MICROBIOLOGY
BASIC AND APPLIED ASPECTS

Dr. Bharat Singh obtained his master's degree in Botany with first division in 1995 and thereafter his Ph.D. degree in 2001 from University of Rajasthan, Jaipur.

Dr. Singh has been teaching Biotechnology to B.Sc. and M.Sc. classes in Stani Memorial P.G. College, Jaipur. So far he has published 18 research papers in reputed national and overseas Journals. He is the author of three books—Immunology, Molecular Biology and Biotechnology, Recombinant DNA and Biotechnology which were well researched and gained a wide popularity. He has presented several papers in national seminars. Besides teaching, Dr. Singh has been engaged with independent research and guiding students for their Ph.D. degrees. He was awarded Junior Research Fellowship of DBT, New Delhi.

The area of his research interest includes production of secondary metabolites in cell cultures. This includes the screening of secondary metabolites against different biological and pharmacological activities.

Currently, he is actively engaged in research on indigenous drugs and drug plants of ancient and traditional medicine systems in relation to their quality control and enhancement of therapeutic spectrum.

MICROBIOLOGY
BASIC AND APPLIED ASPECTS

Dr. Bharat Singh

Pointer Publishers
Jaipur 302 003 (Raj.) India

Distributed by
Prem C. Bakliwal for
Aavishkar Publishers, Distributors
807, Vyas Building, Chaura Rasta
Jaipur 302 003 (Raj.) India
Phone : 0141-2578159
e-mail : aavishkarbooks@hotmail.com

ISBN 978-81-7132-597-9

First Published in 2009 by
Mrs. Shashi Jain for
Pointer Publishers
807, Vyas Building, Chaura Rasta
Jaipur 302 003 (Raj.) India
Phone : 0141-2578159, 2708286
e-mail : pointerpub@hotmail.com
Visit us at : www:pointerpublishers.com

Printed at
Sheetal Printers
Jaipur (Raj.)

PREFACE

Microbiology is an important branch of biological sciences because it deals with many diverse biological disciplines. In addition to studying the natural history of microbes, it also deals with every aspect of microbe—human and environmental interaction. These interactions include genetics, infection mechanism, drug therapy, immunology, genetic engineering, industry, agriculture and ecology. There is growing recognition of the potential of microorganisms in many applied areas. The ability of microorganisms to decompose materials such as insecticides, herbicides, pesticides, oils in oil spills, the potential of microorganisms as food supplements, the exploitation of microbial activity to produce energy such as methane gas for rural consumption, and potential of new therapeutic substances produced by microorganisms are becoming increasingly attractive.

The field of Microbiology has gone through perhaps the most dramatic revolution since Pasteur. In less than five years most of the genomes of pathogenic bacteria became available. Microarray technology has allowed global analysis of gene expression, and proteomics has advanced biochemical studies to universal observation of the protein world. Recombinant DNA technology, makes it feasible to consider genetically manipulated microorganisms for commercial production of new and valuable products for a variety of purposes—as food, medicines and fuel.

Microbiology overlaps with number of scientific disciplines including zoology biochemistry, genetics, ecology, medicine, pharmacology, agriculture science, environmental science, food science etc. The field and challenges for the students of microbiology are virtually limitless and new discoveries are made daily.

Normally the book passes through several proof reading processes. Despite all carefulness, alas, perfection is elusive and errors can still remain. I appreciate constructive critiques from teachers and students alike for improvements, I can make in future editions.

I am highly thankful to my wife Arti Singh for her extensive support during the preparation of this book. I owe a debt of gratitude to Mrs. Barbara Cangan, President, SIIRM and Mr. Elwin Thomas Cangan, Managing Director, SIIRM for their motivation and moral support. I am also thankful to my colleagues and friends of Stani Memorial P.G. College, Jaipur and Department of Botany, University of Rajasthan, Jaipur for their valueable guidance, comments and suggestions.

Lastly, I am grateful to my publisher Mr. Vipin Jain who has masterfully coordinated and extended valuable support during different stages of the production process to allow rapid completion of the project.

Jaipur **Bharat Singh**

CONTENTS

1

HISTORICAL FOUNDATIONS AND SCOPE OF MICROBIOLOGY

The microorganisms are too small to be clearly perceived by the unaided human eye. If an object has a diameter of less than 0.1 mm, or eye can not perceive it at all, and very little detail can be perceived in an object with a diameter of 1 mm. Roughly speaking organisms with a diameter of 1 mm or less are microorganisms and fall into the broad domain of microbiology. So simply microbiology deals with study of microorganisms. At present, there is general agreement to include five major groups as microorganisms : virology, bacteriology, mycology, phycology and protozoology. Despite their small size, microbes are of immense importance to man. They cause diseases, are sources of various foods and medicines and dispose of our wastes. They are infact responsible for the very air we breath since free molecular oxygen was completely absent from the prebiotic atmosphere and this gas has accumulated only as a product of the metabolism of primeval photosynthetic bacteria.

In 6000 BC the ancient Babylonians and Sumerians were brewing beer much as we do today and the Egyptians were baking leaven bread 2000 years later. Despite the antiquity of the microbiological practices, the first documentation of the structure of microorganisms did not occur until the advent of the first microscope in the seventeenth century. Although the spontaneous generation of mice from old rags and of maggots from meat had been disapproved by this time, it was not until last century that Louis Pasteur showed that microbes were not produced *de novo* from mud or decaying organic matter. If Pasteur was the founder of industrial microbiology, Robert Koch was the forefather of medical microbiology. Koch, a German doctor could show in 1876, that the causal agent of bovine anthrax was a bacterium, *Bacillus anthracis*. His remarkable waste remains a landmark in microbiology about the nature of disease and in proving the pathogenicity of a disease-causing agents.in its host through "Koch's postulates". We live in a time when microbiology has come to a age. We use microbes to make beer, wine, cheese, yogurt, soya sauce, antibiotics, vaccines and pesticides. Microbiological reactions are used in sewage treatment, in transforming the chemical structure of drugs, clean cloths and even to extract metals like copper and uranium from their mineral ores. New techniques like gene cloning have been developed within the last decade where these microbes may be used as factory cells for the synthesis of valuable pharmaceutical products like human insulin, hormones, antiviral drugs and vaccines.

HISTORICAL FOUNDATIONS OF MICROBIOLOGY

Microbiology is said to have its roots in a great development of the biological sciences that took place after 1850. It was in this period that Gregor Mendel established the principles of genetics, that Claude Bernard designed his classic experiments in human physiology and that Charles Darwin formulated the process of natural selection inspiring him for publishing on the "Origin of Species". It was also during this period that Louis Pasteur and Robert Koch developed the techniques for isolation, characterisation and identification of microbes of disease. Many of the discoveries in this science have resulted from the prior work of men and women who toiled long hours in dimly light laboratories with the crudest of tools. Each additional insight, whether large or small has added to our current knowledge of living things and processes. This treatment of early history of microbiology will summarize the prominent discoveries made in the past 300 years : microscopy, the rise of scientific methods, the germ theory, and the origins of modern microbiological techniques.

Development of the Microscope

Indeed, several centuries ago there was already a sense that diseases such as the black plague and smallpox were caused by some sort of transmissible matter. But the causes of such phenomena were so vague and obscure because the technology to study them was lacking. Consequently, they remained cloaked in mystery and regarded with superstition—a trend that led even well-educated scientists to believe in spontaneous generation. True awareness of the widespread distribution of microorganisms and some of their characteristics was finally made possible by the development of the first microscope. These devices revealed microbes as discrete entities sharing many of the characteristics of larger, visible plants and animals. Several early scientists fashioned magnifying lenses, but their microscopes lacked the optical clarity needed for examining bacteria and other small, single-celled organisms. The first careful and exacting observations awaited the clever single-lens microscope, hand-fashioned by Antony van Leeuwenhoek, a Dutch linen merchant and self-made microbiologist. The original purpose of the microscopes was to examine cloth for flaws, but Leeuwenhoek turned them to other uses as well. Leeuwenhoek's wide-ranging investigations included observations of tiny organisms he called *animalcules* (little animals), blood, and other human tissues (including his own tooth scrapings), insects, minerals, and plant materials. He constructed over 250 small, powerful microscopes that could magnify up to 300 times. Considering that he had no formal training in science and that he was the first person even to faithfully record this strange new world, his descriptions of bacteria and protozoa were astute and précise. Because of Leeuwenhoek's extraordinary contributions to microbiology, he is known as the father of bacteriology and protozoology. From the time of Leeuwenhoek, microscopes evolved into more complex and improved instruments with the addition of refined lenses, a condenser, finer focusing devices, and built in light sources. The prototype of the modern compound microscope, in use from about the mid-1800s, was capable of magnifications of 1,000 times or more.

The development of an experimental system that answered questions objectively and was not based on prejudice marked the beginning of true scientific thinking. The general approach taken by scientists to explain a certain natural phenomenon is called the scientific method. A primary aim of this method is to formulate a hypothesis, a tentative explanation to account for what has been observed or measured. A good hypothesis must be capable of being either supported or discredited by careful, systematic observation or experimentation. For example, the statement that "bees make honey from pollen" can be experimentally determined by the tools of science, but the statement that "bees buzz because they are happy" cannot.

Discovery Era

As early as 40 BC, Virgil gave a recipe for growing bees artificially, and as late as 1550, von Helmont told how to produce mice by leaving piles of grain and cheese undisturbed in an attic. Even the discovery of single-celled organisms during the latter part of the mid-1600s did not eradicate the idea of abiogenesis, because some scientists assumed that microscopic beings were simply one stage in a process of forming larger organisms from smaller ones. Though some of these earlier ideas seem quaint and ridiculous in light of modern knowledge, we must remember that at the time, mysteries in life were accepted, and the scientific method was not widely practiced. The eventual invalidation of the theory of abiogenesis was a very gradual process. Two hypotheses attempted to explain the origin of the "simpler forms" of life : (a) They arose spontaneously by the massing of vital forces within nonliving matter (abiogenesis; Gr. *a*, not, *bios*, life and *gennan*, to produce), or (b) they arose only from other living things of their same kind (biogenesis). If an investigator was able to verify one, his work automatically falsified the other, because both cannot be true. Among the important variables to be considered in challenging the hypotheses were the effects of nutrients, air, heat, and the presence of preexisting life forms in the environment. One of the first people to doubt the spontaneous generation theory was Francesco Redi of Italy. He conducted a simple test in which he placed meat in a jar and covered it with fine gauze. Flies gathering at the jar were blocked from entering and thus laid their eggs on the outside of the gauze. The maggots subsequently developed without access to the meat, indicating that maggots were the offspring of flies and did not arise from some "vital force" in the meat. This and related experiments laid to rest the idea that more complex animals such as insects and mice developed through abiogenesis, but it did not convince many scientists of the day that simpler organisms could not arise this way.

Frenchman Louis Jablot reasoned that even microscopic organisms must have parents, and his experiments with infusions (dried hay steeped in water) supported this hypothesis. He divided an infusion that had been boiled to destroy any living things into two containers : a heat-treated container that was closed to the air and a heated container that was freely open to the air. Only the open vessel developed microorganisms, which he presumed had entered in air laden with dust. Regrettably, the cause of biogenesis was temporarily set back by John Needham, an Englishman who did similar experiments using mutton gravy. His results were in conflict with Jablot's because both heated and unheated test containers teemed with microbes. Unable to allow Needham's conclusion to remain untested, an Italian naturalist, Lazzaro Spallanzani, took more extensive and sophisticated measures, which included long-term heating of the test vessels and infusions that destroy any lingering resistant forms. He found that a variety of liquid nutrients (broths, urine) thus treated and kept sealed in flasks from air would not develop growth. When he was accused of destroying the vegetative force of the nutrients by overheating, he showed that the heated nutrients could still grow microbes when exposed to air. Undaunted by this argument, other critics objected that Spallanzani's flasks were sealed from oxygen, a gas known to be required in the respiration of animals.

Franz Shultze and Theodor Schwann of Germany felt assured that air was the source of microbes and sought to prove this by passing air through strong chemicals or hot glass tubes into heat-treated infusions in flasks. When the infusions again remained devoid of living things, the supporters of abiogenesis claimed that the treatment of the air had made it harmful to the spontaneous development of life. These studies were followed up by Georg Schroeder and Theodor Van Dusch, who did not treat the air with heat or chemicals, but passed it through cotton wool to filter out microscopic organisms. Again, no microbes grew in the infusions. Although these experiments should have finally laid to rest the arguments for spontaneous generation, they did not. Then, in the mid-1800s, the acclaimed microbiologist Louis Pasteur entered the arena. He had recently been studying the roles of microorganisms in the fermentation of beer and wine, and it was clear to him that these processes were brought about by the activities of microbes introduced into the beverage from air, fruits, and grains. The methods he used to discount abiogenesis were simple yet brilliant. He repeated the experiments using cotton filters to trap dust from air and observed tiny objects (probably spores) in the filters. He also observed that these same filters would initiate growth in previously sterile broths. To further clarify that air was the source of microbes, he filled flasks with broth and fashioned their openings into elongate, swan neck-shaped tubes. The flasks' openings were freely open to the air but curved so that gravity would cause any airborne dust particles to deposit in the lower part of the necks. He heated the flasks to sterilize the broth and then incubated them. As long as the flask remained intact, the broth remained sterile, but if the neck was broken off so that dust fell directly down into the container, microbial growth immediately commenced. Some of these ingenious little flasks are still on display at the Institute Pasteur in Paris in their original sterile form. Pasteur summed up his findings, "For I have kept from them, and am still keeping from them, that one thing which is above the power of man to make; I have kept from them the germs that float in the air, I have kept from them life."

In the Inductive approach, a scientist accumulates specific data or facts and then formulates a general hypothesis that accounts for these facts. The inductive approach asks, "Are various observed events best explained by this hypothesis or by another one?" In the deductive approach, a scientist constructs a hypothesis, tests its validity by outlining particular events that are predicted by the hypothesis, and then performs experiments to test for these events. The deductive process states : "If the hypothesis is valid, then certain specific events can be expected to occur". Natural processes have numerous physical, chemical, and biological factors, or *variables*, that can hypothetically affect their outcome. To account for these variables, experiments are designed : (a) To thoroughly test for or measure the consequences of each possible variable, and (b) to accompany each variable with one or more *control groups*. A control group is designed exactly as the test group but omits only that variable being tested : thus, it may serve as a basis of comparison for the test group. The reasoning is that, if a certain experimental finding occurs only in the test group and not in the control group, the finding must be due to the variable being tested and not to some uncontrolled, untested factor that is not part of the hypothesis.

If experiments do not uphold the hypothesis—that is, if it is found to be flawed—the hypothesis or some part of it is rejected, and it is either discarded or modified to fit the results of the experiment. If the hypothesis is supported by the results from the experiment, it is not (or should not be) immediately accepted as fact. It then must be tested and retested. Indeed, this is an important guideline in the acceptance of a hypothesis. The results of the experiment must be published and then repeated by other investigators. The history of science is littered with the carcasses of apparently ingenious hypotheses that simply could not be repeated and were thus ultimately refuted. In time, as each hypothesis is supported by a growing body of data and survives rigorous scrutiny, it moves to the next level of acceptance—the theory. A theory is a collection of statements, propositions, or concepts that explains or accounts for a natural event. A theory is not the result of a single experiment repeated over and over again but is an entire body of ideas that expresses or explains many aspects of a phenomenon. It is not a fuzzy or weak speculation, as is sometimes the popular notion, but a viable declaration that has stood the test of time and has yet to be disproven by serious scientific endeavors. Often, theories develop and progress through decades of research and are added to and modified by new findings. At some point, evidence of the accuracy and predictability of a theory is so compelling that the next level of confidence is reached and the theory becomes a law or principle.

Golden Era of Microbiology

Louis Pasteur's experiments

By 1860 several workers have begun to realize that there is a causal relationship between the development of microorganisms in organic infusions and the chemical changes that take place in these infusions : microorganisms are agents that bring about major chemical changes. His main contributions to the development of microbiology are as follows : Fermentation and Pasteur's disease concept. The popular belief was that fermentation of grape juice to wine was a natural chemical process, involving the breakdown of the protein albumin. However, Pasteur could observe under his microscope some oval yeast cells and barely, some rods commonly called bacteria. He believed that yeasts played significant role in fermentation. In series of classic experiments, Pasteur first showed that alcohol would be produced from grape yeast mixtures. Using a heat, he then destroyed the yeasts, whereupon alcohol failed to appear in the grape juice. However, when yeasts were returned to the flasks, fermentation took place and wine was formed. Moreover, if care was taken to eliminate the bacteria, the wine would not become sour. Pasteur's work thus seemed to indicate that the microscopic yeasts and bacteria were tiny chemical factories that could bring about important chemical changes. The yeasts appeared to be vital to fermentation process and the bacteria made the wine sour. Pasteur's work appeared to demonstrate that the microorganisms could be a cause of disease, for if they could make the wine sick perhaps they could also make the body ill.

In further experiment, Pasteur believed that bacteria were in the soil and air, and thus their spread could be controlled. If so then, perhaps disease could also be controlled : (a) When the sterile broth containing flask opened in air, then life was appeared, (b) similarly, if sterile broth containing flask neck sealed, then the life was not appeared, and (c) finally if the sterile broth containing swan-necked flask opened in air, then no life was appeared. Others believed that bacteria arose spontaneously from organic matter, as in the sick body, as long as life force was present. Pasteur took the view of spontaneous generation seriously in order to develop his own theory of disease. He first showed that as air become purer, fewer microbes could be located. His swannecked-flasks provided the final defeat for the spontaneous generation theory. The flasks were left open to allow entry of any life force present in untreated air. The curvature of neck prevented the microbes from reaching the broth.

The flasks remained sterile, thus spontaneous generation was eliminated as a viable theory, the idea that microbes existed in the environment was strengthened within a short period of seven years, Pasteur lost his three daughters, who died of typhoid fever, so he thus returned to the study of human diseases.

Robert Koch's experiments

Robert Koch, a German doctor, credited for his contributions to development of microbiology. Koch had observed thread like organisms in blood of animals that had died of anthrax, a disease that was serious threat to farmers killing their sheeps and cattle herds periodically. He injected laboratory animals with a bacteria-ladden blood of dead sheep. He then perform autopsies and noted how the same symptoms appeared regularly. Koch isolated a few of bacilli in the clear sterile vitreous humor of an ox's eye and watched their multiplication. He injected laboratory mice with a silver of wood containing spores and observed as anthrax symptoms developed in the animals. On autopsy, the blood was swarming with the thread like bacteria and he reisolated them in the vitreous humor. The cycle was now complete. In year 1867, Koch postulated his "germ theory of disease". The sequence of procedures by which Koch established his 'germ theory of disease' came to be known as 'Koch's postulates'. Koch accidentally observed that a slice of potato contained masses of bacteria in colonies, each colony separate and distinct from the others, and each composed of billions of rods. It occurred to him that bacteria could grow and multiply on solid surfaces, and he added gelatin to his broth to prepare a solid culture medium. When seeded on to surface, bacteria form visible colonies and each contained only one type of organism. Inoculations of laboratory animals could now be made with the assurance that only one type of microorganisms was involved.

End of Golden Era of Microbiology

The death of Robert Koch in 1910 and advent of world war I led the Pasteur Institute was closed and the German laboratories converted for production of blood components used to treat infections. Thus came to an end what many have called the "Golden Era of Microbiology", a fifty year period from the 1860, work of Pasteur on alcohol fermentation to the death of Koch. This age has witnessed a series of discoveries unparalleled in medicine, most of which involved the identification of agents of diseases. There was the awareness of the relationship of microbes to disease and methods of transmission. This led to sterile practices in hospitals, increased research in pest control and care of preparation and food consumption.

Microbiology of Twentieth Century

The discovery of cell-free alcoholic fermentation by Buchner in 1897 provided the key to chemical analyses of energy yielding processes. In the first two decades, the 20th century parallel studies on the glycolysis by muscle and alcoholic fermentation by yeast gradually revealed that fundamental similarity. There was thus found a common ground for vertebrate physiologists and microbial biochemists. Later vitamins required by animals proved chemically identical with the growth factors required by some yeasts and bacteria. These and some other such discoveries on precursors of coenzymes, etc., from 1920-1935 could demonstrate the basic similarities of all living systems at the metabolic level. The second great advance of biology in early 20th century, the creation of new discipline of genetics had no immediate impact on microbiology. The first important contact between genetics and microbiology occurred in 1941, when Beadle and Tatum succeded in isolating a series of biochemical mutants from the *Neurospora.* In 1943, mutation studies in bacteria by Delbruck and Luria engineered genetic work on these microorganisms. Soon after several mechanisms of genetic transfer were shown to exist in bacteria. In 1944, the work of Avery, McLeod and McCarthy on process of transformation in bacteria revealed that genetic matter in living organisms is DNA.

SCOPE OF MICROBIOLOGY

Microorganisms can cause disease, spoil food, and deteriorate materials like iron pipes, glass lenses, and wood pilings. Most microorganisms are unicellular. In unicellular organisms, all the life processes are performed by a single cell. In the so-called higher forms of life, organisms are composed of many cells that are arranged in tissues and organs to perform specific functions. Regardless of the complexity of an organism, the cell is the basic structural unit of life. The word *cell* was first used more than two centuries ago by an Englishman, Robert Hooke (1635-1703), in his descriptions (1665) of the fine structure of cork and other plant materials. The honeycomb like structure he observed in a thin slice of cork was due to the cell walls of cells that were once living. But the concept of the cell as the structural unit of life—the cell theory—is credited to two Germans, Matthias Schleiden and Theodor Schwann, who in 1838-1839 described cells as the basic structural and functional units of all organisms. Schleiden and Schwann recognized that all cells, no matter what the organism, are very similar in structure. As the concept of the cell as the basic unit of life gained acceptance, investigators speculated on the nature of the substance contained within the cell. Protoplasm (Gr. *proto,* first and *plasm,* formed substance, introduced to characterize the living material of a cell), is a colloidal organic complex consisting largely of protein, lipids and nucleic acids. These substances are enclosed by membranes or cell walls; and the protoplasm always contains nuclei or an equivalent nuclear substance (Fig. 1.1). All biological systems have the following characteristics in common : (a) The ability to reproduce, (b) the ability to ingest or assimilate food substances and metabolize them for energy and growth, (c) the ability to excrete waste products; (d) the ability to react to changes in their environment—sometimes called irritability, and (e) susceptibility to mutation. In the study of microbiology, we encounter "organisms" which may represent the borderline of life. These are the viruses, which are simpler in structure and composition than single cells. Viruses provide an exciting challenge and an opportunity to gain a better understanding of the nature of complex organic substances that may bridge the gap between the living and the nonliving worlds. Viruses are obligate parasites, that is, they are obligated to grow within an appropriate host cell—plant, animal, or microbe. They cannot multiply outside a host cell. However, when a virus enters an appropriate living cell, it is able to direct the synthesis of hundreds of identical viruses, using the cell's energy and biochemical machinery. A virus is made up of substances unique to life : nucleic acids and proteins.

Another manner of subdividing the subject matter of biology is based on the level at which the study is conducted, for example, studies at the level of molecular constituents of the cell (molecular biology), studies at the level of the cell (cell biology), studies at the level of the intact organism (organismal biology), and studies of groups of organisms (population biology). Still another approach is to establish divisions on the basis of form and function, as in morphology or anatomy, physiology, metabolism, and genetics. In some colleges and universities, the study of microbiology is carried out in a department of biology, in others, it is in a department of microbiology or a department of molecular biology.

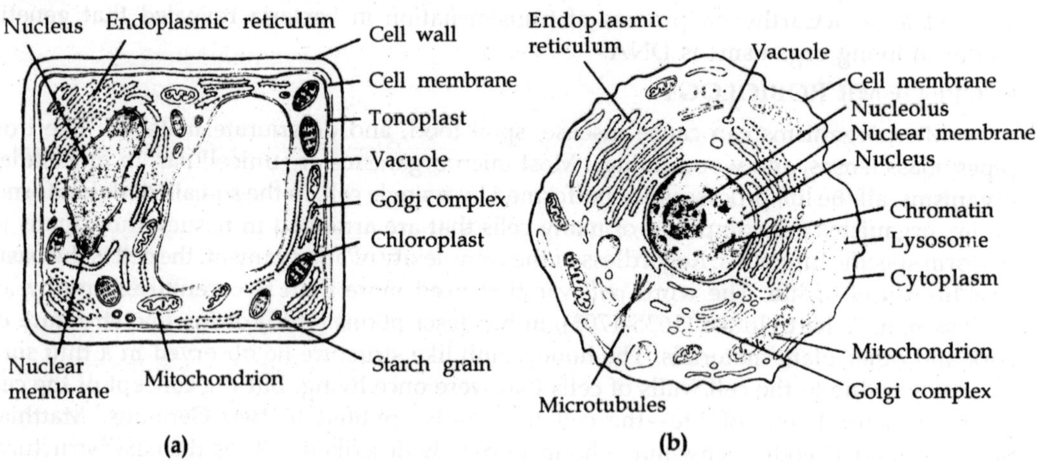

Fig. 1.1 : Typical cell structures : (a) plant cell (b) animal cell.

Microbiology is placed in the broad field of biology, microorganisms have some characteristics which make them ideal specimens for the study of numerous fundamental life processes. This is possible because, at the cellular level, many life processes are performed in the same manner whether they be in microbe, mouse, or human. Microorganisms can be grown conveniently in test tubes or flasks, thus requiring less space and maintenance than larger plants and animals. They grow rapidly and reproduce at an unusually high rate, some species of bacteria undergo almost 100 generations in 24 hours. The metabolic processes of microorganisms follow patterns that occur among higher plants and animals. For example, yeasts utilize glucose in essentially the same manner as cells of mammalian tissue; the same system of enzymes is present in these diverse organisms. The energy liberated during the breakdown of glucose is "trapped" and made available for the work to be performed by the cells, whether they be bacteria, yeasts, protozoa, or muscle cells. In fact, the mechanism by which organisms (or their cells) utilize energy is fundamentally the same throughout the biological world. The source of energy does, of course, vary among organisms. Plants are characterized by their ability to use radiant energy, whereas animals require chemical substances as their fuel. In this respect some microorganisms are like plants, others like animals, and some have the unique ability of using either radiant energy or chemical energy and thus are like both plants and animals. In microbiology we can study organisms in great detail and observe their life processes while they are actively metabolizing, growing, reproducing, aging, and dying. By modifying their environment we can alter metabolic activities, regulate growth, and even change some details of their genetic patterns – all without destroying the organisms.

Bacteriophages have been of inestimable value in elucidating many biological phenomena, including those concerned with genetics. Microorganisms have a wider range of physiological and biochemical potentialities than all other organisms combined. For example, some bacteria are able to utilize atmospheric nitrogen for the synthesis of proteins and other complex organic nitrogenous compounds. Other species require inorganic or organic nitrogen compounds as the initial building blocks for their nitrogenous constituents. Some microorganisms synthesize all their vitamins, while others need to be furnished vitamins. By reviewing the nutritional requirements of a large collection of microorganisms, it is possible to arrange them from those with the simplest to those with the most complex requirements. The increasing complexity of nutritional requirements in such an arrangement is also a reflection of the decreasing synthetic capacity of the organisms so arranged. In addition, this kind of arrangement provides information about the steps in the synthesis of various metabolites, e.g., from atmospheric oxygen to inorganic nitrogen salts to amino acids. The biochemist has used microorganisms having varying degrees of synthetic ability to investigate pathways of synthesis. There is no field of human endeavor, whether it be in industry or agriculture, or in the preparation of food or in connection with problems of shelter or clothing, or in the conservation of human or animal health and the combating of disease, where the microbe does not play an important and often dominant role.

Microbiology is one of the largest and most complex branch of the biological sciences because it deals with many diverse biological disciplines. In addition to studying the natural history of microbes, it also deals with every aspect of microbe-human and environmental interaction. These interactions include genetics, metabolism, infection, disease, drug therapy, immunology, genetic engineering, industry, agriculture, and ecology. Microbiology has numerous practical uses in industry and medicine, some of the more prominent areas that apply microbiological principles are as follows : *Immunology* – studies the system of body defenses that protects against infection. It includes *serology*, a discipline that looks for the products of immune reactions in the blood and tissues, and aids in diagnosis of infectious diseases by this means; and *allergy*, the study of hypersensitive responses to ordinary, harmless materials.

Public health microbiology and epidemiology aim to monitor and control the spread of diseases in communities. The principal American and global institutions involved in this concern are the United States Public Health Service (USPHS) with its main agency, the Centers for Disease Control and Prevention (CDC) located in Atlanta, Georgia, and the World Health Organization (WHO), the medical limb of the United Nations. The CDC collects information on disease from around the United States and publishes it in a weekly newsletter called the Morbidity and Mortality Weekly Report. Food microbiology, dairy microbiology, and aquatic microbiology examine the ecologic and practical roles of microbes in food and water. Agricultural microbiology is concerned with the relationships between microbes and crops, with an emphasis on improving yields and combating plant diseases.

Biotechnology includes any process in which humans use the metabolism of living things to arrive at a desired product, ranging from bread making to gene therapy.

Genetic engineering and recombinant DNA technology involve techniques that deliberately alter the genetic makeup of organisms to mass produce human hormones and other drugs, create totally novel substances, and develop organisms with unique methods of synthesis and adaptation. This is the most powerful and rapidly growing area in modern microbiology.

In fact, many areas of this science have become so specialized that it is not uncommon for a microbiologist to spend his or her whole life concentrating on a single group or type of microbe, biochemical process, or disease. On the other hand, rarely is one person a single type of microbiologist, and most can be classified in several ways. There are, for instance, bacterial physiologists who study industrial processes, molecular biologists who focus on the genetics of viruses, fungal taxonomists interested in agricultural pests, epidemiologists who are also nurses, and dentists who specialize in the microbiology of gum disease. Studies in microbiology have led to greater understanding of many theoretical biological principles. For example, the study of microorganisms established universal concepts concerning the chemistry of life, systems of inheritance, and the global cycles of nutrients, minerals, and gases. The World Health Organization (WHO), a medical agency of the United Nations, estimates that more than 20 million people die each year worldwide from preventable infectious diseases. Many of these diseases occur in developing countries where medical care is unavailable or substandard. Adding to this toll are emerging diseases such as cryptosporidiosis (a protozoan infection) and Lyme disease. Many older diseases, including tuberculosis, measles, and syphilis, are showing a dramatic resurgence in industrialized countries as well. To complicate matters even further, increased numbers of patients with severe compromised immune defenses such as AIDS and cancer patients are being kept alive for extended periods. The scope of microbiology is incredibly diverse. It includes basic microbial research, research on infectious diseases, study of prevention and treatment of disease, environmental functions of microorganisms, and industrial use of microorganisms for commercial, agricultural, and medical purposes.

❑❑❑

2

GENERAL CHARACTERISTICS AND TAXONOMY OF MICROORGANISMS

It is usually not feasible to study the characteristics of a single microorganism, because of its small size; therefore, we study the characteristics of a culture—a population of microorganisms. If we study the characteristics of a culture containing many microorganisms, it is as if we are studying the characteristics of a single organism. A culture that consists of a single kind of microorganism, regardless of the number of individuals, in an environment free of other living organisms is called a pure culture.

Two basic cell lines have appeared during evolutionary history. These lines, termed prokaryotic cells (Gr. *pro,* before, and *karyon,* nucleus) and eukaryotic cells (Gr. *eu,* true or good, and *karyon,* nucleus) differ primarily in the complexity of their cell structure. In general, prokaryotic cells are smaller than eukaryotic cells and they lack special structures such as a nucleus and organelles. All prokaryotes are microorganisms, but only some eukaryotes are microorganisms. The bodies of most microorganisms consist of either a single cell or just a few cells. Because of their role in disease, certain animals such as helminth worms and insects, many of which can be seen with the naked eye, are also considered in the study of microorganisms.

The dimensions of macroscopic organisms are usually given in centimeters (cm) and meters (m), those of most microorganisms fall within the range of micrometers (μm) and to a lesser extent, nanometers (nm) and millimeters (mm). The size range of most microbes extends from the smallest viruses, measuring around 20 am and actually not much bigger than a large molecule, to protozoans measuring 3-4 μm and visible with the naked eye. The majority of microorganisms live a free existence in habitats such as soil and water where they are relatively harmless and often beneficial. A free-living organism can derive all required foods and other factors directly from the nonliving environment. Some microorganisms require interaction with other organisms. One such group, termed parasites, are harbored and nourished by other living organisms, called hosts. Through its actions, a parasite causes damage to its host through infection and disease. Most microbial parasites are some type of bacterium, fungus, protozoan, worm, or virus. Although parasites cause important diseases, they make up only a small proportion of microbes.

GENERAL CHARACTERS

Microorganisms are very small and their size is usually expressed in micrometers (μm). One μm is equivalent to 0.001 millimeter (mm) or about 0.00004 inch; consequently, routine examination of microbial cells requires the use of a high-power microscope, usually at a magnification of about 1,000 diameters. Numerous techniques are available for the microscopic examination of microorganisms. Microbial cells consist of a wide variety of organic compounds. When cells are broken apart and their components subjected to chemical analysis, each kind of microorganism is found to have a characteristic chemical composition. Both qualitative and quantitative differences in composition occur among various species. For example, the occurrence of lipopolysaccharide in cell walls is characteristic of Gram-negative bacteria but not Gram-positive bacteria; on the other hand, many Gram-positive bacteria have cell walls that contain teichoic acids, compounds not found in Gram-negative bacteria. Fungal and algal cell walls are very different in composition from those of bacteria. A major distinction among viruses is made on the basis of the kind of nucleic acid they possess, namely ribonucleic acid (RNA) or deoxyribonucleic acid (DNA). Many microorganisms can be grown in or on a culture medium. Some microorganisms can grow in a medium containing only inorganic compounds, whereas others require a medium containing organic compounds. Some require complex natural substances (peptone, yeast autolysate, blood cells, or blood serum), and some cannot as yet be grown in an artificial laboratory medium and can be propagated only in a living host or living cells. For example, rickettsias require a host in which to grow, such as an animal, a fertilized chicken egg (chick embryo), an arthropod, or a culture of mammalian tissue cells. In addition to specific nutrients, each kind of organism also requires specific physical conditions for growth. For example, some bacteria grow best at high temperatures and cannot grow below 40°C; others grow best in the cold and cannot grow above 20°C; still others, such as bacteria pathogenic to humans, require a temperature close to that of the human body (i.e., 37°C).

The gaseous atmosphere required for growth is also important; for instance, some bacteria require oxygen for growth; oxygen is lethal to others and they can grow only in its absence. Light may be another important physical condition : certain bacteria, such as cyanobacteria, require light as a source of energy, whereas others may be indifferent to light or may even find it deleterious to their growth. Each kind of microorganism grows in a characteristic manner. For example, growth in a liquid medium may be abundant or sparse; it may be evenly dispersed throughout the medium, or it may occur only as a sediment at the bottom or only as a thin film or pellicle at the top. On solid media, microbes grow as colonies—distinct, compact masses of cells that are macroscopically visible. The life processes of the microbial cell are a complex integrated series of chemical reactions collectively referred to as metabolism. The variety of these reactions affords many opportunities to characterize and differentiate various groups of microorganisms. For instance, some organisms may obtain energy by absorbing light, others by oxidizing various organic or inorganic compounds, and others by redistributing the atoms within certain molecules so that the molecules become less stable. Organisms also differ in the ways in which they synthesize their cell components during growth. The various chemical reactions of an organism are catalyzed by proteins called enzymes, and the complement of enzymes possessed by one kind of organism. Certain chemical compounds of microbial cells are called antigens.

Antigenic characterization of a microorganism has great practical importance. If microbial cells enter the animal body, the animal responds to their antigens by forming specific blood serum proteins called antibodies, which bind to the antigens. Antibodies are highly specific for the antigens that induce their formation. Because different kinds of microorganisms have different types of antigens, antibodies are widely used as tools for the rapid identification of particular kinds of microorganisms. For the present this might be explained as a "lock and key system". Because of the highly specific nature of the reaction, if we know the identity of one part of the system (antigen or antibody) we can identify the other. For example, if we take typhoid bacterium antibody and mix it with a suspension of unknown bacterial cells, and a positive reaction occurs, we can conclude that the cells are those of the typhoid organism. The double-stranded chromosomal DNA of each kind of microorganism has certain features that are constant and characteristic for that organism and useful for its classification :

(a) Base composition of DNA—It is important to note that the DNA molecule is made up of base pairs : guanine–cytosine and adenine-thymine. Of the total number of nucleotide bases present in the DNA, that percentage represented by guanine plus cytosine is termed the moles % G + C value (or more briefly, mol % G + C). Values for various organisms range from 23 to 75.

(b) Nucleotide base sequence of DNA—This sequence is unique for each kind of organism and is the most fundamental of all the characteristics of an organism; consequently, it has great significance for microbial classification.

Plasmids are circular DNA molecules that are capable of autonomous replication within bacterial cells, and their presence can confer special characteristics on the cells that contain them, such as the ability to make toxins (toxigenicity), to become resistant to various antibiotics, or to use unusual chemical compounds as nutrients. The ability to cause disease, or pathogenicity, of some microorganisms is certainly a dramatic characteristics and it stimulated much of the early work with microorganisms. Although we now know that relatively few species of microorganisms cause disease, certain microorganisms are pathogenic for animals or plants, and some microorganisms may cause disease in other microorganisms. The habitat of a microorganism is important in characterizing that organism. For example, microorganisms normally found in marine environments generally differ from those in freshwater environments. The microbial population of the oral cavity differs from that of the intestinal tract. Some kinds of microorganisms are widely distributed in nature, but others may be restricted to a particular environment.

MICROBIAL TAXONOMY

A strain is made up of all the descendents of a pure culture; it is usually a succession of cultures derived from an initial colony. Each strain has a specific history and designation. For example, strain ATCC 19554 is a strain of spirilla isolated originally from pond water in Blacksburg, Virginia in 1965 by Wells and Krieg, and cultures of this strain are maintained at the American Type Culture Collection (ATCC), Rockville, Maryland. Cultures of the same species that were isolated from other sources would be considered different strains. Bacterial species consists of a special strain called the type strain together with all other strains that are considered sufficiently similar to the type strain as to warrant inclusion in the species. The type strain is the strain that is designated to be the permanent reference specimen for the species. Unfortunately, it is not always the strain that is most typical of all the strains included in the species, but it is the strain to which all other strains must be compared to see if they resemble it closely enough to belong to the species. Therefore, type strains are particularly important and special attention is given to their maintenance and preservation, particularly by national reference collections such as the ATCC in the United States or the National Collection of Type Cultures in England.

At present there are no specific criteria for a bacterial species that are universally accepted. However, certain criteria based on DNA homology experiments are probably more widely accepted today than any others and eventually may lead to a unifying concept for defining a species. Just as a bacterial species is composed of a collection of similar strains, a bacterial genus is composed of a collection of similar species. One of the species is designated the type species, and this serves as the permanent example of the genus; that is, other species must be judged to be sufficiently similar to the type species to be included with it in the genus. Unfortunately, there is even less agreement about the criteria for a bacterial genus than there is for a bacterial species. Bacteria can be listed in following groups : **Kingdom** (a group of similar divisions), **Division** (a group of similar classes), **Class** (a group of similar orders), **Order** (a group of similar families) and **Family** (a group of similar genera).

Following methods can be used in taxonomic arrangements of bacteria :

The Intuitive Method

Microbiologists who are familiar about characters of the organisms which have been studying for several years may decide that the organisms represent one or more species or genera. The trouble with this method is that the characteristics of an organism that seem important to one person may not be so important to another, and different taxonomists may arrive at very different groupings.

Numerical Taxonomy

By using a computer he or she calculates the % similarity (% S) of each strain to every other strain. For any two strains, this is :

$$\% \text{ S} = \frac{\text{NS}}{\text{NS} + \text{ND}}$$

where NS is the number of characteristics that are the similar (positive or negative) for the two strains, and ND is the number of characteristics that are different. Those strains having a high % S to each other are placed into groups; those groups having a high % S to each other are in turn placed into larger groups, and so on. The degree of similarity needed to rank a group as a species, genus, or other taxon is a matter of judgement on the part of the taxonomist.

Hereditarial Material Relatedness

In the 1960s, the development of that branch of science known as molecular biology provided techniques by which the DNA of one organism could be compared with that of other organisms. At first only crude comparisons could be made, based on mol% G + C values. It is true that two organisms of the same or similar species that are very closely related will have very similar mol% G + C values, and it is also true that two organisms having quite different mol% G + C values are not very closely related. *However, it is important to realize that organisms that are completely unrelated may have similar mol% G + C values.* Therefore, much more precise methods of comparison were needed—namely, methods by which the DNA molecules from various organisms could be compared with respect to the *sequence of their component nucleotides.*

The double-stranded DNA molecules from two organisms are heated to convert them to single strands. The single strands from one organism are then mixed with those from the other organism and allowed to cool. If the two organisms are closely related, heteroduplexes will form. If the two organisms are not closely related, no heteroduplexes will form.

Ribosomes, the small granular-appearing structures within the cell which manufacture proteins, are composed of proteins and RNA. The ribosomal RNA (rRNA) is coded for by only a small fraction of the DNA molecule, the rRNA cistrons. In all bacteria so far studied, the nucleotide sequences of these rRNA genes has been found to be highly conserved; that is, during evolution, the nucleotide sequence has changed more slowly than that of the bulk of the DNA molecule. This means that even if two organisms are only distantly related and show no significant DNA homology, there still may be considerable similarity in the nucleotide sequences of their rRNA cistrons. The degree of similarity that exists can therefore be used as a measure of relatedness between organisms, but at a level beyond that of species. RNA homology experiments and RNA oligonucleotide cataloging are two modern methods used to determine the degree of similarity between the rRNA cistrons of different organisms.

In fact, it is now apparent that present-day bacteria evolved by at least two very different major routes from an early ancestral form and that they now comprise two very large groups : the eubacteria and the archaeobacteria. It has been proposed that these two groups be considered as two separate kingdoms of life, and, indeed, they do seem as distantly related to each other as they are to eukaryotic organisms. Although the kingdom question is still debatable, data obtained from rRNA oligonucleotide cataloging nevertheless make it clear that the archaeobacteria are separated from other bacteria by a great phylogenetic gulf.

Nomenclature

If an organism were to be called *Escherichia coli* in one country and *Coprobacterium intestinale* in another, chaos would result. The name of a species is merely a convenient label. It is not necessarily even descriptive, although some names are. For example, *Micrococcus luteus* means "yellow berry" in Latin, and *Proteus vulgaris* is Latin for "common organism of many shapes". Some species are named after persons : for example, *Escherichia coli*—the organism of the colon, named after Theodor Escherich (a German bacteriologist); : or *Clostridium barkeri*—the spindle-shaped organism, named after H.A. Barker (an American biochemist). Some names are even nonsensical [e.g., *Runella slithyformis*—"the organism whose shape resembles runes (characters of an ancient alphabet) and which is slithy," the latter term being taken from Lewis Carroll's poem "Jabberwocky" from *Alice in Wonderland*]. The important point is that names are only convenient designations. For example, instead of referring to "the rod-shaped, acid-fast bacterium that is slow-growing, is stimulated by glycerol, causes pulmonary tuberculosis in humans, is spread mainly by airborne droplets, forms buff-coloured colonies, synthesizes niacin, reduces nitrate to nitrite, and is pathogenic for guinea pigs", it is much more convenient simply to say "*Mycobacterium tuberculosis*".

Although it might seem that microbial names could be constructed almost at random, the fact is that certain rules must be followed. Bacteria, for example, are named according to rules set down in the *International Code of Nomenclature of Bacteria*; other codes govern the naming of algae, fungi, and viruses. One rule in bacteriological nomenclature is that a name must be written as a Latin or Latinized binomial (two words) and must follow certain rules of Latin grammar. The first word in the binomial is the genus name and is always capitalized. The second word is the specific epithet and is never capitalized. Both the genus name and specific epithet are given in italics. Bacteria are sometimes referred to by common or colloquial names, which have no official standing in nomenclature and are never italicized. Those bacterial names which have official standing in microbiology were published in the *Approved Lists of Bacterial Names* in January, 1980. Any new or additional names must be published in the *International Journal of Systematic Bacteriology* in order to achieve official recognition. The International Code of Nomenclature of Bacteria was developed with reference to the much earlier established International Codes of Zoological and Botanical Nomenclature. All of these codes incorporate certain common principles as listed below :

a. Individual type of organism is designated as a species.

b. The species is designated by a Latin binomial nomenclature.

c. The priority law ensures the use of the oldest available legitimate name.

d. Requirements are given for effective publication of new specific names, as well as guidance in coining new names in reputed Journals.

CLASSIFICATION OF HAECKEL

Present-day classification in biology was established by the work of Carolus Linnaeus (1707-1778), a Swedish botanist. His books on the classification of plants and animals are considered to be the beginning of modern botanical and zoological nomenclature, a system of naming plants and animals. Nomenclature in microbiology, which came much later, was based on the principles established for the plant and animal kingdoms. Until the eighteenth century, the classification of living organisms placed all organisms into one of two kingdoms, plant and animal. Since there are organisms that do not fall naturally into either the plant or the animal kingdom, it was proposed that new kingdoms be established to include those organisms which typically are neither plants nor animals. One of the earliest of these proposals was made in 1866 by a German zoologist, E.H. Haeckel. He suggested that a third kingdom, *Protista*, be formed to include those unicellular microorganisms that are typically neither plants nor animals. These organisms, the protists, include bacteria, algae, fungi, and protozoa. Bacteria are referred to as lower protists; the others—algae, fungi, and protozoa—are called higher protists. Satisfactory criteria were unavailable until late in the 1940s when more definitive observation of internal cell structure was made possible with the aid of the powerful magnification provided by electron microscopy. It was discovered that in some cells, for example typical bacteria, the nuclear substance was not enclosed by a nuclear membrane. In other cells, such as typical algae and fungi, the nucleus was enclosed in a membrane. This discovery— the absence of membrane-bound internal structures in one group of protists (bacteria) and the presence of membrane-bound structures in all the others (fungi, algae, and protozoa)— was a discovery of fundamental significance.

Plantae and Animalia

This assignment of the microbes to either kingdom was made on the basis of the most easily determinable differences between animals and plants : ability of active movement and the ability to photosynthesis. Algae (immobile, photosynthetic) were grouped under plants. Fungi were also lumped with the plants. Microscopic motile forms (including protozoa and bacteria) were lumped together as one group, the Infusoria under animals. The simpler, unicellular forms, were, therefore, sub-divided into two groups : the protozoa with relatively large and complex cells, and the bacteria with much smaller and simple cells. The old Infusoria was thus split into three groups :

a. Metazoa (multicellular) invertebrate animals.

b. Protozoa (unicellular) animals.

c. Bacteria, that were transferred to plants.

This Procrustean distribution of microorganisms into plants and animals presented too many inconsistencies. More information became available on microorganisms, that hardly justified their inclusion in either kingdoms. For instance, unlike plants fungi and most bacteria are non-photosynthetic.

Protists

This kingdom thus included both photosynthetic and nonphotosynthetic forms, some plantlike, some animallike, and some having their own specific properties as well as sharing those of both, plants as well as animals.

Eukaryotes and prokaryotes

Within few years, there were revealed some hitherto unknown interesting features of cell ultrastructure. This led to a fundamental dichotomy among the various groups of organisms with respect to the internal architecture of the cell : two radically different kinds of cells exist in the contemporary living world : (a) The more complex eukaryotic cell is the unit of structure in plants, metazoan animals, protozoa, fungi and algae. (b) The less complex prokaryotic cell is the unit of structure in two microbial groups : bacteria and cyanobacteria. Some authors call the eukaryotic microbes as higher protists and prokaryotic microbes as lower protists. If, the term protists is still to be retained at all, this has been restricted to the eukaryotic microbes i.e., algae, fungi and protozoa just to distinguish them from the Plantae and Animalia. The eukaryotic cells contain a nucleus with a nuclear membrane, multiple chromosomes and a mitotic apparatus to ensure equal partition of the products of chromosomal replication to the two daughter cells. The 'nucleus' of a prokaryotic cell is a single, circular chromosome, without a nuclear membrane and without histones. Prokaryotic cells also lack membrane-bound organelles (mitochondria, lysosomes). There is no endocytosis and intracellular digestive vacuoles. They lack (with few exceptions) steroids. They have unique components (muramate and often diaminopimetate) in cell walls. Some prokaryotes are able to fix atmospheric nitrogen and carry obligate anaerobiosis.

Monera

He proposed a separate kingdom, Monera for the true prokaryotes : the bacteria and cyanobacteria. Higher protists were distributed among two kingdoms. Algae and protozoa were grouped under separate kingdom, Protista, whereas Fungi given the rank of an independent kingdom. The two other kingdoms were the Plantae and Animalia for traditional plants and animals respectively.

WHITTAKER'S FIVE KINGDOM CONCEPT

This system of classification is based on three levels of cellular organization which evolved to accommodate three principal modes of nutrition : photosynthesis, absorption, and ingestion. The prokaryotes are included in the kingdom *Monera*; they lack the ingestive mode of nutrition. Unicellular eukaryotic microorganisms are placed in the kingdom *Protista*; all three nutritional types are represented here. In fact, the nutritional modes are continuous : the mode of nutrition of the microalgae is photosynthetic; the mode of nutrition of the protozoa is ingestive; and the mode of nutrition in some other protists is absorptive with some overlap to the photosynthetic and ingestive modes.

All living organisms are best thought of today as being members of one of the five kingdoms :

Monera

This includes Eubacteria (including rickettsiae, chlamydiae, mycoplasmas), Archaeobacteria, and the Oxyphotobacteria (also called cyanobacteria).

Protista

It consists of algae, protozoa slime molds and water molds. The slime molds and water molds (including other oomycetes) have traditionally been studied by mycologists, and so are usually considered as fungi, but phylogenetically they are said to be more related to protists.

Fungi

It consists of zygomycetes, ascomycetes, basidiomycetes and deuteromycetes.

Plantae

The bryophytes and tracheophytes.

Animalia

The invertebrates and vertebrates.

BERGEY'S CLASSIFICATION

Bergey's Manual of Systematic Bacteriology classified all bacteria in the kingdom Prokaryotae which in turn is divided into 4 divisions.

Division I Gracilicutes

Prokaryotes with a complex cell-wall structure specially Gram-negative bacteria.

Division II Firmicutes

Prokaryotes with a cell-wall structure specially Gram-positive bacteria.

Division III Tenericutes

Prokaryotes that lack a cell wall.

Division IV Mendosicutes

Prokaryotes that show evidence of primitiveness in phylogenetic origin than others.

A comparable manual of classification does not exist for fungi, algae, or protozoa. There are, however, schemes of classification for each group that have wide acceptance and usage. An international system for classification and nomenclature of viruses is in the process of development.

MAIN GROUPS OF MICROBES

We shall now briefly consider the silent features of the main groups of microorganisms.

(A) Bacteria—This group includes the true bacteria (eubacteria) and related organisms i.e., archaeobacteria, blue-green bacteria (cyanobacteria or oxyphotobacteria), mycoplasmas, rickettsiae and chlamydiae.

They are prokaryotes. Reproduction is by simple fission. Nutritional processes are very diverse. Many can thrive in both oxygen-rich and oxygen-free environments. Growth occurs over a wide range of temperatures, usually at a neutral pH. They occur in almost every terrestrial and aquatic environments (Fig. 2.1).

(B) Archaeobacteria— (i) Their 16s rRNA molecules are similar to each other, but differ greatly from those of other bacteria and from eukaryotes.

(ii) Their walls do not contain peptidoglycan, but a range of other unique polysaccharides.

(iii) Their cell membrane is of a single layer of glycero-hydro-carbon-glycerol chains instead of a bilayer of phospholipids arranged tail to tail.

(iv) Their ribosomes are insensitive to chloramphenicol.

(v) They inhabit extreme environments. They include the methanogens; the extreme red halophiles; the salt-loving bacteria of saturated brine and salted fish; and the thermoacidophiles, found in hot sulphur spring or smouldering coal wastes. It is suggested that they represent a very ancient lineage which diverged from the eubacteria very early in the evolutionary process, and have survived only in these specialized ecological niches. The rickettsiae are tiny rods that are transmitted by arthropods and that multiply only in living cells. The chlamydiae are among the smallest bacteria which were formerly considered viruses. The mycoplasmas are the smallest organisms that can be cultivated outside living tissues. Some are involved in lung disorders.

This is a group of small, highly motile bacteria, that adhere to the wall of the host, penetrate to the periplasm where they replicate, causing eventual lysis of cell. Organisms of this type are called bdellovibrios, and perhaps occur in soil. Blue-green bacteria evolve oxygen during photosynthesis and have chlorophyll *a*, a pigment also found in all algae. They are very ancient. They are perhaps major primary producers in the world's oceans, and many are ecologically important as fixers of atmospheric nitrogen. Myxobacteria (also called fruiting bacteria) form fruit bodies, specialized mutlicellular structures like slime molds. Some cells at tips of fruit body produce cysts. They are most complex in behaviour and life cycle patterns found in prokaryotes.

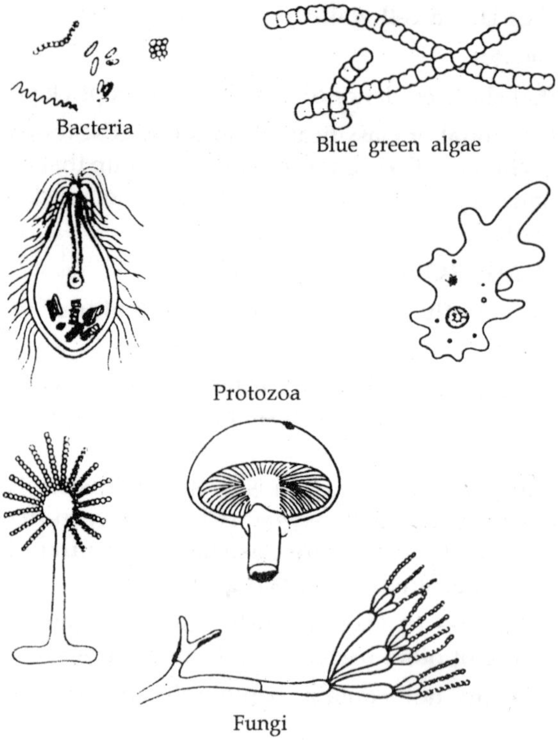

Fig. 2.1 : Microbial complexity and diversity.

(C) Actinomycetes – They are Gram-positive prokaryotes, a large group of filamentous bacteria which characteristically show branching patterns just like those of fungi, to give rise to a spreading mycelium. Also, like fungi, they often readily produce spores when grown in culture.

(i) Mycelial (majority), Gram-positive prokaryotes.

(ii) Some (euactinomycetes) develop only in mycelial state reproducing by unicellular specialized spores (e.g., *Streptomyces, Micromonospora,* mycelial development is transitory and often limited, specialized spores not produced and reproduce primarily by fragmentation into short rod-shaped cells (e.g., *Bifidobacterium, Nocardia*), and the rest are certain Gram-positive, unicellular bacteria which do not form endospores and which are distinguished from the coryneform group by their regular cell shape, and include the lactic acid bacteria *(Pediococcus, Lactobacillus*) and micrococci (*Staphylococcus, Micrococcus*).

(iii) Cell wall has some peptidoglycan as in nearly all Gram-negative prokaryotes. However, a very large number (almost 60) of other peptidoglycan types have been found in these organisms.

(D) Protozoa—To a microbiologist, however, it is not a very helpful definition. It would perhaps be more appropriate to say that they are a group of unicellular non-photosynthetic, eukaryotic microorganisms which normally obtain their food by phagocytosis and which possess no true cell wall. They are involved in many blood and tissue diseases.

Silent features are as follows : The type of movement, used to divide protozoa into major groups—

(i) Amoeboid motion and flagellar movement as in flagellate protozoa, some of which are the colourless counterparts of particular algae. Ciliary movement in ciliates, as in *Paramecium*, whose surface covered with cilia, that are shorter than flagella and have a coordinated motion so that waves of contraction pass over them.

(ii) Food, taken by phagocytosis of solid particles such as bacteria.

(iii) They have generally no cell wall.

(iv) Life cycles less complicated than algae or fungi, exceptions being forms like *Plasmodium*.

(v) Degree of specialization in single cells, as in *Paramecium*. The slime molds form like fungi, macroscopic fruit bodies. But their feeding phase in amoeboid. They live on surface of decaying vegetation. There are two main groups : (i) The cellular slime molds, whose vegetative stage is single amoeboid cells that may aggregate to form a pseudoplasmodium as in *Dictyostelium*, and (ii) acellular slime molds, where a single amoeba produces a multinucleate plasmodium of indefinite size and shape which moves over the surface of the substratum engulfing food particles as it goes. The water molds are of two main types, the oomycetes and the chytrids, originally included among the fungal kingdom. They are shown phylogenetically protists, most obviously because they form flagellated motile cells.

(E) Algae—These are pigmented eukaryotes, ranging in size from microscopic forms to giant kelps of marine waters. The main characters are as follows :

(i) Source of energy is an oxygen-producing photosynthesis occurring in chloroplasts. Within the chloroplast, there are found a variety of chlorophylls, and carotenoids. Chlorophyll *a* always present.

(ii) Found in different morphological types.

(iii) Motility by flagella.

(iv) Cell wall made up of polysaccharide components as pectin, cellulose or xylan, that is sometimes calcified with calcium carbonate.

(v) Reproduction is of sexual and asexual types.

(F) Fungi—Main features are as follows :

(i) Heterotrophic eukaryotic microbes obtaining their food in a soluble form by uptake through plasma membrane.

(ii) Cell wall is usually made up of polysaccharides.

(iii) Motility is totally absent.

(iv) Mycelium made up of individual filaments, the hyphae. Some as yeasts are unicellular i.e., non-mycelial.

(v) Mycelia may be coenocytic, or septate.

(vi) Asexual reproduction by spores.

(vii) Sexual spores are also produced. In some fungi large fruit bodies are produced.

(G) Viruses—Since these have no associated metabolism one can argue whether they are organisms at all. They consist of a fragment of nucleic acid (DNA or RNA, never both) surrounded by a layer of protein and cause important diseases.

Viroids are minute infectious agents of plants which are composed of single naked RNA molecules. Viroid RNAs are very unusual : they are circular, uniquely folded and so small that the largest one so far described (CEV : citrus exocortis viroid) is only 371 nucleotides long-about 1/10th of the size of the smallest RNA virus. Viroids so far are known to cause diseases in plants only.

Prions are associated with certain degenerative diseases of the central nervous system such as kuru, a disease restricted to a few cannibalistic tribes of New Guinea and transmitted by eating uncooked brain, and other diseases such as scrapie of sheep and goats and Crutzfeldt-Jacob disease of humans and animals. Prions appear to be proteins or glycoproteins with no associated nucleic acid of any kind. Since replication of the prion-protein presumably occurs by normal process, this implies that these proteins may be able to bring about the production of their own requisite mRNA. If this were so, prions would have revolutionary impact on molecular biology since they would constitute the only example where genetic information passes from protein to nucleic acid and not the other way around. However, recent evidence shows that scrapie glycoprotein may be a product of a normal host gene, indicating that these infectious agents may be more conventional than thought previously.

CRITERIA TO CLASSIFICATION OF MICROORGANISMS

Several criteria have been used by microbiologists to classify the microbes. These are as follows :

(i) Artificial criteria—The earliest systems of biological classification were artificial. Aristotle, during fourth century BC classified the living organisms into two kingdoms—the animals and the plants. This division was simply based upon (i) ability of active movement and (ii) ability to photosynthesis.

(ii) Natural criteria—In the middle of 18th century Linnaeus (1758) devised a system, that was more useful. This system expressed the biological nature of the objects that it classified.

(iii) Phylogenetic criteria—In the 19th century, the concept of a natural system accordingly changed to phylogenetic ones. However, this approach needs a sound knowledge of fossil records of organisms. We would like to refer here again to Protista. Haeckel separated the unicellular organisms from plants and animals, placing them under a third kingdom, Protista. A modern system of this type is devised by R.H. Whittaker (1969). The microorganisms have been included in the kingdoms Monera, Protista and Fungi in this system.

(iv) Numerical criteria—Michel Adanson in 18th century suggested that the taxonomic arrangement is based upon qualification of the similarities and differences among organisms. There are used as many characters as possible, each of which is of an equal weight. It is assumed that if each phenotypic character is given equal weighting, it should express numerically the taxonomic distances between organisms, in terms of the number of characters they share, relative to the total number of characters examined.

(v) Macromolecular criteria—The growth of molecular biology has now opened up a number of new approaches to microbial taxonomy. For identification and classification of bacteria, a range of criteria are now used. The shape and size of organism, its oxygen, pH and temperature requirements, and laboratory culture characteristics are considered. Photosynthetic ability, capsule production, staining reaction, sporeforming ability and type of movement are other characteristics. Biochemical digestion of certain carbohydrates, proteins and fats are also considered. The melting temperature or buoyant density of DNA (i.e., the temperature at which it becomes denatured, by breakage of the hydrogen bonds holding the two strands together) is directly related to G + C content. Strand separation is accompanied by marked increase in absorbance at 260 nm, the absorption maximum of DNA. When a DNA sample is gradually heated, the absorbance increases as the hydrogen bonds are broken and reaches a plateau at a temperature at which the DNA has all become single-stranded. The midpoint of this rise, the melting temperature (Tm) is a measure of the G + C content. The G + C content may also be determined by subjecting a DNA sample to centrifugation in a CsCl gradient, which affords a precise measure of its density. In both plants and animals, the ranges are relatively narrow and quite similar, centering about a value of 35 to 40 mole percent G + C.

Among the protists, the ranges are much wider, the widest range occurring among the prokaryotes, in which the range extends from about 30 to 75 mole percent G + C. It may be seen that the values vary remarkably among bacteria, from about 30 to 70 mole percent G + C. Various streptococci, pneumococci and lactobacilli have a similar value (38 to 40) which have been traditionally grouped together as lactic acid bacteria, because of their characteristic fermentation. On the other hand, *Lactobacillus bifadus*, very much different with 56 percent G + C, is now renamed *Bifidobacterium*. Within other large groups also, considerable range may be seen which have been grouped into a single genus, mainly on morphological grounds. For example, range in different species of *Proteus, Bacillus,* and *Corynebacterium.* If one examines the mean G + C content of many different stains of a single microbial species, the values are closely similar or identical, as found in several species of *Pseudomonas*. It is very important, as the major genetic divergence reflects a wide evolutionary separation. However, this may not be true, since two organisms with similar mean DNA base composition may differ greatly in genetic constitution.

Similarity of base composition represents only a limited basis for close genetic relatedness, because even distant organisms can by chance have a similar composition. Moreover, as groups diverge in evolution the DNA sequence changes long before the base composition. Homology of sequence (DNA-DNA homology) can be measured quantitatively in terms of the ability of DNA stands from two different sources to form molecular hybrids *in vitro*. Surveys among higher organisms have revealed close parallelism between the results of such hybridization and the phylogenetic relations concluded on other grounds. When DNA preparations from two related strains of bacteria were mixed and treated at

10 to 30°C below the Tm value, hybrid DNA molecules are formed. However, DNA preparations from two unrelated bacteria, could not hybridise. This discovery of reassociation of single-stranded DNA molecules from different biological sources to form hybrid duplexes laid the foundations of DNA-DNA homology in bacterial genetic relatedness. Among bacteria this is useful only within closely related groups because it quickly disappears in the wider range of variation. Since duplexes can also be formed between single-stranded DNA and complementary RNA strands, analogous DNA-RNA reassociations can be performed. Ribosomal RNA hybridization to DNA is useful for estimating more distant kinship among bacteria. RNA sequences have proved more useful than DNA sequences because they involve relatively small segments of the chromosome : those that code the base sequence either of the tRNAs or of the rRNAs. An alternative is thus to sequence the smaller molecules (RNA or protein) derived from DNA. This technique is thus used for comparative sequencing of specific RNA molecules. The molecules are highly conserved in an evolutionary sense and provide a very useful measure of the phylogenetic relationships of the wider groupings of microorganisms.

SPECIFIC NAMES ASSIGNMENTS

Some species of microorganisms (especially pathogens) are also called by informal names such as the gonococcus (*Neisseria gonorrhoeae*) or the leprosy bacillus (*Mycobacterium leprae*), but this is not the usual practice. If we were to adopt common names such as the "little yellow coccus" or the "club-shaped diphtheria bacterium" the terminology would become even more cumbersome and challenging than scientific names. Even worse, common names are notorious for varying from region to region, even within the same country. The method of assigning the scientific or specific name is called a binomial (two-name) system of nomenclature. The scientific name is always a combination of the generic (genus) name followed by the species name. The generic part of the scientific name is capitalized, and the species part begins with a lowercase letter. For example, *Campylobacter jejuni*.

To save space, an organism's scientific name is sometimes given in shorthand form, such as *C. jejuni*. The source for nomenclature is usually Latin or Greek, but if other languages such as English or French are used, the endings of these words are revised to have Latin endings. In general, the name first applied to a species will be the one that takes precedence over all others. An international group oversees the naming of every new organism discovered, making sure that standard procedures have been followed and that there is not already an earlier name for the organism or another organism with that same name. The inspiration for names is extremely varied and often rather imaginative. A number of species have been named in honour of a microbiologist who originally discovered the microbe or who has made outstanding contributions to the field. Some names are given on basis of origin :

(a) *Homo sapiens* (hoh'-moh say'-pee-enz) Gr. *Homo*, man, and *sapiens*, wise.

(b) *Micrococcus luteus* (my"-kroh-kok'-us loo'-tee-us) Gr. *Micros*, small, and *kokkus*, berry; L. *luteus*, yellow. "The little yellow berry".

(c) *Pseudomonas tomato* (soo"-doh-mon'-us toh-may'-toh) Gr. *Pseudo*, false, *monas*, unit, and *tomato*, the fruit. A bacterium that infects the common garden tomato.

(d) *Corynebacterium diphtheriae* (kor-eye''-nee-bak-ter'-ee-yum dif'-theer-ee-aye) Gr. *Coryne,* club, *bacterium,* little rod, and *diphtheriae,* the causative agent of the disease diphtheria. "The club-shaped diphtheria bacterium".

(e) *Campylobacter jejuni* (cam-pee'-loh-bak-ter jee-joo'-nye) Gr. *Kampylos,* curved, *bakterion,* little rod, and *jejunum,* a section of intestine. One of the most important causes of intestinal infection worldwide.

(f) *Lactobacillus sanfrancisco* (lak''-toh-bass-ill'-us san-fran-siss'-koh) L. *Lacto,* milk, and *bacillus,* little rod. A bacterial species used to make sourdough bread.

(g) *Vampirovibrio chlorellavorus* (vam-pye'-roh-vib'-ree-oh klor-ell-ah'-vor-us) F. *Vampire;* L. *vibrio,* curved cell; *Chlorella,* a genus of green algae; and *vorus,* to devour. A small, curved bacterium that sucks out the cell juices of *Chlorella.*

(h) *Giardia lamblia* (jee-ar'-dee-uh lam'-blee uh) for Alfred Giard, a French microbiologist, and Vilem Lambl, a protozoan that causes a severe intestinal infection.

EVOLUTIONARY TRENDS IN MICROORGANISMS

Evolution (evolution, to rollout) is an important theme that underlies all of biology, including microbiology. From its simplest standpoint, evolution states that living things change gradually through hundreds of millions of years, and that these evolvements are expressed in various types of structural and functional changes through many generations. The process of evolution is selective : those changes that most favour the survival of a particular organism or group of organisms tend to be retained and those that are less beneficial to survival tend to be lost. Space does not permit a detailed analysis of evolutionary theories, but its occurrence is supported by a tremendous amount of evidence from the fossil record and from the study of morphology (Gr. *Morphos,* form and *logos,* to study), physiology, (Gr. *Physis,* nature) and genetics (inheritance). Evolution accounts for the millions of different species on earth and their adaptation to its many and diverse habitats. Evolution is founded on two preconceptions : (a) That all new species originate from preexisting species, and (b) that closely related organisms have similar features due to having evolved from common ancestral forms. Usually, evolution progresses toward greater complexity, and evolutionary stages range from simple, more primitive forms that are closer to an ancestral organism to more complex, advanced forms.

Although we use the terms *primitive* and *advanced* to denote the degree of change from the original set of ancestral traits, it is very important to realize that all species presently residing on earth are modern, but some have arisen more recently in evolutionary history than others. The evolutionary patterns of organisms are often drawn as a family tree, with the trunk representing the main ancestral lines and the branches showing offshoots into specialized groups of organisms. This sort of arrangement places the more ancient groups at the bottom and the more recent ones at the top. The branches may also indicate closeness in relationships among the offshoot lines and, if a time scale is added, the beginning and length of an evolutionary period (Fig. 2.2).

In time, it became clear that certain organisms did not truly fit either of these categories, so a third kingdom for simpler organisms that lacked tissue differentiation (protists) was recognized. Eventually, when significant differences became evident even among the protists Robert Whittaker proposed a fourth kingdom for the bacteria and a fifth

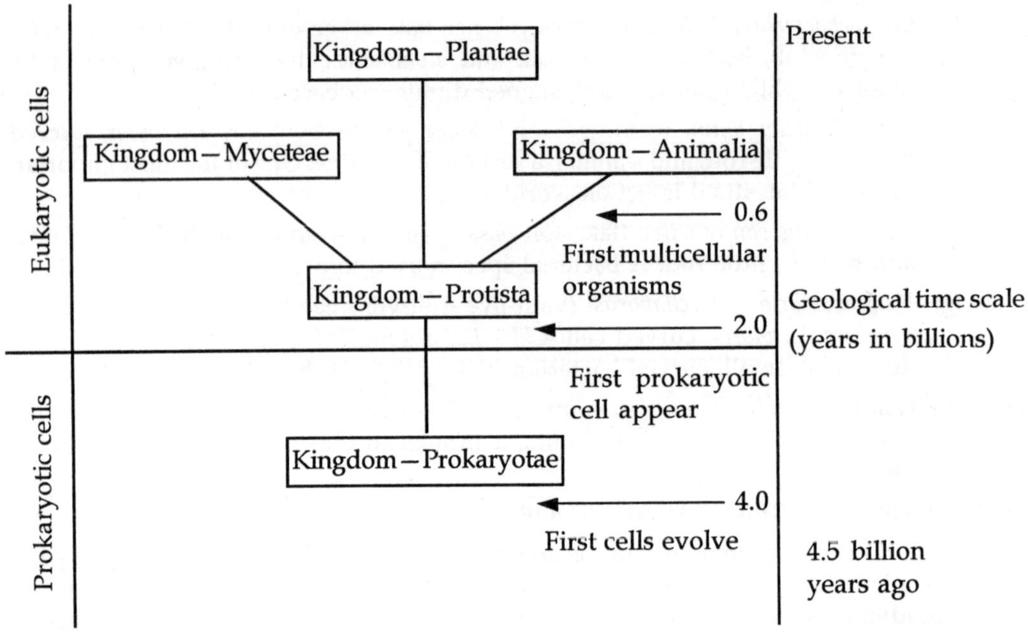

Fig. 2.2 : These kingdoms are designated on basis of structure and type of cells.

one for the fungi. This system places all living things in one of five basic kingdoms : (a) The Prokaryotae or Monera, (b) the Protista, (c) the Myceteae or Fungi, (d) the Plantae, and (e) the Animalia. The simple, single-celled organisms at the base of the family tree are in the kingdom Prokaryotae (also called Monera). Since only those organisms with prokaryotic cells are placed in it, the nature of cell structure is the main defining characteristics for this kingdom. It includes all of the microorganisms commonly known as bacteria. Two main subgroups of the kingdom are the eubacteria (Gr. *Eu*, true, and *bakterion*, little rod), bacteria with typical prokaryotic cell structure and the archaeobacteria (Gr. *Archaios,* ancient), bacteria with a typical cell structure and function, primitive bacteria were the earliest cells to appear on earth and were the original ancestors of both more advanced bacteria and eukaryotic organisms.

The kingdom – Protista (Gr. *Protos,* first) contains mostly single-celled microbes that lack more complex levels of organization, such as tissues. Its members include both the microscopic algae, defined as independent photosynthetic cells with rigid walls; and the protozoans, animal-like creatures that feed upon other live or dead organisms and lack cell walls. The kingdom – Myceteae (Gr. *Mycos,* the fungi) contains the fungi, single or multicelled eukaryotes that are encased in cell walls and absorb nutrients from other organisms. With the exception of certain infectious worms and arthropods, the final two kingdoms, Animalia and Plantae, are generally not included in the realm of microbiology because most are large, multicellular organisms with tissues, organs, and organ systems. In general, animals move freely and feed on other organisms, while plants grow in an attached state and exhibit a nutritional scheme based on photosynthesis.

3

METHODS IN MICROBIOLOGY

Since microbiologists cannot rely as much as other scientists on senses other than sight, they are confronted by some unique problems. First, since most habitats harbor microbes in complex associations, it is often necessary to separate the species from one another. In addition, to maintain and keep track of such elusive research subjects, it is usually necessary to grow them under artificial conditions. A third difficulty in working with microbes is that they are invisible and widely distributed, and undesirable ones can be introduced into an experiment and cause misleading results. These impediments were largely responsible for the development of sterile, pure culture, and aseptic techniques. Microbiologists have five basic techniques to manipulate, grow, examine, and characterize microorganisms : inoculation, incubation, isolation, inspection, and identification. Some or all of these procedures are performed by microbiologists, whether the beginning laboratory student, the researcher attempting to isolate drug-producing bacteria from soil, or the clinical microbiologist working with a specimen from a patient's infection.

ISOLATION OF MICROORGANISMS

Introduction of a tiny sample into a container of nutrient medium (L. *Middle*), which provides an environment in which they multiply. This process is called inoculation (L. *In, in*, and *oculum*, bud). The observable growth that appears in or on the medium is known as a culture (Gr. *Cultus,* to tend). The nature of the sample being cultured depends upon the objectives of the analysis. Clinical specimens for determining the cause of an infectious disease are obtained from body fluids (blood, cerebrospinal fluid), discharges (sputum, urine, feces), or diseased tissue. Culture media are contained in test tubes, flasks, or Petri dishes, and they are inoculated by such tools as loops, needles, pipettes, and swabs. Media are extremely varied in nutrient content and consistency, and can be specially formulated for a particular purpose. Culturing obligate parasites (including viruses and certain bacteria) that cannot grow on artificial media requires live cell cultures or host animals.

For an experiment to be properly controlled, sterile technique is necessary. This means that the inoculation must start with a sterile medium and use inoculating tools with sterile tips. Measures must be taken to prevent introduction of nonsterile materials such as room air and fingers directly into the culture. In some ways, culturing microbes is analogous to gardening on a microscopic scale. Cultures can be compared to carefully separated plots of tiny plants in which extreme care is taken to exclude weeds. A pure culture is a container of medium that grows only a single known species or type of microorganism. This type of culture is most frequently used for laboratory study, because it allows the systematic examination and control of one microorganism by itself. Instead of the term, *pure culture* some microbiologists prefer the terms axenic (Gr. *A*, no, and *xenos*, stranger), meaning that the culture is free of other living things except for the one being studied.

A mixed culture is a container that holds two or more *identified*, easily differentiated species of microorganisms, not unlike a garden plot containing both carrots and onions. A contaminated (Gr. *Con*, together, and L. *tangere*, to touch) culture was once pure or mixed but has since had contaminants introduced into it, like weeds into a garden, because contaminants have the potential for causing disruption, constant vigilance is required to exclude them from microbiology laboratories, as you will do doubt witness from your own experience. Isolation techniques are based on the concept that if an individual microbial cell is separated from other cells and provided adequate space on a nutrient surface, it will grow into a discrete mound of cells called a colony. Because it was formed from a single cell, a colony consists of just that single species and no other. To achieve proper isolation, a small number of cells must be inoculated into a relatively large volume or over an expansive area of medium. It generally requires the following materials : A medium that has a relatively firm surface, a Petri plate and an inoculating loop. In the streak plate method, a small droplet of culture or sample is spread over the surface of the medium according to a pattern that gradually thins out the sample and separates the cells spatially over several sections of the plate. In the loop dilution, or pour plate technique, the sample is inoculated serially into a series of cooled but still liquid agar tubes so as to dilute the number of cells in each successive tube in the series. Inoculated tubes are then plated out (poured) into sterile Petri dishes and allowed to solidify (harden). The end result (usually in the second or third plate) is that the number of cells per volume is so decreased that cells have ample space to grow into separate colonies.

NUTRITIONAL REQUIREMENTS OF MICROORGANISMS

Numerous media (singular, medium) have been developed for bacterial cultivation. Because the nutritional requirement of bacteria vary widely, there are great differences in the chemical compositions of the media used in the laboratory. Bacteria also exhibit wide differences with respect to the physical conditions favouring their growth, such as temperature, pH, and gaseous environment. All forms of life, from microorganisms to human beings, share certain nutritional requirements for growth and normal functioning. All organisms require a source of energy. Some rely on chemical compounds for their energy and are designated as chemotrophs. Others can utilize radiant energy (light) and are called phototrophs. Both chemotrophs and phototrophs exist among bacteria. All organisms require a source of electrons for their metabolism. Some organisms can use reduced inorganic compounds as electron donors and are termed lithotrophs. Other organisms use organic compounds as electron donors and are called organotrophs. All organisms require carbon in some form for use in synthesizing cell components. All organisms require at least small amounts of CO_2. However, some can use CO_2 as their major, or even sole, source of carbon; such organisms are termed autotrophs. Others require organic compounds as their carbon source and are termed heterotrophs.

Bacteria are extremely versatile in this respect. Unlike eukaryotes, some bacteria can use atmospheric nitrogen. Others thrive on inorganic nitrogen compounds such as nitrates, nitrites, or ammonium salts and still others derive nitrogen from organic compounds such as amino acids. All organisms require oxygen, sulphur and phosphorus for cell components. Sulphur is needed for synthesis of certain amino acids (cysteine, and methionine). Some bacteria require organic sulphur compounds, some are capable of utilizing inorganic sulphur compounds, and some can even use elemental sulphur. Phosphorus, usually supplied in the form of phosphate, is an essential component of nucleotides, nucleic acids, phospholipids, teichoic acids, and other compounds. All living organisms require metal ions, such as K^+, Ca^{2+}, Mg^{2+}, and Fe^{2+} for normal growth. Other metal ions are also needed but usually only at very low concentrations, such as Zn^{2+}, Cu^{2+}, Mn^{2+}, Mo^{6+}, Ni^{2+}, B^{3+}, and Co^{2+}; these are often termed trace elements and often occur as contaminants of other components of culture media in amounts sufficient to support bacterial growth. Most bacteria do not require Na^+, but certain marine bacteria, cyanobacteria, and photosynthetic bacteria do require it. For those members of the archaeobacteria known as the 'red extreme halophiles', the requirement is astonishing : they cannot grow with less than 12 to 15 percent NaCl. All living organisms contain vitamins and vitamin like compounds. These function either as coenzymes for several enzymes or as the building blocks for coenzymes. Some bacteria are capable of synthesizing their entire requirement of vitamins from other compounds in the culture medium, but others cannot do so and will not grow unless the required vitamins are supplied preformed to them in the medium. All living organisms require water, and in the case of bacteria all nutrients must be in aqueous solution before they can enter the cells. Water is a highly polar compound that is unequaled in its ability to dissolve or disperse cellular components and to provide a suitable milieu for the various metabolic reactions of a cell.

Phototrophs

The phototrophic bacteria are species that use inorganic compounds as their source of electrons (i.e., photolithotrophs). For example, *Chromatium okenii* uses H_2S as its electron donor, oxidizing it to chemical sulphur :

$$H_2S \rightarrow S + 2e^- + 2H^+$$

Some phototrophic bacteria use organic compounds such as fatty acids and alcohols as electron donors and are therefore photoorganotrophs. For example, *Rhodospirillum rubrum* can use succinate as an electron donor :

$$\text{Succinate} \rightarrow \text{fumarate} + 2e^- + 2H^+$$

Certain phototrophic bacteria are not restricted to being phototrophic. As indicated before, chemotrophs rely on chemical compounds rather than light for their energy, and under same circumstances a phototrophic bacterium can grow as a chemotroph.

Chemotrophs

The chemotrophic bacteria are species that use inorganic compounds as their source of electrons (i.e., chemolithotrophs). For example, bacteria of the genus *Nitrosomonas* use ammonia as their electron source, obtaining energy by oxidizing ammonia to nitrite :

$$NH_4^+ + {}^3/_2\, O_2 + H_2O \rightarrow NO_2^- + 2H_2O^+$$

This reaction involves a net transfer of 6 electrons, causing a valence change of the nitrogen atom from -3 to +3. Certain bacteria can grow as either chemolithotrophs or chemoorganotrophs. For example, *Pseudomonas pseudoflava* can use either the organic compound glucose or the inorganic compound H_2 as its source of electrons :

$$\text{Glucose} + 6H_2O \rightarrow 6CO_2 + 24H^+ + 24e^-$$

$$H_2 \rightarrow 2H^+ + 2e^-$$

Autotrophs

Some autotrophs are facultative autotrophs; i.e., they can either live as autotrophs, deriving their carbon from CO_2, or they can live as heterotrophs, deriving their carbon from organic compounds. For example, *P. pseudoflava* can live as a heterotroph, using glucose as a source of carbon for assimilation (and also as its source of electrons, as mentioned above); however, if H_2 is provided as the electron source, then it can use CO_2 as its sole carbon source and can grow as an autotroph. The fact that an organism can grow and reproduce in such a mixture of inorganic compounds indicates that it has an elaborate capacity for synthesis. That is, the organism can transform these compounds into the carbohydrates, proteins, nucleic acids, lipids, vitamins, and other complex organic substances that constitute the living cell.

Heterotrophs

Heterotrophic bacteria have been studied more widely in comparison to autotrophs because heterotrophs, in a sense, are of more immediate concern to us : it is here that we find all the species that cause diseases of human beings, other animals and plants, as well as those that constitute the greater part of the microbial population in our immediate environment. The heterotrophic bacteria, although they constitute one major nutritional group, vary considerably in the specific nutrients required for growth, particularly with respect to their organic carbon sources, nitrogen sources, and vitamin requirements. Organisms such as lactobacilli that have elaborate requirements for specific nutrients, i.e., vitamins and other growth-promoting substances, are designated fastidious heterotrophs.

Obligate Parasites

At present, such bacteria can be propagated only in association with a living host which, in a sense, serves as the medium. One example is the bacterium that causes leprosy, *Mycobacterium leprae,* which can be cultivated by infecting mice or armadillos. Other examples include the rickettsias, the chlamydias, and the spirochete that causes syphilis, *Treponema pallidum.*

MEDIA : BACTERIAL NUTRITIONAL SUPPLY

Media are needed for cultivation of autotrophs and also useful for nutritional requirements for heterotrophs. However, for routine cultivation of heterotrophs, chemically defined media are not generally used, so certain complex raw materials such as peptones, meat has available numerous media.

Media Types

Media can be classified in three types :

(a) Liquid forms of media

Liquid media are defined as water-based solutions that do not solidify at temperatures above freezing and that tend to flow freely when the container is titled. These media, termed broths, milks, or infusions, are made by dissolving various solutes in distilled water. Growth occurs throughout the container and can then present a dispersed cloudy or particulate appearance. A common laboratory medium, nutrient broth, contains beef extract and peptone dissolved in water. Methylene blue milk and litmus milk are opaque liquids containing whole milk and dyes. Fluid thioglycollate is a slightly viscous broth used for determining oxygen utilization of bacteria. At ordinary room temperature, semisolid media exhibit a clotlike consistency. This is because they contain an amount of solidifying agent (agar or gelatin) that thickens them but does not produce a firm substrate. Semisolid media are used to determine the motility of bacteria and to localize a reaction at a specific site. Both motility test medium and SIM contain a small amount (0.3-0.5%) of agar. The medium is stabbed carefully in the center and later observed for the pattern of growth around the stab line.

(b) Solid forms of media

Solid media provide a firm surface on which cells can form discrete colonies and are advantageous for isolating and subculturing bacteria and fungi. They come in two forms : liquefiable and nonliquefiable. Liquefiable solid media, sometimes called reversible solid media, contain a solidifying agent that is thermoplastic—its physical properties change in response to temperature. By far the most widely used and effective of these agents is agar (The Malay word-sea weed), a complex polysaccharide isolated from the red alga *Gelidium*. The benefits of agar are numerous. It is solid at room temperature and most incubation temperatures, and it melts (liquefies) at the boiling temperature of water (100°C). Once liquefied, agar does not resolidify until it cools to 42°C, allowing it to be inoculated and poured in liquid form at temperatures (45-50°C) that will not harm the microbes or the handler, Agar is flexible and moldable, and it provides a basic framework to hold moisture and nutrients, though it is not itself a digestible nutrient for the vast majority of microorganisms. Any medium containing 1% to 5% agar usually has the word agar in its name. Nutrient agar is a common one. Like the broth, it contains beef extract and peptone, as well as 1.5% agar by weight. Many of the examples covered in the section on functional categories of media contain agar. Although gelatin is not nearly as satisfactory as agar, it will create a reasonably solid surface in concentrations of 10% to 15%.

(c) Nonliquefiable forms of media

Nonliquefiable solid media have less versatile applications than agar media because they are not thermoplastic. They include materials such as rice grains (used to grow fungi), cooked meat media (good for anaerobes), and potato slices; all of these media start out solid and remain solid after heat sterilization. Other solid media containing egg and serum start out liquid and are permanently coagulated or hardened by moist heat.

Chemical Composition of Media

Synthetic media come in many forms. Some media such as minimal media for fungi contain nothing more than a few essential compounds such as salts and amino acids dissolved in water. Others contain a variety of defined organic and inorganic chemicals (Fig. 3.2). Such standardized and reproducible media are most useful in research and cell culture, where the exact nutritional needs of the test organisms are known. If even one component of a given medium is not chemically definable, the medium belongs in the next category. Complex or nonsynthetic media contain at least one ingredient that is not chemically definable—not a simple, pure compound and not representable by an exact chemical formula. Most of these substances are extracts of animals, plants, or yeasts, including such material as ground-up cells, tissues, and secretions. Examples are blood, serum, and meat extracts or infusions. Infusions are high in vitamins, minerals, proteins, and other organic nutrients. Other nonsynthetic ingredients are milk, yeast extract, soyabean digests, and peptone. Peptone is a partially digested protein, rich in amino acids, that is often used as a carbon and nitrogen source. Nutrient broth, blood agar, and EMB agar, though different in function and appearance, are all nonsynthetic media. A specific example can be used to compare what differentiates a synthetic medium from a nonsynthetic one. Both synthetic *Euglena* medium (Fig. 3.1) and nonsynthetic nutrient broth contain

Chemical constituents	Quantity
Glycine	5.0 g/l
Glutamic acid	6.0 g/l
Aspartic acid	4.0 g/l
Sucrose	30 g/l
Malic acid	2.0 g/l
Succinic acid	1.04 g/l
Boric acid	1.14 mg/l
Thiamine HCl	12 mg/l
Monopotassium PO_4	0.6 g/l
$MgSO_4 . 7H_2O$	0.8 g/l
$CaCO_3$	0.16 g/l
Ammonium carbonate	0.72 g/l
$FeCl_3$	60 mg/l
Zinc sulphate	40 mg/l
$CuSo_4 . 5H_2O$	0.62 mg/l
Manganese sulphate	6.0 mg/l
Cobalt sulphate	5.0 mg/l
Ammonium molybdate	1.34 mg/l

Fig. 3.1 : Chemical composition of growth medium for *Euglena*.

amino acids. But *Euglena* medium has three known amino acids in known amounts, while nutrient broth contains amino acids (in peptone) in variable types and amounts. Special media used for cultivation of *Azotobacter* (Fig. 3.2), *Rhizobium* (Fig. 3.3) and members of actinomycetes (Fig. 3.4).

Chemical constituents	Quantity
Sucrose	20 g/l
K_2HPO_4	1.0 g/l
$MgSO_4 \cdot 7H_2O$	0.50 g/l
NaCl	0.50 g/l
$FeSO_4$	1.0 g/l
Na_2MoO_4	0.005 g/l
$CaCO_3$	2.0 g/l
Agar-Agar	15.0 g/l

Fig. 3.2 : Jensen's medium for *Azotobacter* bacteria.

Chemical constituents	Quantity
Mannitol	10.0 g/l
K_2HPO_4	0.50 g/l
$MgSO_4 \cdot 7H_2O$	0.20 g/l
NaCl	0.10 g/l
Yeast extract	1.0 g/l
Agar-Agar	20 g/l

Fig. 3.3 : YEMA medium for *Rhizobium* species.

Chemical constituents	Quantity
Dextrose	1.0 g/l
KH_2PO_4	0.10 g/l
$NaNO_3$	0.10 g/l
KCl	0.10 g/l
$MgSO_4 \cdot 7H_2O$	0.10 g/l
Agar-Agar	15.0 g/l

Fig. 3.4 : Medium for members of actinomycetes.

Media are used for primary isolation, to maintain cultures in the lab, to determine biochemical and growth characteristics, and for numerous other functions. Note that some media can serve more than one function. For example, a medium such as brain-heart infusion is general purpose and enriched; mannitol salt agar is both selective and differential; and blood agar is both enriched and differential. General purpose media are designed to grow as broad a spectrum of microbes as possible. As a rule, they are nonsynthetic and contain a mixture of nutrients that could support the growth of pathogens and nonpathogens alike. Examples include nutrient agar and broth, brain-heart infusion, and trypticase soy agar (TSA). TSA contains partially digested milk protein (casein), soyabean digest, NaCl and agar. An enriched medium contains complex organic substances such as blood, serum, hemoglobin, or special growth factors (vitamins, amino acids) that certain species must have in order to grow. Bacteria that require growth factors and complex nutrients are termed fastidious (L. *Fastidium*, loatting or disgust), For example, the fastidious species *Streptococcus pneumoniae* (a cause of pneumonia) is cultured on blood agar, which is made by adding sterile sheep, horse, or rabbit blood to a sterile agar base. Pathogenic *Neisseria* (one species causes gonorrhea) are grown on Thayer-Martin medium or chocolate agar, which is essentially cooked blood agar.

Selective and Differential Media

Selective media are very important in primary isolation of a specific type of microorganism from samples containing a highly mixed population—for example, feces, saliva, skin, water, and soil. They hasten isolation by suppressing the background organisms and favouring growth of the desired ones. Mannitol salt agar (MSA) contains a concentration of NaCl (7.5%) that is quite inhibitory to most human pathogens. One exception is the genus *Staphylococcus*, which grows well in this medium and consequently can be amplified in very mixed samples. Bile salts, a component of feces, inhibit most Gram-positive bacteria while permitting many Gram-negative rods to grow. Media for isolating intestinal pathogens (MacConkey agar, eosin-methylene blue [EMB] agar) contain these salts as a selective agent. Dyes such as methylene blue and crystal violet also inhibit certain Gram-positive bacteria. Other agents that have selective properties are antimicrobic drugs and acid. Some selective media contain strongly inhibitory agents to favour the growth of a pathogen that would otherwise be overlooked because of its low numbers in a specimen. Selenite and brilliant green dye are used in media to isolate *Salmonella* from feces, and tellurite is used to isolate oral streptococci from saliva.

A differential medium can grow several types of microorganisms, shows up as variations in colony size and colour, in media colour changes, and in the formation of gas bubbles and precipitates. The disparities in appearance are due to the type of agents added and the way the cells react to them. For example, when microbe X metabolizes a certain substance not used by organism Y, then X will cause a visible change in the medium and Y will not. The simplest differential media demarcate two reaction types—showing the use or nonuse of a particular nutrient, or one colony reacting with a dye and the other one not. Dyes serve as differential agents by acting as pH indicators, changing colour in response to the production of an acid or base. For example, MacConkey agar contains lactose and neutral red, a dye that is yellow when neutral and pink or red when acidic. A common intestinal bacterium like *Escherichia coli* that gives off acid when it metabolizes lactose develops red to pink colonies, and one like *Salmonella* that does not give off acid remains its natural colour (off-white). Mitissalivarius agar differentiates oral *Streptococcus* species by means of sucrose. Whereas *S. salivarius* can use the sucrose to synthesize a slime layer and produce large "gumdrop" colonies, *S. mitis* does not do this and remains tiny and flat.

Miscellaneous Media

Reducing media are important for growing anaerobic bacteria or determining oxygen requirements. Carbohydrate fermentation media contain sugars that can be fermented (converted to acids) and a pH indicator to show this reaction. Transport media are used to maintain and preserve specimens that have to be held for a period of time prior to clinical analysis, or to sustain delicate species that die rapidly if not held under stable conditions. Stuart's and Amies transport media contain salts, buffers, and absorbants to prevent cell destruction by enzymes, pH changes, and toxic substances, but will not support growth. Assay media are used by technologists to test the effectiveness of antimicrobial drugs and by drug manufacturers to assess the effect of disinfectants, antiseptics, cosmetics, and preservatives on the growth of microorganisms. Enumeration media are used by industrial and environmental microbiologists to count the numbers of organisms in milk, water, food, soil, and other samples.

Preparation of Culture Media

Media of the nutrient broth or nutrient agar type are prepared by compounding the required individual ingredients or, more conveniently, by adding water to a dehydrated product which contains all the ingredients. Practically all media are available commercially in powdered form.

The preparation of bacteriological media usually involves the following steps :

(a) Individual chemical constituent or the complete dehydrated medium, is dissolved in the appropriate volume of distilled water.

(b) Medium pH is determined with a pH meter.

(c) For solidification of medium, agar is added and the medium is boiled to dissolve the agar.

(d) The medium is poured into tubes or flasks.

(e) The medium is sterilized, generally by autoclaving (15 lbs/inch^2). Some media (or specific ingredients) that are heat-labile are sterilized by filtration.

ASEPTIC CONDITIONS REQUIRED FOR GROWTH OF MICROBES

Temperature

The temperature that allows for most rapid growth during a short period of time (12 to 24 h) is known as the optimum growth temperature. It can be seen that the maximum temperature at which growth occurs is usually quite close to the optimum temperature, whereas the minimum temperature for growth is usually much lower than the optimum. Many microbiologists restrict the term *psychrophile* to organisms that can grow at 0°C but have an optimum temperature of 15°C or lower and a maximum temperature of about 20°C; the term *psychrotroph* or *facultative psychrophile* is used for those organisms able to grow at 0°C but which grow best at temperatures in the range of about 20 to 30°C. During isolation of strict psychrophiles it is usually necessary to maintain the source samples at cold temperatures from the time, they are collected and also to chill all media before attempting isolation. This is because strict psychrophiles usually die if they are even temporarily exposed to room temperature. The physiological factors responsible for the low temperature maxima for strict psychrophiles are not entirely clear, but some factors that have been implicated are heat instability of ribosomes and various enzymes, increased leakage of cell components, and impaired transport of nutrients above the maximum temperature. Mesophiles grow best within a temperature range of approximately 25 to 40°C. For example, all bacteria that are pathogenic for humans and warm-blooded animals are mesophiles, most growing best at about body temperature (37°C). Thermophiles grow best at temperatures above 45°C. The growth range of many thermophiles extends into the mesophilic region; these species are designated *facultative thermophiles.* Factors that have been implicated in the ability to grow at high temperatures are an increased thermal stability of ribosomes, membranes, and various enzymes. Loss of the fluidity that exists within the lipid bilayer of the cytoplasmic membrane may be a factor governing the minimum temperature. It is important to note that a bacterial species may not manifest the same characteristics in every detail when grown at different temperatures. For example, *Serratia marcescens* forms a blood-red to orange pigment when cultured at 25°C but produces little or no pigment when cultured at 37°C.

Gases

Bacteria display such a wide variety of responses to free oxygen that it is convenient to divide them into four groups on the following bases :

(a) Aerobic bacteria require oxygen for growth and can grow when incubated in an air atmosphere (i.e., 21 percent oxygen).

(b) Anaerobic bacteria do not use oxygen to obtain energy; moreover, oxygen is toxic for them and they cannot grow when incubated in an air atmosphere. Some can tolerate low levels of oxygen, but others cannot tolerate even low levels and may die upon brief exposure to air.

(c) Facultatively anaerobic bacteria do not require oxygen for growth, although they may use it for energy production if it is available. They are not inhibited by oxygen and usually grow as well under an air atmosphere as they do in the absence of oxygen.

(d) Microaerophilic bacteria require low levels of oxygen for growth but cannot tolerate the level of oxygen present in an air atmosphere.

Gaseous Toxicity

Oxygen is both beneficial and poisonous, because its strong oxidizing ability makes it an excellent terminal electron acceptor for the energy-yielding process known as respiration. However, oxygen is also a toxic substance. Aerobic and facultative organisms have developed protective mechanisms that greatly mitigate this toxicity, but microaerophiles and anaerobes are deficient in these mechanisms and are restricted to habitats where little or no oxygen is present. Molecular oxygen can directly oxidize certain essential reduced groups, such as thiol (-SH) groups, or enzymes, resulting in enzyme inactivation. For instance, the enzyme complex known as nitrogenase, responsible for nitrogen fixation, is irreversibly destroyed by even small amounts of oxygen. Various cellular enzymes catalyze chemical reactions involving molecular oxygen; some of these reactions can result in addition of a single electron to an oxygen molecule, thereby forming a superoxide radical (O_2^-.). Superoxide radicals can inactivate vital cell components. However, recent studies suggest that their greatest detrimental action is through production of even more toxic substances such as hydrogen peroxide (H_2O_2) and hydroxyl radicals (OH.).

Hydrogen peroxide is not a free radical, but it is a powerful oxidizing agent that is highly toxic to many kinds of cells. Another toxic derivative of oxygen is an energized form known as singlet oxygen $(^{1}\Delta_{g})O_{2}$, which is produced in biological systems by certain photochemical reactions. Aerobic and facultative organisms, have developed various protective mechanisms against the toxic forms of oxygen. In general, anaerobic bacteria have either no superoxide dismutase or only relatively low levels compared to aerobes. Many anaerobes are also deficient in catalase and/or peroxidase.

Conditions for Aerobic Bacteria

When aerobic organisms are to be grown in large quantities, it is advantageous to increase the exposure of the medium to the atmosphere. This can be accomplished by dispensing the medium in shallow layers, for which special containers are available. Aeration can also be increased by constantly shaking the inoculated liquid cultures.

Conditions for Anaerobic Bacteria

During preparation, the culture medium is boiled for several minutes to drive off most of the dissolved oxygen. A reducing agent, e.g., cysteine, is added to further lower the oxygen content. Oxygen-free N_2 is bubbled through the medium to keep it anaerobic. The medium is then dispensed into tubes which are being flushed with oxygen-free N_2, stoppered tightly, and sterilized by autoclaving. Such tubes can be stored for many months before being used.

Use of prereduced media for cultivation of stringent anaerobes. (A) Tube of prereduced medium containing an atmosphere of oxygen-free N_2. (B) To inoculate, the stopper is removed and a gas cannula inserted to flush the tube continuously with oxygen-free CO_2 and maintain anaerobic conditions. The medium is inoculated with a few drops of culture by means of a Pasteur pipette. (C) After inoculation, the tube is restoppered and incubated.

pH of Media

For most bacteria the optimum pH for growth lies between 6.5 and 7.5, and the limits generally lie somewhere between 5 and 9. However, a few bacteria prefer more extreme pH values for growth. For example, *Thiobacillus thiooxidans* has an optimum pH of 2.0 to 3.5 and can grow in a range between pH 0.5 and 6.0. On the other hand, an unclassified bacterium isolated from an alkaline spring in California was found to grow best at a pH of 9.0 to 9.5 and could grow within a range from 8.0 to 11.4. When bacteria are cultivated in a medium originally adjusted to a given pH, for example, 7.0, it is very likely that this pH will change as a result of the chemical activities of the organism. If a carbohydrate is present, it may be fermented or oxidized to organic acids, thus decreasing the pH of the medium.

METHODS OF SELECTION OF PURE CULTURES

In order to achieve its isolation into pure culture, it is helpful—and often necessary— —to first achieve an increase in the relative number of the species, preferably to the point where the species becomes the numerically dominant component of the population. This can be accomplished by the use of selective methods. These methods favour the growth of the desired species while discouraging, or even killing, the other organisms present in the mixed culture. Chemical, physical, or biological methods are used in order to achieve selection of a particular kind of bacterium.

Chemical Methods

This particular kind of selection is often referred to by a special name, enrichment. For example, if we wish to isolate, from soil, bacteria capable of utilizing a very complex organic compound like α-conidendrin, a constituent of wood, we find that when we inoculate a medium such as nutrient agar directly with the soil sample, our chances of finding α-conidendrin—utilizing bacteria will be very limited. There are so many other rapidly growing bacteria present that the more slowly growing kind we wish to obtain will be soon overgrown. Consequently, we prepare a liquid enrichment medium in which α-conidendrin, is the sole source of carbon. Under these conditions, only organisms capable of utilizing this compound will be able to grow well. However, it is important to recognize that other bacteria may be able to grow to some extent by utilizing organic compounds made by the conidendrin-utilizing organisms and that the method is not completely specific. As another example, if we wish to select for nitrogen-fixing bacteria, nitrogen gas (N_2) can be supplied as the sole nitrogen source, since only nitrogen-fixing bacteria will be able to grow well. Certain aquatic bacteria, such as *Caulobacter* species, are capable of growing with very low levels of carbon or nitrogen source. Consequently, one way to select for such bacteria is to inoculate a mixed culture into a very dilute medium, e.g., a broth containing only 0.01 percent peptone. The medium must have low enough levels of nutrients that other kinds of organisms will not be grow well in it.

Intestinal bacteria can grow in the presence of bile salts such as sodium deoxycholate, whereas nonintestinal bacteria are usually inhibited. Consequently, a medium containing crystal violet dye plus sodium deoxycholate will allow Gram-negative intestinal bacteria to grow but will inhibit most other kinds of bacteria. An example of such a medium is MacConkey agar, which is widely used to select for Gram-negative intestinal pathogens such as *Salmonella* and *Shigella* species. *Campylobacter jejuni* is a frequent cause of diarrhea in humans, yet diarrheic stool samples contain many other kinds of bacteria that interfere with the isolation of this species. By incorporating certain antibiotics or other chemotherapeutic agents, such as vancomycin, polymyxin, and trimethoprim, into the culture medium, most of these contaminants can be inhibited without affecting the growth of *C. jejuni*.

Physical Methods

Vegetative cells will be killed (80°C for 15 min) by this treatment but endospores will survive and subsequently germinate and grow. To select for psychrophilic or psychrotrophic bacteria, cultures are incubated at low temperatures e.g., 0 to 5°C. For selection of thermophiles, a high incubation temperature is used e.g., 55°C. To select for acid-tolerant bacteria, a low-pH medium can be used. For example, to select for the lactobacilli present in cheddar cheese, the pH of the medium is maintained at 5.35 with an acetic acid/acetate buffer; other organisms in the cheese cannot grow well at such a low pH. Similarly, to select for alkali-tolerant organisms, a high-pH medium can be used. For example, to select for the cholera-causing bacterium, *Vibrio cholerae*, from a stool sample, we can use a medium with a pH of 8.5; most other intestinal bacteria are unable to grow at this pH.

A membrane filter having a pore size of 0.15 μm is placed on the surface of an agar plate. The unusually small size of treponemes favours them to penetrate the pores of the filter to reach the underlying agar. Moreover, treponemes have the ability to swim through solid agar media; consequently, they migrate away from the filter and grow to form a hazy zone within agar, from which they can be subcultured. Other bacteria from the oral cavity are either too large to penetrate the membrane filter or, if they can penetrate are unable to migrate away through the agar.

Biological Methods

A sputum sample containing *Streptococcus pneumoniae* is ordinarily contaminated by many other bacterial species. However, laboratory mice are extremely susceptible to infection by *S. pneumoniae*, and if the sputum sample is injected into a mouse, the pathogen will multiply extensively. Nonpathogenic bacteria present in the sample will either be inhibited or killed by the defense mechanisms of the animal. In a sensitive animal serves as the selective medium. Moreover, you can often devise a satisfactory selective procedures comparing the characteristics of the species sought with those of the accompanying contaminants. Differences in these characteristics, e.g., in susceptible to certain antibiotics, can provide the basis for a suitable selective procedure.

MAINTENANCE OF PURE CULTURES

Most major biological companies maintain large culture collections. The strains are used for screening of new, potentially effective chemotherapeutic agents; as assay tools for vitamins and amino acids; as agents for the production of vaccines, antisera, antitumor agents, enzymes, and organic chemicals; and as reference cultures that are cited in company patents. For these and other purposes it is extremely important to have properly identified and cataloged strains of bacteria available. Consequently, a considerable amount of research has been performed to develop methods whereby bacterial strains can be preserved and stored until they are needed. Several different methods have been developed, since not all bacteria respond in a similar manner to a specific method. Moreover, there are various practical considerations such as the amount of labour involved and the amount of storage space required. However, all the methods which we will now describe have the same objective : to maintain strains alive and uncontaminated and to prevent any change in their characteristics.

The culture medium, the storage temperature, and the time interval at which the transfers are made vary with the species and must be ascertained beforehand. The temperature and the type of medium chosen should support a slow rather than a rapid rate of growth so that the time interval between transfers can be as long as possible. Many of the more common heterotrophs remain viable for several weeks or months on a medium like nutrient agar. The transfer method has the disadvantage of failing to prevent changes in the characteristics of a strain due to the development of variants and mutants. Many bacteria can be successfully preserved by covering the growth on an agar slant with sterile mineral oil. The oil must cover the slant completely; to ensure this, the oil should be about ½ in above the tip of the slanted surface. Maintenance of viability under this treatment varies with the species. This method of maintenance has the unique advantage that you can remove some of the growth under the oil with a transfer needle, inoculate a fresh medium, and still preserve the original culture. Most bacteria die if cultures are allowed to become dry, although spore- and cyst-formers can remain viable for many years. However, freeze-drying can satisfactorily preserve many kinds of bacteria that would be killed by ordinary drying. In this process a dense cell suspension is placed in small vials and frozen at -60 to -78°C. The vials are then connected to a high-vacuum line. The ice present in the frozen suspension sublimes under the vacuum, i.e., evaporates without first going through a liquid water phase. This results in dehydration of the bacteria with a minimum of damage to delicate cell structures. The vials are then sealed off under a vacuum and stored in a refrigerator. Many species of bacteria preserved by this method have remained viable and unchanged in their characteristics for more than 30 years. Only minimal storage space is required; hundreds of lyophilized cultures can be stored in a small area. Furthermore, the small vials can be sent conveniently through the mail to other microbiology laboratories when packaged in special sealed mailing containers. Lyophilized cultures are revived by opening the vials, adding liquid medium, and transferring the rehydrated culture to a suitable growth medium.

The cell suspension is sealed into small ampoules or vials and then frozen at a controlled rate to-150°C. The ampoules or vials are then stored in a liquid nitrogen refrigerator either by immersion in the liquid nitrogen (-196°C) or by storage in the gas phase above the liquid nitrogen (-150°C). The liquid nitrogen method has been successful with many species that cannot be preserved by lyophilization, and most species can remain viable under these conditions for 10 to 30 years or more without undergoing change in their characteristics. However, the method is relatively expensive, since the liquid nitrogen in the refrigerators must be replenished at regular intervals to replace the loss due to evaporation.

COLLECTION OF PURE CULTURES

Subcultures of some strains were often sent to other microbiologists; other subcultures were received and added to the scientist's own collection. Certain strains had taxonomic importance because they formed the basis for descriptions of species and genera. Others had special properties useful for various purposes. However, many important strains became lost or were inadequately maintained. Thus, it became imperative to establish large central collections whose main purpose would be the acquisition, preservation, and distribution of authentic cultures of living microorganisms. In France a collection of bacteria is maintained at the Institute Pasteur in Paris, in England the National Collection of Type Cultures is in London; the Federal Republic of Germany maintains the Deutsche Sammlung von Microorganisms in Darmstadt; and Japan maintains a large collection at the Institute for Fermentation in Osaka. Many other such collections exist. In the United States the major collection is the American Type Culture Collection (ATCC), located in Rockville, Maryland. In 1980 the collection included the following numbers of strains : bacteria, 11,500; bacteriophages, 300; fungi and fungal viruses, 13,700; protozoa, 720; algae, 130; animal-cell cultures, 500; animal viruses, rickettsiae, and chlamydiae, 1,135, and plant viruses, 220. More than 1 million ampoules of lyophilized or frozen living strains are inventoried and stored at the ATCC. Other large collections in the United States are more specialized in scope. For example, the Northern Utilization Research and Development Division, USDA, at Peoria, Illinois, maintains a collection of yeasts, molds, and bacteria especially for use in fermentations. The Quartermaster Research and Development Center, U.S. Army, Natick, Massachusetts, maintains a collection of microbial strains that are associated with deterioration processes. A number of smaller collections of a specialized nature also exist, such as the collection of anaerobic bacteria maintained by the Department of Anaerobic Microbiology, Virginia Polytechnic Institute and State

University, Blacksburg. One of the major functions of a large national collection is the preservation of type strains. Microbiologists who propose a new species are expected to deposit the type strain with one or more national collections so that it can be preserved and so that subcultures can be distributed to other workers for study and comparison with other microorganisms.

PROPERTIES OF PURE CULTURES

Colonies range in size from extremely small (pinpoint), measuring only a fraction of a millimeter in diameter, to large colonies measuring 5 to 10 mm in diameter. Although the colonies of a given species have a characteristic diameter, one must be aware of certain factors affecting colony diameter. For instance, only well separated colonies should be measured, since such colonies tend to have a larger diameter than those which are crowded together. This is because widely separated colonies are subject to less competition for nutrients and less inhibition by toxic products of metabolism. Moreover, young colonies are smaller than older colonies; therefore the time at which measurements are made must be stated. There is generally an upper limit to the final size of the colonies of a given species; i.e., a point is reached where further incubation no longer results in a corresponding increase in size.

Certain species have colonies possessing a highly wrinkled surface. For a pure culture, all the colonies on the plate should have a similar type of surface; however you should bear in mind that some pure cultures may exhibit surface variation. One of the commonest variations is known as the S→R variation. This is due to the presence of mutant cells that give rise to some rough (R) colonies in a population that otherwise produces smooth (S) colonies. Some R mutants produce rough colonies because they lack the ability to make capsules, or, if the species is Gram-negative, they may no longer be able to form O antigens. For several species of pathogenic bacteria, the surface texture of colonies may bear a relation to virulence. For instance, S colonies of *S. pneumoniae* or of *Salmonella* species are usually virulent, whereas R colonies are not. On the other hand, for strains of *Mycobacterium tuberculosis* a rough surface showing serpentine cords is usually a good indicator of virulence.

MICROSCOPY

We have already seen how Antony van Leeuwenhoek is honoured for providing the first accurate report of the occurrence of bacteria with the help of his single lens microscope of simplest possible design. Later on Robert Hooke had used a compound microscope in 18th century, but these were incapable of good performance, due to defects such as chromatic and spherical aberration inherent in their basic design. The microbial existence is indeed another world, but it would remain largely uncharted without an essential tool; the microscope. Your efforts in exploring microbes will be more meaningful if you understand some essentials of microscopy (study of microscope techniques) and specimen preparation. A discovery by early microscopists that spurred the advancement of microbiology was that clear, glass spheres could act as a lens to magnify (L. *Magnus*, great, and *ficere,* to make) small objects. Magnification in most microscopes results from a complex interaction between visible light waves and the curvature of the lens. When a beam or ray of light transmitted through air strikes and passes through the convex surface of glass, it experiences some degree of refraction (L. *Refringere,* to break apart), defined as the bending or change in the angle of the light ray as it passes into the lens. The greater the differences in the composition of the two substances the light passes between, the more pronounced is the refraction. When an object is placed a certain distance from the spherical lens and illuminated (L. *Illuminatus,* to light up) with light, an optical replica, or image, of it is formed by the refracted light. Depending upon the size and curvature of the lens, the image appears enlarged to a particular degree, which is called its power of magnification and is usually identified with a number combined with × (read "times"). This behaviour of light is evident if one looks through an everyday object such as a glass ball, a water-filled flask, or a magnifying glass. It is basic to the function of all optical, or light, microscopes, though many of them have additional features that define, refine, and increase the size of the image. The first microscopes were simple, meaning they contained just a single magnifying lens and a few working parts. Among the refinements that led to the development of today's compound microscope were the addition of a second magnifying lens system, a lamp in the base to give off visible light and illuminate the specimen, and a special lens called the condenser that converges or focuses the rays of light to a single point on the object.

Lens of Light Microscopy

Magnification of the object or specimen by a compound microscope occurs in two phases. The first lens in this system (the one closes to the specimen) is the objective lens, and the second (the one closest to the eye) is the ocular lens, or eyepiece. The objective forms the initial image of the specimen, called the *real image.* When this image is projected up through the microscope body to the plane of the eyepiece, the ocular lens forms a second image, the *virtual image.* The virtual image is the one that will be received by the eye and converted to a retinal and visual image. The magnifying power of the objectives alone usually ranges from 4 × to 100 ×, and the power of the ocular alone ranges from 10× to 20×. The total power of magnification of the final image informed by the combined lenses is a product of the separate powers of each lens :

Objective power ×	**Ocular power**	=	**Total magnification**
10× low power	20×	=	200×
40× high power	10×	=	400×
100× oil immersion	10×	=	1,000×

The power of the ocular usually remains constant for a given microscope depending on the power of the ocular, the total magnification of standard light microscopes can vary from 40× with the lowest power objective (called the scanning objective) to 2,000× with the highest power objective (the oil immersion objective).

Resolution of Object

Resolution is the capacity of an optical system to distinguish or separate two adjacent objects or points from one another, for example, at a certain fixed distance (about 10 inches), the lens in the human eye can resolve two small objects as separate points just as long as the two objects are no closer than 0.2 mm apart. The eye examination given by optometrists is in fact a test of the resolving power of the human eye for various-sized letters read at a distance of 20 feet.

Resolution of object can be determined by following formula :

$$\text{Resolving power (R.P.)} = \frac{\text{Light wavelength (in nm)}}{2 \times \text{Objective lens's numerical aperture}}$$

The light source for optical microscopes consists of a band of coloured wavelengths in the visible spectrum. The shortest visible wavelengths are in the violet-blue portion of the spectrum (400 nm), and the longest are in the red portion (750 nm). Since the wavelength must pass between the objects that are being resolved, shorter wavelengths (in the 400-500 nm range) will provide between resolution. Some microscopes have a special blue filter placed over the lamp to limit the longer wavelengths of light from entering the specimen. Without going into the mathematical derivation of this constant, it is sufficient to say that each objective has a fixed numerical aperture reading that is determined by the microscope design and ranges from 0.1 in the lowest power lens to approximately 1.25 in the highest power (oil immersion) lens. The most important thing to remember is that a higher numerical aperture number will provide better resolution. In order for the oil immersion lens to arrive at its maximum resolving capacity, a drop of oil must be inserted between the tip of the lens and the specimen on the glass slide. Since oil has the same optical qualities as glass, it prevents refractive loss that normally occurs as peripheral light passes from the slide into the air; this property effectively increases the numerical aperture. There is an absolute limitation to resolution in optical microscopes, which can be demonstrated by calculating the resolution of the oil immersion lens using a blue-green wavelength of light :

$$\text{R.P.} = \frac{500\,\text{nm}}{2 \times 1.25}$$

$$= 200 \text{ nm (or } 0.2\ \mu\text{m)}$$

This means that the oil immersion lens can resolve any cell or cell part as long as it is at least 0.2 μm in diameter, and that it can resolve two adjacent objects as long as they are no closer than 0.2 μm. In general, organisms that are 0.5 μm or more in diameter are readily seen. This includes fungi and protozoa and some of their internal structures and most bacteria. However, a few bacteria and most viruses are far too small to be resolved by the optical microscope and require electron microscopy. In summary then, the factor that most limits the clarity of a microscope's image is its resolving power. Even if a light microscope were designed to magnify several thousand times, its resolving power could not be increased, and the image it produced would simply be enlarged and fuzzy.

A typical problem is spherical aberration, a distortion in the image caused by irregularities in the lens, which creates a curved, rather than flat, image. Another is chromatic aberration, a blurry, rainbowlike image that is caused by the lens acting as a prism and separating visible light into its coloured bands. To correct for these, the manufacturer incorporates several compensating lenses in the objective and ocular. Brightness and direction of illumination also affect image formation. Because too much light can reduce contrast and burn out the image, an iris diaphragm on most microscopes controls the amount of light entering the condenser. The lack of contrast in cell components is compensated for by using special lenses (the phase-contrast microscope) and by adding dyes.

Bright-field Microscopy

The light microscope is a two-lens system in which the lens nearest, the object is known as the objective lens, and that nearest the eye is called as ocular lens. Visible light rays are projected through a condenser that focuses them into a sharp cone, then sends them through the opening in the stage on which the slide rests. The light passes through the slide, bounces of the object, then passes through the objective lens to form a magnified image darken than the background around it. The bright-field microscope is the most widely used type of light microscope. Although we ordinarily view objects like the words on this page with light reflected off the surface, a bright-field microscope forms its image when light is transmitted through the specimen. The specimen, being denser and more opaque than its surroundings, absorbs some of this light, and the rest of the light is transmitted directly up through the ocular into the field. As a result, the specimen will produce an image that is darker than the surrounding brightly illuminated field. The bright-field microscope is a multipurpose instrument that can be used for both live, unstained material and preserved, stained material.

Dark-field Microscopy

In this case a special condenser scatters the light and causes the light to hit the object from different directions. Some light is reflected from the object into the lens and the object is seen clearly, but since there is not direct background light, the surrounding area appears as dark. This microscopy is used in a diagnosis of certain diseases when small live organisms near the limit of resolution of the microscope must be observed. A bright-field microscope can be adapted as a dark-field microscope by adding a special disk called a *stop* to the condenser. The stop blocks all light from entering the objective lens except peripheral light that is reflected off the sides of the specimen itself. The resulting image is a particularly striking one : brightly illuminated specimens surrounded by a dark (black) field. Some of Leeuwenhoek's more successful microscopes probably operated with dark-field illumination. The most effective use of dark-field microscopy is to visualize living cells that would be distorted by drying or heat or cannot be stained with the usual methods.

Phase-contrast Microscopy

Internal components of a live, unstained cell also lack contrast and can be difficult to distinguish. But cell structures do differ slightly in density, enough that they can alter the light that passes through them in subtle ways. The phase-contrast microscope has been constructed to take advantage of this characteristic. This microscope contains devices that transform the subtle changes in light waves passing through the specimen into differences in light intensity. For example, denser cell parts such as organelles alter the pathway of light more than less dense regions (the cytoplasm). Light patterns coming from these regions will vary in contrast. The amount of internal detail visible by this method is greater than by either bright-field or dark-field methods. The phase-contrast microscope is most useful for observing intracellular structures such as bacterial spores, granules, and organelles, as well as the locomotor structures of eukaryotic cells. Like the phase-contrast microscope, the *differential interference contrast (DIC) microscope* provides a detailed view of unstained, live specimens by manipulating the light. But this microscope has additional refinements, including two prisms that add contrasting colours to the image and two beams of light rather than a single one. DIC microscopes produce extremely well-defined images that are vividly coloured and appear three dimensional.

Fluorescence Microscopy

The name of this type of microscopy originates from the use of certain dyes (acridine, fluorescein) and minerals that show fluorescence. This means that the dyes emit visible light when bombarded by shorter ultraviolet rays. For an image to be formed, the specimen must first be coated or placed in contact with a source of fluorescence. Subsequent illumination by ultraviolet radiation cause the specimen to give off light that will form its own image, usually an intense yellow, orange, or red against a black field. Fluorescence microscopy has its widest applications in diagnosing infections caused by certain bacteria, protozoans, and viruses. A staining technique with fluorescent dyes is commonly used to detect *Mycobacterium tuberculosis* (the agent of tuberculosis) in patients' specimens. In a number of diagnostic procedures, fluorescent dyes are affixed to specific antibodies. These *fluorescent antibodies* can be used to detect the causative agents in such diseases as syphilis, chlamydiosis, trichomoniasis, herpes, and influenza. A newer technology using fluorescent nucleic acid stains can differentiate between live and dead cells in mixtures.

Electron Microscopy

Although this microscope was originally conceived and developed for studying nonbiological materials such as metals and small electronic parts, biologists immediately recognized the importance of the tool and began to use it in the early 1930s. Unlike the light microscope, the electron microscope forms an image with a beam of electrons that can be made to travel in wavelike patterns when accelerated to high speeds. These waves are 100,000 times shorter than the waves of visible light. Because resolving power is a function of wavelength, electrons have tremendous power to resolve minute structures. Indeed, it is possible to resolve atoms with an electron microscope, though the practical resolution for biological applications is approximately 0.5 nm. Because the resolution is so substantial, it follows that magnification can also be extremely high—usually between 5,000 × and 750,000 × for biological specimens and up to 5,000,000 × in some applications. Its capacity for magnification and resolution makes the EM an invaluable tool for seeing the finest structure (called the ultrastructure) of cells and viruses.

It employs components analogous to, but not necessarily the same as, those in light microscopy. For instance, it magnifies in stages by means of two lens systems, and it has a condensing lens, a specimen holder, and focusing apparatus. Otherwise, the two types have numerous differences. An electron gun aims its beam through a vacuum to ring-shaped electromagnets that focus this beam on the specimen. Specimens must be pretreated with chemicals or dyes to increase contrast and cannot be observed in a live state. The enlarged image is displayed on a viewing screen or photographed for further study rather than being observed through an eyepiece.

Transmission electron microscopes are the method of choice for viewing the detailed structure of cells and viruses. This microscope produces its image by transmitting electrons through the specimen. Because electrons cannot readily penetrate thick preparations, the specimen must be sectioned into extremely thin slices (20-100 nm thick) and stained or coated with metals that will increase image contrast. The darkest areas of TEM micrographs represent the thicker (denser) parts, and the lighter areas indicate the more transparent and less dense parts. The TEM can also be used to produce negative images and shadow casts of whole microbes.

The scanning electron microscope is designed to create an extremely detailed three-dimensional view of anything from a fly's eye to AIDS viruses escaping their host cells. To produce this image, the SEM does not transmit electrons; it bombards the surface of a whole, metal-coated specimen with electrons while scanning back and forth over it. A shower of electrons deflected from the surface is picked up with great fidelity by a sophisticated detector, and the electron pattern is displayed as an image on a television screen. The contours of the specimens resolved with scanning electron micrography are very revealing and often surprising. Areas that look smooth and flat with the light microscope display intriguing surface features with the SEM. Improved technology has continued to refine electron microscopes and to develop variations on the basic plan.

STAINING TECHNIQUES

The smear technique, developed by Robert Koch more than 100 years ago, consists of spreading a thin film made from a liquid suspension of cells on a slide and air drying it. Next, the air-dried smear is usually heated gently by a process called fixation that simultaneously kills and secures the specimen to the slide. For some microbial cells, fixation is performed with chemicals such as alcohol and formalin to preserve more delicate cell structures. Like images on undeveloped photographic film, the unstained cells of a fixed smear are quite indistinct, no matter how great the magnification or how fine the resolving power of the microscope. The process of "developing" a smear to create contrast and make inconspicuous features stand out, requires staining techniques. Staining is any procedure in which coloured chemicals called dyes are applied to specimens. Dyes impart a colour to cells or cell parts by becoming affixed to them through a chemical reaction. In general, they are classified as basic (cationic) dyes, which have a positive charge, or acidic (anionic) dyes, which have a negative charge. Since chemicals of opposite charge are attracted to each other, cell parts that are negatively charged will attract basic dyes and those that are positively charged will attract acidic dyes. Many cells, especially those of bacteria, have numerous negatively charged acidic substances and thus stain more readily with basic dyes. Two basic types of staining technique are used, depending upon how a dye reacts with the specimen. Most procedures involve a positive stain, in which the dye actually sticks to the specimen and gives it colour. A negative stain, on the other hand, is just the reverse (like a photographic negative). The dye does not stick to the specimen but settles around its outer boundary, forming a silhouette. In a sense, negative staining "stains" the glass slide to produce a dark background that highlights the cells in the specimen. Nigrosin (blue-black) and India ink (a black suspension of carbon particles) are the most common dyes used for negative staining. The cells themselves do not stain because these dyes are negatively charged and are repelled by the negatively charged surface of the cells. The value of negative staining is its relative simplicity and the reduced shrinkage or distortion of cells, as the smear is not heat-fixed. A quick assessment can thus be made regarding cellular size, shape, and arrangement.

Simple stains require only a single dye and an uncomplicated procedure, differential stains use two different coloured dyes, called the *primary dye* and the *counterstain*, to distinguish between cell types or parts. These staining techniques tend to be more complex and sometimes require additional chemical reagents to produce the desired reaction. Most simple staining techniques take advantage of the ready binding of bacterial cells to dyes like malachite green, crystal violet, basic fuchsin, and safranin. Simple stains cause all cells in a smear to appear more or less the same colour, regardless of type, but they reveal such bacterial characteristics as shape, size, and arrangement. A simple stain with Loeffler's methylene blue is distinctive. The blue cells stand out against a relatively unstained background, so that size, shape, and grouping show up easily.

Gram Stain Technique

Christan Gram (1984), a Danish researcher gave a concept of this stain technique, which differentiates between two kinds of bacteria—Gram-negative and Gram-positive bacteria. A thin smear of bacterium is prepared on slide. To the smear crystal violet solution is applied for 30-35 seconds. The slide later gently rinsed in distilled water and put on Gram-iodine solution for 30 seconds. Lateron the slide is washed with absolute alcohol. This usually takes from 20 seconds to one minute. Microscopic examination of slide will reveal the Gram-positive retain the violet-iodine colour while negative ones loose the blue-purple colour after alcohol wash, and will be of original colour. Some bacteria retain colour while others loose out, why? To explain this concept, it has been postulated that since Gram-negative bacteria have relatively a high lipid content in their cell walls, the alcohol dissolves the lipids, that allows the leakage of crystal violet iodine complex. Gram-positive bacteria with less lipids in their cell walls are less susceptible to the action of alcohol. Another theory suggests that peptidoglycans found in high concentration in cell walls of Gram-positive bacteria perhaps traps the crystal violet-iodine complex in its many cross linkage. Gram-negative bacteria with less peptidoglycan and fewer cross linkages loose the stain readily.

Acid-fast Stain Technique

The basic principle of this procedure originated with Paul Ehrlich in 1882 as an improved method for detecting hard to stain *Mycobacterium tuberculosis* in specimens. It was determined that these bacterial cells have a particularly impervious outer wall that holds fast (tightly or tenaciously) to the dye (carbol fuchsin) even when washed with a solution containing acid or acid alcohol. This stain is used for other medically important mycobacteria such as the leprosy bacillus and for *Nocardia*, an agent of lung or skin infections.

Endospore Stain Technique

This stain is designed to distinguish between spores and the cells that they come from. Of significance in medical microbiology are the Gram-positive, spore-forming members of the genus *Bacillus* (the cause of anthrax) and *Clostridium* (the cause of botulism and tetanus)—dramatic diseases of universal fascination.

MICROMETRY

Micrometry is the measurement of microorganisms. For this an ocular micrometer serves as scale ocular micrometer is simply a disc of glass upon which are etched lines. They are usually etched 100 equally spaced divisions, marked 0 to 10 upon an ocular micrometer. When placed on the ocular, a ruled lines superimpose certain distance markers on the microscope field. However, the scale on ocular micrometer does not have any standard value. We can find out the value of one division of this unknown scale by calibrating it with a known scale. Thus actual value of one division of this unknown scale by calibrating it with a known scale. Thus actual value of one division of ocular micrometer is found by using another known scale.

Method of Micrometry

(a) Remove eye piece lens and insert the ocular micrometer on the metal diaphragm. Replace the ocular lens and mount in the microscope and observe. There will be seen scale lines of ocular micrometer.

(b) Stage micrometer mounted on the stage of microscope and bring its scale in the centre of microscope field under a sharp focus. Firstly this is done with lower power objective and later with high-power finally with oil-immersion objectives.

(c) Later ocular micrometer and stage micrometer scales are adjusted in such a way that lines of former superimpose upon those of the latter.

(d) One divisional values of ocular micrometer is calculated for different objectives.

(i) 100 divisions of stage micrometer = 0.1 mm = 100 μm

1 division of stage micrometer = 1.0 μm

Suppose 30 divisions of ocular = 3 divisions of stage micrometer

$$\text{i.e., } 3 \times 100 = \frac{300\ \mu m}{30\ \mu m} = 10\ \mu m$$

1 divisions of ocular = 10 μm

(ii) High-power objective

40 divisions of ocular = 1 division of stage micrometer

$$\text{1 division of ocular} = \frac{100\ \mu m}{40\ \mu m} = 2.5\ \mu m$$

(iii) Oil-immersion objective

60 divisions of ocular = 1 division of stage micrometer

$$\text{one division of ocular} = \frac{100\ \mu m}{60\ \mu m} = 1.66\ \mu m$$

(e) Replace the stage micrometer with the slide bearing the microorganism whose measurements are to be taken. For instance, total length of chain of bacterium covers 18 divisions of ocular micrometer, when examined under oil-immersion, thus total length of chain would be 18 × 1.66 = 29.88 μm.

Counting of Spores and Cells

The simplest technique is to use a special counting chamber, of a type that used in a hemocytometer for counting the blood cells. The counting chamber is simply a ruled slide with a supported glass cover that holds a definite volume of fluid. This allows us to determine the exact number of cells, spores for a given volume of fluid. The counting chamber on glass slide has nine squares (each area of 1 mm × 1 mm) etched over it. Nine large squares (each 1 mm × 1mm), medium sized squares (each 0.2 mm × 0.2 mm) and small squares (each 0.05 mm × 0.05 mm). Each large square is 1 mm × 1 mm = 1.0 mm^2 in area. Since the cover glass is 0.1 mm above the counting chamber, the depth of chamber is 0.1 mm. Thus each large square has a volume of 1 mm × 1 mm × 0.1 mm or 1/10 cm × 1/10 cm × 1/100 cm = 1/10,000 cm^3 or 10^{-4} cm^3.

Method

(a) In central large square, cells are counted, thus 25 medium sized squares (0.2 mm × 0.2 mm) in this large square are to be used. Count 100-150 cells, or spores in total.

(b) 100-150 cells per large square i.e., 1/10,000 cm^3. If so multiply this count by 10^4 (10,000) to find out the number of cells, spores per cm^3.

(c) Less than 100-150 cells per large square. In this case count cells, spores in some (3 or 4) corner large squares to have a total count of 100 or 50. Divide this total number by the number of large squares to find out average number of cells, spores per large square. Multiply this average number by 10^4 to find out their number per cm^3.

Suppose 100 cells in large squares

100/4 = 25 cells per large square

25 × 10^4 = 250,000 cells per cm^3

(d) It is tedius and time consuming to count in all large squares. So count the cells, spores in all the medium-sized squares and multiply this average value by 25 for number of cells per large square. Later on multiply this number by 10^4 for total number of cells per cm^3.

150 cells in three medium-sized squares

150/3 = 50 cells per medium size square

50 × 25 = 1250 cells per large square (1/10,000 cm^3)

1250 × 10^4 = 12,500,000 or 1.25 × 10^7 cells per cm^3

(e) If number may be too high. In this case suspension is diluted. The dilution is to be done to a level, suitable for count. If 9 ml of water are added to 1 ml of suspension, it gives 1/10 dilution. If so, the cell count is also to be multiplied by dilution factor to have a final count of cells, or spores.

❑❑❑

4

Prokaryotic Cells : The Bacteria and Mycoplasma

In certain species of bacteria the cells are arranged in groups, the most common of which are pairs, clusters, chains, trichomes, and filaments. It is important to recognize these patterns of shape and arrangement, since they are often characteristic of a taxonomic group, e.g., a genus. Some bacteria also possess appendages, which can be made visible by special staining techniques or by electron microscopy. The discovery of this internal structure was made possible by the development of electron-microscope techniques and of instruments for slicing a bacterial cell into extremely thin sections. The terms microbial cytology and bacterial anatomy have become commonplace in microbiological literature. The various structures of a bacterial cell differ from one another not only in their physical features but also in their chemical characteristics and in their functions. Thus biologists today seek to integrate the structural, chemical, and functional properties of bacterial cell.

SHAPE AND SIZE OF BACTERIAL CELL

Bacteria are very small, most being approximately 0.5 to 1.0 μm in diameter. A relatively large surface through which nutrients can enter compared to a small volume of cell substance to be nourished accounts for the unusually high rate of growth and metabolism of bacteria. Moreover, because of the high surface area volume ratio, the mass of cell substance to be nourished is very close to the surface; therefore, no circulatory mechanism is needed to distribute the nutrients that are taken in, and there is thought to be little or no cytoplasmic movement within a bacterial cell. Typical bacterial cells are spherical (cocci; singular, coccus); straight rods (bacilli; singular, bacillus); or rods that are helically curved (Fig. 4.1). Although most bacterial species have cells that are of a fairly constant and characteristic shape, some have cells that are pleomorphic, i.e., that can exhibit a variety of shapes (Fig. 4.2). Bacterial cells are usually arranged in a manner characteristics of their particular species. Although it is rare that all the cells of a species are arranged in the same manner, it is the predominant arrangement that is the important

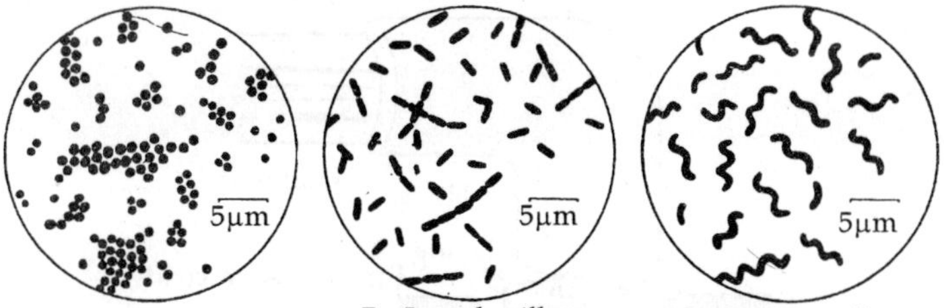

Fig. 4.1 : Shape diversity in bacteria.

feature. Cocci appear in different arrangements depending on plane of cell division. Bacilli are not arranged as complex as those of other Cocci. But some species, such as *Bacillus subtilis*, form chains (streptobacilli); others, such as *Beggiatoa* and *Saprospira* species, form trichomes, which are similar to chains but have a much larger area of contact between the adjacent cells. In other bacillus species, such as *Corynebacterium diphtheriae*, the cells are lined side by side like matchsticks (palisade arrangement) and at angles to one another. Still others, such as *Streptomyces* species, form long, branched, multinucleate filaments called hyphae (singular, hypha) which collectively form a mycelium. Curved bacteria are usually curved with a twist. Bacteria with less than one complete twist have a vibrioid shape, where as those with one complete twist have a helical shape. In addition to the common bacterial shapes, many others also occur; pearshaped cells (e.g., *Pasteuria*); lobed spheres (e.g., *Sulfolobus*); rods with squared rather than the usual hemispherical ends (e.g., *Bacillus anthracis*); disks arranged like stacks of coins (e.g., *Caryophanon*); rods with helically sculptured surfaces (e.g., *Seliberia*); and many others.

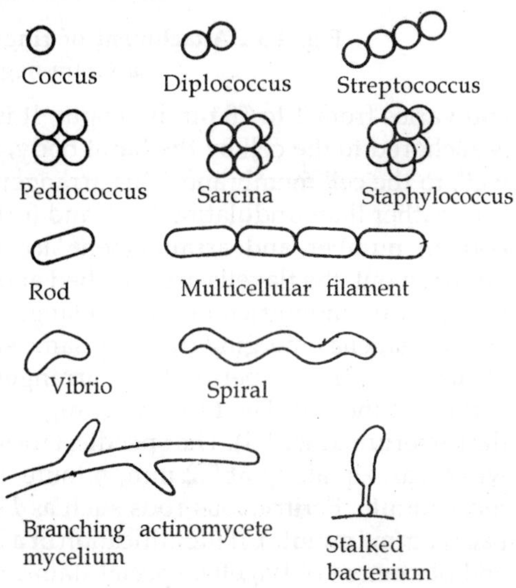

Fig. 4.2 : Range of structures of bacteria.

STRUCTURE OF BACTERIAL CELL

The evolutionary history of bacteria extends back at least 3.5 billion years. The fact that bacteria have endured for so long in such a variety of habitats indicate a cellular structure and function that are amazingly versatile and adaptable. The general cellular organisation of a prokaryotic cell can be represented with the following micromap :

Flagella

The primary function of flagella is to confer motility, or self-propulsion – that is, the capacity of a cell to swim freely through an aqueous habitat. The extreme thinness of a bacterial flagellum necessitates high magnification to reveal its special architecture, which occurs in three distinct parts : the filament, the hook (sheath), and the basal body (Fig. 4.3). The filament, a helical structure composed of proteins, is approximately 20 nm in diameter

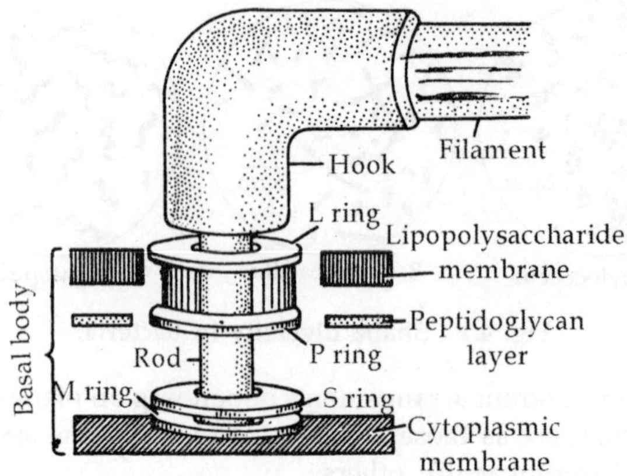

Fig. 4.3 : Attachment of rings to cell wall and cell membrane of a Gram-negative bacterial cell.

and varies from 1 to 70 nm in length. It is inserted into a curved, tubular hook. The hook is anchored to the cell by the basal body, a stack of rings firmly anchored through the cell wall, to the cell membrane. This arrangement permits the hook with its filament to rotate 360°, rather than undulating back and forth like a whip as was once thought. Flagella vary both in number and arrangement according to two general patterns : (1) In a *polar* arrangement, the flagella are attached at one or both ends of the cell. Three subtypes of this pattern are : monotrichous with a single flagellum; lophotrichous with small bunches or tufts of flagella emerging from the same site; and amphitrichous with flagella at both poles of the cell. (2) In a peritrichous arrangement, flagella are dispersed randomly over the surface of the cell. The type of arrangement has some bearing on the swimming speed of the bacterium (Fig. 4.4). The speediest forms are polar, flagellated cells such as *Thiospirillum*, which can zip along at 5.2 mm/minute, and *Pseudomonas aeruginosa*, which can swim 4.4 mm/minute. Peritrichous rods such as *Escherichia coli* tend to swim at a relatively slower pace (1 mm/minute). If identification of a bacterium requires detection of the actual number and placement of flagella, special stains or electron microscope preparations are required, since flagella are too minute to be seen in unstained live preparations with an ordinary light microscope. Often it is sufficient to know simply whether a bacterial species is motile. One way to detect motility is to place a tiny mass of cells into a soft (semisolid) medium.

Flagellated bacteria can perform some rather sophisticated feats. They can detect and move in response to chemical signals—a type of behaviour called chemotaxis (Gr. *Chemo*, chemicals, and *taxis*, an arrangement). Positive chemotaxis is movement of a cell in the direction of a favourable chemical stimulus (usually a nutrient); negative chemotaxis is movement away from a repellent (potentially harmful) compound. The flagellum is effective in guiding bacteria through the environment primarily because the system for detecting chemicals is linked to the mechanisms that drive the flagella. Located in the cell membrane are clusters of receptors that bind specific molecules coming from the immediate environment. The attachment of sufficient numbers of these molecules transmits signals to the flagellum and sets it into rotary motion. If several flagella are present, they become aligned and rotate as a group. As a flagellum rotates counterclockwise, the cell itself is driven clockwise. This combination results in a run, defined as smooth, linear swimming in the direction of the stimulus. Large motile bodies such as boats and fish make use of the inertia of water for their propulsion. When pushed against with, for example, an oar, a propeller blade, or fins,

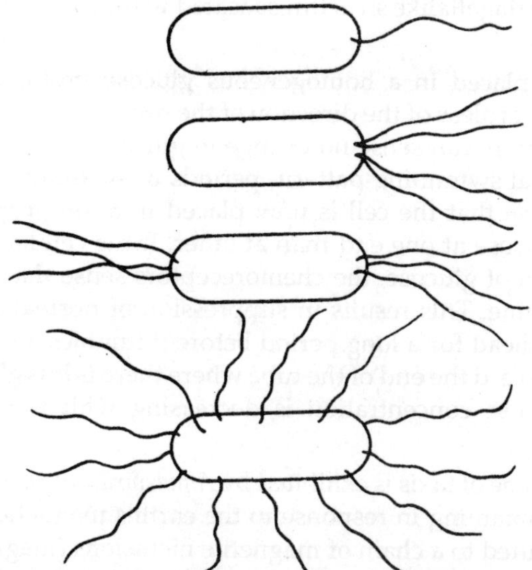

Fig. 4.4 : Types of flagellation in bacterial cells.

the water temporarily acts as a solid, thereby enabling the boat or fish to generate a forward propulsive force. However, the small size of bacteria prohibits their use of the inertia of water to gain propulsive force, because the drag forces due to the viscosity of water become thousands of times greater than any forces that can be generated from inertia. The difficulty would be similar to what we would encounter if we attempted to row a boat on a lake filled with thick molasses. Bacteria propel themselves by rotating their helical flagella. The principle involved can be illustrated by imagining the penetration of a piece of cork by a corkscrew. If one tries to ram the corkscrew directly through the cork, great force will probably be needed. On the other hand if one merely rotates the corkscrew, the cork can be easily penetrated. In the case of bacteria, the cork is analogous to the viscous medium and the corkscrew to the helical flagellum. It is apparent from this analogy that a mutant bacterium having straight rather than helical flagella would be unable to swim. The nature of the rotary motor that spins each corkscrew-shaped flagellum is still not understood, but the rings found in the basal body are probably involved. It is known that the flagellar motor is driven by the protonmotive force, i.e., the force derived from the electrical potential and the hydrogen-ion gradient across the cytoplasmic membrane. Moreover, recent studies suggest that the concentration of cGMP (guanosine 3′-5′-cyclic phosphoric acid) within the cell governs the direction in which the rotation occur. Bacteria having polar flagella swim in a back-and-forth fashion; they reverse their direction of swimming by reversing the direction of flagella rotation.

However, when the flagella motors reverse, conformational changes occur along the flagella, the bundle flies apart, and the cell tumbles wildly. Finally, the flagellar motors resume their normal direction, the flagellar bundle again forms, and the cell begins to swim—but now in a different direction. Certain helical bacteria (spirochetes) exhibit swimming motility, particularly in highly viscous media, yet they lack external flagella.

However, they have flagellalike structures located within the cell, just beneath the outer cell envelope.

If the cell is placed in a homogeneous glucose broth, the glucose concentration remains constant regardless of the direction of the bacterium's swimming, and the glucose-specific chemoreceptors can sense no change in glucose concentration. Consequently, the cell exhibits a normal swimming pattern- periods of swimming with intermittent periods of tumbling. Suppose that the cell is now placed in a long capillary tube with a higher concentration of glucose at one end than at other. If the cell happens to swim toward the higher concentration of glucose, the chemoreceptors sense that the glucose concentration is increasing with time. This results in suppression of normal tumbling, causing the cell to swim smoothly ahead for a long period before it tumbles. On the other hand, if the cell happens to swim toward the end of the tube where there is less glucose, the chemoreceptors sense that the glucose concentration is decreasing with time, and no suppression of tumbling occurs.

Still another type of taxis is exhibited by *Aquaspirillum magnetotacticum*; this organism exhibits directed swimming in response to the earth's magnetic field or to local magnetic fields. This is attributed to a chain of magnetite inclusions (magnetosomes) within the cell, which allows the cell to become oriented as a magnetic dipole. Because of the downward inclination of the Earth's magnetic field in the regions where these bacteria have been found, magnetotaxis may serve to direct the cells downward in aquatic environments toward oxygen-deficient areas more favourable for growth.

Pili (Fimbriae)

Fimbriae are small, bristlelike fibers sprouting off the surface of many bacterial cells. Their exact composition varies, but most of them contain protein. Fimbriae (L. *Fringe*) have an inherent tendency to sick to each other and to surfaces. For example, they are often responsible for the mutual clinging of cells that leads to films and other thick aggregates of cells on the surface of liquids, and for the microbial colonization of inanimate solids such as rocks and glass. Some pathogens can colonize and infect host tissues because of a tight adhesion between their fimbriae and epithelial cells. For example, the gonococcus (agent of gonorrhea) invades the genitourinary tract and *Escherichia coli* invades the intestine by this means. Mutant forms of these pathogens that lack fimbriae are unable to cause infections. A pilus (L. *Hair*) is an elongate, rigid tubular structure made of a special protein, *pilin*. So far, true pili have been found only on Gram-negative bacteria, where they are involved primarily in a mating process between cells called conjugation (L. *Conjugatus*, linked together) which involves partial transfer of DNA from one cell to another. A pilus from the donor cell unites with a recipient cell, thereby providing a cytoplasmic connection for making the transfer. Production of pili is controlled genetically, and conjugation takes place only between compatible Gram-negative cells. In some Gram-positive bacteria spinae have been reported. These are tubular pericellular non prosthecate rigid appendages made up of single protein-moiety-spinin. They are said to help adjust cells to some environmental conditions as salinity, pH, temperature etc.

Capsule

Some bacterial cells are surrounded by a viscous substance forming a covering layer or envelope around the cell wall. If this layer can be visualized by light microscopy using special staining methods, it is termed a capsule. If the layer is too thin to be seen by light microscopy, it is termed a microcapsule; if it is so abundant that many cells are embedded in a common matrix, the material is called slime. By light microscopy, capsules appear to be amorphous gelatinous areas surrounding a cell; however, special techniques designed to preserve delicate structures for observation by electron microscopy have revealed that capsules consist of a mesh or network of fine strands. Capsules can serve a number of functions, depending on the bacterial species. (a) They may provide protection against temporary drying by binding water molecules. (b) They may block attachment of bacteriophages. (c) They may be antiphagocytic, i.e., they inhibit the engulfment of pathogenic bacteria by white blood cells and thus contribute to invasive or infective ability (virulence). (d) They may promote attachment of bacteria to surface; for example, *Streptococcus mutens*, a bacterium associated with producing dental caries, firmly adheres to the smooth surfaces of teeth because of its secretion of a water-insoluble capsular glucan. (e) If capsules are composed of compounds having an electrical charge, such as sugar-uronic acids, they may promote the stability of bacterial suspension by preventing the cells from aggregating and settling out, because cells bearing similarly charged surfaces tend to repel one another. Capsules composed of a single kind of sugar are termed homopolysaccharides; are usually synthesized outside the cell from disaccharides by exocellular enzymes. The synthesis of glucan (a polymer of glucose) from sucrose by *S. mutans* is an example. Other capsules are composed of several kinds of sugars and are termed heteropolysaccharides; these are usually synthesized from sugar precursors that are activated (energized) within the cell, attached to a lipid carrier molecule, transported across the cytoplasmic membrane, and polymerized outside the cell. A few capsules are polypeptides. For example, the capsule of the anthrax organism, *B. anthracis*, is composed entirely of a polymer of glutamic acid.

Prostheca

They are characteristic of a number of aerobic bacteria from freshwater and marine environments. Some bacterial genera such as *Caulobacter* have a single prostheca; others such as *Stella* and *Ancalomicrobium* have several. Prosthecae increase the surface area of the cells for nutrient absorption, which is advantageous in dilute environments. Some prosthecate bacteria may form a new cell (bud) at the end of a prostheca; others have an adhesive substance at the end of a prostheca that aids in attachment to surfaces. Although the term stalk is sometimes used interchangeably with the terms *prostheca* or *hypha*, it is perhaps better to restrict its use to certain nonliving ribbonlike or tubular appendages that are excreted by the cell, such as those found in the genera *Gallionella or Planctomyces.*

Cell Wall

Immediately below the glycocalyx lies a second layer, the cell wall. This structure accounts for a number of important bacterial characteristics. In general, it determines the shape of a bacterium, and it also provides the kind of strong structural support necessary to keep a bacterium from bursting or collapsing due to changes in osmotic pressure. In this way, the cell wall functions like a bicycle tire that maintains the necessary shape and prevents the more delicate inner tube from bursting when it is expanded.

The cell walls of most bacteria gain this relatively rigid quality from a unique macromolecule called peptidoglycan (PG). This compound is composed of a repeating framework of long *glycan* chains cross-linked by short peptide fragments to provide a strong but flexible support framework (Fig. 4.5). Peptidoglycan is only one of several materials found in cell walls, and its amount varies among the major bacterial groups.

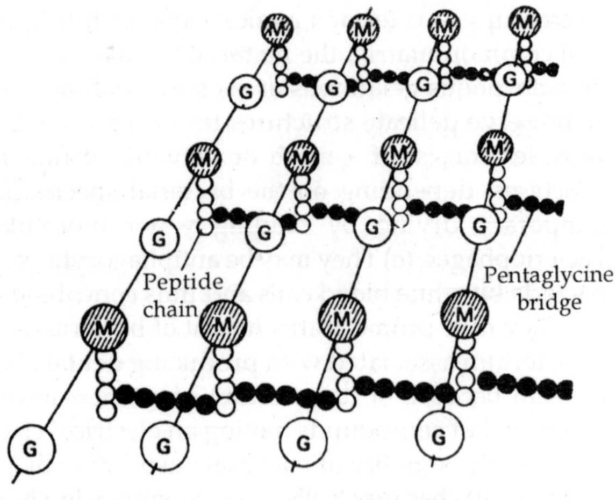

Fig. 4.5 : Diagrammatic sketch to show the structure of glycopeptide of cell wall of bacteria. The basic heteropolymer consists of alternating molecules of N-acetyl-glucosamine (G) and N-acetylmuramic acid (M) with a tetrapeptide attached to the latter.

Because many bacteria live in aqueous habitats with a low solute concentration, they are constantly absorbing excess water by osmosis. Were it not for the strength and relative rigidity of the peptidoglycan in the cell wall, they would rupture from internal pressure. Understanding this function of the cell wall has been a tremendous boon to the drug industry. Several types of drugs used to treat infection (penicillin, cephalosporins) are effective because they can destroy the peptidoglycan layer of bacteria. With their cell walls incomplete or missing, such cells have very little protection from lysis (Gr., *Loosen*) some disinfectants (alcohol, detergents) also kill bacterial cells by damaging the cell wall. Lysozyme, an enzyme contained in tears and saliva, provides a natural defense against certain bacteria by hydrolyzing peptidoglycan. More than a hundred years ago, long before the detailed anatomy of bacteria was even remotely known, a Danish physician named Hans Christian Gram developed a staining technique, the Gram stain, that delineates two generally different groups of bacteria. Although the reasons were not determined for many years, the contrasting staining reactions that occur with this stain are due entirely to some very fundamental differences in the structure of bacterial cell walls. The two major groups shown by this technique are the Gram-positive bacteria and Gram-negative bacteria. Since the Gram stain does not actually reveal the nature of these physical differences, we must turn to the electron microscope and to biochemical analysis. The extent of the differences between Gram-positive and Gram-negative bacteria is evident in the physical appearance of their cell envelopes. In Gram-positive cells, a microscopic section resembles an open-faced sandwich with two layers; the thick outer cell wall, composed primarily of peptidoglycan, and the cell membrane. A similar section of a Gram-negative cell envelope shows a complete sandwich with three layers; the cell wall, composed of an outer membrane and a thin layer of peptidoglycan, and the cell membrane.

The bulk of the Gram-positive cell wall is a thick, homogeneous sheath of peptidoglycan ranging from 20 to 80 nm in thickness. It also contains tightly bound acidic polysaccharides, including techoic acid and lipotechoic acid. Techoic acid is a polymer of ribitol or glycerol and phosphate embedded in the peptidoglycan sheath. Lipotechoic acid is similar in structure but is attached to the lipids in the plasma membrane. These molecules appear to function in cell wall maintenance and enlargement during cell division, and they also contribute to the acidic charge on the cell surface. In some cases, the cell wall of Gram-positive bacteria is pressed tightly against the cell membrane with very little space between them, but in other cells, a thin periplasmic space is evident between the cell membrane, and cell wall (Fig. 4.6).

	Character	Gram-negative	Gram-positive
(a)	Chemical composition	Lipoprotein Lipopolysaccharide Peptidoglycan	Peptidoglycan Techoic acid Lipotechoic acid
(b)	Porin proteins	Yes	No
(c)	Molecular permeability	Less penetrable	More penetrable
(d)	Periplasmic space	Present in all	Present in some
(e)	Overall thickness	Thinner (10-15 nm thick)	Thicker (25-30 nm thick)
(f)	Outer membrane	Yes	No

Fig. 4.6 : Comparison of cell walls of Gram-negative and Gram-positive bacteria.

The Gram-negative cell wall is more complex in morphology because it contains an outer membrane (OM), a thinner shell of peptidoglycan, and has an extensive space surrounding the peptidoglycan. The outer membrane is somewhat similar in construction to the cell membrane, except that it contains specialized types of polysaccharides and proteins. The uppermost layer of the OM contains *lipopolysaccharide* (LPS), which is essentially a polysaccharide fragment integrated into the membrane lipids. The innermost layer of the OM is another lipid layer anchored by means of proteins to the peptidoglycan layer below. The outer membrane serves as a partial chemical sieve by allowing only relatively small molecules to penetrate. Access is provided by special membrane channels formed by *porin proteins* that completely span the outer membrane. The size of these porins can be altered so as to block the entrance of harmful chemicals, making them one defense of Gram-negative bacteria against certain antibiotics. The bottom layer of the Gram-negative wall is a single, thin (1-3 nm) sheet of peptidoglycan. Surrounding the peptidoglycan between the outer membrane and the cell membrane is a well-developed periplasmic space.

The outer membrane contributes an extra barrier in Gram-negative bacteria that makes them more impervious to some antimicrobic chemicals such as dyes and disinfectants so they are generally more difficult to inhibit or kill than are Gram-positive bacteria. One exception is for alcohol-based compounds, which can dissolve the lipids in the outer membrane and disturb its integrity. The cell wall or its parts can interact with human tissues and contribute to disease. Gram-negative lipopolysaccharides called *endotoxins* are a complicating factor, causing fever and shock in meningitis and typhoid fever. Proteins attached to the outer portion of the cell wall of several Gram-positive species, including *Corynebacterium diphtheriae* (the agent of diphtheria) and *Streptococcus pyogenes* (the cause of strep throat), also have toxic properties. The lipids in the cell walls of certain *Mycobacterium species* are harmful to human cells as well.

Several bacteria groups lack the cell wall structure of Gram-positive or Gram-negative bacteria, and some bacteria have no cell wall at all. Although these exceptional forms can stain positive or negative in the Gram stain, examination of their fine structure and chemistry shows that they do not really fit the descriptions for typical Gram-negative or-positive cells. For example, the cells of *Mycobacterium* and *Nacardia* contains peptidoglycan and stain Gram-positive, but the bulk of their cell wall is comprised of unique types of lipids. One of these is a very long chain fatty acid called *mycolic acid,* or cord factor, that contributes to the pathogenicity of this group. The thick, waxy nature imparted to the cell wall by these lipids is also responsible for a high degree of resistance to certain chemicals and dyes. Such resistance is the basis for the acid-fast stain used to diagnose tuberculosis and leprosy. Since they are from a more ancient and primitive line of prokaryotes, the archaebacteria exhibit unusual and chemically distinct cell walls. In some, the walls are composed almost entirely of polysaccharides, and in others, the walls are pure protein; but as a group, they all lack the true peptidoglycan structure described previously. Since a few archaebacteria and all mycoplasmas lack a cell wall entirely, their cell membrane must serve the dual functions of protection as well as transport. Some bacteria that ordinarily have a cell wall can lose it during part of their life cycle. These wall-deficient forms are referred to as L forms or L-phase variants (for the Lister Institute where they were discovered). L forms arise naturally from a mutation in the wall-forming genes, or they can be induced artificially by treatment with a chemical such as lysozyme or penicillin that disrupts the cell wall. When a Gram-positive cell is exposed to either of these two chemicals, it will lose the cell wall completely and become a *protoplast* a fragile cell

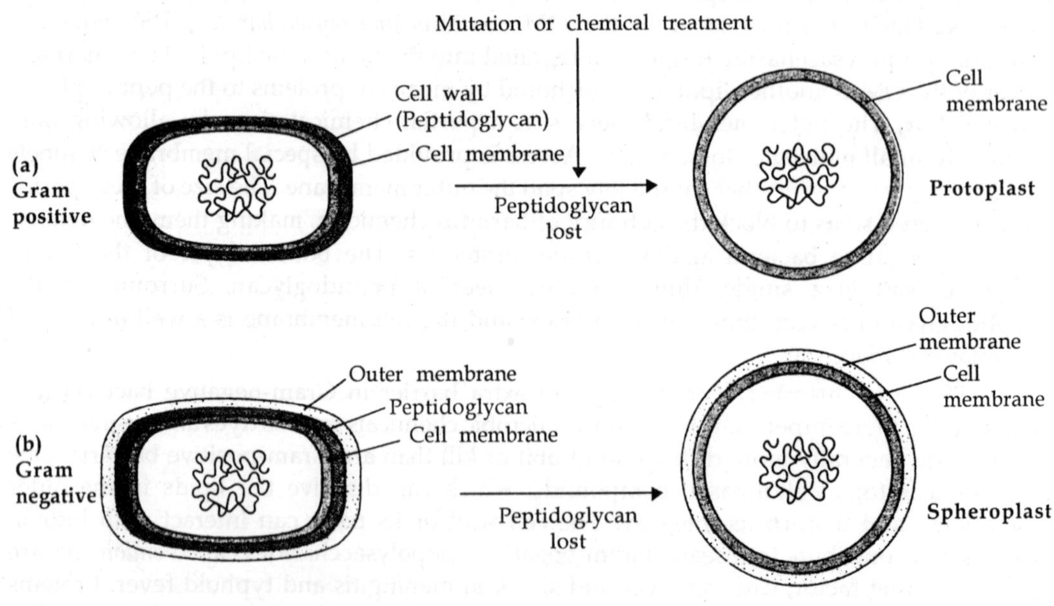

Fig. 4.7 : Walled bacterial cells converted to L forms.

bounded only by a membrane that is highly subject to lysis (Fig. 4.7). A Gram-negative cell exposed to these same substances loses its peptidoglycan but retains its outer membrane, leaving a less fragile but nevertheless weakened *spheroplast*.

Cell Membrane

Cytoplasmic membrane, a very thin (5-10 nm), flexible sheet molded completely around the cytoplasm. Bacterial cell membranes have this typical structure, containing primarily phospholipids (making up about 30-40% of the membrane mass) and proteins (contributing 60-70%). Major exceptions to this description are the membranes of mycoplasmas that contain high amounts of sterols—rigid lipids that stabilize and reinforce the membrane—and archaebacterial membranes, which contain unique branched hydrocarbons rather than fatty acids. In some locations, the cell membrane extends inwardly into coiled passages, or sacs, in the cytoplasm called mesosomes. These are prominent in Gram-positive bacteria but harder to see in Gram-negative bacteria because of their relatively small size. Mesosomes presumably increase the internal surfaces area available for membrane activities, although whether they are artifacts of fixations or true functioning structures continues to be a controversial topic. Some of the proposed functions of mesosomes are to participate in cell wall synthesis and to guide the duplicated chromatin bodies (bacterial chromosomes) into the two daughter cells during cell division.

A major action of the cell membrane is to regulate *transport*, that is, the passage of nutrients into the cell and the discharge of wastes. Although water and small solute molecules can diffuse across the membrane unaided, the membrane is a selectively permeable structure with special carrier mechanisms for passage of most molecules. The glycocalyx and cell wall can bar the passage of large molecules, but they are not the primary transport apparatus. The cell membrane is also involved in *secretion*, or the discharge of a metabolic product into the extracellular environment. The membranes of prokaryotes are an important site for a number of metabolic activities. It is here that most enzymes of respiration and other energy-processing activities are located. Enzyme systems located in the cell membrane also synthesize structural macromolecules to be incorporated into the cell envelope and appendages.

A significant difference exists between the phospholipids of eubacteria and those of archaeobacteria. In eubacteria the phospholipids are phosphoglycerides, in which straight-chain fatty acids are ester-linked to glycerol (Fig. 4.8). In archaeobacteria, the lipids are polyisoprenoid branched-chain lipids, in which long-chain branched alcohols (phytanols) are ether-linked to glycerol. The cytoplasmic membrane is a hydrophobic barrier to penetration by most water-soluble molecules (Fig. 4.9). However, specific proteins in the membrane allow, indeed facilitate, the passage of small molecules across the membrane. The cytoplasmic membrane also contains various enzymes involved in respiratory metabolism and in synthesis of capsular and cell-wall components; moreover, because of its impermeability to protons (hydrogen ions), the cytoplasmic membrane is the site of generation of the protonmotive force—the force that drives ATP synthesis in many organisms, certain nutrient transport systems and flagellar motility. Consequently the cytoplasmic membrane is an extremely important functional structure, and damage to it by physical or chemical agents can result in the death of the cell. Proteins are synthesized within the cell, but some can pass across the cytoplasmic membrane barrier to the outside;

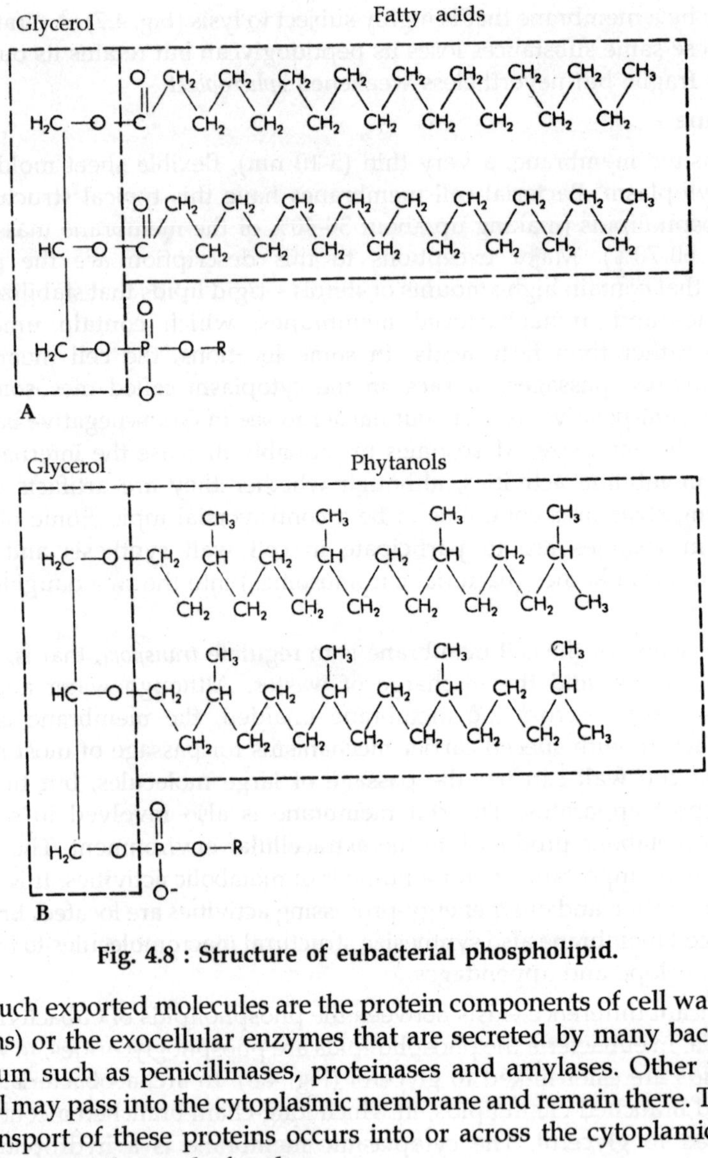

Fig. 4.8 : Structure of eubacterial phospholipid.

examples of such exported molecules are the protein components of cell walls (e.g., porins or lipoproteins) or the exocellular enzymes that are secreted by many bacteria into their culture medium such as penicillinases, proteinases and amylases. Other proteins made within the cell may pass into the cytoplasmic membrane and remain there. The mechanism by which transport of these proteins occurs into or across the cytoplamic membrane is unknown. The genes that code for these proteins carry a message that results in the addition of a sequence of about 20 extra amino acids (the signal peptide) to the proteins during their synthesis within the cell. Unlike ordinary proteins, proteins carrying a signal peptide are destined to be transported into or across the cytoplasmic membrane.

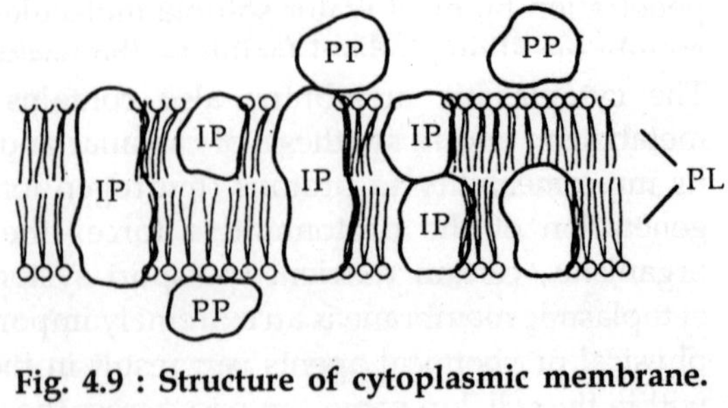

Fig. 4.9 : Structure of cytoplasmic membrane.

Protoplast

Protoplasts can be prepared from Gram-positive bacteria by treating the cells with an enzyme such as lysozyme, which selectively dissolves the cell wall, or by cultivating the bacteria in the presence of an antibiotic such as penicillin, which prevents the formation of the cell wall. In either case, the osmotic pressure of the medium must be sufficiently high to protect the organisms from bursting. Bacteria normally occur in hypotonic environments and they continually take up water by osmosis; thus, they tend to expand, pressing the cytoplasmic membrane tightly against the rigid cell wall. In the absence of a rigid cell wall, there is nothing to prevent the continued expansion and eventual bursting of a protoplast. This bursting can be prevented by preparing protoplasts in an isotonic medium, i.e., in a medium that has an osmotic pressure similar to that of the protoplast.

Spheroplast

However, the cell walls of Gram-negative bacteria differ from those of Gram-positive bacteria by possessing an outer membrane. Although the peptidoglycan of the cell wall may be destroyed by lysozyme or its synthesis inhibited by antibiotics, the flexible outer membrane of the cell wall remains. Because the treated cell has two membranes, the cytoplasmic membrane of the protoplast plus the outer membrane of the cell wall, the cell is called a spheroplast rather than a protoplast. Some bacteria, the mycoplasmas, never has cell walls and are bounded by only a cytoplasmic membrane; therefore, they have many of the propertes of protoplasts, yet they manage to thrive nonetheless. Most mycoplasmas are parasities of animals, plants, or arthropods, and therefore live in osmotically favourable or isotonic environments. Some are able to attain a degree of rigidity by incorporating cholesterol into their cytoplasmic membranes. Most mycoplasmas have a more or less spherical shape, but one genus, *Spiroplasma,* consists of helical cells.

Membranous Intrusions

Many bacteria, especially Gram-positive bacteria, possess membrane invaginations in the form of systems of convoluted tubules and vesicles termed mesosomes. Those known as central mesosomes penetrate deeply into the cytoplasm, are located near the middle of the cell, and seem to be attached to the cell's nuclear material; they are thought to be involved in DNA replication and cell division. In contrast, peripheral mesosomes show only a shallow penetration into the cytoplasm, are not restricted to a central location, and are not associated with nuclear material; they seem to be involved in export of exocellular enzymes such as penicillinase. Extensive intracellular membrane systems occur in methane-oxidizing bacteria, in certain chemoautotrophic bacteria and in nearly all phototrophic bacteria. For example, in phototrophic bacteria they are the site of the photosynthetic apparatus of the cell; the infoldings provide a large surface area to accommodate a high content of light-absorbing pigments.

Cytoplasm

The cell material bounded by the cytoplasmic membrane may be divided into (a) the cytoplasmic area, granular in appearance and rich in the macromolecular RNA-protein bodies known as ribosomes, on which proteins are synthesized; (b) the chromatinic area, rich in DNA; and (c) the fluid portion with dissolved substances. Unlike animal or plant cells, there is no endoplasmic reticulum to which ribosomes are bound; some ribosomes are free in the cytoplasm, and others, especially those involved in the synthesis of proteins to be transported out of the cell, are associated with the inner surface of the cytoplasmic membrane. When the ribosomes of prokaryotes undergo sedimentation in a centrifuge, they have a sedimentation coefficient of 70 Svedberg units (70S) and are composed of two subunits, a 50S and a 30S subunit. Its major component is water (70-80%), which serves as a solvent for the cell pool, a complex mixture of nutrients, including sugars, amino acids, and salts. The components of this pool serve as building blocks for cell synthesis or as sources of energy.

Cytoplasm Inclusions

Volutin granules, also known as metachromatic granules, are composed of polyphosphate. They stain an intense reddish-purple colour with dilute methylene blue and can be observed by light microscopy. By electron microscopy they appear as round, dark areas. Volutin serves as a reserve source of phosphate. Another polymer often found in aerobic bacteria, especially under high-carbon, low-nitrogen culture conditions, is a chloroform-soluble, lipidlike material, poly-β-hydroxybutyrate, (PHB), which can serve as a reserve carbon and energy source. PHB granules can be stained with lipid-soluble dyes such as Nile blue. Polysaccharide granules, i.e., glycogen, can be stained brown with iodine. By electron microscopy they appear as dark granules. Some bacteria that live in aquatic habitats form gas vacuoles that provide buoyancy. By light microscopy these are bright, refractile bodies; by electron microscopy they are seen to have a regular shape; hollow, rigid cylinders with more or less conical ends and having a striated protein boundary. This boundary is impermeable to water, but the various dissolved gases in the culture medium can penetrate it to fill the cavity. The identifying feature of gas vacuoles is that they can be made to collapse under pressure and thereby lose their refractility.

Chromatin Bodies/Plasmids

The hereditary material of bacteria exists in the form of a single circular strand of DNA designated as the chromatin (Gr. *Chromato*-colour, and *in*, fiber) body, or bacterial chromosome. By definition, bacteria do not have a nucleus—that is, their DNA is not enclosed by a nuclear membrane, but instead is aggregated in a dense area of the cell called the nucleoid (L. *Nucis*, nut, and *oid*, formor like). The chromatin body is actually an extremely long molecule of DNA that is tightly coiled around special basic protein molecules so as to fit inside the cell compartment. Arranged along its length are genetic units (genes) that carry information required for bacterial maintenance and growth. When exposed to special stains or observed with an electron microscope, chromatin bodies have a granular or fibrous appearance. Since bacteria have a single chromosome, they are haploid. Although the chromatin body is the minimal genetic requirement for bacterial survival, many bacteria contain other, non-essential pieces of DNA called plasmids (Gr. *Plasma*, a form, and *id*, belonging to). These tiny, circular extra-chromosomal strands can be free or integrated into the chromosome; they are duplicated and passed on to offspring. They are not essential to bacterial growth and metabolism, but they often confer protective traits such as resisting drugs, and producing toxins and enzymes.

Ribosomes

A bacterial cell contains thousands of tiny, discrete units called ribosomes (Gr. *Ribose*, a pentose sugar, and *soma*, body). When viewed even by very high magnification, ribosomes show up as fine, spherical specks dispersed through the cytoplasm that often occur in chains (polysomes). They are also attached to the cell membrane and to mesosomes. Chemically, a ribosome is a combination of a special type of RNA called ribosomal RNA, or rRNA (about 60%), and protein (40%). One method of characterizing ribosomes is by S, or Svedberg, units, which rate the molecular sizes of various cell parts that have been spun down and separated by molecular weight and shape in a centrifuge. Heavier, more compact structures sediment faster and are assigned a higher S rating. Combining this method of analysis with high resolution electron micrography has revealed that the prokaryotic ribosome, which has an overall rating of 70S, is actually composed of two smaller subunits. The 30S unit looks something like a heart; the 50S unit bears a resemblance to a crown.

Bacterial Endospores

Ample evidence indicates that the anatomy of bacteria helps them adjust rather well to adverse habitats. But of all microbial structures, nothing can compare to the bacterial endospore (Gr. *Endo,* inside, and *sporos,* seed) for withstanding hostile conditions and facilitating survival. It is a generic term that refers to any tiny cell-like bodies that are produced by vegetable or reproductive structures of microorganisms (Fig. 4.10). Spores can be quite variable in origin, form, and function. The bacterial type discussed here is called an endospore because it is produced inside a cell. It functions in *survival,* not reproduction, because no increase in cell numbers is involved in its formation. Endospores are dormant bodies produced by Gram-positive genera *Bacillus, Clostridium,* and *Sporosarcina.* These bacteria have a two-phase life cycle—a vegetative cell and an endospore. The vegetative cell is a metabolically active and growing entity that can be induced by genetic signals to undergo spore formation or sporulation. Once formed, the spore exists in an inert, resting conditions that shows up prominently in a spore or Gram stain. Spores also stand out in a fresh hanging drop preparations because of their *refractility* (A property of causing light rays to bend, thus creating differences in contrast).

Once this stimulus has been received by the vegetative cell, it undergoes a conversion to a committed sporulating cell called a sporangium (NL *Sporos,* seed, and *angeion,* vessel). Complete transformation of a vegetative cell into a sporangium and then into a spore

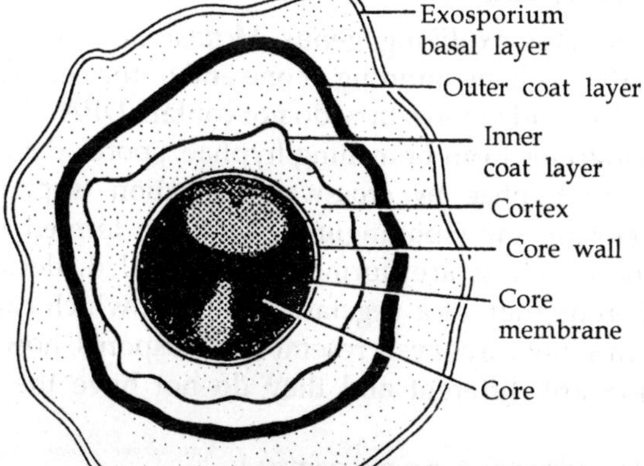

Fig. 4.10 : Sectioned endospore of a *Bacillus* sp.

requires 6 to 8 hours in most spore-forming species. Bacterial endospores are the hardiest of all life forms, capable of withstanding extremes in heat, drying, freezing, radiation, and chemicals that would readily kill vegetative cells. Their survival under such harsh conditions is due to several factors. The heat resistance of spores has been linked to their high content of calcium and *dipicolinic acid,* although the exact role of these chemicals is not yet clear. We know, for instance, that heat destroys cells by inactivating proteins and DNA, and that this process requires a certain amount of water in the protoplasm. Since the deposition of calcium dipicolinate in the spore removes water and leaves the spore very dehydrated, it is less vulnerable to the effects of heat. Moreover, since the spore is already very dehydrated, it is also metabolically inactive and highly resistant to damage from further drying. The thick, impervious cortex and spore coats also protect against radiation and chemicals. The longevity of bacterial spores verges on immortality. After lying in a state of inactivity for a varying period of time, spores can be revitalized when favourable conditions arise. The breaking of dormancy, or germination, happens in the presence of water and a specific chemical or environmental stimulus (germination agent). Once initiated, it proceeds to completion quite rapidly (1 ½ hours). While the specific germination agent varies among species, it is generally a small organic molecule such as an amino acid or an inorganic salt. This agent stimulates the formation of hydrolytic (digestive) enzymes by the spore membranes. These enzymes digest the cortex and expose the core to water. As the core rehydrates and takes up nutrients, it begins to grow out of the spore coats. Although the majority of spore-forming bacteria are relatively harmless, several bacterial pathogens are sporeformers. In fact, some aspects of the diseases they cause are related to the persistence and resistance of their spores. *Bacillus anthracis* is the agent of anthrax, a skin and lung infection of domestic animals transmissible to humans. The genus *Clostridium* includes even more pathogens, including *C. tetani,* the cause of tetanus (lockjaw), and *C. perfringens,* the cause of gas gangrene. When the spores of these species are embedded in a wound that contains dead tissue, they can germinate, grow, and release potent toxins. Another toxin-forming species, *C. botulinum,* is the agent of botulism, a deadly form of food poisoning. Because they inhabit the soil and dust, spores are a constant intruder where sterility and cleanliness are important.

Exospores, Conidiospores and Cysts

Cells of the methane-oxidizing genus *Methylosinus* form exospores, i.e., spores external to the vegetative cell, by budding at one end of the cell. These are desiccation and heat-resistant, but unlike endospores they do not contain DPA. The large group of bacteria known as the actinomycetes form branching hyphae; spores develop, singly or in chains, from the tips of these hyphae by crosswall formation (septation). If the spores are contained in an enclosing sac (sporangium), they are termed sporangiospores; if not, are called conidiospores. Cysts are dormant, thick-walled, desiccation-resistance forms that develop by differentiation of a vegetative cell and which can later germinate under suitable conditions. In some ways cysts resemble endospores; however, their structure and chemical composition are different and they do not have the high heat resistance of endospores.

IDENTIFICATION METHODS OF BACTERIA

Isolation and laboratory growth are also necessary for discovery and verification of new species. The methods that a microbiologist uses to identify bacteria to the level of genus and species fall into the main categories of morphology (microscopic and macroscopic), bacterial physiology or biochemistry, serological analysis, and genetic techniques. Data from a cross section of such tests can produce a unique profile of each bacterium. Final differentiation of any unknown species is accomplished by comparing its profile to the characteristics of known bacteria in tables, charts, and keys. Many of the identification systems are automated and incorporate computers to process data and provide a "best fit" identification. However, not all methods are used on all bacteria. A few bacteria can be identified by placing them in an automated machine that analyzes only the kind of fatty acids they contain; in contrast, some are identifiable by a Gram stain and a few physiological tests; others may require a diverse spectrum of morphologic, biochemical, and genetic tests.

Microscopic morphology—Traits that can be valuable aids to identification are combinations of cell shape and size, Gram stain reaction, acid-fast reaction, and special structures, including endospores, granules, and capsules.

Macroscopic morphology—Appearance of colonies, including texture, size, shape, pigment, speed of growth and patterns of growth in broth and gelatin media.

Physiological-biochemical characteristics—Enzymes and other biochemical properties of bacteria are a fairly reliable and stable expression of the chemical identity of each species. Dozens of diagnostic test exist for determining the presence of specific enzymes and to assess nutritional and metabolic activities. Examples include tests for fermentation of sugars; capacity to digest or metabolize complex polymers such as proteins and polysaccharides; production of gas; presence of enzymes such as catalase, oxidase and decarboxylases; and sensitivity to antimicrobic drugs.

Chemical analysis—Analyzing the types of specific structural substances that the bacterium contains, such as the chemical composition of peptides in the cell wall and lipids in membranes.

Serological analysis—Bacteria have surface and other molecules called antigens that are recognized by the immune system. One immune response to antigens is the production of molecules called antibodies that are designed to bind tightly to the antigens. This response is to specific that antibodies can be used as a means of identifying bacteria in specimens and cultures.

Genetic and molecular analysis—Examining the genetic material itself has revolutionized the identification and classification of bacteria.

G + C base composition—The overall percentage of guanine and cytosine (the G + C content as compared to A+T content) in DNA is a general indicator of relatedness because it is a trait that does not change rapidly. Bacteria with a significant difference in G+C percentage are likely to be genetically distinct species or genera. For example, although superficially similar in Gram reaction, shape, and other morphologic characteristics, *Escherichia* has a G + C base composition of 48-52% and *Pseudomonas* has a composition of 58-70%, indicating that they probably are not closely related.

DNA analysis using genetic probes—The exact order and arrangement of the DNA code is unique to each organism with a technique called *hybridization*, it is possible to identify a bacterial species by analyzing segments of its DNA. This requires small fragments of single-stranded DNA (or RNA) called probes that are known to be complementary to the specific sequences of DNA from a particular microbe. The test is conducted by extracting unknown test DNA from cells in specimens or cultures and binding it to special blotter paper. After several different probes are added to the blotter, it is observed for visible signs that the probes have become fixed (hybridized) to the test DNA.

Nucleic acid sequencing and rRNA analysis—One of the most viable indicators of evolutionary relatedness and affiliation are comparisons in the sequence of nitrogen bases in ribosomal RNA, a major component of ribosomes. Ribosomes have the same function (protein synthesis) in all cells and they tend to remain more or less stable in their nucleic acid content over long periods. Thus, any major differences in the sequence or "signature" of the rRNA is likely to indicate some distance in ancestry.

RICKETTSIAS

Ricketts located a similar organism in blood of animals infected with Mexican typhus and discovered that in this case lice were important to transmission. The fundamental differences from ordinary bacteria are (a) smaller size with smaller genome providing fewer enzymes (b) longer generation time and (c) requirement of exogenous energy supply for growth.

Structure

The rickettsia measure about 0.3-0.5 μm in diameter and 0.3-0.4 μm in length and usually appear to rods with rounded edges, a form known as the *Coccobacillus*. They also exist in alternate shapes and hence pleomorphic. They have no flagella, pili, capsules, or spores. The cell wall is chemically similar to that of Gram-negative bacteria. Reproduction by binary fission. Rickettsia infects both humans and arthropods, the latter serving as vector. Arthropod is primary host and the human is secondary host. Ticks and mites pass the rickettsia to their young through the eggs in a process called transovarian infection. When the arthropod feeds in the skin of infected person, it takes a blood meal and becomes infected particularly along its intestinal tract. As the infected cells of the arthropod burst, the organism fill the intestinal cavity and are deposited during defecation on the skin of next individual. By scratching the site, the person himself inoculate the rickettsia into the blood streak, and symptoms of disease soon appear. The symptoms are a high fever lasting for several days and skin rash. The rash begins as pink spots called macules. They progress to pink-red spots, like pimples, called papules. They soon become dark red and then fade without leaving evidence of scaring. This type of rash is maculopapular rash. Rickettsial infections usually respond to treatment with tetracyclins.

Weil-Felix test—This test is used to identify to presence of rickettsia without specifying a particular species. A patient's blood sample is first obtained, and then clean serum is separated from red blood cells. A sample of serum is mixed with the cells of bacterium, *Proteus ox19*. The bacteria will clump together if the serum contains antibodies against the rickettsia indicating that a person has the disease or once had it. If no clumping takes plate, rickettsial antibodies are probably not present. The Weil-Felix test works because, rickettsia posses antigens, also located in *Proteus* cells, and therefore, both organisms stimulate the immune system to produce the same antibodies.

CHLAMYDIA

This is a sub group of rickettsia, and are among the smallest recognized bacteria that cause diseases in humans and animals. They measure about 0.25 μm in diameter. These were originally thought as viruses. However, later it was found that they posses both DNA and RNA as well as number of enzymes. They can be treated as some antibodies. They grow only in living tissues such as fertilized chicken eggs and tissue cultures. They have complex reproductive cycle, that occurs in the cytoplasm of an infected cell. The chlamydia, called an elementary body (EB) enlarge to form an initial body or reticulate body that divides by binary fission to form a series of small, dense particles, the developmental cycle alternates between two forms—(A) the elementary body (EB), 300 nm in diameter is specialized for survival when released from the cell, and (B) reticulate body (RB) or initial body upto 1000 nm in diameter and engaged in intracellular multiplication within a few hours after entry into the host cells, the EB begins to undergo profound changes in their cell envelopes and characteristics central condensate begins to disperse to form a more homogeneous cytoplasm in which strands of nucleic acids and ribosomes are seen. The resulting RBs continue to grow in size and 10-15 hours after infection binary fission begins. At 20-30. After infection some of the RBs develop a central condensation of cytoplasmic contents, decrease in size and become typical EBs. Most of the RBs, however, continue to multiply until the host cell cytoplasm is almost filled by the colony. The process is completed within about 24 hour and the particles may be seen in cells cytoplasm, as granule like inclusions, which are characteristics of chlamydial diseases.

Chlamydia trachomatis, has more than one strain called, TRIC agent. TRIC agent is responsible for trachoma (a eye disease), inclusion conjunctivitis (eye disease) and nongonococcal urethritis (reproductive system disease). When TRIC agent infects the reproductive system and is pass in sexual contacts the disease is called nongonococcal urethritis. This name was coined as the symptoms are remarkably similar to those in gonorrhea (disease caused by *Neisseria gonorrhoeae,* a Gram-negative diplococcus, susceptible to penicillin but responds only to tetracycline. Chlamydial ophthalmia, disease also found in small children.

MYCOPLASMA

Nocard and Roux (1898) investigated the pleural fluids of cattle suffering from bovine pleuropneumonia, discovered the organisms that were entirely different from any other microorganisms that are known similar pleomorphic (Gr. *Pleo,* many, and *morphe,* form) organisms were later isolated from other animals as goats, dogs and human beings. Similar forms were also found growing as saprophytes in decaying in organic matter. These were named as pleuropneumonia like organisms (PPLO). The species discovered by Roux and Nocard was given the first binomial *Asterococcus myloides* by Borrel *et. al.* (1910), meaning rounded and stellate forms with radial, mold like filaments. It was later on put under the genus *Mycoplasma* by Nowak (1929) and these organisms are now commonly called as mycoplasma.

Characters of Mycoplasma

1. Mycoplasma are free living, parasites, and saprophytic organisms.
2. These are unicellular, prokaryotic, usually non-motile, and form fried egg shaped colonies.
3. These can be easily filtered by bacterial filters.
4. Both RNA and DNA fragments are found in mycoplasma.

5. These are sensitive to tetracyclins but resistant to penicillins.
6. The cell wall is totally absent.
7. Cell membrane is triple layered, made up of lipoproteins, flexible, and 7-11 nm thick.
8. These are highly pleomorphic forms.
9. The mycoplasma vary in size from 300 nm to 0.2 μm in diameter. The volume of smallest PPLO is about 1×10^{-3} μm^3.
10. The cell membrane encloses the cytoplasm which contains numerous ribosomes, network of fibrillar DNA, electron dense area, soluble RNA and soluble proteins.
11. However, mycoplasma divide unevenly into minute bodies called elementary bodies.
12. The elementary bodies range in size from about 330 nm to 450 nm.
13. Growth rate of mycoplasma is very high, generation time is 1-3 hours.

Plant-Pathogenic Mycoplasma

Doi et al., (1967) discovered pleomorphic mycoplasma like organisms in phloem cells of plants affected by different yellow type diseases. Mycoplasma occur from temperate to tropical regions but it is in warmer areas that serious loses occur in crops as coconut, citrus, rice, maize, cotton. They also occur in insects and as saprophytes in soil and sewage. Monroe-Eaton isolated a tiny virus like agent from the respiratory tract of patients. This was named as *Mycoplasma pneumoniae*, the Eaton agent, *Mycoplasma hominis* causes a sexually transmitted disease, mycoplasmal urethritis.

❑❑❑

5

EUKARYOTIC CELLS : THE MICROORGANISMS

The structure of these new cells was so versatile that eukaryotic microorganisms soon spread out into available habitats and adopted greatly diverse styles of living. The first primitive eukaryotes were probably single-celled and independent, but, over time, some forms began to aggregate, forming colonies and filaments. With further evolution, some of the cells within colonies became *specialized,* or adapted to perform a particular function advantageous to the whole colony, such as locomotion, feeding, or reproduction. Complex multicellular organisms evolved as individual cells in the organism lost the ability to survive apart from the intact colony. Although a multicellular organism is composed of many cells, it is more than just a disorganized assemblage of cells like a colony. Rather, it is composed of distinct groups of cells that cannot exist independently of the rest of the body. Looking at modern eukaryotic organisms, we find examples of many levels of cellular organization. All protozoa, as well as numerous algae and fungi, live a unicellular or colonial existence. Truly multicellular organisms are found only among plants and animals, and some of the fungi and algae. Only certain eukaryotes are small enough to fall into the realm of microbiology—namely, the protozoa and some of the algae, fungi, and animal parasites. All eukaryotic microbial cells have a cytoplasmic membrane, nucleus, mitochondria, endoplasmic reticulum, Golgi apparatus, vacuoles, and cytoskeleton. A cell wall, locomotor appendages, chloroplast, and glycocalyx are found only in some groups.

STRUCTURE OF EUKARYOTIC CELL

The cells of eukaryotic organisms are so varied that no one member can serve as typical example. A eukaryotic cells have following organelles :

Cilia and Flagella

Locomotion by means of flagella or cilia is a common property of protozoa, many algae, and a few fungal and animal cells. Although they share the same name, the structure and operation of eukaryotic flagella are much different from those of prokaryotes. The eukaryotic flagellum is thicker (by a factor of 10), structurally more complex, and covered by an extension of the cell membrane. A single flagellum is a long, sheathed cylinder containing regularly spaced hollow tubules—microtubules—that extend along its entire length. A cross section reveals nine pairs of closely attached microtubules surrounding a single central pair. This scheme, called the 9 + 2 arrangement, is a universal pattern of flagella and cilia. During locomotion, the adjacent microtubules slide past each other, which whips the flagellum back and forth. Flagella can move the cell by pushing it forward like a fishtail or by pulling it by a lashing or twirling motion. The placement and number of flagella can be useful in identifying flagellated protozoa and certain algae. Cilia (L. *Siliam*, eyelid) are very similar in overall architecture to flagella, but they are shorter and more numerous (some cells have several thousand). They are found only on a single group of protozoans and certain animal cells. In the ciliated protozoans, the cilia occur in rows over the cell surface, where they beat back and forth in regular oarlike strokes. Such protozoans are among the most rapid of all motile cells. The fastest ciliated protozoan can swim up to 2,500 μm/sec—a meter and a half per minute. On some cells, cilia also function as feeding and filtering structures. A notable example of the function of animal cilia occurs in the epithelial cells in the respiratory tract.

Glycocalyx

Most eukaryotic cells have a glycocalyx, an outermost boundary that comes in direct contact with the environment. This structure is usually composed of polysaccharides and appears as a network of fibers, a slime layer, or a capsule much like the glycocalyx of prokaryotes. Because of its positioning, the glycocalyx contributes to protection, adherence of cells to surfaces, and reception of signals from other cells and from the environment. The nature of the layer supporting the glycocalyx varies among the several eukaryotic groups. Fungi and most algae have a thick, rigid cell wall; whereas protozoa, a few algae, and all animal cells lack a cell wall and have only a cytoplasmic membrane beneath.

Cell Wall

The cell walls of algal and fungal cells are rigid and provide structural support and shape, but they are different in chemical composition from prokaryotic cell walls. Fungal cell walls have a thick, inner layer of polysaccharide fibers composed of chitin or cellulose and a thin outer layer of mixed glycans. The cell walls of algae are quite varied in chemical composition. Substances commonly found among various algal groups are cellulose, pectin, mannans, and minerals such as silicon dioxide and calcium carbonate.

Cell Membrane

The cytoplasmic membrane of eukaryotic cells is a typical bilayer of lipids in which protein molecules are embedded. In addition to phospholipids, eukaryotic membranes also contain *sterols* of various kinds. Sterols are different from phospholipids in both structure and behaviour. Their relative rigidity confers stability on eukaryotic membranes. This strengthening feature is extremely important in cells lacking a cell wall. Cytoplasmic membranes of eukaryotes are functionally similar to those of prokaryotes, serving as selectively permeable barriers in transport. Unlike prokaryotes, eukaryotic cells also contain a number of individual membrane-bound organelles that are extensive enough to comprise 60% to 80% of their volume.

Nucleus

The nucleus is a compact sphere that is the most prominent organelle of eukaryotic cells. It is separated from the cell cytoplasm by an external boundary called a nuclear envelope. The envelope has a unique architecture. It is composed of two parallel membranes separated by a narrow space, and it is perforated with small, regularly spaced openings, or pores, formed at sites where the two membranes unite. The nuclear pores are passage ways through which macromolecules migrate from the nucleus to the cytoplasm and vice versa. The nucleus contains a matrix called the nucleoplasm and a granular mass, the nucleolus (L. *Diminutive of nucleus*), that can stain more intensely than the immediate surroundings due to its RNA content. The nucleolus synthesizes the RNA portions of ribosomal subunits, which are transported through the nuclear pores into the cytoplasm for final assembly. A prominent feature of the nucleoplasm in stained preparations is a network of dark fibers known as chromatin (Gr. *Chromas,* colour) because of its attraction for dyes. Analysis has shown that chromatin actually comprises the eukaryotic chromosomes, large units of genetic information in the cell. The chromosomes in the nucleus of most cells are not readily visible because they are long, linear DNA molecules bound in varying degrees to *histone* proteins, and they are far too fine to be resolved as distinct structures without extremely high magnification. During mitosis (Gr. *Mitos,* thread and *osis,* a condition of) however, when the duplicated chromosomes are separated equally into daughter cells, the chromosomes themselves finally become readily visible as discrete, X- or Y-shaped structures. To ensure its continued existence, a eukaryotic cell maintains a specified number of chromosomes that is typical for its species. The set of chromosomes exist in the haploid (Gr. *Haplos,* single, and *eidos,* form) single, unpaired state or in the diploid (Gr. *Di,* two, and *eidos,* form) state. Most fungi, many algae, and some protozoa are examples of organisms whose cells occur in the haploid state during most of the life cycle. For examples, the chromosome number of one type of ameba is 25, and one species of the fungus *Penicillium* possesses 5 chromosomes. The cells of animals and plants, and some protozoa and algae, spend the greater part of their life cycle in the diploid state. This state comes about when the female haploid gamete (egg) fuses with the male haploid gamete (sperm) and produces a double (diploid) set of chromosomes in the offspring during sexual reproduction. Examples are humans with 46 chromosomes (23 pairs), a parasitic blood fluke with 16 chromosomes (8 pairs), and a redwood tree with 22 chromosomes (11 pairs). During normal growth or cell division (asexual reproduction) of all eukaryotic organisms, the chromosome number, whether haploid or diploid, is maintained by mitosis. When it comes to sexual reproduction in diploid organisms, however, the diploid chromosome number must be reduced to haploid, since gametes must be haploid for proper fertilization. This is accomplished through reduction division, or meiosis (Gr. *Meios,* diminution). During this process, diploid cells in the sex organs undergo two sets of divisions that result in haploid sex cells. Formation of a diploid zygote then restores the chromosome number characteristic of that organism. Another example of the function of meiosis occurs in the sexual reproductive cycles of haploid fungi. Normal mitosis alone can produce their gametes, but after these gametes unite, they form a diploid zygote.

Endoplasmic Reticulum

The endoplasmic (Gr. *Endo,* within, and *plasma,* something form) reticulum (L. *Rete,* net or network) can be likened to a labyrinth on a microscopic scale. In sections, it appear as a parallel series of thin, flat pouches bounded by membranes. Two kinds of endoplasmic reticulum are, the rough endoplasmic reticulum (RER) and the smooth endoplasmic reticulum (SER). Electron micrographs show that the RER originates from the outer membrane of the nuclear envelope and extends in a continuous network through the cytoplasm, even out to the cell membrane. This architecture permits the RER to transport materials from the nucleus to the cytoplasm and ultimately to the cell's exterior. The RER appears rough because of large numbers of ribosomes partly attached to its membrane to act as secretary factories. Proteins synthesized on the ribosomes are shunted into the cavity of the reticulum and held there for later packaging and transport. The smooth ER is a closed tubular network without ribosomes that function in nutrient processing and synthesis, and storage of nonprotein macromolecules such as lipids.

Golgi Complex

The Golgi complex (*Golgi,* named for C. Golgi) also called the Golgi body or apparatus, is a discrete organelle consisting of a stack of several flattened, disc-shaped sacs, or *cisternae.* These sacs have outer limiting membranes and cavities like the endoplasmic reticulum, but they do not form a continuous network. This organelle is always closely associated with the endoplasmic reticulum both in its location and function. At a site where it meets the Golgi complex, the endoplasmic reticulum buds off tiny membrane bound packets of protein called transitional *vesicles* (L. *Vesios,* bladder) that are picked up by the forming face of the Golgi complex. Once in the complex itself, the proteins are often modified by the addition of polysaccharides and lipids. The final action of this apparatus is to pinch off finished condensing vesicles that will be conveyed to organelles such as lysosomes or transported outside the cell as secretary vesicles. The production of antibodies and receptors in human white blood cells involve this packaging and transport system.

Nucleus, Endoplasmic Reticulum, Golgi Apparatus : Assembly Unit

This network includes ribosomes, which originate in the nucleolus, and the rough endoplasmic reticulum, which is continuously connected with the nuclear envelope. Initially, a genetic code on DNA is copied into RNA and passed out through the nuclear pores directly to the ribosomes on the endoplasmic reticulum. Here, specific proteins are synthesized according to the RNA code and deposited in the lumen of the endoplasmic reticulum. After being transported by a transitional vesicle to the Golgi complex, the protein products are chemically modified and packaged into vesicles that can be used by the cell in a variety of ways. Some of the vesicles contain enzymes to digest food inside the cell, other vesicles are secreted to digest materials outside the cell, and yet others are important in the enlargement and repair of the cell wall and membrane.

Lysosome

A lysosome (Gr. *Lysis,* dissolution, and *soma,* body) is one type of vesicle originating from the Golgi complex that contains a variety of enzymes. Lysosomes are involved in intracellular digestion of food particles and in protection against invading microorganisms. They also participate in digestion and removal of cell debris in damaged tissue. Other types of vesicles include vacuoles (L. *Vacuus,* empty) which are membrane-bound sacs containing fluids or solid particles to be digested, excreted, or stored. They are formed in phagocytic cells (certain white blood cells and protozoans) in response to food and other substances that have been engulfed. The contents of a food vacuole are digested through the merger of the vacuole with a lysosome. Other types of vacuoles are used in storing reserve food such as fats and glycogen.

Mitochondria

Although the nucleus is the cell's control center, none of the cellular activities it oversees could proceed without a constant supply of energy, the bulk of which is generated in most eukaryotes by mitochondria (Gr. *Mitos,* thread, and *chondrion,* granule). When viewed with light microscopy, mitochondria appear as round to elongate particles scattered through the cytoplasm. The internal ultrastructure reveals that a single mitochondrion consists of a smooth, continuous outer membrane that forms the external contour, and an inner, folded membrane nestled neatly within the outer membrane. The folds on the inner membrane, called cristae (L. *Crista,* a comb), appear fingerlike in cross section, but are actually more shelflike or platelike. The cristae membranes hold the enzymes and electron carriers of aerobic respiration. This is an oxygen-using process that extracts chemical energy contained in nutrient molecules and stores it in the form of high-energy molecules, or ATP. The spaces around the cristae are filled with a chemically complex fluid called the matrix (L. *Matter,* mother or origin) that holds ribosomes, DNA, and the pool of enzymes and other compounds involved in the metabolic cycle. Mitochondria (along with chloroplasts) are unique among organelles in that they divide independently of the cell, contain circular strands of DNA, and have prokaryotic sized 70S ribosomes.

Chloroplasts

Chloroplasts (Gr. *Chloros,* green, and *plastos,* to form) are remarkable packets found in algae and plant cells that are capable of converting the energy of sunlight into chemical energy through photosynthesis. The photosynthetic role of chloroplasts makes them the primary producers of organic nutrients upon which all other organisms (except certain bacteria) ultimately depend. Another important photosynthetic product of chloroplasts is oxygen gas. Although chloroplasts resemble mitochondria, chloroplasts are larger, contain special pigments, and are much more varied in shape. Although there are differences among various algal chloroplasts, most are generally composed of two membranes, one enclosing the other. The even, outer membrane completely covers an inner membrane folded into small, disclike sacs called thylakoids (Gr. *Thylakon,* a small sac, and *eidos,* form) that are stacked upon one another into grana. These structures carry the green pigment chlorophyll and sometimes additional pigments as well. Surrounding the thylakoids is a ground substances called the stroma (Gr. *Stroma,* mattress or bed). The role of the photosynthetic pigments is to absorb and transform solar energy into chemical energy that is then converted by reactions in the stroma to carbohydrates.

Cytoskeleton

This framework appears to have several functions, such as anchoring organelles, providing support, and permitting shape changes and movement in some cells. The two main types of cytoskeletal elements are microfilaments and microtubules. Microfilaments are thin protein strands that attach to the cell membrane and form a network through the cytoplasm. Some microfilaments are responsible for movements of the cytoplasm, often made evident by the streaming of organelles around the cell in a cyclic pattern. Other microfilaments are active in *ameboid motion,* a type of movement typical of cells such as amebas and phagocytes that produces extensions of the cell membrane (pseudopods) into which the cytoplasm flows. Microtubules are long hollow tubes that maintain the shape of eukaryotic cells without walls and transport substances from one part of a cell to another. The spindle fibers that play an essential role in mitosis are actually microtubules that attach to chromosomes and separate them into daughter cells.

Ribosomes

Ribosomes are distributed in two ways : some are scattered freely in the cytoplasm and cytoskeleton, while others are intimately associated with the rough endoplasmic reticulum as previously described. Multiple ribosomes are often found arranged in short chains called polyribosomes. The basic structure of eukaryotic ribosomes is similar to that of prokaryotes, both being composed of large (60S) and small (40S) subunits of ribonucleoprotein. By contrast, however, the eukaryotic ribosome is larger (80S) and originates in the nucleolus.

MYCETEAE

The position of the fungi (Gr. *Fungos,* mushroom) in the biological world has been debated for many years. Although they were originally classified with the green plants, they were later separated from plants and placed in a group with algae and protozoa (the Protista). However, even at that time, many mycologists were struck by several unique qualities of fungi that warranted placement into their own separate kingdom, which eventually was done. The kingdom Myceteae is large and filled with forms of great variety and complexity. For practical purposes, the approximately 100,000 species of fungi can be divided into two groups : The *macroscopic fungi* (mushrooms, puffballs, gill fungi) and the *microscopic fungi* (molds, yeasts). Although the majority of fungi are either unicellular or colonial, a few complex forms such as mushrooms and puffballs do have a limited form of cellular specialization. Cells of the microscopic fungi exist in two basic morphological types : yeasts and hyphae. A yeast cell is distinguished by its round to oval shape and by its mode of asexual reproduction. It grows swellings on its surface called buds. Some species form pseudohyphae (Gr. *Pseudo,* false, and *hyphe,* a web), a chain of yeasts formed when buds remain attached in a row. Hyphae are long, threadlike cells found in the bodies of filamentous fungi, or molds. Some fungal cells exist only in a yeast form; others occur primarily as hyphae; and a few, called dimorphic (Gr. *Di,* two, and *morphe* form), can take either form, depending upon growth conditions, especially changing temperature.

Mode of Nutrition in Fungi

All fungi are heterotrophic (Gr. *Hetero,* other, and *troph,* to feed). They acquire nutrients from a wide variety of organic materials called substrates. Most fungi are saprobes (Gr. *Sapros,* rotten, and *bios,* to live) meaning that they obtain these substrates from the remnants of dead plants and animals in soil, fresh water, or salt water. Parasitic fungi actually use the bodies of living animals or plants as a substrate, although very few fungi absolutely require a living host. In general, the fungus penetrates the substrate and secretes enzymes that reduce it to small molecules that can be absorbed by the cells. Fungi have enzymes for digesting an incredible array of substances, including feathers, hair, cellulose, petroleum products, wood, and rubber. It has been said that every naturally occurring organic material on earth can be attached by some type of fungus. Fungi are often found in nutritionally poor or adverse environments. In our laboratory, we occasionally discover fungi in mineral and weak acid solutions ("bottle imps") and on specimens preserved in formaldehyde. Various fungi thrive in substrates with high salt or sugar content, at relatively high temperatures, and even in snow and glaciers.

Fungal Organisation

The colonies of yeasts are much like those of bacteria in that they have a soft, uniform texture and appearance. The colonies of filamentous fungi are noted for the striking cottony, hairy, or velvety textures that arise from their microscopic organization and morphology. The woven, intertwining mass of hyphae that makes up the body or colony of a mold is called a mycelium (Gr. *Mykes,* fungus, and *helos,* nail). Although hyphae contain the usual eukaryotic organelles, they also have some unique organizational features. Nonseptate hyphae consist of one long, continuous cell not divided into individual compartments by septa (L. *Seepio,* to fence or wall) or cross walls with this construction, the cytoplasm and organelles move freely from one region to another, and each hypha can have several nuclei. In most fungi, the hyphae are divided into segments by cross walls, a condition called septate. The nature of the septa varies from solid partitions with no communication between the compartments to partial walls with small pores that allow the flow of organelles and nutrients between adjacent compartments.

Vegetative hyphae (mycelia) are responsible for the visible mass of growth that appears on the surface of a substrate and penetrates it to digest and absorb nutrients. During the development of a fungal colony, the vegetative hyphae ultimately give rise to structures called reproductive, or aerial, hyphae that orient vertically from the vegetative mycelium. These hyphae are responsible for the production of fungal reproductive bodies called spores.

Mode of Reproduction in Fungi

Most can propagate by the simple outward growth of existing hyphae or by fragmentation, in which a separated piece of mycelium can generate a whole new colony. But the primary reproductive mode of fungi involves the production of various types of spores. Do not confuse fungal spores with the more resistant, nonreproductive bacterial spores. Fungal spores are responsible not only for multiplication but also for survival, producing genetic variation, and dissemination. Due to their compactness and relatively light weight, spores are dispersed widely through the environment by air, water, and living things. Normal outdoor air can contain up to 100,000 fungal spores per cubic meter, and moist air containing nutrients can have as many as one billion spores per cubic meter. The fungi exhibit such a marked diversity in spores that they are largely classified and identified by their spores and spore-forming structures. The most general subdivision is based on the way the spores arise. Asexual spores are the products of mitotic division of a single parent cell, and sexual spores are formed through a process of two parental nuclei fusing, followed by meiosis. Asexual spores are of two types :

(a) Sporangiospores are formed by successive cleavages within a saclike head called a sporangium (Gr. *Sporos,* seed, and *angeion,* vessel) which is attached to a stalk, the sporangiophore. These spores are initially enclosed but are released when the sporangium ruptures.

(b) Conidia (Gr. *Konidion,* dust particle) bears the conidiospores, are free spores not enclosed by a spore-bearing sac. They develop by being pinched off the tip of a special fertile hypha, or by the segmentation of a preexisting vegetative hypha. Conidia are the most common asexual spores, and they occur in the following forms :

Arthrospore (Gr. *Arthron,* joint) – A rectangular spore formed when a septate hypha fragments at the cross walls.

Chlamydospore (Gr. *Chlamys,* cloak) – A spherical conidium formed by the thickening of a hyphal cell.

Blastospore – A spore produced by budding from a parent cell that is a yeast or another conidium; also called a bud.

Phialospore (Gr. *Phialos,* a vessel) – A conidium that is budded from the mouth of a vase-shaped spore-bearing cell called a phialide or sterigma, leaving a small collar.

Microconidium and macroconidium – The smaller and larger conidia formed by the same fungus under varying conditions. Microconidia are one-celled, and macroconidia have two or more cells.

Porospore – A conidium that grows out through small pores in the spore-bearing cell; some are composed of several cells.

Sexual Reproduction

Just as in plants and animals, this linking of genes from two parents creates offspring with combinations of genes different from either parent. The offspring from such a union can have slight variations in form and function that are potentially advantageous in the adaptation and survival of their species. Since fungal cells spend most of their life cycle in the haploid state, their cells already have a chromosome number compatible for sexual union. In general, during sexual reproduction, the haploid nuclei from two compatible parental cells fuse, forming a diploid nucleus that subsequently participates in sexual spore development the majority of fungi produce sexual spores at some point. The nature of this process varies from the simple fusion of fertile hyphae of two different strains to a complex union of differentiated male and female structures and the development of special fruiting structures. Four different types of sexual spores have been identified, but we will consider the three most common : zygospores, ascospores, and basidiospores.

Zygospores (Gr. *Zygon*, yoke, to join) are sturdy diploid spores formed when hyphae of two opposite strains (called the plus and minus strains) fuse and create a diploid zygote that swells; becomes covered by strong, spiny walls; and is released into the environment. When its wall is disrupted, and moisture and nutrient conditions are suitable, the zygospore germinates and forms a sporangium. Meiosis of diploid cells of the sporangium results in haploid nuclei that develop into sporangiospores. Both the sporangia and sporangiospores that arise from sexual processes are outwardly identical to the asexual type, but because the spores arose from the union of two separate fungal parents, they are not genetically identical. Through differentiation, each of these cells enlarges to form an ascus (ascospores), and its diploid nucleus undergoes meiosis (often followed by mitosis) to form four to eight haploid nuclei that will mature into ascospores. Late in the cycle, the asci burst and release the ascospores.

Basidospores (Gr. *Basidi*, a pedestal) are haploid sexual spores formed on the outside of a club-shaped cell called a basidium. In general, spore formation follows the same pattern of two mating types coming together, fusing and forming terminal cells with diploid nuclei. Each of these cells become a basidium, and its nucleus produces, through meiosis, four haploid nuclei. These nuclei are extruded through the top of the basidium, where they develop into basidiospores.

Classification of Fungi

This is due to the fact that the organisms do not always perfectly fit the neat categories made for them, and even experts cannot always agree on the nature of the categories. Since the fungi exhibit exceptions to the rule, there are several ways to classify them. For our purposes, we will adopt a classification scheme with a medical mycology emphasis, in which the kingdom Myceteae is subdivided into two subkingdoms, the Myxomycota (slime molds) and the Eumycota (true fungi). From the beginnings of fungal classification, any fungus that lacked a sexual state was called 'imperfect' and placed in a catchall category, the Fungi Imperfecti or Deuteromycota. Gradually, many species of Fungi Imperfecti were found to make sexual spores, and they were assigned to the taxonomic grouping that best fit those spores. This created a quandary, because several very important species, especially pathogens, had to be regrouped and renamed. In many cases, the older names were so well entrenched in the literature that it was easier to retain them and assign a new generic name for their sexual stage. For instance, *Blastomyces* and *Histoplasma* have a sexual state assigned to the genus *Ajellomyces*.

Significance of Fungi

I. Nearly all fungi are free-living and do not require a host to complete their life cycles. Even among those fungi that are pathogenic, most human infection occurs through accidental contact from an environmental source such as soil, water, or dust. Humans are generally quite resistant to fungal infection, except for two main types of fungal pathogens : the true pathogens, which can infect even healthy persons, and the opportunistic pathogens, which attack persons who are already weakened in some way.

II. The mycelium of a single colony of *Armillaria ostoye* covers an area of $2\,^1/_2$ square miles and could easily weigh 100 tons. Most of the fungus lies hidden beneath the ground where it periodically fruits into edible mushrooms. Experts with the forest service took random samples over a large area and found that the mycelium is from genetically identical stock. They believe this fungus was started from a single mating pair of sexual hyphae around 1,000 years ago. Another colossal species is from Michigan—*A. bulbosa* is estimated to weigh 250,000 pounds and is perhaps 1,500 years old. Both fungi are able to grow on the stumps and roots of trees, and can become pests.

III. This fungus *Taxomyces andreanae* synthesizes taxol, a substance that has shown great promise in treating breast and ovarian cancers. What makes this discovery of particular importance is that the original source of taxol was the pacific yew itself, yet this tree is somewhat uncommon and slow growing. Environmentalists feared that the few stands of trees would have to be wiped out to obtain the drug and even then, supplies would not meet the demands for therapy. An additional scientific surprise is that the fungus actually acquired its genes for synthesizing taxol from the live cells of its host tree. The next stage is already underway : to culture and possibly genetically engineer the fungus in the laboratory to improve its taxol output in preparation for clinical drug testing and marketing.

IV. When the fungus *Arabis holboellii* invades the tissues of the rock cress, it induces the plant to produce a tall spike of leaves that looks like a buttercup flower. The surface of this 'flower' becomes a substrate for the fungus to give off yellow reproductive structures and a sweet saplike substance that attracts bees and flies. The fake flower is so attractive to these insects that they prefer it to real flowers.

V. This tenacious fungus *Cladosporium* is to blame for the black mildew spots that appear on walls, shower grouting, and other hard surfaces. The mold is also known to inhabit air vents and internal walls of air conditioners and coolers. It grows tightly against these surfaces, deriving miniscule amounts of nutrients and absorbing periodic infusions of moisture. The dozens of cleaners on the marked attest to the difficulty in suppressing and removing this fungus. Researchers of 'sick building syndrome' are beginning to collect data on the possible role of this and other fungi in the odd collection of symptoms—headache, rash, fatigue—that some people experience in certain buildings.

VI. The list of opportunistic fungal pathogens has been increasing in the past few years because of newer medical techniques that keep compromised patients alive. Even so-called harmless species found in the air and dust around us may be able to cause opportunistic infections in patients who already have AIDS, cancer, or diabetes. Fungi are involved in other medical conditions besides infections. Fungal cell walls give off chemical substances that can cause allergies. The toxins produced by poisonous mushrooms can induce neurological disturbances and even death. The mold *Aspergillus flavus* synthesizes a potentially lethal poison called aflotoxin, which is the cause of a disease in domestic animals that have eaten grain infested with the mold and is also a suspected cause of liver cancer in humans. Fungi pose an ever-present economic hindrance to the agricultural industry. It has been estimated that as much as 40% of the yearly fruit crop is consumed not by humans but by fungi. On the beneficial side, however, fungi play an essential role in decomposing organic matter and returning essential minerals to the soil. Industry has tapped the biochemical potential of fungi to produce large quantities of antibodies, alcohol, organic acids, and vitamins. Some fungi are eaten or used to impart flavourings to food. The yeast *Saccharomyces* produces the alcohol in beer and wine and the gas that causes bread to rise.

ALGAE

The algae (L. Sea weeds) are a group of photosynthetic organisms usually recognized by their large members, such as seaweeds and kelps. In addition to being beautifully coloured and diverse in appearance, they vary in length from a few μm to 200 meters. Microscopic forms are included in the kingdom Protista; macroscopic forms such as kelps are grouped with the plants. Algae occur in unicellular, colonial, and filamentous forms, and the larger forms can possess tissues but would not be classified among the protists. Algal cells as a group contain all of the eukaryotic organelles. The most noticeable of these are the chloroplasts, which contain, in addition to the green pigment chlorophyll, a number of other pigments that create the yellow, red, and brown colouration of some groups. The chloroplasts can be of such unique character that they are used in identification. Most algal cells are enclosed by an intricately patterned cell wall, which accounts for the distinctive appearance of unicellular members such as diatoms and dinoflagellates. The outermost structure in a group called the euglenoids is a thick, flexible membrane called a *pellicle*. Motility by flagella or gliding is common among the algae, and many members contains tiny, light-sensitive areas (eyespots) that coordinate with the flagella to guide the cell toward the light it requires for photosynthesis.

Algae are one of the main components of the large floating community of microscopic organisms called *plankton*. In this capacity they play an essential role in the aquatic food chain and produce most of the earth's oxygen. Other algal habitats include the surface of soil, rocks, and plants, and several species are even hardy enough to live in hot springs or snowbanks. A more technical system divides the microscopic algae into four divisions, based on the types of chlorophyll and other pigments, the type of cell covering, and the nature of their stored foods. The common names for these divisions are : (a) Euglenophyta or euglenoids, (b) Chlorophyta or green algae, (c) Chrysophyta or golden brown algae, and (d) Pyrrophyta or dinoflagellates. Algae reproduce asexually through fragmentation, binary fission, and mitosis, and some produce motile spores. Their sexual reproductive cycles can be highly complex, with stages similar to fungi. The capacity of algae to photosynthesize is extremely important to the earth. They form the basis of aquatic food chains and play an essential role in the earth's oxygen and carbon dioxide balance. Some products of algae have industrial applications. For example, fossilized marine diatoms yield an abrasive powder called diatomaceous earth that is used in polishes, bricks, and filters, and certain seaweeds are a source of agar and algin used in microbiology, dentistry, and the food and cosmetics industries.

Prototheca, an unusual non-photosynthetic alga, which has been associated with skin and subcutaneous infections in humans and animals. The primary medical threat from algae is due to a type of food poisoning caused by the toxins of certain marine dinoflagellates. During particular seasons of the year, the overgrowth of these motile algae imparts a brilliant red colour to the water, which is referred to as a 'red tide'.

PROTOZOA

Habit and Habitat of Protozoa

Free-living protozoa have even been found in the polar regions and at very high altitudes. Parasitic protozoa may be found in association with most animal groups. Many protozoa survive dry conditions by the formation of a resistant cyst, or dormant stage. For example, the soil ameba *Naegleria* is a resistant cyst in dry weather, is a naked ameba in moist soil, and becomes flagellated when flooded with water. For example, the malarial parasite *Plasmodium* produces male gametes in response to a drop in temperature on transfer from a warm-blooded mammalian host to a mosquito. The distribution of trophic (vegetative) forms of protozoa in the sea and freshwater and of cyst forms in the atmosphere has resulted in the spread of free-living species throughout the world. From the ecological standpoint, protozoa may be divided into free-living forms and those living on or in other organisms. The latter group is referred to as the symbiotic protozoa. Some of the symbiotic ones are parasitic and may cause disease. The factors which influence the distribution and number of free-living protozoa in a habitat are moisture, temperature, light, available nutrients, and other physical and chemical conditions.

The vegetative or trophic stages of free-living protozoa occur in every type of salt water, freshwater, sand, soil, and decaying organic matter. Obviously, for those protozoa which bear chromatophores and carry out photosynthesis, sunlight is essential. It follows also that those protozoa which feed on photosynthetic microorganisms also require sunlight, albeit indirectly. Some protozoa can tolerate a wide range of pH, for example, pH 3.2 to 8.7. However, for the majority a pH range of 6.0 to 8.0 is optimal for maximum metabolic activity. The protozoan population in an aquatic environment is influenced by the chemical constituents of the water. Some protozoa thrive in water rich in oxygen but low in organic matter (mountain springs, brooks, or ponds); others require water rich in minerals. Still others prefer water with little oxygen but many decomposition products (e.g., black bottom slime and sewage). Some species have been found to live in both salt water and freshwater. The nutrient supply in a habitat is a major determining factor in the distribution and number of protozoa within it. Species of *Paramecium* and other holozoic protozoa must have a supply of bacteria or other protozoa. As a general rule, holozoic protozoa which feed on a variety of organisms are widely distributed; those that are more selective and feed only a few species are limited in their distribution. Most protozoa have an optimum temperature of between 16 and 25°C; the maximum is between 36 and 40°C. The minimum temperature is less detrimental. The temperature tolerance varies with different environmental conditions. Even warm waters (30 to 56°C) of hot springs have been known to contain protozoa. The so-called red snow of high altitudes is due to the presence of several hematochrome-bearing flagellates.

The association between protozoa and their host can differ in various ways. In commensalism, the host in neither injured nor benefitted, but commensal is benifitted. Ectocommensalism is often represented by protozoa which attach themselves to a host's body. Endocommensalism is the association when the protozoan is inside the host's body, e.g., the protozoa which live in the lumen of the alimentary tract. *Mutualism* occurs between some protozoa and their hosts. For example, certain flagellates are present in the gut of termites and digest the woody material eaten by the termite to a glycogenous substance which can be used by the host cells. If deprived of these flagellates, the termite dies; if the flagellates are removed from the termite gut, they too perish. In parasitism, one organism—the parasite—lives at the expense of the other. The parasite feeds on the host cells or cell fragments by pseudopodia or cytostome, or enters the host tissues and cells, living upon the cytoplasm and even the nuclei. As a result, the host may develop pathological conditions. The sporozoa are strictly parasitic and are among the most important of the disease-producing protozoa.

Significance of Protozoa

In marine waters, zooplankton are protozoa that feed on the photosynthetic phytoplankton. They in turn become food for larger marine organisms. This food chain can be represented as follows :

Solar energy → Producers → Primary consumers → Secondary consumers

Also of particular importance in the ecological balance of many communities, in wetlands as well as aquatic environments, are the saprophytic and bacteria—feeding protozoa. They make use of the substances produced and organisms involved in the final decomposition stage of organic matter. It follows that microorganisms are important in the degradation of sewage. Although bacteria participate in a prominent way in the process, the role of protozoa is becoming more completely understood and appreciated. Biological sewage treatment involves both anaerobic digestion and/or aeration. Anaerobic protozoa such as species of *Metopus, Saprodinium,* and *Epalxis* are active in the anaerobic steps, while those treatment steps requiring aeration and flocculation include the aerobic protozoa such as *Bodo, Paramecium, Aspidisca,* and *Vorticella.* In the treatment of industrial wastes, where there is an accumulation of nitrates and phosphates, the settling tanks are illuminated to promote the growth of algae and protozoa. These protists remove the inorganic material from the water for their own synthesis. Some protozoa cause disease in animals, including humans. They have caused untold misery. Such parasitic protozoa multiply within the host much as bacteria do. Some live only as obligate parasites and may produce chronic or acute diseases in humans. Some well-known protozoan diseases in humans are intestinal amebiasis. Protozoa have also become important research organisms for biologists and biochemists for the following reasons. Many protozoa are easily cultured and maintained in the laboratory.

Morphology of Protozoa

Leishmania donovani, the cause of the human disease kala azar, measures 1 to 4 μm in length. *Ameba proteus* measures 600 μm or more. Certain common ciliates reach 2,000 μm, or 2 mm, and the tests (a kind of protective envelope) of some extinct (fossilized) members of *Foraminiferida,* the nummulites, measure up to 15 cm in diameter.

Forms of Protozoa

Most protozoan cells are single units containing the major eukaryotic organelles except chloroplasts. Their organelles can be highly specialized for feeding, reproduction, and locomotion. The cytoplasm is usually divided into a clear outer layer called the *ectoplasm* and a granular inner region called the *endoplasm.* Ectoplasm is involved in locomotion, feeding, and protection. Endoplasm houses the nucleus, mitochondria, and food and contractile vacuoles. Some ciliates and flagellates even have organelles that work somewhat like a primitive nervous system to coordinate movement. Since protozoans lack a cell well, they have a certain amount of flexibility. Their outer boundary is a cell membrane that regulates the movement of food, wastes, and secretions. Cell shape can remain constant (as in most ciliates) or change constantly (as in amebas). Certain amebas (foraminiferans) encase themselves in hard shells made of calcium carbonate. Some notable exceptions are giant amebas and ciliates that are large enough (3-4 mm in length) to be seen swimming in pond water.

Locomotion in Protozoa

Protozoa may move by three types of specialized organelles : pseudopodia, flagella, and cilia. In addition, a few protozoa without such organelles can carry out a gliding movement by body flexion. A pseudopodium is a temporary projection of part of the cytoplasm of those protozoa which do not have a rigid pellicle. Pseudopodia are therefore characteristics of the amebas (Sarcodina). These organelles are also used for capturing food substances. The flagellum is an extremely fine filamentous extension of the cell. As a rule, the number of flagella present in an individual protozoan varies from one to eight; one or two is the most frequent number. A flagellum is composed of two parts : an elastic filament called an axoneme and the contractile cytoplasmic sheath that surrounds the axoneme. In certain parasitic *Mastigophora,* such as *Trypanosoma,* there is a very delicate membrane that extends out from the side of the body with a flagellum bordering its outer margin. Cilia, in addition to their locomotor function, also aid in the ingestion of food and serve often as a tactile organelle. They are fine and short threadlike extensions from the cell. They may be uniform in length or may be of different lengths depending on their location. Generally, cilia are arranged in longitudinal, oblique, or spiral rows, inserted either on the ridges or in the furrows. Electron microscopy has shown that the fine structure of the flagella and cilia of all eukaryotes follows the same basic design. Sections show two central and nine double peripheral microtubules ("9 + 2" structure) along most of the length of the shaft, which is enveloped by a membrane continuous with the pellicle.

Reproduction in Protozoa

Protozoa multiply by asexual reproduction; the majority of higher animals reproduce by sexual means. This is not to say that sexual processes are absent in the protozoa.

Asexual Reproduction

Asexual reproduction occurs by simple cell division, which can be equal or unequal—the daughter cells are of equal or unequal sizes, respectively. If two daughter cells are formed, then the process is called binary fission: if many daughter cells are formed, it is multiple fission.

Binary fission—The pseudopodia are withdrawn before the nucleus divides. After the nucleus divides, the organism elongates and constricts in the center in order to form two daughter cells. Amebas with special protective coverings are more complex in their manner of binary fission, which is directly related to the type of covering they possess. In those with soft coverings, the division plane is longitudinal along the body axis and the covering constructs into two halves. In those with more rigid coverings, part of the cytoplasm protrudes from the aperture to secrete a new covering over its surface.

In flagellates, fission is longitudinal along the major body axis. Since the flagella themselves are incapable of division, they must be regenerated from basal bodies (the blepharoplasts which arise in the vicinity of the old basal bodies). Thus multiplication of basal bodies usually precedes cell division. In dinoflagellates, division is at right angles to the cell axis because the flagella which determine the plane of division are not at the front end but at the side of the cell. Transverse fission is characteristic for ciliates.

Ciliates have two types of nuclei—the macronucleus, which determines vegetative processes, and the micronucleus, which is involved in sexual processes. During asexual binary fission the diploid micronucleus divides normally by mitotic division. The macronucleus undergoes DNA synthesis and divides into two portions without the regular reduplication of the chromosomes (amitosis). However, each daughter macronucleus contains the full complement of genes (in fact, in multiple sets—polyploidy). In some primitive ciliates found in marine sands of the intertidal zone, the macronucleus is incapable of division; after fission of the organism a new one is formed from the micronucleus.

Multiple fission—Division is usually preceded by formation of multiple nuclei within the mother cell, which then cleaves rapidly to form a corresponding number of daughter cells. Multiple fission is not as widespread as binary fission but it often takes place in addition to the later process. In ciliates and flagellates, this type of fission is found in relatively few species. Multiple fission occurs commonly in the foraminifera, the radiolaria, and the heliozoa. Perhaps the best-known examples of multiple fission are found in the sporozoa, e.g., in the malarial parasite *Plasmodium* where it is known as schizogony and serves to spread the parasite quickly in the host.

Budding—Instead, in protozoology it is often used to describe the varied processes by which sessile protozoa produce motile offspring. That is, the mother cell remains sessile and releases one or more swarming daughter cells. The swarmer differs from the parent cell not only in a lower degree of differentiation but also in the possession of special locomotor organelles. Some form of budding is found in all sessile ciliates and is used to disseminate the species while the mother cell remains *in situ*. Budding can be exogenous or endogenous. The former involves formation and separation of the bud toward the outside. Definite exogenous budding is seen in suctorians when a portion or portions of an adult sessile suctorian bud from the parent, develop cilia, and swim away.

Sexual Reproduction

Sexual fusion of two gametes (syngamy or gametogamy) occurs in various groups of protozoa. Conjugation, which is generally a temporary union of two individuals for the purpose of exchanging nuclear material, is a sexual process found exclusively in the ciliates. After exchange of nuclei, the conjugants separate and each of them gives rise to its respective progeny by fission or budding. When the gametes are morphologically alike, they are called isogametes. When they are unlike in morphology, they are anisogametes and can be either microgametes or macrogametes. That is, they are like the spermatozoa and the ova of metazoa, respectively. Thus microgametes are motile, relatively small, and usually numerous in comparison to macrogametes. Anisogametes are common among the sporozoa.

Classification of Major Groups of Protozoa

The most recent system divides them into seven phyla with several subphyla and classes, but this method may be more complex than is necessary for our survey. We will simplify this system by dividing the subkingdom Protozoa into four medically important groups, based on method of motility, mode of reproduction, and stages in the life cycle, summarized as follows :

The Flagellata

Motility is primarily by flagella alone or both flagella and ameboid movement; single nucleus; sexual reproduction, when present, by syngamy; division by longitudinal fission. Several parasitic forms lack mitochondria and Golgi apparatus; most species form cysts and are free-living; the group also includes several parasites. Examples : *Trypanosoma* and *Leishmania, Giardia,* an intestinal parasite, *Trichomonas,* a parasite of the reproductive tract of humans spread by sexual contact.

The Sarcodina

Cell form is primarily an ameba, major locomotor organelles are pseudopodia, although some species have flagella during reproductive states. Asexual reproduction by fission; two groups have an external shell; mostly uninucleate; usually encyst. Examples : *Entamoeba* is a pathogen or parasite of humans.

The Ciliata

Trophozoites are motile by cilia; some have cilia in tufts for feeding and attachment; most develop cysts; have both macronuclei and micronuclei; division by transverse fission; show relatively complicated and advanced behaviour. Examples : *Balantidium coli*, inhabit in vertebrate intestines.

The Sporozoa

Life cycles are complex, with well-developed asexual and sexual stages. Motility is absent in most cells except male gametes. Sporozoans produce special sporelike cells called sporozoites (Gr. *Sporos,* seed, and *zoon,* animal) following sexual reproduction, which are important in transmission of infections; most form thick-walled zygotes called oocysts;

entire group is parasitic. Examples : *Plasmodium*, an intracellular parasite with a complex cycle alternating between humans and mosquitos. *Toxoplasma gondii* causes an acute infection (toxoplasmosis) in humans.

Pathogenic Protozoa

In this survey, we first introduce some concepts of the host-protozoan relationship and then look at examples that illustrate some of the main features of protozoan diseases. Parasitic protozoa are traditionally studied along with the helminths in the science of parasitology. Although a parasite is generally defined as an organism that obtains food and other needs at the expense of a host, the range of host-parasite relationships can be very broad. At one extreme are the so-called 'good' parasites, which occupy their host with little harm. Most human parasites go through three general stages : (a) The microbe is transmitted to the human host from a source such as soil, water, food, other humans, or animals. (b) The microbe invades and multiplies in the host, producing more parasites that can infect other suitable hosts. (c) The microbe leaves the host in large numbers by a specific means and, to survive, must find and enter a new host. There are numerous variations on this simple theme. For instance, the microbe can invade more than one host species, and undergo several changes as it cycles through these hosts, such as sexual reproduction or encystment.

I. Trypanosomes—Trypanosomes are parasites belonging to the genus *Trypanosoma* (Gr. *Trypnon,* borer, and, *soma,* body). The two most important representatives are *T. brucei* and *T. cruzi,* species that are closely related but geographically restricted. *Trypanosoma brucei* occurs in Africa, where it causes approximately 10,000 new cases of sleeping sickness per year. *Trypanosoma cruzi,* the cause of Chagas' disease, is endemic to South and Central America, where it infects approximately 10 to 20 million people a year and causes 50,000 deaths. Both species have long, crescent-shaped cells with a single flagellum that is sometimes attached to the cell body by an undulating membrane.

The mammalian hosts are numerous, including dogs, cats, oppossums, armadillos, and foxes. The vector is the *reduviid bug,* an insect that is sometimes called the "kissing bug" because of its habit of biting its host at the corner of the mouth. Transmission occurs from bug to mammal and from mammal to bug, but usually not from mammal to mammal, except across the placenta during pregnancy. The trypanosome trophozoite multiplies in the intestinal tract of the reduviid bug and is harbored in the feces. During the night, the bug seeks a host and bites the mucous membranes, usually of the eye, nose, or lips. As it fills with blood, the bug soils the bite with feces containing the trypanosome. Ironically, the victims themselves inadvertently contribute to the entry of the microbe by scratching the bite wound. The trypanosomes, ultimately become established and multiply in muscle and white blood cells. Periodically, these parasitized cells rupture, releasing large numbers of new trophozoites into the blood. Eventually, the trypanosome can spread to many systems, including the lymphoid organs, heart, liver, and brain. Manifestations of the resultant disease range from mild to very severe, and include fever, inflammation, and heart and brain damage.

II. *Entamoeba* —Several species of amebas cause disease in humans, but probably the most common disease is amebiasis, or amebic dysentery (an inflammatory disease), caused by *Entamoeba histolytica.* This microbe is widely distributed in the world, from northern zones to the tropics, and is nearly always associated with humans. Amebic dysentery is the fourth most common protozoan infection in the world. This microbe has a life cycle quite different from the trypanosomes in that it does not involve multiple hosts and a blood-sucking vector. It lives part of its cycle as a trophozoite and part as a cyst. Since the cyst is the more resistant form and can survive in water and soil for several weeks, it is the more important stage for transmission.

The viable, heavy-walled cyst passes through the stomach unharmed. Once inside the intestine, the cyst germinates into a large, multinucleate ameba that subsequently divides to form small amebas. These trophozoites migrate to the large intestine, colonize the intestinal surface, and begin to feed and grow. From this site, they can penetrate the lining of the intestine and invade the liver, lungs, and skin. Common symptoms include gastrointestinal disturbances such as nausea, vomiting, and diarrhea, leading to weight loss and dehydration. Untreated cases with extensive damage to the organs experience a high death rate. The cycle is completed in the infected human when certain trophozoites in the feces begin to form cysts with several nuclei, which then pass out of the body with fecal matter.

III. Helminths—Adult animals are usually large enough to be seen with the naked eye, and they range from the longest tapeworms, measuring up to 25 yards in length, to roundworms less than one millimeter in length. Nevertheless, they are included among microorganisms because the microscope is necessary to see the details of smaller helminths and to identify their eggs and larvae. The two major groups of parasitic helminths based upon morphological form are the flatworms with a very thin, often segmented body plan, and the roundworms (Nematodes; Gr. *Nemato,* thread, and *eidos,* form) with an elongate, cylindrical, unsegmented body plan. The flatworm group is subdivided into the cestodes (L. *Cestus,* abelt, and *ode,* like), or tapeworms, named for their long, ribbonlike arrangement, and the trematodes (Gr. *Trema,* hole), or flukes, characterized by flat, ovoid bodies.

All helminths are multicellular animals equipped to some degree with organs and organ systems. In parasites, the most developed organs are those of the reproductive tract, with some degree of reduction in the digestive, excretory, nervous, and muscular systems. Reproduction of individuals is primarily sexual, involving the production of eggs and sperm in the same worm or in separate male and female worms. Fertilized eggs are usually released to the environment and are provided with a protective shell and extra food to aid their development into larvae. Even so, most eggs and larvae are vulnerable to heat, cold, drying, and predators, and are destroyed or unable to reach a new host. To counteract this formidable mortality rate, certain worms have adapted a reproductive capacity that borders on the incredible. A single female *Ascaris* can lay 200,000 eggs per day, and a large female can contain over 25,000,000 eggs at varying stages of development.

IV Pinworm—*Enterobius vermicularis,* the pinworm or seatworm. This worm causes a very common infestation of the large intestine. Worms range from 2 to 12 mm long, and have a tapered, curved cylinder shape. The condition they cause, enterobiasis, is usually a simple, uncomplicated infection that does not spread beyond the intestine. A cycle starts when a person swallows microscopic eggs picked up from another infected person by direct contact or by touching articles that person has touched. The eggs hatch in the intestine and then release larvae that mature into adult worms in the appendix. There, male and female worms mate, and the female migrates out to the anus to deposit eggs, which causes intense itchiness that is relieved by scratching. Herein lies a significant means of dispersal—scratching contaminates the fingers, which, in turn, transfer eggs to bedclothes and other inanimate objects.

□□□

6

MICROBIAL GENETICS

The molecular footprints of ancient DNA exchanges between diverse genetic species are evident in most, if not all, of the bacteria for which complete DNA sequences are known. In addition to their evolutionary significance, these processes are crucial for bacterial survival in natural environments where chemical insults constantly threaten to damage DNA and destroy chromosome integrity. Genetic recombination is the production of new combinations of genes derived from two different parental cells. Several processes have been described by which prokaryotic cells can exchange genetic information to yield recombinants. They are conjugation , transformation, and transduction. Conjugation requires physical contact between two cells of opposite mating type, resulting in the transfer of DNA from the donor cell to the recipient cell. The transfer of cell-free DNA into a recipient cell is called transformation, whereas transduction involves bacteriophage-mediated transfer of genetic information from a donor cell to a recipient cell. The recipient cell temporarily becomes a partial diploid (merodiploid) for the short region of homology between the transferred DNA and the DNA of the recipient chromosome.

THE PLASMID : AN EXTRACHROMOSOMAL MATERIAL

Transfer of genetic information between different *E. coli* strains was found to depend on the presence in some cells (called donor cells) of a small "extra" chromosome called F (fertility) factor. Subsequent to the discovery of F, many other extrachromosomal DNA elements, called plasmids, were discovered. One exception to circularity is the linear plasmids found in the Lyme disease bacterium *Borrelia burgdorferi.* Some plasmids, called episomes, commonly, integrate into the bacterial chromosome. F factor is one example of an episome. The size of plasmids can range from 1 to 2000 kb. Compare this to the size of the *E. coli* bacterial chromosome, which is approximately 4639 kb. A plasmid that can mediate its own transfer to a new strain is called a conjugative plasmid, whereas one that cannot is referred to as nonconjugative. Plasmids that have no known identifiable function other than self-replication are often referred to as cryptic plasmids. It is important to examine and understand plasmids because of the enormous role they play in biotechnology and medicine. Plasmids are the workhorses of recombinant DNA technology, serving as the vehicles by which individual genes from diverse organisms can be maintained separate from their genomic origins.

Partitioning assures that after replication each daughter cell gets a copy of the plasmid. For plasmids present in high copy numbers (50 to 100 copies per cell; e.g., pBR322), random diffusion may be enough to get at least one copy of the plasmid to each daughter cell. However, random segregation of low-copy-number plasmids (only 2 to 4 copies per cell) would likely mean that, following cell division, one of the daughter cells would not receive a plasmid. The plasmid would eventually be diluted from the population. Consequently, regulated partitioning mechanisms are essential for these plasmids. The mechanism used for partitioning differs depending on the plasmid. Partitioning, especially of low-copy-number plasmids such as F factor and P1, usually involves genes called *par* organized on plasmids as gene cassettes. Cis-acting means that, for a plasmid to replicate, the DNA sequence in question must be part of that plasmid. It does not encode a diffusible product. In contrast, a trans-acting replication gene encodes a protein that does diffuse through the cytoplasm. The upstream gene in a typical *par* operon encodes an ATPase essential to the DNA segregation process. The organization of the partitioning loci in F factor and P1 are the same, but the genes are called *par* and *sop*, respectively. ParB/SopB proteins bind as dimers to the downstream *parS/sopS* centromere-like regions. ParA/Sop A (the ATPases) then bind through protein–protein interactions with the B proteins. Genetic fusions between the Par proteins and green fluorescent protein (GFP) from the jellyfish *Aequorea victoria* have been used to visualize the location and movement of plasmids in the cell during segregation. The data suggest that the plasmids are intrinsically located at midcell where they replicate. After replication, the daughter plasmids pair with ParA and B proteins at the *par C–sopC* site, then move from midcell to quarter-cell positions, ensuring each daughter cell receives a plasmid copy.

Two plasmids that share some regulatory aspect of their replication are said to be incompatible. For example, if two different plasmids produce similar repressors for replication initiation, then the repressor of one could regulate the replication of the other and vice versa. The choice as to which plasmid will actually replicate is random. So, in any given cell, copies of one plasmid type could outnumber copies of a second plasmid.

Mobilizable Plasmids

The colicin E1 (ColE1) plasmid is an example of a nonconjugative, mobilizable plasmid. It has a DnaA-independent origin of replication but contains a site called *bom* (basis of mobility) that functions much like *oriT* of the F factor. While some ColE1 genes are required for mobilization (*mob* genes), transfer will not occur unless some functions are provided by F factor. Not all conjugative plasmids will fill this role, however, where members of the incompatibility groups IncF, IncI, or IncP conjugative plasmids will efficiently mobilize ColE1, members of IncW will do so inefficiently.

Plasmid Resistance

One important factor, resistance (R) factor, was first recognized by the fact that organisms in which it was present were resistant to a number of chemotherapeutic agents. A single R factor my carry traits for resistance to as many as seven or more chemotherapeutic or chemical agents. There is evidence that R factors did not arise as a direct result of the widespread use of antibiotics. Individuals living in highly sequestered geographic areas who have not been subjected to antimicrobial agents possess normal bacterial flora harboring R factors. Also, organisms maintained in the lyophilized state from time periods prior to the widespread use of antimicrobial agents have been shown to carry R factors. Thus, the origin of these plasmids probably occurred in response to encounters with antimicrobials produced in the natural environment. Regardless of their origins, the fact that R factors are transferred from normal flora to pathogenic microorganisms under natural conditions (e.g., the gastrointestinal tract) generates considerable concern with respect to the widespread and indiscriminate use of antimicrobial agents that could foster the further spread of these plasmids.

It is also possible to demonstrate intergeneric transmission of plasmids among members of this closely related group of organisms. Plasmid-associated resistance factors have been well characterized in most bacteria (one notable exception is *Rickettsia*). For example, resistance of *Enterococcus hirae* (previously called *E. faecalis*) to erythromycin and lincomycin is plasmid associated.

Replication of Plasmid

Theta (θ)-type replication has been examined most extensively among circular plasmids of Gram-negative bacteria. The mechanism involves melting of the parental strands at a plasmid origin, synthesis of a primer RNA, and then DNA synthesis by covalent extension of that primer. θ–type DNA-synthesis can start from one or multiple origins, and replications can be either uni- or bidirectional, depending on the plasmid. Electron microscopy revealed these replication intermediates look like θ-shaped molecules.

The current model (rolling circle replication) posits that replications initiates by the plasmid-encoded Rep protein, introducing a site-specific nick on one strand (the plus strand) of the DNA. The resulting 3′-OH end serves as a primer for leading strand synthesis involving host replication proteins (DNA pol III, ssb, and a helicase). Elongation from the 3′-OH end accompanied by displacement of the parental strand continues until the replicome reaches the reconstituted origin at which point a DNA strand transfer reaction takes place to terminate leading strand replication. The result is a double-stranded DNA molecule comprised of the parental minus strand and the newly synthesized plus strand, and a single-stranded DNA intermediate corresponding to the parental plus strand.

Strand displacement—Replication is promoted by the joint activity of three plasmid-encoded proteins, RepA, B, C, with respectively, 5′-3′ helicase, primase, and initiator activities. RepC binds to iteron sequences within the origin and probably with the RepA helicase, promoting exposure of two small palindromic sequences (*ssi*) located on opposite strands at the origin. DNA synthesis occurs with oppsosite polarities from these two sites. Stem-loop structures generated by the *ssi* sequences are probably required for assembly of the RepB primase, an event required to initiate replication. Linear plasmids have two basic structures. The spirochete *Borrelia* contains linear plasmids with a covalently closed hairpin loop at each end. This group replicates via concatameric intermediates. Linear plasmids in the Gram-positive filamentous *Streptomyces* have a covalently bound protein attached at each end. F factor is a low-copy-number plasmid whose replication is stringently controlled. ColE1, in contrast, is a high-copy-number plasmid that can be present at 50 to 100 copies per cell. The ColE1 plasmid undergoes unidirectional replication (one replication fork) and can replicate in the absence of *de novo* protein synthesis. It does this by utilizing a variety of stable host replication proteins (e.g., Pol I, DNA-dependent RNA polymerase, Pol III, DnaB, DnaC and DnaG) rather than making its own. Because replication initiation of the bacterial chromosome requires *de novo* protein synthesis, the addition of chloramphenicol to ColE1-containing cells results in the amplification of the plasmid copies relative to the chromosome. But even though the ColE1 replication system allows for a fairly high copy number per cell, the system is regulated so as not to exceed a given number (Fig. 6.1).

As the transcript progresses through *ori*, an RNA-DNA hybrid molecule forms that is a suitable substrate for the bacterial host enzyme RNaseH. RNaseH cleaves the RNA

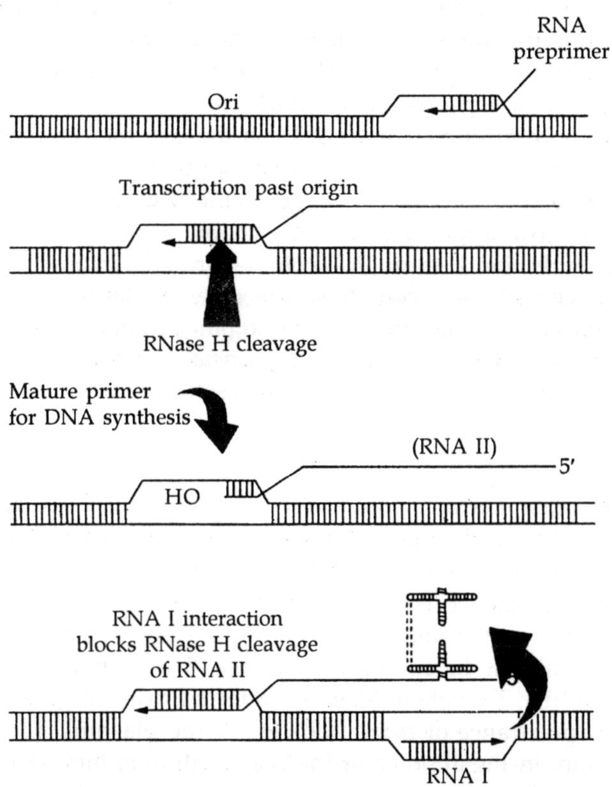

Fig. 6.1 : Plasmid ColE1 controlled by counter transcript replication.

primer, leaving a 3′-OH end suitable for DNA Pol III and replication. The processed RNA primer is called RNA II. The RNA I and RNA II precursor molecules each contain a significant amount of secondary stemloop structure, and because the two RNA molecules are complementary, the loop areas can base pair with the help of the plasmid Rom protein. In order for RNA II to prevent replication, it must achieve a concentration that will favour its interaction with the RNA II precursor. It is this aspect of the interaction that allows the plasmid to achieve 20 to 50 copies per cell before RNA I concentrations are high enough to prevent further replication. Following the initial molecular "kiss" of the loop areas, the molecules melt into each other, which extends the base pairing. Formation of the RNA-RNA complex disrupts the RNA-DNA hybrid at *ori* and interferes with processing by RNaseH, thereby aborting replication initiation. Lack of RNA I degradation decreases plasmid copy number because the resulting higher RNA I level reduces the number of possible replication events. So, under normal circumstances, when the plasmid copy number in a cell decreases, less RNA I is made and preexisting RNA I is degraded. This allows a greater number of initiation events, and the plasmid copy number will rise.

RepA protein must bind to sites within a region of the P1 genome called *oriR* in order to trigger P1 replication. Within *oriR* there are five copies of a 19 bp sequence, called an iteron, to which RepA can bind. Another nine iterons are found in a separate region called *incA*. Replication will only initiate if RepA binds to the iterons in *oriR*. The nine iterons present at *incA* were initially thought to titrate the level of available RepA to a point insufficient for binding to *oriR*, especially when multiple copies of P1 are present in the cell. The theory was that with multiple copies of P1 in the cell, the number of competing iterons would increase and thus decrease the probability that RepA would fully occupy the *oriR* iterons and initiate replication. However, artificially increasing RepA did not increase copy number. It now appears that the monomeric forms of RepA that bind to iterons can subsequently bind through protein–protein interactions with other RepA monomer–iteron complexes, even those on sister plasmids. The result, called handcuffing, effectively ties up the plasmids, preventing movement of the replication fork. As the cell volume increases, the RepA dimers disassociate and handcuffing decreases. Since RepA autorepresses its own expression, the increase in cell volume also derepresses *repA*.

Addiction Modules

Low-copy-number plasmids such as F factor, as well as R1, R100, and P1, all have the remarkable ability to kill cells that have lost the plasmid. In F, two plasmid-associated genes are responsible : *ccdA* and *ccdB* (coupled cell division). The *ccdB* product (11.7 kDa) is an inhibitor of cell growth and division. Its activity is countered by the unstable *ccdA* product (8.7 kDa). If plasmid copy number falls below a certain threshold, the loss of ccdA through degradation will unleash the killing activity of ccdB. Affected cells form filaments, stop replicating DNA, and die. This phenomenon is referred to as programmed cell death. The population has the appearance of being addicted to the plasmid. The target of ccdB is DNA gyrase. ccdB toxin, in the absence of the ccdA antidote, binds to GyrA, thereby inhibiting DNA gyrase activity.

The role of antidote is played by a small, unstable RNA transcript that binds to toxin-encoding RNA and inhibits its translation. The best-characterized system of this type is the *hok/sok* locus of plasmid R1. Translation of mRNA encoding the Hok (host-killing) protein is regulated by *sok* (suppressor of killing) antisense RNA. The unchecked production of Hok (52 amino acids) causes a collapse of the transmembrane potential. In the orthodox view of bacteria as single-celled organisms, programmed cell death does not make adaptive sense. However, it is becoming increasingly clear that in nature bacteria seldom behave as isolated cells. They often take on characteristics normally associated with multicellular organisms. Biofilms are one obvious example.

CONJUGATION IN BACTERIA

Conjugation involves direct cell-to-cell contact to achieve DNA transfer. For this process, certain types of extrachromosomal elements called plasmids are usually required. The prototype conjugative plasmid is the F, or fertility, factor of *E. coli.* F factor is an *E. coli* plasmid (100 kb) with genes coding for autonomous replication, sex pili formation, and conjugal transfer functions. In addition, there are several insertion sequences situated at various sites. F is considered an episome, since it replicates either independently of the host chromosome or as part of the host genome. Cells containing an autonomous F are referred to as F^+ cells. Replication of the F factor in this situation requires host proteins but is independent of the *DnaA* gene product.

In the Enterobacteriaceae, the presence of specific structural appendages (sex pili) on the cell surface is correlated with the ability of the cell to serve as a donor of genetic material. Of the 25 known transfer genes, 14 are involved with F-pilus formation (*traA, -L,-E,-K,-B,-V,-W,-C,-U,-F,-H,-G,-Q,X*). TraA is the F-pilin protein subunit while all the other genes are involved in the complex process of assembling the sex pilus (Fig. 6.2). Sex pilus assembly occurs in the membranous layers of the cell envelope. F-donor cells possess only one to three sex pili. The tip of the pilus is involved in stable mating pair formation (TraN and TraG involvement), interacting with the OmpA protein on the outer membrane of the recipient. Once initial contact between the donor pilus and the recipient is established, the pilus is thought to contract, bringing the cell surfaces of the donor and recipient cells into close proximity. Following mating-aggregate formation, a signal that a mating pair has formed is relayed from the TraD pore protein to TraM, a protein that binds the TraD pore, and ultimately to TraY, a site-specific DNA-binding protein that binds *oriT* (origin of conjugal transfer). The transfer of F^+ DNA initiates from *oriT* as opposed to *oriV*, the vegetative replication origin used for plasmid maintenance. TraY and integration host factor (IHF, an *E. coli* DNA-binding protein) change the architecture of *oriT* to a structure suitable for binding the bifunctional TraI endonuclease/helicase. The DNA transfer machinery, known as a transferosome or relaxosome, is then assembled at this origin. The

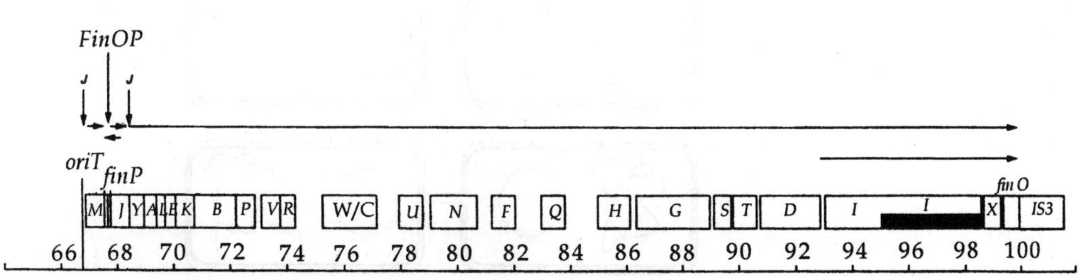

Fig. 6.2 : Transfer regions of plasmid's physical and genetic map.

plasmid-encoded TraI endonuclease/helicase nicks the F plasmid at *oriT,* and the 5′ end of the DNA is attached to the protein (Fig. 6.3). The nicked DNA, containing TraI as a pilot protein, travels through the TraD pore at the membrane bridge, not through the pilus itself, as originally believed (Fig. 6.4). The 5′-ended strand is transferred to the recipient via a rolling circle type of replication with the intact trand serving as a template. The *traI* gene product (DNA helicase I) unwinds the plasmid duplex (1000 bp/s), pumping one strand into the recipient cell. DNA pol III synthesizes a replacement strand in the donor. The 5′ end of the strand, upon entering the recipient, becomes anchored to the membrane. As the donor strand is transferred into the recipient cell, it, too, will undergo replication and become circularized. The circularization process does not rely on the host RecA synaptase but may require a plasmid-encoded recombination system.

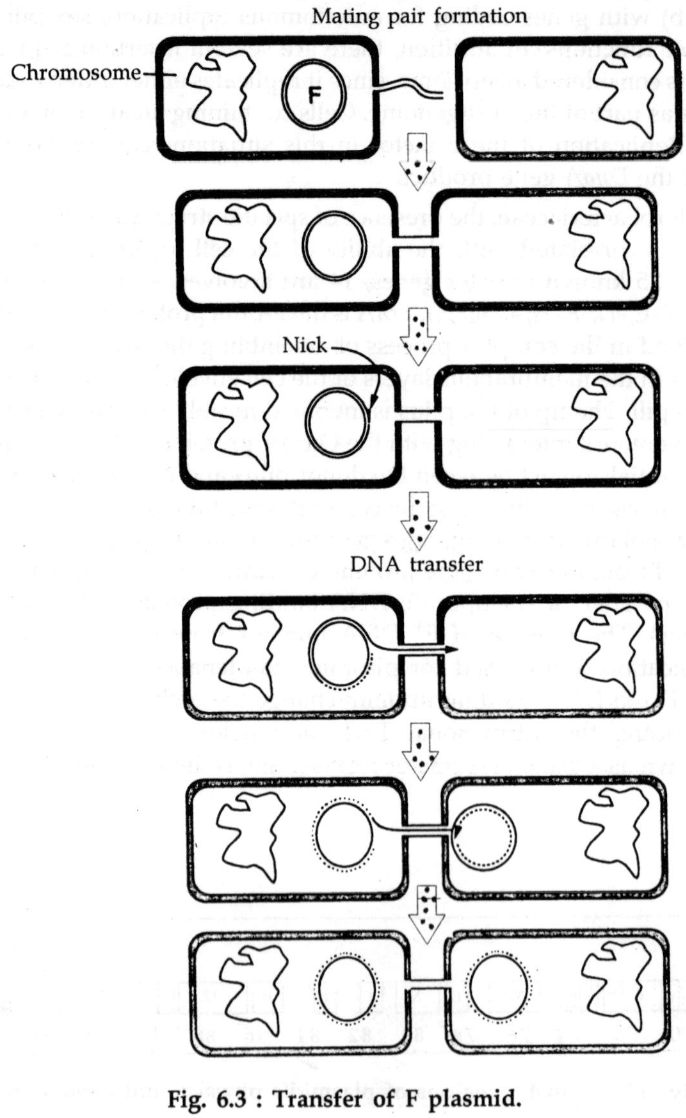

Fig. 6.3 : Transfer of F plasmid.

Two genes, *traS* and *traT*, are required for surface exclusion, with TraT being an outer membrane protein. It is believed that the TraT protein might block mating pair stabilization sites or, alternatively, might affect the synthesis of structural proteins (other than OmpA) necessary for stabilization. If mating pair formation occurs in spite of the efforts of TraT, TraS is thought to block DNA-strand transfer, although how it does this is not known. An odd phenomenon never fully explained is that growth of F^+ cells into late stationary phase imparts a recipient ability almost equivalent to an F^- cell. These cells are called F^- phenocopies and are nonpiliated.

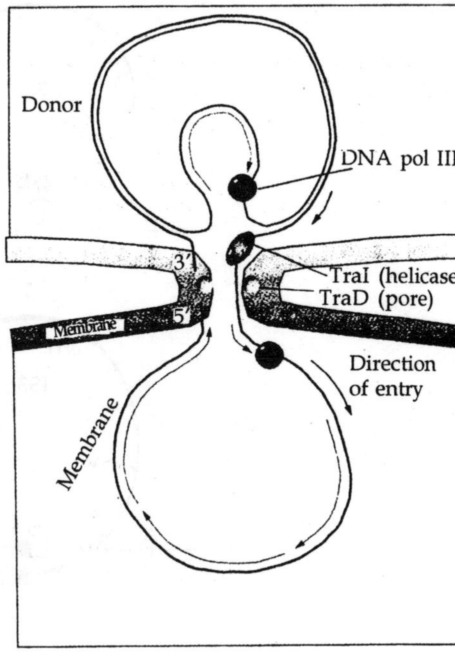

Fig. 6.4 : Conjugative transfer of F.

The *finO* and *finP* products interact to form a FinOP inhibitor of *tra* gene expression. In contrast to these other plasmids, the F factor is *finO finP*$^+$ and so is naturally derepressed for conjugation. However, if another plasmid that is *finO*$^+$ resides in the same cell as F, then the F *finP* gene product and the coresident plasmid's *finO* gene product can interact, producing the $FinOP_F$ inhibitor. This will act as a negative regulator of the F factor's *traJ*, which itself is the positive regulator for the other *tra* genes. Consequently, the fertility of F will be inhibited. FinP is a 79 nucleotide untranslated RNA molecule that is complementary (antisense) to the untranslated leader region of *traJ* mRNA. FinP antisense RNA is thought to form an RNA/RNA duplex with *traJ* mRNA, occluding the *traJ* ribosome-binding site and thereby preventing translation of the TraJ protein. The FinO protein (21 kDa) in F-like plasmids other than F stabilizes FinP antisense RNA by preventing degradation by RNAse E. This increases the *in vivo* concentration of the antisense RNA.

A cell that contains an integrated F factor is referred to as an Hfr cell (high frequency of recombination). The frequency of insertion occurs at about 10^{-5} to 10^{-7} per generation—that is, among a population of 10^7 F^+ cells, 1 to 100 cells will have an integrated F. Integration involves homologous recombination between two covalently closed circular DNA molecules forming one circular molecule that contains both of the original DNA structures (Fig. 6.5). It is thought that the insertion sequences (IS) present in the F genome (F carries two IS3 and one IS2 sequence) and those in the host chromosome (*E. coli* contains five each of IS2 and IS3) serve as regions of homology for the insertional event. F integration is predominantly RecA dependent, but rare RecA-independent Hfr's can be formed based on the transposition functions of the insertion elements. Once integrated, the F DNA is replicated along with the host chromosome. However, in situations where the host *DnaA* gene is inactive, replication of the entire chromosome can initiate from an integrated F in a process called integrative suppression. An integrated F factor still has active transfer functions such that an Hfr to F^- cross will transfer host DNA to the recipient where the donor DNA can recombine with the recipient DNA.

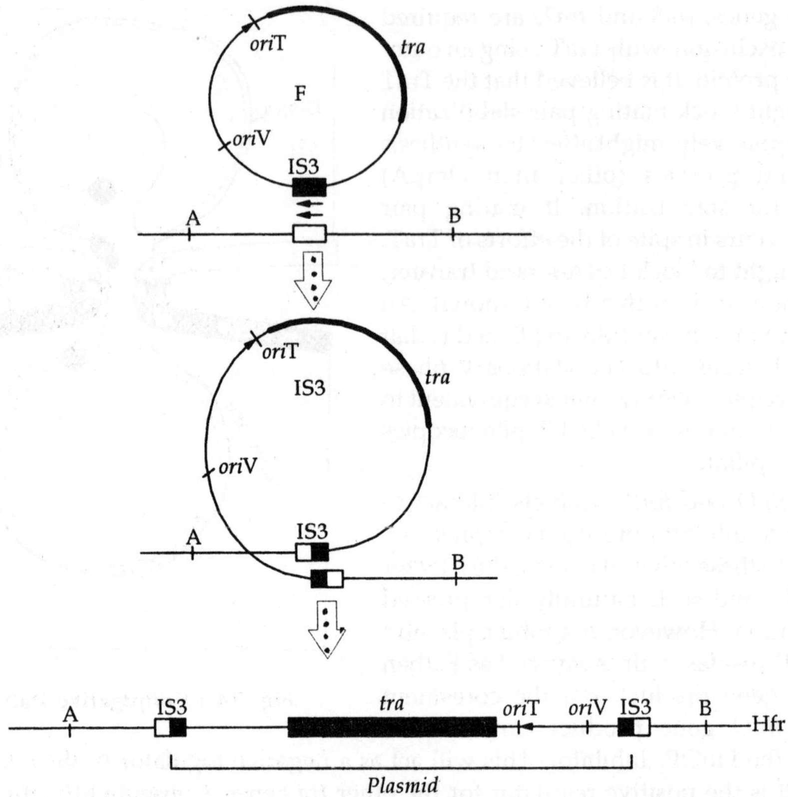

Fig. 6.5 : F factor and bacterial chromosome integration.

If the orientation of the chromosomal IS element was in the opposite direction, then the resulting Hfr would transfer gene A as a proximal marker and gene B would be one of the last genes transferred. Since conjugal transfer of the host chromosome in an Hfr cell is time dependent, a gene can be mapped relative to the position of the integrated F factor simply by determining how long it takes for the gene to be transferred to a recipient.

Normally, excision of the F factor restores the host chromosome to its original state. However, improper or aberrant excision can occur at a low frequency, forming a plasmid containing both F and bacterial DNA. This type of plasmid is called an F-prime (F′). There are two types of faulty genetic exchanges that can result in F′ formation. The first involves recombination between a region on the bacterial chromosome and one within the integrated F factor. The resulting F′ has lost some F sequences but now carries some host DNA originally located at one or the other side of the integrated F. This is a type I F′. As an alternative, host sequences located on both sides of the integrated F can undergo genetic exchange. The F′ formed in this situation (type II) contains all of F plus some host DNA from both sides of the point where F was integrated. In both situations the host DNA contained in the F′ is deleted from the host chromosome. However, when F′ (type I or type II) is transferred to a new host, a partially diploid (merodiploid) situation occurs for the host genes contained on the F′.

Cis/Trans Complementation Test

If the mutations lie in two different genes, then they should complement each other in trans—that is, an F′ with the genotype *trpA*$^+$ *trpB*$^-$ will produce a functional TrpA protein that can complement a host with the genotype *trpA*$^-$ *trpB*$^+$. This premise depends, of course, on whether the gene produces a diffusible gene product—that is, protein or RNA. Regulatory genes such as operators or promoters do not produce diffusible products and so mutations in these genes cannot be complemented in trans. An exception occurs with intragenic complementation. Two different mutations in a gene whose product functions as a multimeric enzyme (homodimer or tetramer) can sometimes complement each other in trans. The resulting active enzyme is composed of two different mutant polypeptides where the defect in one polypeptide sequence is compensated by the presence of the correct sequence on the other polypeptide. A wonderful example of this is called α complementation with β-galactosidase.

Enterococcal Conjugation and Pheromones

In *Enterococcus hirae (E. faecalis)* conjugative plasmids can be placed into two general categories. Members of the first group (e.g., pAD1, pOB1, pPD1, pJH2, pAM1, pAM2, pAM3) transfer at a relatively high frequency in broth (10^{-3} to 10^{-1} per donor cell). Members of the second group transfer poorly in broth but are fairly efficient when matings are carried out on filter membranes (e.g., pAC1, pIIP501, and p5M15346). The reason for this difference involves the production of sex pheromones by streptococci. Plasmids that transfer efficiently in broth use the pheromones to generate cell-to-cell contact. Recipient cells excrete small, soluble peptides (seven to eight amino acids long) that induce certain donor cells to become adherent. This adherence property facilitates formation of donor—recipient mating aggregates that arise from random collisions. Once the aggregate forms, transfer of the plasmid begins. The model for induction of pAD1 conjugation starts with a product of pAD1, TraA, repressing many other pAD1 genes (Fig. 6.6). The cAD1 pheromone produced from a potential recipient binds to the PrgZ surface component of the donor cell's oligopeptide permease. The peptide is then transported into the donor cell through a transmembrane channel formed by two other members of the permease : OppB and OppF. Once inside, cAD1 is thought to bind TraA, causing the repressor to release from its DNA-binding site. Transcription then proceeds through TraE, a positive regulator of many pCD1 transfer genes including the *asa 1* gene (a.k.a. *prgB* in pCF10) encoding aggregation substance. Aggregation substance (AS) is a cell-surface microfibrillar protein that will recognize a compound known as binding substance (BS) produced on all potential recipient cells.

Autostimulation is prevented because the plasmid also contains a cAD1-inducible gene (*iad; prgY* in pCF10) that encodes an inhibitor of AD1 peptide. The iAD1 peptide binds to extracellular cAD, effectively neutralizing its effect. However, this only works at low concentrations of cAD1. In addition, cAD1 autoinduces genes required for cell surface exclusion, so a cell already containing a pAD1 plasmid will not receive another copy from a second cell.

Crown Gall Disease

The Ti plasmid *Agrobacterium tumefaciens* is very large (1.2 □× 10^8 daltons) but can be transferred from this bacterium to dicotyledonous plants, resulting in production of a

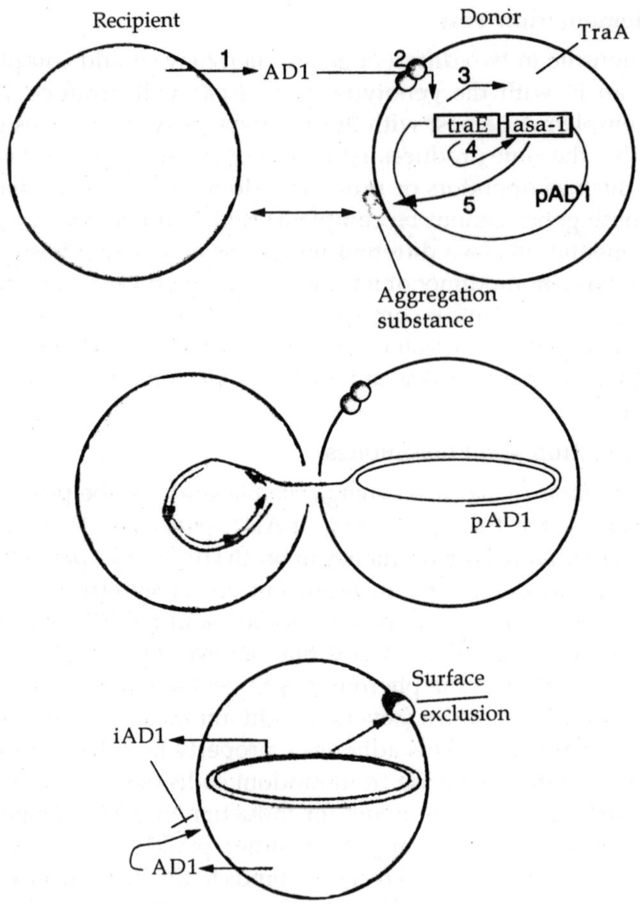

Fig. 6.6 : Pheromone-induced conjugation of pAD1 in streptococci.

plant tumor (e.g., crown gall disease). When wounded, plant cells release specific phenolic signal molecules called opines (e.g., acetosyringone) that are sensed by a two-component signal transduction system, VirA/VirG. In addition, monosaccharide and acid pH in the environment potentiate expression by the VirA/VirG system. Activation of the VirA/VirG system signals transcription of other *vir* genes whose products transfer a defined segment of the Ti plasmid to the plant nuclear genome, transforming the plant cell and leading to tumorous growth. Conjugal transfer of Ti between bacteria or between bacteria and plants is also regulated by cell–cell signaling. Donor bacteria induced by the conjugal opines release a soluble homoserine lactone signal molecule (autoinducer) that interacts with TraR (a homolog of the *Vibrio fischeri* LuxR) and induces conjugal functions.

TRANSFORMATION IN BACTERIA

Griffith first discovered the phenomenon, called transformation, in 1928 in the course of his investigations of *Streptococcus pneumoniae* (pneumococcus). Capsule-producing pneumococci were shown to be virulent for mice, while nonencapsulated strains were a virulent. Griffith discovered that if mice were injected with mixtures of heat-killed encapsulated (smooth = S) and live uncapsulated (rough = R) cells, a curious phenomenon occurred :

$$\underset{\text{(Live)}}{\text{R cells}} + \underset{\text{(Dead or heat-killed)}}{\text{S cells}} \xrightarrow[\text{into mice}]{\text{Injected}} \text{Dead mice}$$

The R cells recovered the ability to produce capsules and regained the capacity of virulence. There are different antigenic types of capsular material produced by different strains of pneumococci. Consequently, Griffith showed that if the avirulent R cells injected into the mice were derived from capsular type II cells and the heat-killed cells were of capsular type III, the viable, encapsulated cells recovered from the dead mice were of capsular type III. This indicated a transformation took place, so type II R cells now produced capsules of antigenic type III.

Avery, MacLeod, and McCarty (1944) demonstrated that is was DNA from the heat-killed encapsulated strain that was responsible for the transformation. They found that if living, rough type II cells were exposed to DNA isolated from type III encapsulated pneumococci, viable type III encapsulated organisms that were virulent for mice could be recovered. These findings were of exceptional importance because they showed that DNA had the ability to carry hereditary information.

Transformation processes in different microorganisms can be categorized into two main DNA uptake routes known as the *Streptococcus – Bacillus* model (Gram-positive bacteria) and the *Haemophilus – Neisseria* model (Gram-negative bacteria). Keep in mind that separation into these two basic models is artificial, since some of the bacteria in one group share features of the transformation process typically associated with the other group.

Streptococcal Transformation

The competent state in pneumococci is induced by a specific protein, the competence-stimulating peptide (CSP, 17 amino acids). Binding of this activator protein to receptors on the plasma membrane triggers the synthesis of at least a dozen new proteins within minutes. After induction by CSF, cells develop the capacity of bind DNA molecules. After binding to recipient cell membranes, donor DNA molecules are acted upon by a translocasome (also known as transformasome) complex located at the cell surface of competent recipient cells. The translocasome includes the endonuclease EndA. This nuclease attacks and degrades one strand of DNA while facilitating the entry of the complementary strand into the cell (Fig. 6.7). After entry, the single-stranded DNA can be recombined into the chromosome of the transformed cell. The development of competence in *Streptococcus pneumoniae* is a wonderful example of a phenomenon known as quorum sensing. Under appropriate conditions, the 17 amino acid CSP (encoded by *comC*) is processed, exported from the cell, and accumulates in the growth medium until a certain cell density is reached. Once CSP accumulates to a certain level, it binds to the membrane-localized ComD histidine kinase, which autophosphorylates and subsequently transphosphorylates its cognate response regulator protein ComE. ComE-P then activates transcription of competence-specific genes, including the *comCDE* operon in an autocatalytic feedback loop. ComE-P also appears to trigger a regulatory cascade that induces expression of an alternative σ factor ComX (σH), which in turn drives the expression of numerous other operons. Some of these new expressed gene products form the DNA uptake transformasome while other products are thought to remodel the cell wall around the transformasome.

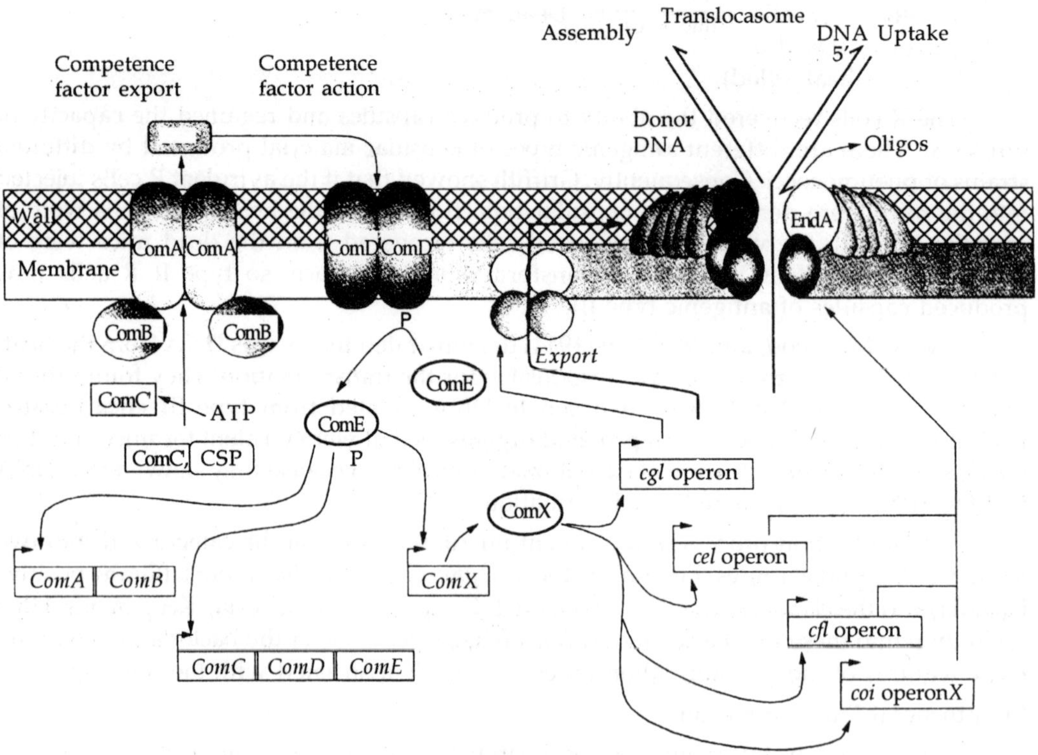

Fig. 6.7 : Streptococcal transformation.

A mismatch repair system called Hex repairs different base—base mismatches with different efficiencies. Thus, the efficiency of integration of genetic markers into the genome of the recipient cell varies drastically with different types of mismatches. Transition mutations (G/T or exhibit a low efficiency of transformation (LE markers) because the Hex system is very efficient at repairing the mutations. The HexA protein recognizes the mismatch in heteroduplex DNA resulting from transformation and selectively removes the entire donor strand, which can be as large as up to several kilobases. It is also proposed that HexB protein is activated by binding HexA-DNA mismatch complexes and in some manner increase the efficiency of the system. In contrast to LE transition mutations, transversion mutations (G/C or C/C mismatches) are not recognized by the Hex system. DNA fragements containing these mutations exhibit 10- to 20-fold higher transformation efficiency and are referred to as high-efficiency (HE) markers.

Transformation in *Bacillus subtilis*

Competence development in *B. subtilis* is subjected to three types of regulation : nutritional, growth stage specific, and cell-type specific. The regulatory pathway of competence intersects with those of several other stress responses including sporulation, motility, and secretion of macromolecule-degrading enzymes. It is useful to think of competence development as consisting of two modules. Module 1 is a quorum-sensing machine whose major input is population density signaled by two pheromones (CSF and ComX). Once the pheromones are sensed, a small peptide, ComS, is produced. The second module senses ComS and produces the transcriptional regulator ComK in response. ComK activates transcription of the late competence genes required to import DNA and the recombination genes required to incorporate DNA into the chromosome. As cell density increases, so does the extracellular concentration of ComX (Fig. 6.8). Once it reaches a critical concentration, ComX will interact with a membrane histidine kinase sensor protein called ComP. This triggers a regulatory cascade in which ComP autophosphorylates itself on a conserved histidine residue and then transphosphorylates a cytoplasmic regulatory protein called ComA. Phosphorylated ComA (ComA-P) is a transcriptional activator that activates the *comS* gene.

ComK is subject to degradation by the ClpCP protease. An increase in ComS will interfere with this degradation, allowing ComK to accumulate. Regulation of ComK proteolysis is actually very similar to what we described previously for the *E. coli* σ S

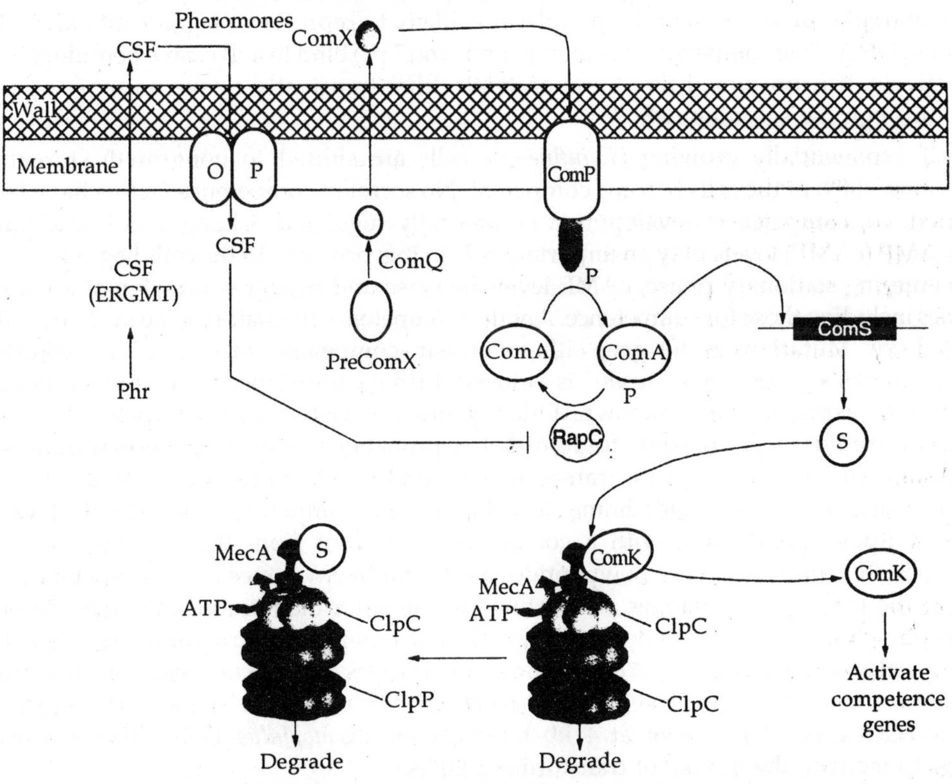

Fig. 6.8 : Transformation in *B. subtilis*.

protein. In *Bacillus*, an adaptor protein called MecA binds to ComS and to the ClpC ATPase chaperone of the ClpCP protease, thus expanding the substrate specificity of ClpC. ComK is then fed into the ClpP barrel and degraded. ComS, however, competes with ComK for binding MecA. Thus, at high enough concentrations, ComS will preferentially bind ClpC and effectively prevent the degradation of ComK. ComK levels increase as a result and induce expression of the genes required for uptake and processing of DNA.

Transformation in *Neisseria* and *Haemophilus influenza*

DNA first nonspecifically binds to the neisserial cell surface by electrostatic forces. Then an unknown protein or protein complex recognizes a specific sequence (GCCGTCTGAA) within the bound DNA. This sequence, resembling a bar code, indicates what DNA is suitable for uptake. Transport of the tagged DNA across the membrane involves an amazing type IV pilus structure similar to that described previously for type II protein export. The pilus biogenesis factors involved are PilD, T, F, and G, located at the inner membrane. The pilus, comprised of PilE protein, is assembled and protrudes through an outer membrane pore formed by PilQ (a.k.a. Omc). Another protein, PilC, helps gate the polymerized pili through the pore. One model for DNA uptake is that depolymerization of the pilus would somehow drag DNA into the periplasm. Once in the periplasm, the net-like peptidoglycan sacculus in the periplasm is likely to represent a significant barrier for incoming DNA. Two competence proteins (ComL and Tpc) bind to and cleave peptidoglycan, respectively. It is presumed that these proteins either transiently cleave murein or guide DNA through the sacculus net.

If exponentially growing *H. influenzae* cells are shifted to nongrowth starvation conditions, 100% of the cells become competent. No secreted competence factors have been reported, so, competence development is internally regulated. Changes in intracellular cyclic AMP (cAMP) levels play an important role in this process. As the cells begin to starve upon entering stationary phase, cAMP levels increase and trigger expression of a variety of genes including those for competence. Another competence regulatory gene in *Haemophilus* is called *sxy*. Mutations in this gene either eliminate competence or render it constitutive. An operon of six genes, *comA – comF*, is induced during competence development, but the activator of expression is unknown. Other genes associated with competence in this organism are Por, a periplasmic foldase that is probably involved in the maturation of periplasmic proteins required for transformation, and DprA, a protein required for DNA transport across the inner membrane. Envelopes from competent cells exhibit elevated levels of lipopolysaccharide with a composition different from that of log-phase cell envelopes. Six apparently new polypeptides are found in envelopes from competent cells. Most of the polypeptide changes are confined to the outer membrane, although one new polypeptide is associated with the inner cytoplasmic membrane. Structural changes in the envelope also occur in competent cells. Numerous vesicles called *transformasomes* bud from the surface and contain proteins that react specifically with conserved sequences (5′AAGTGCGGTCA3′) present at 4 kb intervals on *Haemophilus* DNA. These vesicles appear to mediate the uptake of transforming DNA.

TRANSDUCTION IN BACTERIA

There are two types of transduction reported in bacteria : generalized and specialized.

Generalized Transduction

During normal loading of nucleic acid into virus protein heads, the packaging apparatus occasionally makes a mistake, packaging chromosomal DNA into the phage rather than phage DNA. When a transducing bacteriophage binds to a bacterial cell, the donor DNA is injected into the bacterium by the phage and becomes integrated into the genome of the new cell through generalized recombination. Thus, all genes from a donor cell can potentially be transduced to a recipient cell population. Phages that can mediate generalized transduction include P1 (*E. coli*) and P22 (*Salmonella*). Phage DNA in P22 is linear but circularly permuted and terminally redundant. Thus, the termini of a P22 DNA molecule contain duplicate DNA sequences, but different P22 molecules within a population contain different terminal sequences. This seemingly odd chromosome structure is due to the packaging method P22 uses to insert DNA into its head. After infecting a cell, the linear P22 DNA circularizes through recombination between the terminal redundant ends. The circular molecule replicates via a rolling circle type of DNA synthesis, forming a long concatemer.

Once the P22 head is full (headful packaging), a second, nonsequence-specific cut is made. This second cut also defines the start of packing for the next phage head. However, to assure that each head receives a complete genome, the P22 system packs a little more than one genome length of P22 DNA. This is the source of the terminal redundancy of the P22 genome. One consequence of packing extra DNA is that the packing start site changes for each subsequent phage DNA molecule packaged from the concatemer. When the packaging system encounters a sequence in the bacterial chromosome that is similar to a *pac* site, it does not distinguish this site from P22 *pac* sites. It will use this homologous site to package chromosomal DNA such that progressive packaging will generate a series of phage particles that carry different parts of the chromosome. The size of packaged DNA is approximately 44 kb, which is about one-one hundredth the size of the *Samonella* chromosome.

Cotransduction of two or more genes requires that the genes be close enough to each other on the host chromosome such that both genes can be packaged into the same phage head. The closer the two genes are to each other, the higher the probability they will be cotransduced. The only thing that could separate the two genes would be a recombinational event occurring between them. Thus, the closer the two genes are to each other, the smaller the recombinational target that could separate them.

In abortive transduction, DNA that is transferred to the recipient cell does not become integrated into the genome of the recipient cell. Since this DNA is not replicated, it is transmitted unilaterally from the original cell to only one of the daughter cells and only transiently expresses the function of the genetic information in these cells.

Specialized Transduction

Specialized transduction is mediated by bacteriophage that integrate (lysogenize) into a specific site (*att*) on the bacterial chromosome. This specificity limits transfer of genetic material to host markers that lie within the immediate vicinity of this site. Lambda (λ) phage is the most actively studied phage of this type. λ phage almost invariably integrates at an *att* site near the galactose (*gal*) region of the chromosome and therefore can mediate specialized transduction of genes in that area to recipient cells. An infecting λ DNA integrates into the bacterial chromosome (A). In low-frequency transduction (LFT), the integrated phage (prophage) may be induced through some forms of stress to enter a cycle of lytic infection (Fig. 6.9). As a relatively rare event (10^{-5} – 10^{-6} per cell), a portion of the genome of the phage is replaced by a specific segment of the host chromosome (B). This occurs due to improper excision of integrated prophage DNA in a manner similar to the formation of type I F′ factors. These are defective phages because they lack some portion of the phage genome, but they are capable of transducing and integrating into a new host, carrying the original host genes with them (C). This establishes a merodiploid situation. At this point, the defective λ cannot produce more phage particles on its own. Nevertheless, a high-frequency transducing lysate can be made following an LFT event. This happens because a cell doubly lysogenized with a defective λ (carrying donor DNA) and a normal λ can be induced to yield new phage progeny (D).

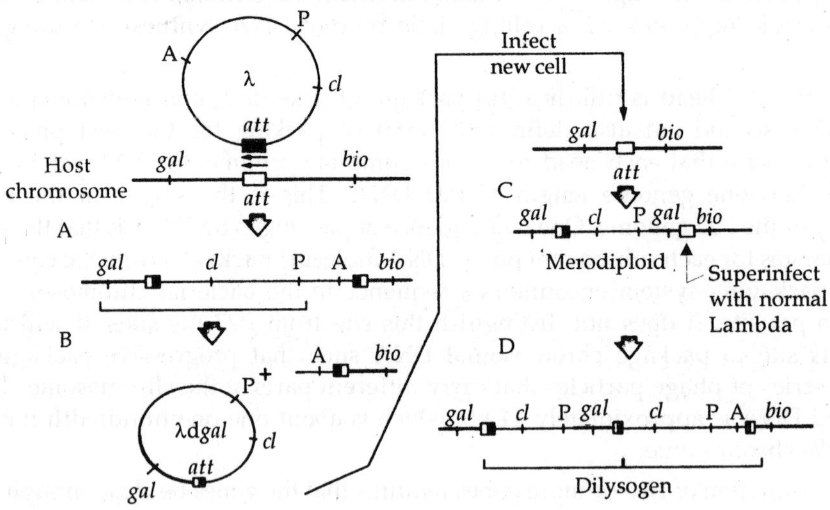

Fig. 6.9 : Specialized transduction.

Intergeneric transduction has been demonstrated between closely related members of the enteric group of organisms—for example, between *E. coli* and *Salmonella* or *Shigella* species. Various genetic traits such as fermentation capabilities, antigenic structure, and resistance to chemotherapeutic agents are transducible. Keep in mind that transduction is not limited to the bacterial chromosome. Genetic information residing on plasmids may also be transferred by transduction. In lysogenic conversion, a new phenotypic trait acquired by the host cell is due to a phage gene. Because the gene is part of the normal phage genome, every cell in a population that has been lysogenized acquires the genetic property. This mass conversion of the cell population distinguishes lysogenic conversion from transduction and other genetic transfer mechanisms. Cells that are lysogenized by β phage are designated *tox*⁺. The ability to produce toxin is inherent in the genome of the bacteriophage rather than in any trait that may have been transduced from the original host cell. Thus, cells that are "cured" of the lysogenic state no longer produce toxin. The *tox*⁺ gene is considered to be a part of the prophage genome; however, it is not essential for any known phage function and may be modified or eliminated without any effect on β phage replication.

RECOMBINATION IN BACTERIA

There are basically two types of recombination that can occur in *E. coli* and many other bacteria. These are RecA-dependent general recombination (or homologous recombination) and RecA-independent or nonhomologous recombination. General recombination requires a large region of homology between the donor and recipient DNAs but does not produce a net gain of DNA by the recipient genome. Homologous sequences are merely exchanged. In contrast, the nonhomologous recombinational mechanisms require very little sequence homology (as little as 5 or 6 bp) but actually produce a net gain of DNA by the recipient. Nonhomologous recombination may be divided into two types : site specific, where exchange occurs only at specific sites located on one or both participating DNAs, and illegitimate, which includes other RecA-independent events such a transposition.

Recombination of General Type

The *recA* product is a synaptase that facilitates the alignment of homologous DNA sequences. Loss of *recA* through mutation reduces recombination frequencies by 99.9%. The process of recombination can be viewed in six steps : (a) strand breakage, (b) strand pairing, (c) strand invasion/assimilation, (d) chiasma or crossover formation, (e) breakage and reunion, and (f) mismatch repair. A specific DNA sequences called chi *(χ)* is the primary target for initial recombination events. χ sequences, which are naturally present in *E. coli,* clearly enhance recombination and define recombinational hotspots. The χ-consensus sequence is an octomer, 5′-GCTGGTGG-3′, and occurs every 5 to 10 kb. χ sequences are important for accomplishing the first goal in the process of recombination, which is to take a double-stranded donor molecule and produce a single-strand from it that can invade and displace one strand of the recipient DNA (Fig. 6.10). The process begins with the *recBCD* or *recJ* products entering at the ends of a double-stranded DNA molecule and unwinding the DNA using their helicase activities. RecJ is part of an alternative recombination pathway called the RecF pathway. When the sliding RecBCD complex nears a χ site, it introduces a single-strand nick a few nucleotides from the 3′ side of χ. RecBCD then degrades the 3′-terminated strand during further DNA unwinding.

RecBCD then loads the new single-stranded 3′ end with the RecA synaptase. RecD actually inhibits recombination at sites other than χ by inhibiting the loading of RecA on single-stranded DNA. It has been proposed that upon encountering χ RecD undergoes an as-yet undocumented change that either ejects RecD from the complex or exposes a domain for RecA loading. Once the resultant single-strand tail is bound to RecA, the complex is poised to invade the donor DNA and begin the recombination process. The RecA protein (active species is a tetramer of 38,000 dalton monomers) is known to promote rapid renaturation of complementary single strands, hydrolyzing ATP in the process. It is believed that the RecA synaptase binds to ssDNA, which spontaneously increases RecA ability to bind dsDNA. RecA protein bound to ssDNA then aids in a search for homology between the donor strand and the recipient molecule.

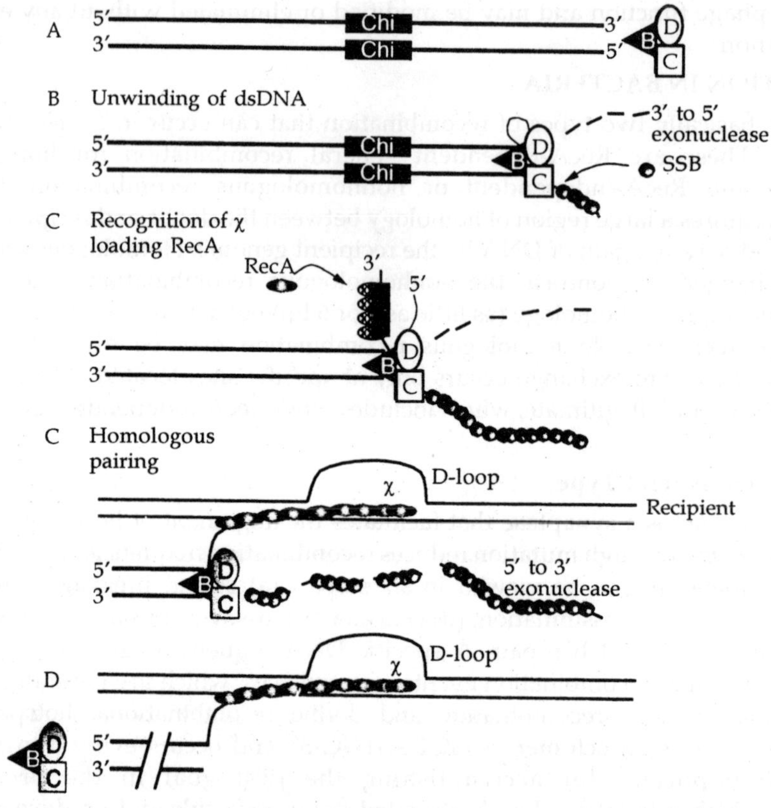

Fig. 6.10 : D loop formation in recombination mechanism.

The search for homology continues as the RecA protein either reiteratively forms such complexes or translates the two DNA molecules relative to each other until areas of homology are found. Once homologous regions are encountered and the single- and double-stranded DNAs are complexed, a stable displacement loop (D loop) is formed.

Some data suggest that RecA protein will partially denature the recipient DNA duplex and allow the assimilating strand to track in with a 5′ to 3′ polarity. Extensive assimilation seems to involve RecA binding cooperatively to additional areas in the crossover region to form a "protein–DNA filament". Subsequent to RecA binding, the RuvAB complex is responsible for the branch migration phase of recombination that occurs after the crossover site, known as the Holliday junction, has formed (Fig. 6.11). RuvA targets the complex to the Holliday junction, enabling RuvB to assemble as a hexamer around two opposite arms of the Holliday junction. The DNA is then pumped out of the RuvAB complex in opposite directions to drive branch migration.

The fact that certain *Dna* mutations cause a decrease in recombination frequencies tends to support the involvement of replication in genetic exchange. While strand assimilation occurs, it is proposed that an endonuclease such as ExoV *(recBCD* gene product) attacks the D loop. ExoV possesses both exo-and endonuclease activity. D loops are known to be susceptible to ExoV. The resolution of recombination intermediates is catalyzed by RuvC protein, which binds the Holliday junction as a dimer. This process effects the crossover. Cleavage requires homology and occurs at a preferred sequences [5′ (A/T)TT ↓ (G/C)3′]. Nicks are introduced symmetrically across the junction to give ligatable products.

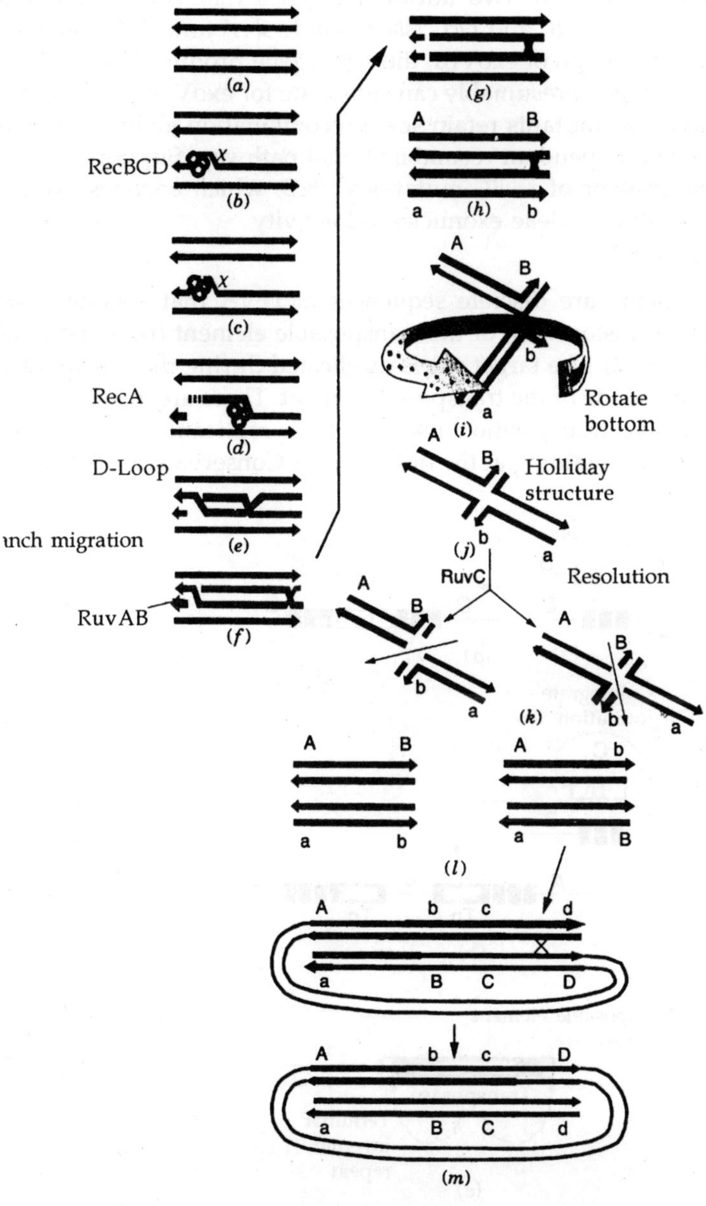

Fig. 6.11 : Recombination of genetic material.

Recombination Genetics

The identification of these genes has led to the discovery of multiple recombinational pathways. Mutations in *recA* reduce recombination frequencies to 0.1% of normal levels commensurate with its central role in many homologous recombination systems. One of these pathways was defined by mutations in *recBC,* which reduce the frequency to 1%. Other pathways were discovered following a search for revertants of these *recBC* mutants. Among the revertants found were two additional genes that can suppress the Rec$^-$ phenotype. The *suppressors* of *rec****B*** and *recC (sbc)* include *sbcA* and *sbcB*. The *sbcA*$^+$ locus negatively controls the expression of exoVIII, the *recE*$^+$ gene product. The *sbcA* mutations increase exoVIII activity, which presumably can substitute for exoV *(recBCD)* in the RecBC pathway. The fact that *recBC* mutants retain some recombination ability suggested there was at least one RecBCD-independent recombinational pathway. Support for this theory came from another suppressor of *recBC* mutations, *sbcB,* which encodes exonuclease I. Mutations in *sbcB* decrease exoncleae exonuclease I activity.

TRANSPOSABLE ELEMENT

Transposable elements are discrete sequences of DNA that encode functions to catalyze the movement (translocation) of the transposable element from one DNA site to a second, *target* site (Fig. 6.12). The target site is duplicated during the transposition event with a copy found to either side of the transposed element. There are essentially two types of transpositon. Replicative transposition involves both replication and recombination with a copy of the element remaining at the original site. Conservative transposition does

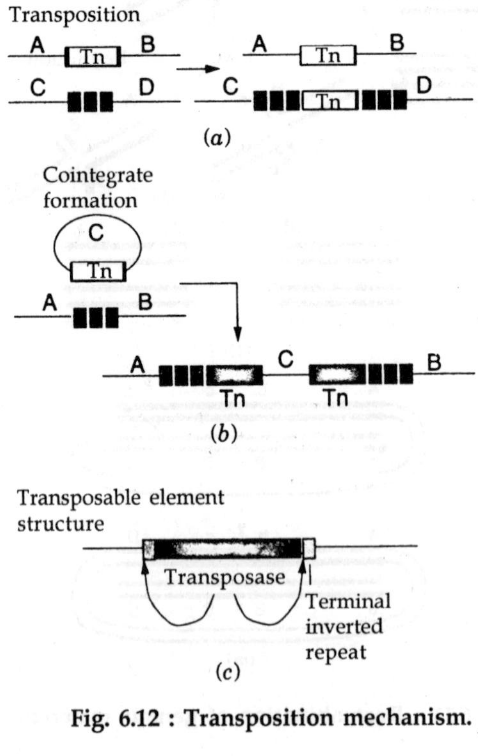

Fig. 6.12 : Transposition mechanism.

not involves replication. The element is simply moved to a new location. When the target site occurs within a gene, either type of transposition will generate insertion mutations. In addition to causing insertion mutations, transposition of these elements can cause deletions, inversions, and cointegrate formation in which two distinct replicons (e.g., plasmids) are joined (replicon fusion). Single stranded ends, and the short ends are filled in by repair synthesis, leading to the flanking duplications.

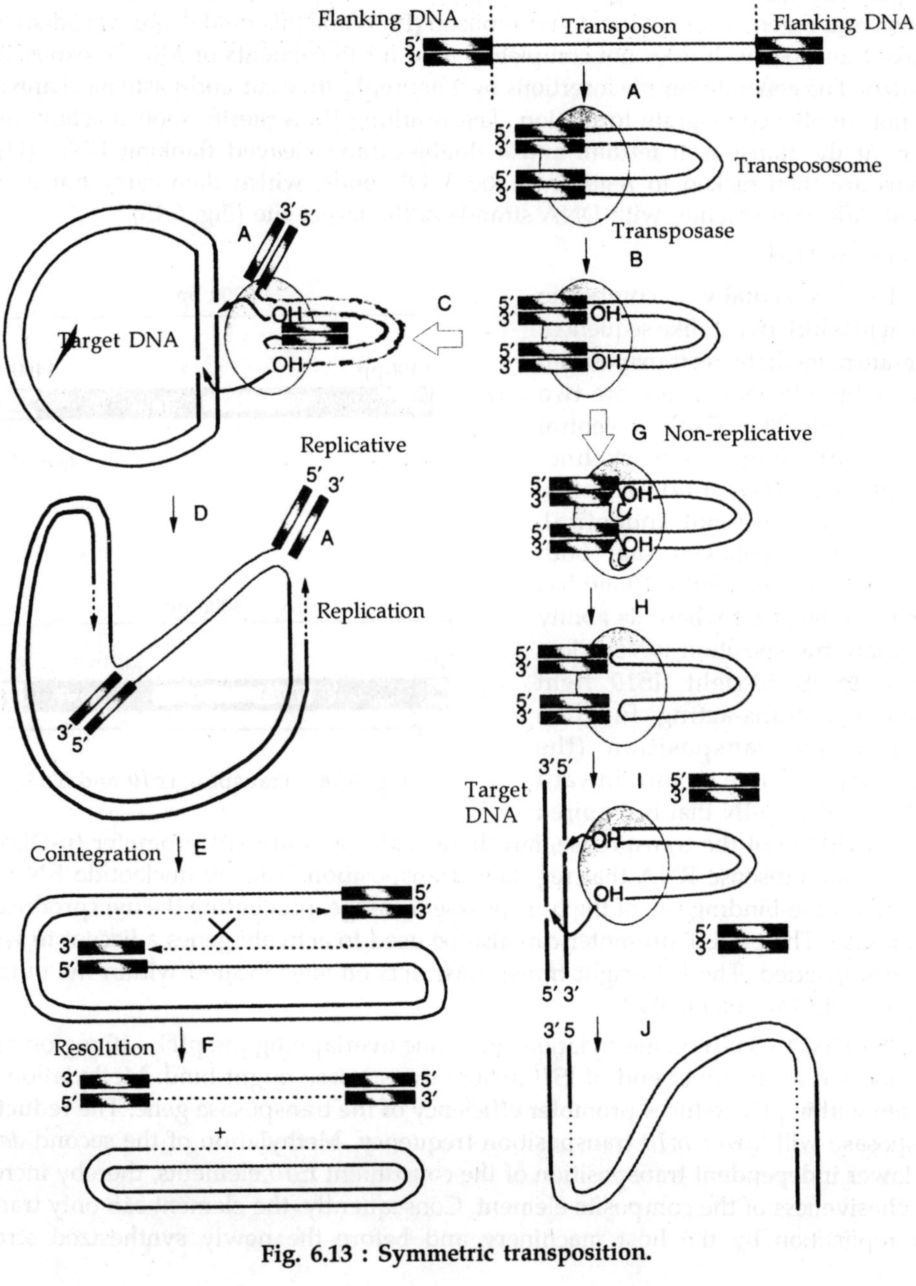

Fig. 6.13 : Symmetric transposition.

The transposase hydrolytically nicks one strand at each end of the transposon (Fig. 6.13). In replicative transposition, the 3′-OH ends of the nicked element attack the target DNA, resulting in a transesterification reaction covalently joining the transposon ends to the target DNA (C). Attachment of both ends of the transposon to the target immediately forms two replication forks (D). The transposon is completely replicated, with final sealing of the replicated DNA to flanking sequences generating a cointegrate (E). Resolution of the cointegrate by genetic exchange between the two transposon copies results in a simple insertion and regeneration of the donor replicon (F). While this model appears adequate for Tn*3*-like transposons, it does not completely hold for IS elements or Mu. Transposon such as Tn*10* or Tn*5* generate simple insertions by a nonreplicative cut-and-paste mechanism that does not involve cointegrate formation. The resulting transesterification reaction yields a hairpin at the transposon termini and a double-strand cleaved flanking DNA (H). The hairpins are then nicked to regenerate the 3′-OH ends, which then carry out a second transesterification reaction with DNA strands at the target site (Fig. 6.13).

Transposon Tn*10*

Tn*10* is actually a composite element in which two IS-like sequences cooperate to mediate the transposition of the entire element. There are two IS*10* elements that flank a central region containing a tetracycline-resistant locus (Fig. 6.14). The two IS*10* elements are not indentical, however, it is believed that both started out the same, but IS*10*-left has evolved to the point where its ability to mediate transposition is very low relative to IS *10*-right. IS*10* right produces a trans-acting function required for transposition (the transposase). There is an inward promoter called p-IN that is required for transcription of the transposase, but there is also an outward promoter (p-OUT) that produces an antisense RNA that regulates transposition. This 69 nucleotide RNA binds to the ribosome-binding site of the transposase message, preventing the overproduction of transposase. The p-OUT promoter can also be used to activate genes adjacent to where a Tn*10* has inserted. The IS*10*-right transposase acts on sites located within the outermost 70 bp of both IS*10* elements.

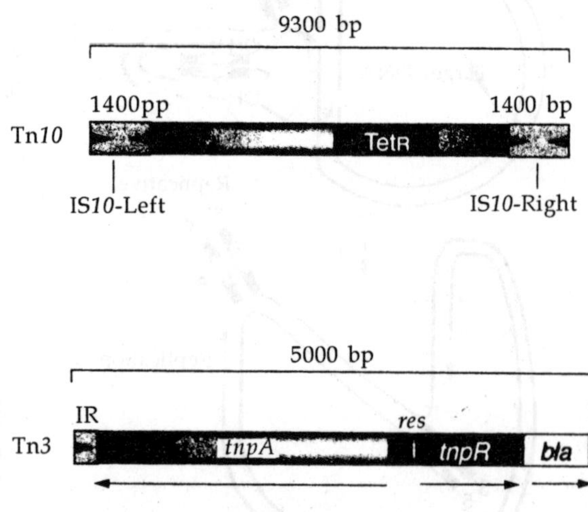

Fig. 6.14 : Transpon Tn*10* and Tn3.

IS*10* has two GATC methylation sites : one overlapping the pIN – 10 region and the other located at the inner end of IS*10* where transposase might bind. Methylation of the *dam* site within pIN reduces promoter efficiency of the transposase gene. The reduction in transposase will lower *tn10* transposition frequency. Methylation of the second *dam* site will lower independent transposition of the constituent IS*10* elements, thereby increasing the cohesiveness of the composite element. Consequently, the element can only transpose after replication by the host machinery and before the newly synthesized strand is methylated. Although Tn*10* can insert at many different sites in both *E. coli* and *S. enterica* chromosomes, the sites of insertion are not totally random. There is approximately one Tn*10* insertion "hot spot" per thousand base pairs of DNA. The specificity sequence is a 6 bp consensus sequence (GCTNAGC). This consensus specificity is located within the 9 bp target site, which is cleaved by staggered nicks during Tn*10* insertion. As noted earlier, in addition to promoting the movement of a transposon from one site to another, transpositional recombination can result in a variety of genetic rearrangements such as inversions or deletions (Fig. 6.15).

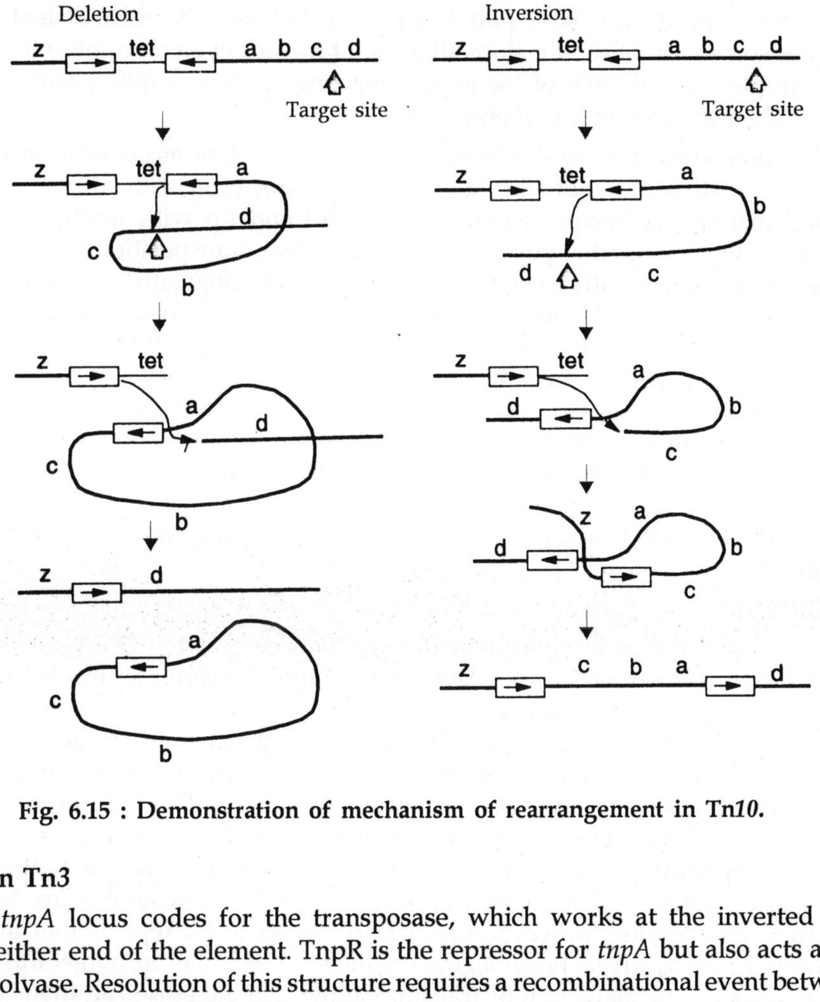

Fig. 6.15 : Demonstration of mechanism of rearrangement in Tn*10*.

Transposon Tn3

The *tnpA* locus codes for the transposase, which works at the inverted repeats located at either end of the element. TnpR is the repressor for *tnpA* but also acts as a site-specific resolvase. Resolution of this structure requires a recombinational event between the two transposons. For Tn3, the *res* site is the point where recombination, carried out by the TnpR resolvase, occurs.

Conjugative Transposition

This organism *Enterococcus hirae* contains an 18 kb Tet^R element designated Tn916 that mediates mating and self-transfer to other enterococci. Now known to be more widespread in nature, conjugative transposons can excise from the chromosome, form a circular intermediate, and be transferred to a new cell via conjugation. Thus, conjugative transposons are plasmid-like in that they have a covalently closed circular transfer intermediate and they are transferred by conjugation, but unlike plasmids, the circular intermediate of a conjugative transposon does not appear to replicate. An interesting feature of Tn916 is that the excision process involves dissimilar sequences at either end of the transposon. During excision, staggered cuts are made at both ends of the element, leaving six nucleotide single-strand overhangs called coupling sequences. The Tn916 circle is formed by covalently coupling the ends, but since the coupling sequences are not homologous, they do not base pair but form a bubble. The mismatched joint region undergoes some form of repair, since all the joint regions of circular intermediates end up as homoduplexes, with 80% of the joints reflecting the left coupling sequence and 20% reflecting the right coupling sequence.

The precise excision of Tn916 suggests a conservative mechanism of transposition in which replication does not leave a copy of the transposon at the donor site. However, following mating, the recipient chromosome will end up with multiple copies of the transposon, suggesting the possibility of replicative transposition. Although initially discovered in Gram-positive organisms, examples of conjugative transposons are also found in the Gram-negative genera *Bacteroides* and *Neisseria*. The transfer of many conjugative plasmids is regulated, but only in a few cases is the signal known. The signal for a *Bacteroides* conjugative plasmid named Tc^R Em^R Dot is tetracycline. Like the bacterial equivalent of an aphrodisiac, exposure to a low level of tetracycline stimulates mating 1000-fold. Curiously, *Bacteroides* conjugative transposons transfer only when the donor and recipient reside on solid surfaces. Mating in liquid culture does not occur. How mating became regulated by tetracycline is an evolutionary mystery. Tet has only been in use as an antibiotic for a short time—not long enough for such a complex system to develop in response.

INTEGRONS

There are numerous unexplained resemblances between DNAs of evolutionarily distinct species. Genetic elements called integrons are the reason for this high level of gene trafficking. Integrons are elements that encode a site-specific recombination system that recognizes and captures mobile gene cassettes. Integrons include an integrase (Int) and an adjacent recombination site (*attI*). A glimpse into how this type of gene trafficking occurs comes from studies with *Vibrio cholerae,* the cholera agent. This organism has a versatile integron acquisition system embedded in its own genome that appears capable of capturing genes serendipitously encountered from other organisms. The proposed integron contains an integrase gene and adjacent gene(s) flanked by DNA repeats that are targets for the integrase. The integrase, however, can act on similar DNA repeats that may be found flanking genes in DNA from heterologous organisms. *Vibrio* may encounter this heterologous DNA as a result of conjugation or transformation. The integrase can then "capture" this new DNA by recognizing these DNA repeats and can integrate the heterologous gene at one of *Vibrio's* own DNA repeats.

MUTATION

Any heritable change in the nucleotide sequence of a gene is called a mutation regardless of whether there is any observable change in the characteristics (phenotype) of the organism. But before delving into the molecular details of this process, the terminology of mutations must be defined. For example, a bacterial strain that contains all of the genetic information required to grow on a minimal salts medium is called wild-type or prototrophic, whereas a mutant strain requiring one or more additional nutrients is auxotrophic. Mutations themselves come in a variety of different forms. A change in a single base is a point mutation. The most common type of point mutations are transition mutations that involve changing a purine to a different purine (A ↔ G) or a pyrimidine to a different pyrimidine. A transversion mutation occurs when a purine is replaced by a pyrimidine or vice versa. A point mutation can change a specific codon, resulting in an incorrect amino acid being incorporated into a protein. The result is a missense mutation. The codon change could also result in a translational stop codon being inserted into the middle of a gene. Since a mutation of this sort does not code for an amino acid, it is a nonsense mutation. A missense mutation allows the formation of a complete polypeptide, whereas a nonsense mutation results in an incomplete (truncated) protein. Nonsense mutations (also referred to as amber, UAG; ochre, UGA; or opal, UAA mutations), if they occur in an operon involving several genes transcribed from a single promoter, may have ploar (distal) effects on the expression of genes downstream from the mutation. A mutation of this type is a polar mutation (Fig. 6.16).

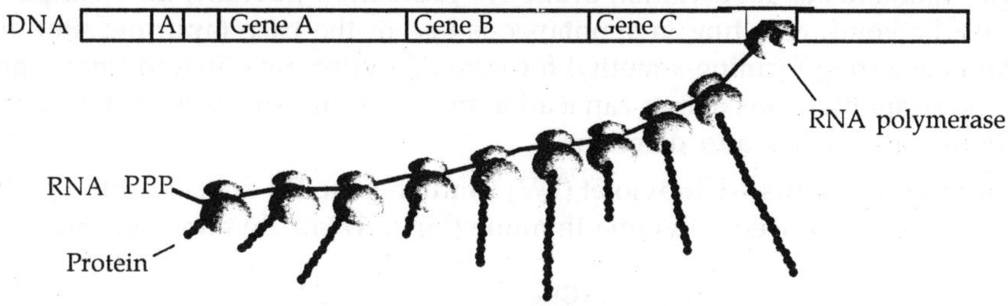

(a) Normal operon expression

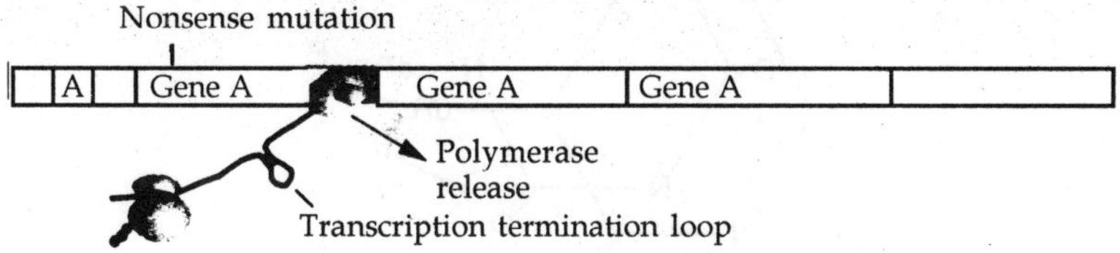

(b) Polar mutation

Fig. 6.16 : Polar mutation acts as termination of transcription.

The key to understanding polar mutations is that some nonsense mutations will cause premature transcription termination, lowering the amount of downstream message produced and available for translation. Premature transcription termination occurs when the nontranslated mRNA downstream from the nonsense codon possesses a secondary structure (i.e., stem-loop) that mimics transcription termination signals. If a process causes the removal of a series of bases in a sequence, the result is a deletion mutation. Likewise, the addition of extra bases into a sequence is an addition or insertion mutation. Both additions and deletions can result in changing the translational reading frame, causing all

,of the amino acids situated downstream of the mutation to be incorrect. The offending mutation in this event is a frameshift mutation.

Spontaneous Mutations

Spontaneous mutations are rare, ranging from 10^{-6} to 10^{-8} per generation depending on the gene and organism. Mutation rate is calculated as the number of mutations formed per cell doubling according to the formula $a = m/\text{cell generations} = m \ln 2/n - n_0$, where a = mutation rate and m is the number of mutations that occur as the number of cells increases from n_0 to n. There are many causes of spontaneous mutations. Even though the replication apparatus is very accurate with various "proofreading" and repair functions, such as the $3' \rightarrow 5'$ exonuclease activity of DNA polymerase III, mistakes do occur at a very low rate. One reason that misincorporation by DNA polymerase can go undetected involves tautomeric considerations discussed below. Besides misincorporation mistakes, endogenously occurring DNA damage is an important source of spontaneous mutations. For example, cytosine deaminates spontaneously to yield uracil ($G:C \rightarrow A:T$). In addition, purines are particularly susceptible to spontaneous loss from DNA by breakage of the glycosidic bond between the base and sugar. The result is the formation of an apurinic site in the DNA.

The spontaneous methylation of the N7 position of guanine, for example, weakens the glycosidic bond, resulting in spontaneous loss of the base (apurinic site) or opening of the imidazole ring forming a methyl formamide pyrimidine. In addition to mispairing, some of these spontaneous events can lead to major chromosomal rearrangements such as duplications, inversions, and deletions.

Mutational events—Ultraviolet (UV) irradiation at 254 nm causes the production of pyrimidine dimers between thymine-thymine (Fig. 6.17) or thymine-cytosine or cytosine-

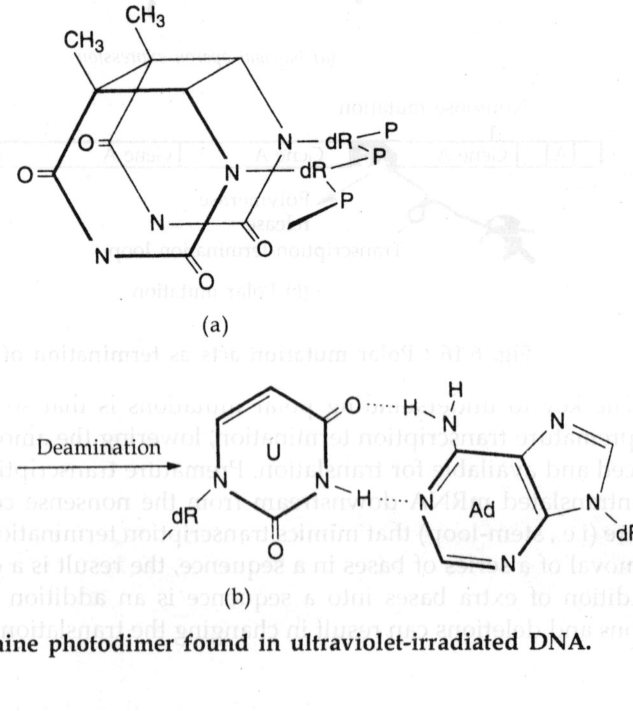

Fig. 6.17 : Thymine photodimer found in ultraviolet-irradiated DNA.

cytosine pairs. Ultraviolet irradiation may also result in distortion of the backbone of the DNA helix, causing replication errors. Ionizing radiations cause instability in the DNA molecule, resulting in single-strand breaks.

Local distortions in the helix may result in replication or recombination errors. Compounds such as nitrous acid cause the deamination of adenine and cytosine, and alkylating agents such as ethylethane sulphonate produce base analogs that result in transitions in base pairing in the nucleic acid structure. Acridine dyes may intercalate

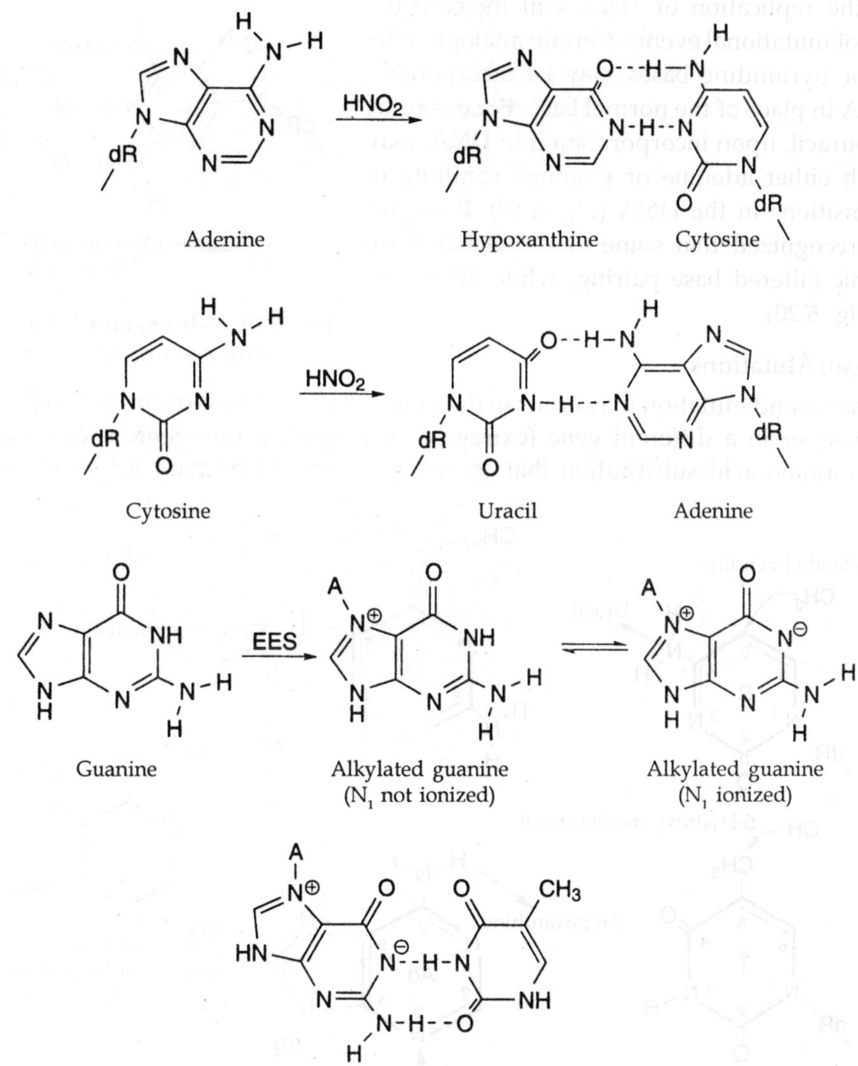

Hydrogen bonding of alkylated guanine (in ionized state) with thymine

Fig. 6.18 : Tautomeric shifting.

between the stacked bases, distorting the structure of the DNA and causing frameshift errors during replication. Abnormal base pairing may also result from tautomeric shifts in the chemical structure of the bases (Fig. 6.18).

Normally, the amino and keto forms predominate (ca. 85%). Tautomeric shifts occurring during the replication of DNA will increase the number of mutational events. Certain analogs of the purine or pyrimidine bases may be incorporated into DNA in place of the normal base. For example, 5-bromouracil, upon incorporation into DNA, may pair with either adenine or guanine, resulting in base transitions in the DNA (Fig. 6.19). It should also be recognized that some modified bases are mutagenic (altered base pairing) while others are lethal (Fig. 6.20).

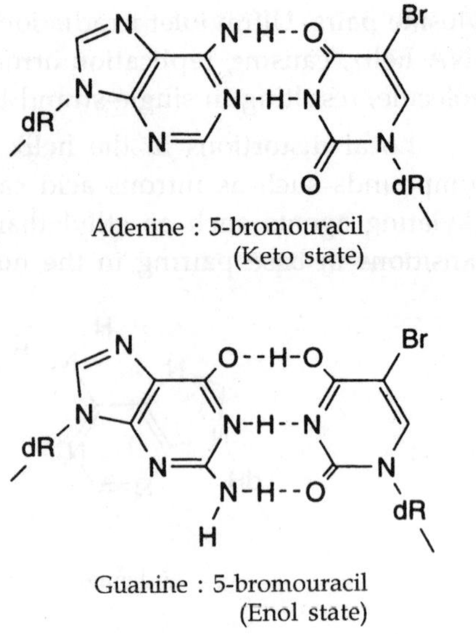

Fig. 6.19 : 5-bromouracil base pairing with adenine and guanine.

Suppressor Mutations

The second mutation may occur in the same gene as the original mutation (intragenic suppressor) or in a different gene (extragenic suppressor). Intragenic suppressors may cause an amino acid substitution that compensates for the primary missense mutation,

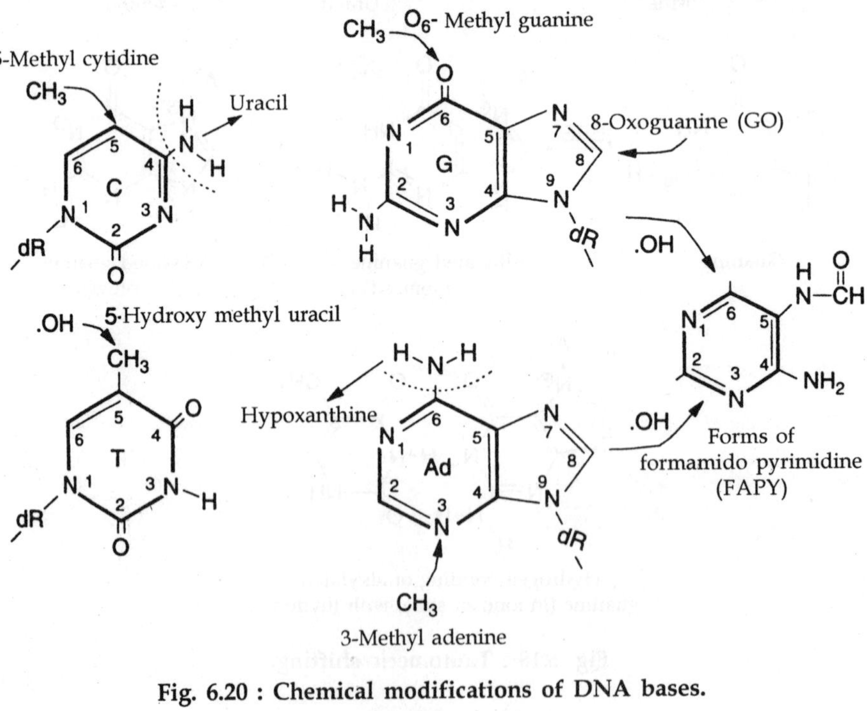

Fig. 6.20 : Chemical modifications of DNA bases.

thereby partially restoring the lost function, or they may be the result of an insertion or deletion that compensates for the original frameshift mutation. Extragenic nonsense suppressors are usually mutations that alter the genes encoding tRNAs. The revertant phenotypes may be found to contain tRNAs that recognize nonsense condons (e.g., UAA and UAG) and result in the insertion of a particular amino acid at the site of the nonsense mutation.

REGULATION AND EXPRESSION OF GENE ACTIVITY IN BACTERIA

Several adjacent genes code for a single, long mRNA molecule that directs the synthesis of several enzymes of a specific metabolic pathway. The consequence of such an arrangement is that the amount of synthesis of gene products is coordinately regulated. Therefore, if a cell is stimulated to synthesize a large amount of one of the enzymes of a group, it will also make large amounts of the other enzymes of the same group. Maintenance of induction requires the continued synthesis of *m*RNA to balance its degradation. Thus this *m*RNA instability, coupled with transcriptional control, assumes that only necessary proteins are synthesized by the cell. We can better understand the regulation of gene expression in prokaryotes by discussing the lactose (*lac*) operon of *E. coli.* This operon is now by far the best understood part of any cellular genome. There are other bacterial operons, but they are less well understood and are different in detail from the *E. coli lac* operon.

When inducers such as lactose or other β-galactosides are added to a culture of *E. coli,* there is a 1,000-fold increase in the rate of synthesis of the enzymes β-galactosidase (which hydrolyzes lactose to glucose and galactose), β-galactoside permease (which transports lactose into the cell), and thiogalactoside transacetylase (which plays no role in lactose utilization but may play a role in detoxifying certain thiogalactosides). The genes for these proteins are linked together on the *E. coli* chromosome. In the absence of control, the rate of enzyme production would be constant and depend only on the structural genes (such as *z, y,* and *a*), amino acid levels, activating enzymes, and other substances. However, the control of the rate of enzyme synthesis is directed by the regulatory genes designated *i, p,* and *o,* where *i* is the repressor gene, *p* the promoter gene, and *o* the operator gene. The i gene codes for a repressor protein which binds to the DNA of the operator o gene, thus preventing transcription, that is, the synthesis of *m*RNA (Fig. 6.21). The promoter gene p is considered to be the site on the DNA where the RNA polymerase enzyme, catalyzing the synthesis of *m*RNA, binds, and is thus the site where the specific *lac m*RNA.

(a) Using the protein-synthesizing machinery of the cell (ribosomes), the *m*RNA directs the synthesis of polypeptides in a process called translation.

(b) The genes *z, y,* and *a* operate as a single unit of transcription, which is initiated at *p.*

(c) Transcription of the operon is both negatively and positively controlled.

Negative control is mediated by the *lac* repressor which binds to the *o* gene and blocks transcription. Inducers, such as lactose, stimulate *lac m*RNA synthesis by binding to the repressor and reducing its affintiy for the operation (Fig. 6.21). Both repression and induction of enzyme synthesis are negative control systems because, in either case, the synthesis of enzyme can proceed only when the repressor is removed from its blocking site on the *o* gene.

Positive control of enzyme synthesis is said to occur when an association between a protein and a part of the regulatory region of an operon is essential for expression of

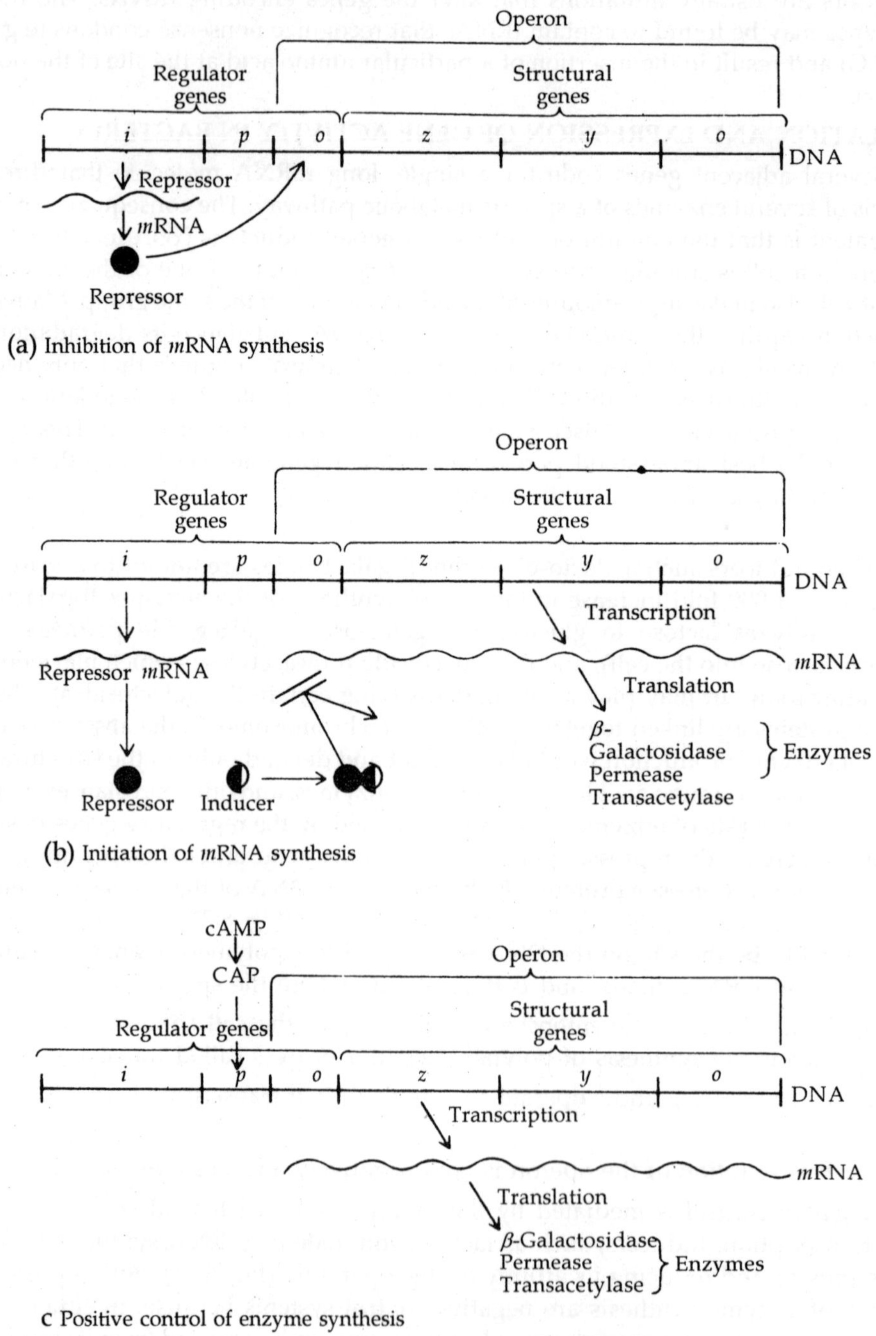

Fig. 6.21 : *Lac* operon model for gene regulation in *E. coli.*

related structural genes in the operon. Expression of the *lac* operon is inhibited when a more efficient source of energy, such as glucose, is present in the medium. The presence of glucose results in a decreased concentration of intracellular cyclic AMP. Cyclic AMP is necessary for efficient expression of the *lac* operon since it activates the catabolite gene activator protein (CAP), which in turn activates transcription of *lac* *m*RNA by RNA polymerase at the promoter site. Thus both cyclic AMP and a specific inducer acting in concert are necessary for the synthesis of many inducible enzymes in *E coli*. Little enzyme is made if either is absent.

7

MICROBIAL PHYSIOLOGY AND METABOLISM

Although metabolism (Gr. *Metabollein,* change) entails thousands of different reactions, most of them fall into one of two general categories. Anabolism (Gr. *Anabole,* a throwing up), sometimes also called biosynthesis, is any process that results in synthesis of cell molecules and structures. It is a building and bond-making process that forms larger molecules from smaller ones, and it usually requires the input of energy. Catabolism (Gr. *Katabole,* a throwing down) is the opposite, or complement, of anabolism. Catabolic reactions are degradative; they break bonds, convert larger molecules into smaller components, and often produce energy. The linking of anabolism to catabolism ensures the efficient completion of many thousands of cellular processes. Metabolism is a cyclical self-regulatory process that maintains the stability of the cell. Its actions provide a dynamic pool of chemical building blocks and give rise to enzymes and structural components of the cell. Metabolism of nutrients can extract energy in the form of adenosine triphosphate (ATP), or other high-energy compounds, that can be channeled into such processes as biosynthesis, transport, growth, and motility. Metabolism is highly organized and responsive to fine controls. For instance, there are mechanisms for diminishing or ceasing the production of a substance that is not in demand and for diverting excess nutrients into storage.

ENZYMES

The chemical reactions of life, even when highly organized and complex, cannot proceed without a special class of molecules called enzymes (Gr. *En,* in, and *syme,* leaven). Enzymes are a remarkable example of catalysts (Gr. *Katalysis,* dissolution), chemicals that increase the rate of a chemical reaction without becoming part of the products. Do not make the mistake of thinking that an enzyme creates a reaction. Because of the greater energy of some molecules, a reaction would occur spontaneously at some point. But such uncatalyzed metabolic reactions do not occur fast enough to sustain life processes, and thus enzymes, which speed up the rate of reactions, are indispensable to life.

Characters of Enzymes

1. Have unique characteristics such as shape, specificity and function.
2. Enable metabolic reactions to proceed at a speed compatible with life.
3. Provide a reactive site for target molecules called substrates.
4. Act as organic catalysts to speed up the rate of cellular reactions.
5. Are composed of protein and may require cofactors.
6. Associate closely with substrates, but do not become integrated into the reaction products.
7. Can be recycled, thus function in extremely low concentrations.
8. Are limited by particular conditions of temperature and pH.
9. Lower the activation energy required for a chemical reaction to proceed.
10. Are not used up or permanently changed by the reaction.

Activity of Enzymes

During a chemical reaction, reactants are converted to products by bond formation or breakage. A certain amount of energy is required for every such reaction, which limits its rate. This resistance to a reaction, which must be overcome for a reaction to proceed, is measurable and is called the energy of activation. In the laboratory, overcoming this initial resistance can be achieved by : (a) Increasing thermal energy (heating), which increases molecular velocity. (b) Increasing the concentration of reactants, or (c) adding a catalyst. In living systems, the first two alternatives are not feasible, since elevating the temperature is potentially harmful and increasing the density of reactants is not practical. This leaves only the action of catalysts, and enzymes fill this need efficiently and potently. At the molecular level, an enzyme physically promotes a reaction by serving as a physical site upon which one or more reactant molecules—the substrate (or substrates)—can be positioned for various interactions. Although an enzyme becomes physically attached to the substrate and participates directly in bonding, it does not become a part of the products, is not used up by the reaction, and can function over and over again. In fact, studies have shown that a single molecule of an enzyme can catalyze the reactions of several million substrate molecules.

Structure of Enzyme

Enzymes can be classified as simple or conjugated. Simple enzymes consist of protein alone, while conjugated enzymes (Fig. 7.1) contain protein and other nonprotein molecules. A conjugated enzyme, sometimes referred to as a holoenzyme (Gr. *Holos*, whole), is a combination of a protein, now called the apoenzyme, and one or more cofactors. Cofactors are either organic molecules, called coenzymes, or inorganic elements (metal ions). In some enzymes, the cofactor is loosely associated with the apoenzyme by noncovalent bonds, while in others it is tenaciously linked by covalent bonds.

Apoenzymes range in size from small polypeptides with about one hundred amino acids and a molecular weight of 12,000 to large polypeptide conglomerates with thousands of amino acids and a molecular weight of over one million. Like all proteins, an apoenzyme

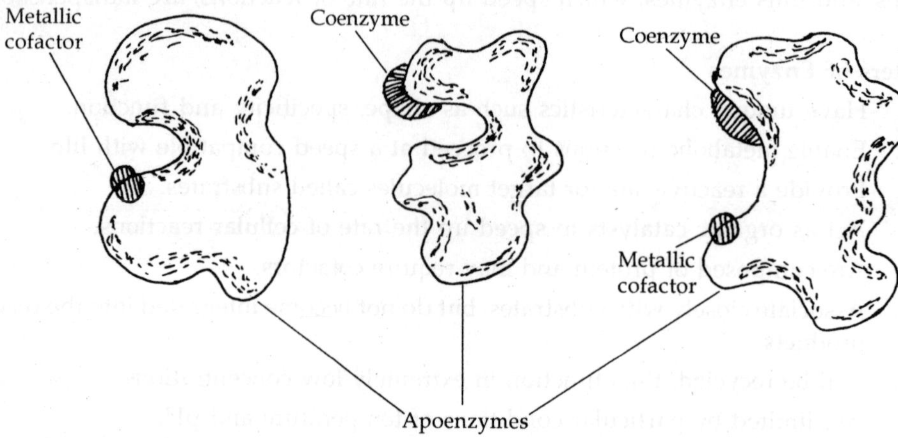

Fig. 7.1 : Structure of conjugative enzyme.

exhibits levels of molecular complexity called the primary, secondary, tertiary, and in larger enzymes, quaternary organization. The first three levels of structure arise from a single polypeptide chain that has undergone an automatic folding process and achieved stability by forming disulfide bonds. Folding causes the surface of the apoenzyme to acquire three-dimensional surface features. This relates to enzyme specificity in that : (a) The apoenzyme of each enzyme differs from others in its primary structure (the type and sequence of amino acids); (b) nuances in polypeptide folding produce a particular three-dimensional configuration (tertiary structure); and (c) surface features of the tertiary structure provide a unique and specific site (usually a crevice or groove) for attachment of substrate molecules. The site that accepts a substrate is called the active, or catalytic site, and there can be from one to several such sites.

Enzymes of Unconventional Nature

Just when it seemed that the protein nature of biological catalysts and enzymes had been well established, researchers working on bacteria discovered a novel type of RNA that also acts as a catalyst. Even more startling is the fact that the target of these molecules, called *ribozymes,* is RNA itself. In their basic mechanisms, ribozymes are really very similar to protein-based enzymes : they have a very specialized active site that binds to a certain place on a substrate and changes that substrate in some way. It appears that the precise function of RNA-based enzymes is somewhat more restricted. They help process the genetic code by cleaving RNA molecules, but they do not handle the diversity of substrates that regular protein enzymes do. Meanwhile, back at the lab, immunologists were studying enzymatic possibilities from yet another perspective. They were already well acquainted with the concept of a lock-and-key specific fit because of the way that antibodies lock onto their targets. The catalytic antibodies, or *abzymes,* have extreme specificity for their substrate and can speed up a reaction, form an abzyme-substrate complex, participate in bond breaking, and release a product. What is very exciting about the use of abzymes is that, unlike enzymes, antibodies are very easy to fine tune to fit almost any possible substrate configuration and can be manufactured in large amounts using monoclonal methods. The applications for medicine and industry are considerable, for instance, an abzyme could be designed to do double duty—both attaching to a target molecule and enzymatically destroying it. This technology has the potential for treating cancers, infections, abnormal blood clots, and other medical problems very specifically. The possible use of abzymes to purify drugs and inactivate viruses is already being explored.

Interactions of Enzymes and Substrates

For a reaction to take place, a temporary enzyme-substrate union must occur, fittingly described as a complementary look-and-key fit, at the active site (Fig. 7.2). The bonds formed between the substrate and enzyme are weak and, of necessity, easily reversible. Once the temporary enzyme-substrate complex has formed, appropriate reactions occur on the substrate, often with the aid of a cofactor, and a product is formed and released. The enzyme can then attach to another substrate molecule and repeat this action.

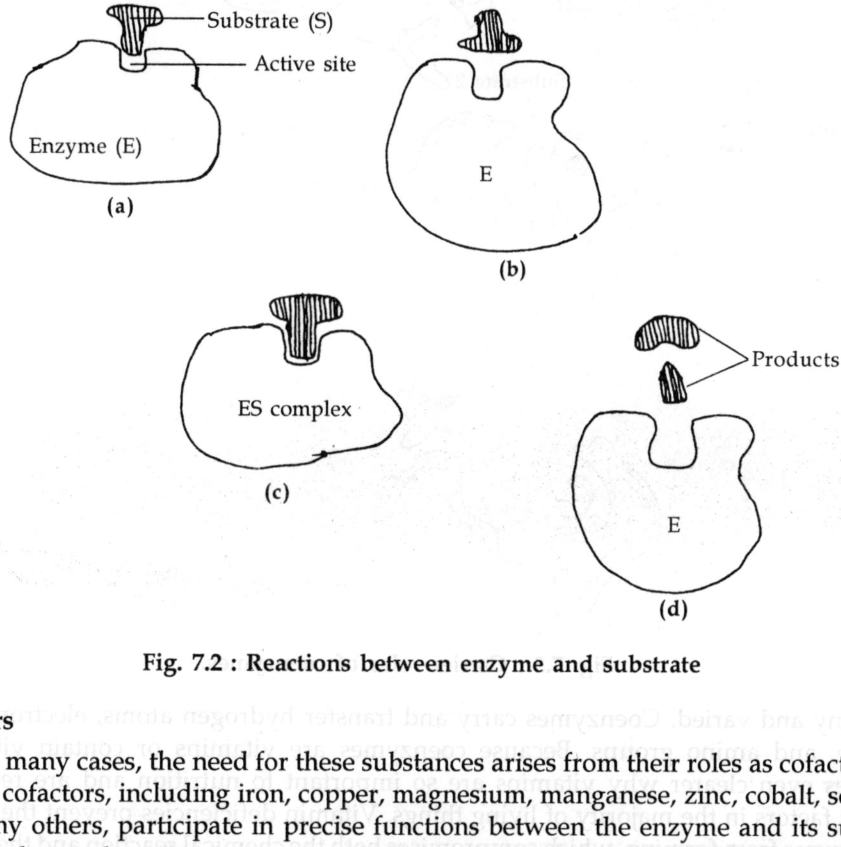

Fig. 7.2 : Reactions between enzyme and substrate

Cofactors

In many cases, the need for these substances arises from their roles as cofactors. The metallic cofactors, including iron, copper, magnesium, manganese, zinc, cobalt, selenium, and many others, participate in precise functions between the enzyme and its substrate. In general, metals activate enzymes, help bring the active site and substrate closely, together, and participate directly in chemical reactions with the enzyme-substrate complex. Coenzymes are organic compounds that work in conjunction with an apoenzyme to perform a necessary alteration of a substrate. The general function of a coenzyme is to remove a functional group from one substrate molecule and add it to another, thereby serving as a transient carrier of this group (Fig. 7.3). The specific activities of coenzymes

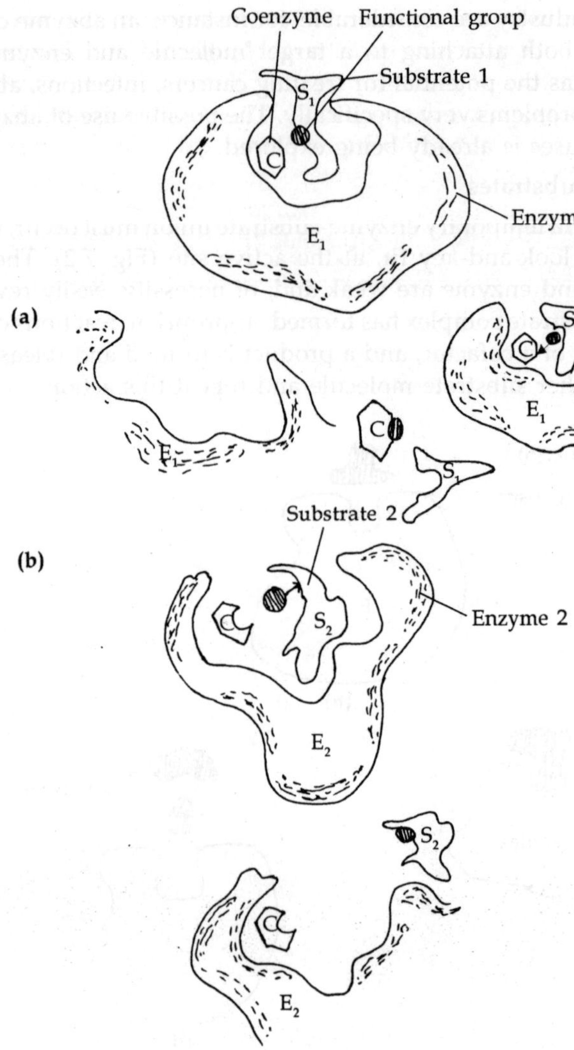

Fig. 7.3 : Carrier roles of coenzymes.

are many and varied. Coenzymes carry and transfer hydrogen atoms, electrons, carbon dioxide, and amino groups. Because coenzymes are vitamins or contain vitamins, it becomes even clearer why vitamins are so important to nutrition and are required as growth factors in the majority of living things. Vitamin deficiencies prevent the complete holoenzyme from forming, which compromises both the chemical reaction and the structure or function dependent upon that reaction.

Functions of Enzymes

Location and regularity of enzyme action—Enzymes can be classified according to their site of activity. After initial synthesis in the cell, exoenzymes are transported extracellularly, where they break down (hydrolyze) large food molecules or harmful chemicals. Examples of exoenzymes are cellulase, amylase, and penicillinase. By contrast, endoenzymes are retained intracellularly and function there. Most enzymes of the metabolic pathways are of this variety. In terms of their presence in the cell, enzymes are not all produced in equal amounts or at equal rates. Some, called constitutive enzymes (Fig. 7.4), are always present and in relatively constant amounts, regardless of the amount of substrate. The enzymes involved in utilizing glucose, for example, are very important in metabolism and thus are constitutive. An induced (or inducible) enzyme is not constantly present and is produced only when its substrate is present. Induced enzymes are present in amounts ranging from a few molecules per cell to several thousand times that many, depending on the metabolic requirement.

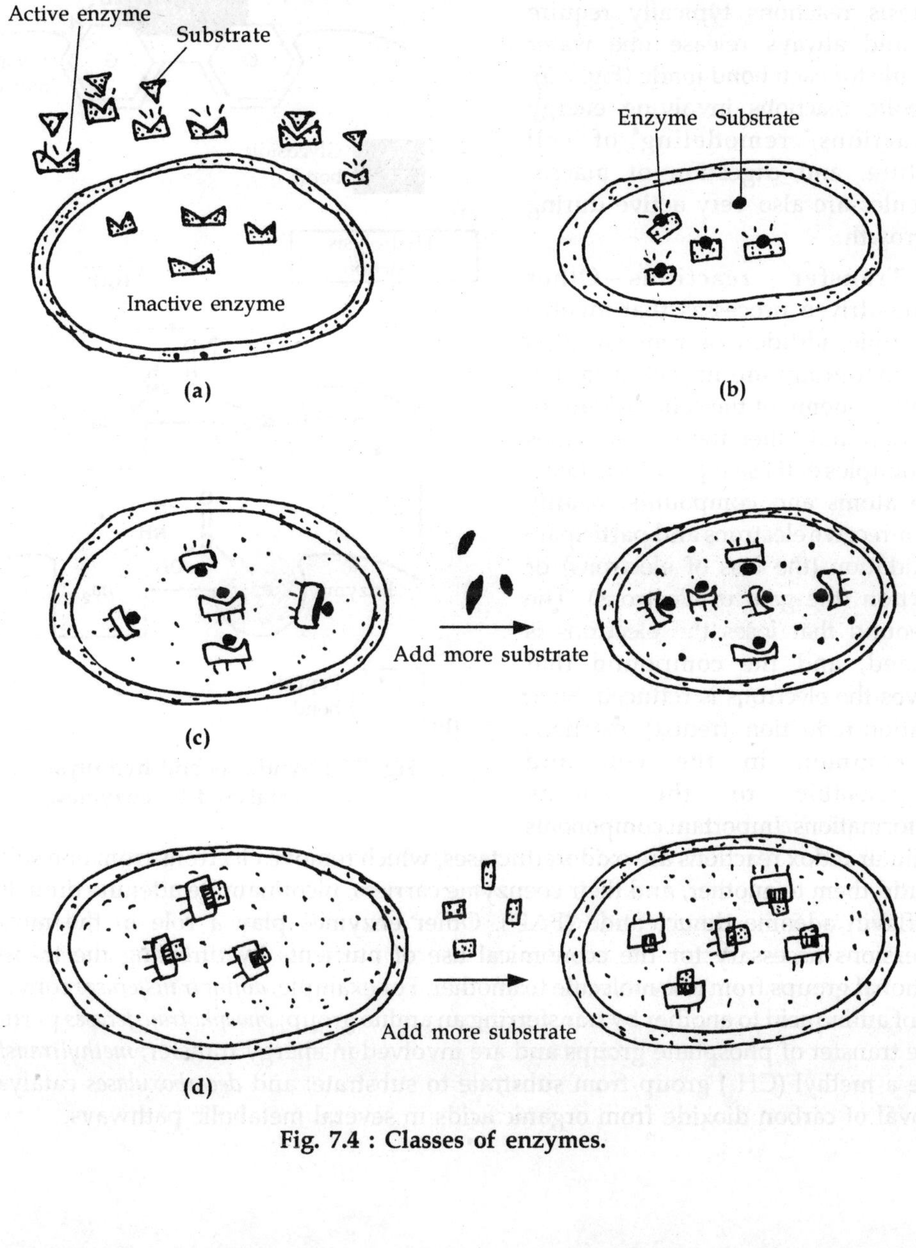

Fig. 7.4 : Classes of enzymes.

Synthesis and hydrolysis—A growing cell is in a frenzy of activity, constantly synthesizing proteins, DNA, and RNA; forming storage polymers such as starch and glycogen; and assembling new cell parts. Such anabolic reactions require enzymes (ligases) to form covalent bonds between smaller substrate molecules. Also known as condensation reactions, synthesis reactions typically require ATP and always release one water molecule for each bond made (Fig. 7.5). Catabolic reactions involving energy transactions, remodeling of cell structure, and digestion of macromolecules are also very active during cell growth.

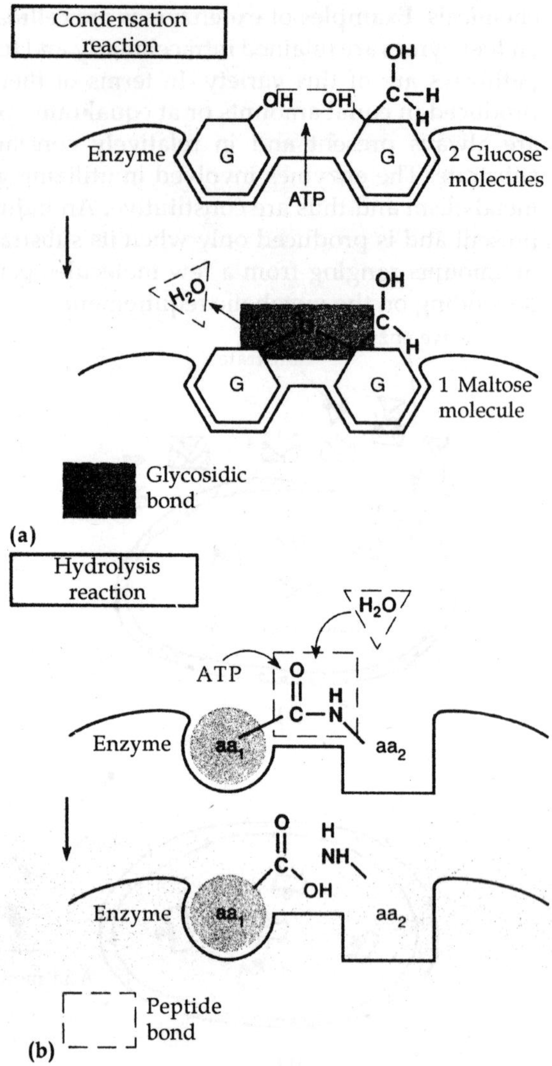

Fig. 7.5 : Synthesis and hydrolysis reactions catalyzed by enzymes.

Transfer reactions—Other enzyme-driven processes that involve the simple addition or removal of a functional group are important to the overall economy of the cell. Oxidation-reduction and other transfer activities are examples of these types of reactions. Some atoms and compounds readily give or receive electrons and participate in oxidation (the loss of electrons) or reduction (the gain of electrons). The compound that loses the electrons is oxidized, and the compound that receives the electrons is reduced. Such oxidation-reduction (redox) reactions are common in the cell and indispensable to the energy transformations. Important components of cellular redox reactions are oxidoreductases, which remove electrons from one substrate and add them to another, and their coenzyme carriers, nicotinamide adenine dinucleotide and flavin adenine dinucleotide (FAD). Other enzymes play a role in the molecular conversions necessary for the economical use of nutrients by directing the transfer of functional groups from one molecule to another. For example, *aminotransferases* convert one type of amino acid to another by transferring an amino group; *phosphotransferases* participate in the transfer of phosphate groups and are involved in energy transfer; *methyltransferases* move a methyl (CH_3) group from substrate to substrate; and *decarboxylases* catalyze the removal of carbon dioxide from organic acids in several metabolic pathways.

The role of microbial enzymes in disease—Many bacterial pathogens secrete unique exoenzymes that help them avoid host defenses or promote their multiplication in tissues. Because these enzymes contribute to pathogenicity, they are referred to as virulence factors, or toxins in some cases, *Streptococcus pyogenes* produces a streptokinase that digests blood clots and apparently assists in invasion of wounds. It also produces a protease that accounts for a severe form of 'flesh eating' disease. The lipases of *Staphylococcus aureus* (the cause of skin boils) increase the virulence of this species by promoting its invasion of oil-producing glands on the skin. *Pseudomonas aeruginosa,* a respiratory and skin pathogen, produces elastase and collagenase, which digest elastin and collagen. These increase the severity of certain lung diseases and burn infections. *Clostridium perfringens,* an agent of gas gangrene, synthesizes lecithinase C, a lipase that profoundly damages cell membranes and accounts for the tissue death associated with the disease.

Nomenclature of Enzymes

Given the complex chemistry of the cell and the extraordinary diversity of living things, many more enzymes probably remain to be discovered. To prevent discrepancies, a standardized system of nomenclature and classification was developed. The naming of enzymes is really quite straightforward. A name is composed of two parts : a prefix or stem word derived from a certain characteristic—usually the substrate acted upon or the type of reaction catalyzed, or both—followed by the ending-*ase*. The system assigns the enzyme to one of the following six classes, based on its general biochemical action : (a) *Oxidoreductases* transfer electrons from one substrate to another, and *dehydrogenases* transfer a hydrogen from one compound to another. (b) *Transferases* transfer C, N, P or S groups from one substrate to another. (c) *Hydrolases* cleave bonds on molecules with the addition of water. (d) *Lyases* add groups to or remove groups from double-bonded substrates. (e) *Isomerases* change a substrate into its isomeric form. (f) *Ligases* catalyze the formation of bonds with the input of ATP and the removal of water. Each enzyme is also assigned a systematic name that indicates the specific reaction it catalyzes, but some of these names are too cumbersome to be used routinely, and the common name, based on the substrate or some other notable feature, is preferred. Examples are *hydrogen peroxide oxidoreductase,* known more commonly as *catalase,* and *mucopeptide N-acetylmuramoylhydrolase,* called *lysozyme.* With this system, an enzyme that digests a carbohydrate substrate is a *carbohydrase;* more specifically, *amylase* acts on starch (amylose is a major component of starch). A *disaccharidase* such as *maltase* digests 2-unit sugars such as maltose. An enzyme that hydrolyzes peptide bonds of a protein is a *proteinase, protease,* or *peptidase,* depending on the size of the protein substrate. Some fats and other lipids are digested by *lipases.*

Nature of Enzymes

In general, enzymes operate only under the natural temperature, pH, and osmotic pressure of an organism's habitat. When enzymes are subjected to changes in these normal conditions, they tend to be chemically unstable, or labile. Low temperatures inhibit catalysis, while high temperatures denature the apoenzyme. Denaturation is a process by which the weak bonds that collectively maintain the native shape of the apoenzyme are broken. This disruption causes extreme distortion of the enzyme's shape and prevents the substrate from attaching to the active site such nonfunctional enzymes block metabolic reactions, which can lead to cell death.

Regulation of Enzyme Activity

The cell responds to environmental conditions by adopting those metabolic reactions that most favour growth and survival. Because enzymes are critical to these reactions, the regulation of metabolism is largely the regulation of enzymes by an elaborate system of checks and balances. Before we examine some of these control systems, let us take a look at some general features of metabolic pathways. Metabolic reactions rarely consist of a single action or step. More often, they occur in a multistep series or pathway, with each step catalyzed by an enzyme. An individual reaction is shown in various ways, depending on the purpose at hand. The product of one reaction is often the reactant (substrate) for the next, forming a linear chain of reactions. Many pathways have branches that provide alternate methods for nutrient processing. Others take a cyclic form, in which the starting molecule is regenerated to initiate another turn of the cycle. Every pathway has one or more enzyme pacemakers that sets the rate of a pathway's progression. These enzymes respond to various control signals and, in so doing, determine whether or not a pathway proceeds. Regulation of pacemaker enzymes proceeds on two fundamental levels. Either the enzyme itself is directly inhibited or activated, or the amount of the enzyme in the system is altered (decreased or increased).

Enzymatic Behaviour Controlled by Direct Methods

Competitive inhibition—In competitive inhibition, other molecules with a structure similar to the normal substrate can occupy the enzyme's active site. Although these molecular mimics have the proper fit for the site, they cannot be further acted upon. This effectively prevents an enzyme from attaching to its usual substrate and blocks its activity, output, and possibly the rest of that pathway. This mechanism is common in natural metabolic pathways.

Feedback control—The term *feedback* is borrowed from the field of engineering, where it describes a process in which the product of a system is *fed back* into the system to control it. Feedback is positive or negative, depending upon whether the process is stimulated (+) or inhibited (-) by the product. Negative feedback is common in electrical appliances such as hot plates or furnaces with automatically controlled temperatures. These devices possess a means of controlling the temperature (a heating coil or flame) and a temperature sensor (thermostat) that is set to a desired point and regulates the turning on or off of the heat source. In the case of a hot plate, when the set temperature has been reached, the thermostat switches the heating elements off. Many enzymatic reactions are controlled by a similar negative mechanism, whereby the end product being fed back into the system *negates* (cancels) an enzyme's activity :

$$A + B \xrightarrow{\text{Enzyme}} C$$

builds up,
cancels action of

At high levels of substrate and scant levels of end product, the enzyme works freely, but as the end product builds up, it stops the action of that enzyme. It usually targets the first enzyme in a metabolic pathway. Allosteric (Gr. *Allos*, other, and *steros*, solid) enzymes have an additional regulatory site for the attachment of molecules other than substrate. When an end product molecule fits into this regulatory site, the enzyme's active site is so distorted that it can no longer bind to its substrate. Although this does not denature the enzyme and is quite reversible, allosteric inhibitors can temporarily stop the action of that enzymes (Fig. 7.6). This mechanism is termed feedback inhibition. When at some point the end product is used up and more of it is needed, the enzyme will be released from inhibition and can resume catalysis.

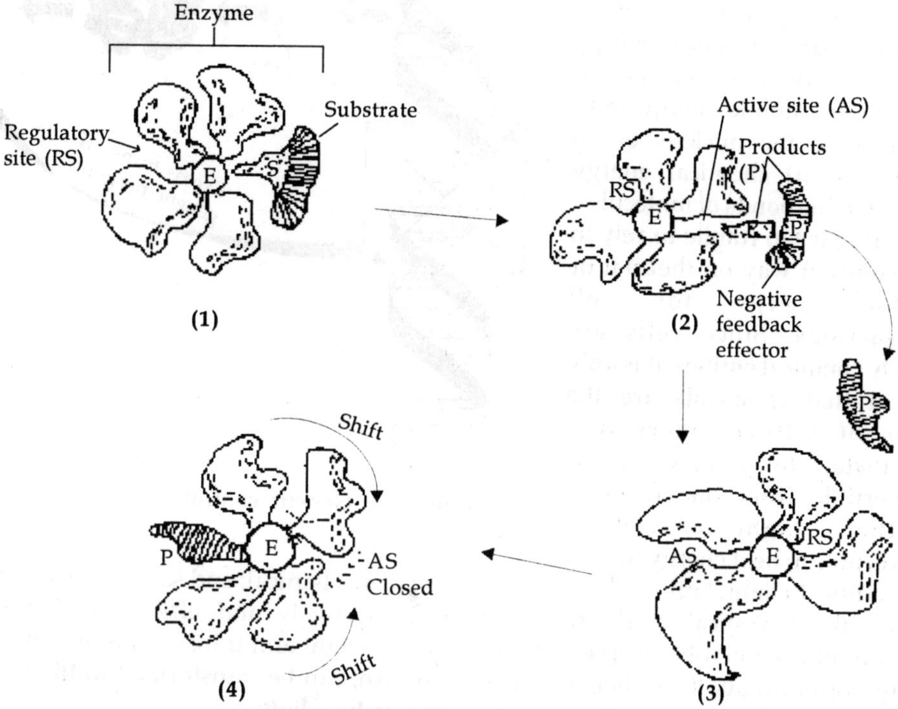

Fig. 7.6 : Negative feedback mechanism controls the enzyme activity.

Enzyme Synthesis Controlled by Different Methods

For catalysis to continue, enzymes eventually must be replaced. This works into the scheme of the cell, where replacement of enzymes can be regulated according to cell demand. Enzyme repression is a means to stop further synthesis of an enzyme somewhere along its pathway. As the level of the end product from a given enzymatic reaction has built to excess, the genetic apparatus responsible for replacing these enzymes is automatically suppressed (Fig. 7.7). The response time takes longer than feedback inhibition, but its effects are more enduring. A response that resembles the inverse of feedback repression is enzyme induction. A classic model of enzyme induction occurs in the response of *Escherichia coli* to certain sugars. For example, if a particular strain of *E. coli* is inoculated into a medium whose principal carbon source is lactose, it will produce lactase to hydrolyze it into glucose and galactose. If the bacterium is subsequently inoculated into a medium containing only sucrose as a carbon source, it will ceases synthesizing lactase and begin synthesizing sucrase.

ENERGY UTILIZATION

Energy is the capacity to do work or cause change. Energy commonly exists in

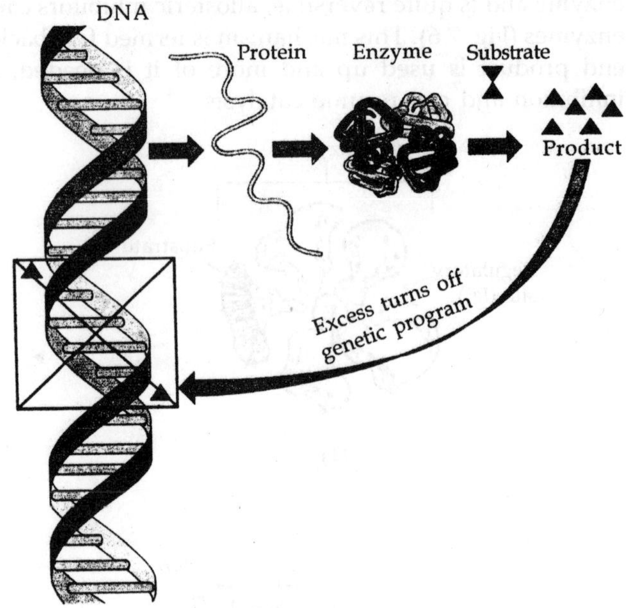

Fig. 7.7 : Genetic control of enzyme synthesis.

various forms : (a) Thermal, or heat, energy from molecular motion; (b) radiant (wave) energy from visible light or other rays; (c) electrical energy from a flow of electrons; (d) mechanical energy from a physical change in position; (e) atomic energy from reactions in the nucleus of an atom; and (f) chemical energy present in the bonds of molecules. Cells are far too fragile to rely in any constant way on thermal or atomic energy for cell transactions. Since cells are largely chemical entities, it is only logical that chemicals are the basis of cellular energetics. Fortunately for cells, energy is convertible from one form to another. Light energy is convertible to heat energy; heat is convertible to light; light to chemical; chemical to mechanical; and so on. This phenomenon allows the biological world to use energy effectively and efficiently. For example, photosynthetic microbes trap the energy in visible light and transform it into the chemical energy of nutrients. This chemical energy, in turn, can be transformed to the mechanical energy of flagellar movement, or it can even produce light.

This capacity of photosynthesis to convert solar energy into chemical energy provides both a nutritional and an energy basis for all heterotrophic living things. Although energy exists in several general forms, only chemical energy can routinely operate cell transactions at the level of the protoplasm. The general topic of this section is energy as it relates to the oxidation of fuels, the role of redox carriers, and the generation of ATP. The energy of the cell resides generally in the bonds between atoms and specifically in valence electrons that atoms can exchange. When these bonds remain untapped or unchanged and are not spent doing cellular work, they have what is called potential energy when bond energy is freed for cellular work, it is termed kinetic energy. Not all cellular reactions are equal with respect to energy. Some release energy, and others require it to proceed. If the cellular reaction :

$$A + B \xrightarrow{\nearrow ATP} C$$

proceeds with the release of energy, it is termed exergonic (Gr. *Exo*, without, and *ergon*, work). Energy released from some exergonic reactions can be stored as bond energy in ATP. If the reaction :

$$X + Y \xrightarrow{ATP} Z$$

requires the input of energy, it is endergonic (Gr. *Endo*, within, and *ergon*, work). Endergonic reactions can be driven forward by energy given off by ATP.

Indeed, a cell can be likened to a gasoline engine that releases energy as it burns the fuel and uses the energy to perform work. The engine does not actually produce energy, but it converts the potential energy of the fuel to kinetic energy of work. The processes by which cells handle energy are well understood, and most of them are explainable in basic biochemical terms. At the simplest level, cells possess specialized enzyme systems that entrap the energy present in the bonds of nutrients as they are progressively broken (Fig. 7.8). During exergonic reactions, energy released by electrons is stored in various high-energy phosphate molecules such as ATP.

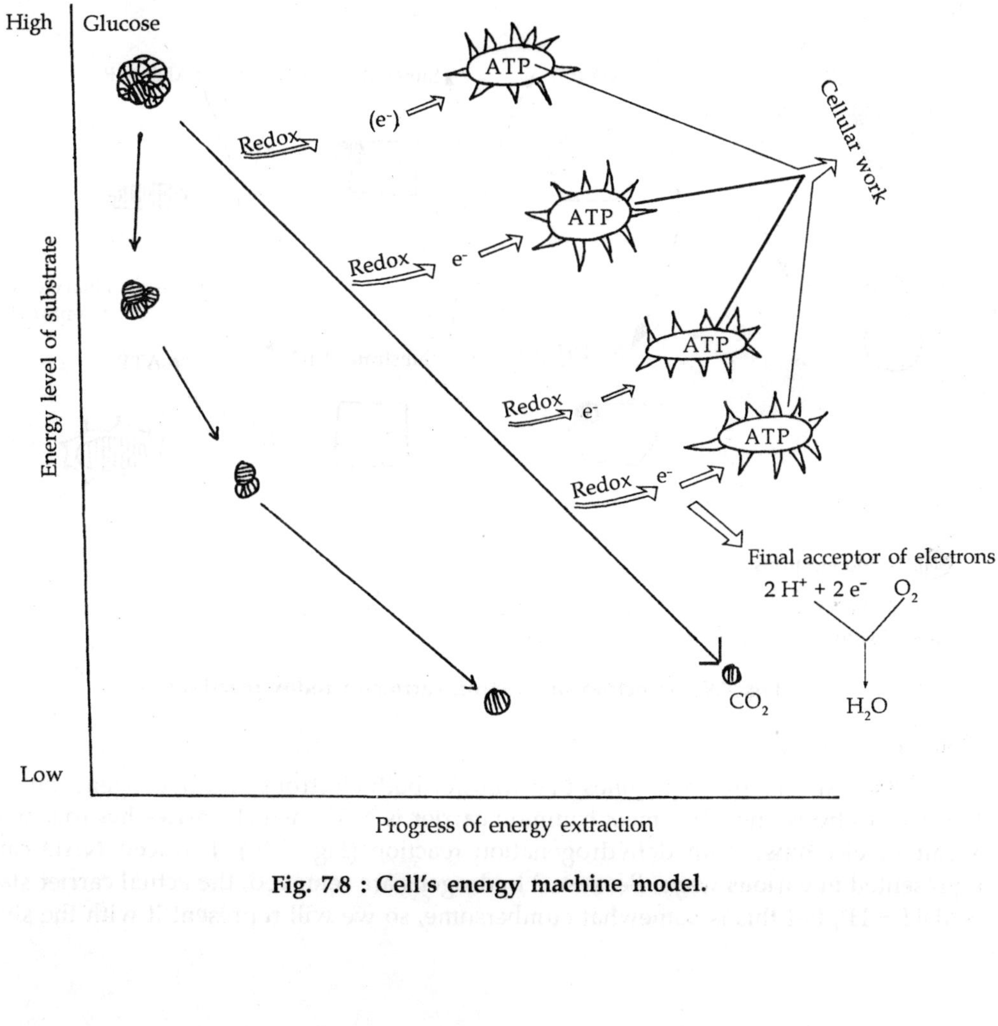

Fig. 7.8 : Cell's energy machine model.

BIOLOGICAL OXIDATION AND REDUCTION

Redox reactions always occur in pairs, with an electron donor and an electron acceptor, which constitute a *conjugate pair*, or *redox pair*. Written in equation form :

Electron donor (oxidized) $\rightarrow$ e^- Electron acceptor (reduced) + Energy

This process salvages electrons along with their inherent energy, but it also releases some kinetic energy, leaving the reduced compound with less energy than the oxidized one. The released energy can phosphorylate, or add an inorganic phosphate (P_i), to ADP or to some other compound, a process that captures this energy and stores it in a high-energy molecule (ATP). In many cases, the cell does not handle electrons as discrete entities but rather as parts of an atom such as hydrogen. For simplicity's sake, we will continue to use the term *electron transfer,* but keep in mind that hydrogen ions are often part of the transfer process. The removal of hydrogens (one hydrogen atom consists of a single proton and a single electron) from a compound during a redox reaction is called dehydrogenation. The job of handling these hydrogens and electrons falls to one or more carriers, which function as short-term repositories for the electrons until they can be used in ATP synthesis (Fig. 7.9).

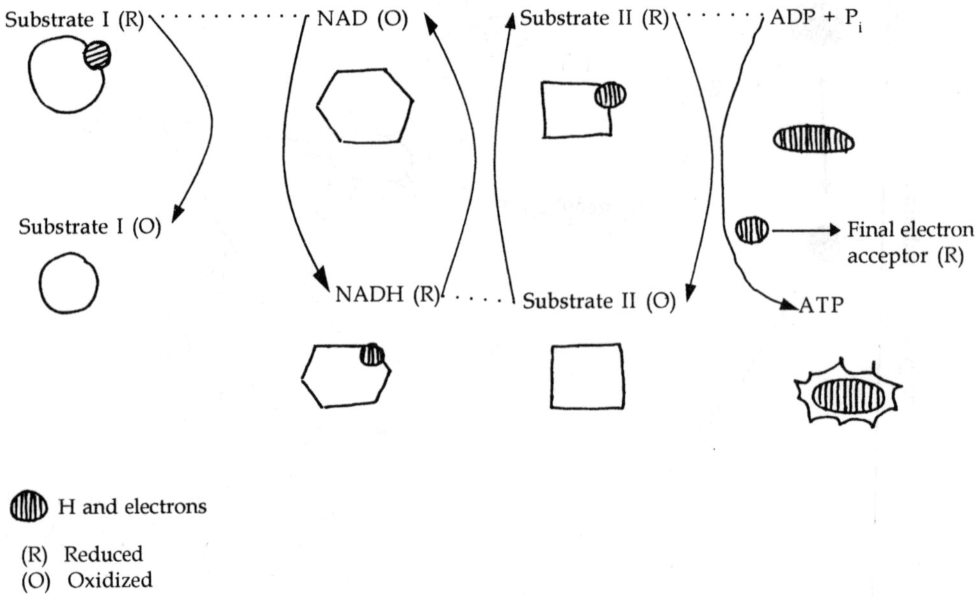

Fig. 7.9 : Function of electron carrier in redox reactions.

Molecular Shuttle

Most carriers are coenzymes that transfer both electrons and hydrogens, but some transfer electrons only. The most common carrier is NAD, which carries hydrogens (and a pair of electrons) from dehydrogenation reaction (Fig. 7.10). Reduced NAD can be represented in various ways. Because 2 hydrogens are removed, the actual carrier state is NADH + H^+, but this is somewhat cumbersome, so we will represent it with the shorter

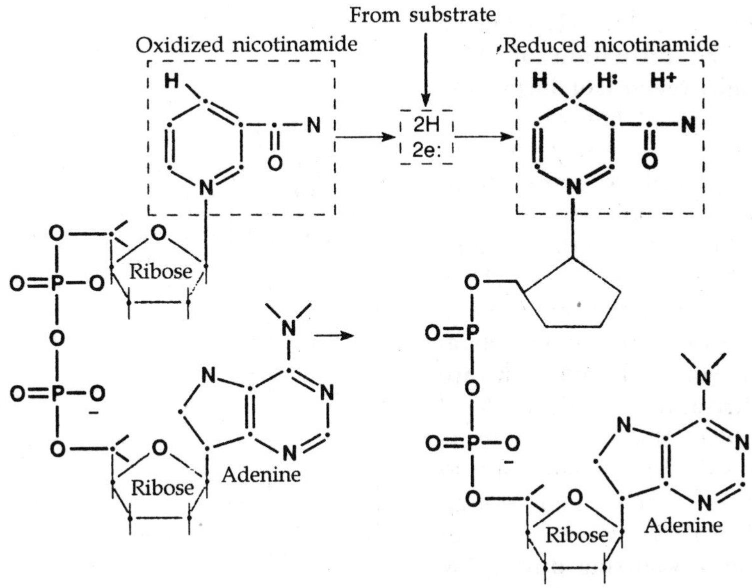

Fig. 7.10 : Mechanism of reduction of NAD.

NADH. In catabolic pathways, electrons are extracted and carried through a series of redox reactions until the final electron acceptor at the end of a particular pathway is reached. In aerobic metabolism, this acceptor is molecular oxygen, and in anaerobic metabolism, it is some other inorganic or organic compound. Other common redox carriers are FAD, NADP (NAD phosphate), coenzyme A, and the compounds of the respiratory chain, which are fixed into membranes.

ATP

ATP has also been described as metabolic money because it can be earned, banked, saved, spent, and exchanged. As a temporary energy repository, ATP provides a connection between energy-yielding catabolism and all other cellular activities that require energy. Some clues to its energy-storing properties lie in its unique molecular structure (Fig. 7.11). ATP is a three-part molecule consisting of a nitrogen base (adenine) linked to a 5-carbon sugar (ribose), with a chain of three phosphate groups bonded to the ribose. The type, arrangement, and especially the proximity of atoms in ATP combine to form a compatible but unstable high-energy molecule. The high energy of ATP originates in the orientation of the phosphate groups, which are relatively bulky and carry negative charges. The proximity of these repelling electrostatic charges impose a strain that is most acute on the bonds between the last two phosphate groups. When ATP is formed, this instability is accommodated by the bond stress being distributed through the rest of the molecule. Breaking the bonds between two successive phosphates yields adenosine diphosphate (ADP) and adenosine monophosphate (AMP).

In many instances, the energy released during ATP hydrolysis powers biosynthesis by activating individual subunits before they are enzymatically linked together. ATP is also used to prepare a molecule for catabolism such as the double phosphorylation of a 6-

carbon sugar during the early stages of glycolysis. The formation of ATP occurs through a reversal of the process by which it was spent. In heterotrophs, the energy infusion that regenerates a high-energy phosphate comes from certain steps of catabolic pathways, in which nutrients such as carbohydrates are degraded and yield energy. ATP is formed when substrates or electron carriers add a high-energy phosphate bond to ADP. Some ATP molecules are formed through *substrate level phosphorylation*—that is, energy is released directly from a substrate to ADP. Other ATPs are formed through *oxidative phosphorylation,* a series of redox reactions occurring during the final phase of the respiratory pathway. Phototrophic organisms have a system of *photophosphorylation,* in which the ATP is formed through a series of sunlight-driven reactions.

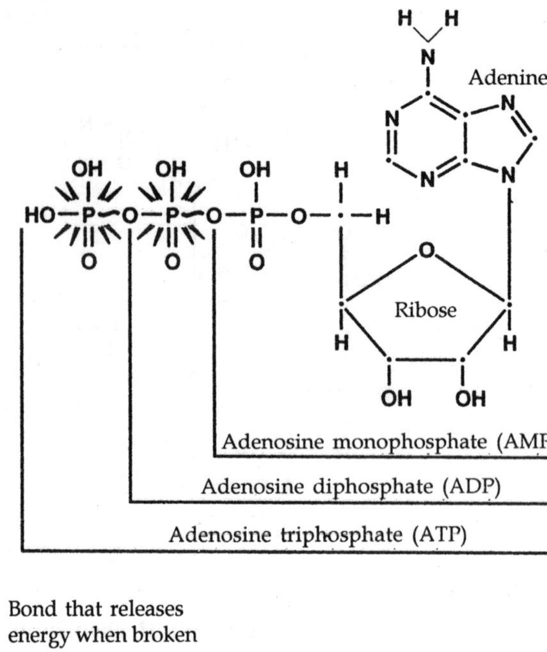

Fig. 7.11 : ATP molecule's structure.

BIOENERGETICS

Initial work with bacteria and yeasts, followed by studies with animal and plant cells, clearly demonstrated metabolic similarities and strongly supported the concept of the universality of metabolism. The study of the production and use of energy by cells is called bioenergetics. The two major types of metabolic pathways covered by this science are catabolic routes that gradually degrade nutrients and anabolic or biosynthetic routes that are involved in cell growth. Although these pathways are interconnected and interdependent and most individual reactions are reversible, anabolic pathways are not simply reversals of catabolic ones. At first glance it might seem more economical to use identical pathways but having different enzymes and divided pathways allows anabolism and catabolism to proceed simultaneously without interference.

Catabolism

The primary catabolism of fuels that results in energy release in many organisms proceeds through a series of three coupled pathways : (a) Glycolysis (Gr. *Glykys,* sweet, and *lysis,* a loosening), also called the Embden-Meyerhof-Parnas (EMP) pathway, (b) the tricarboxylic acid cycle (TCA), also known as the citric acid or Krebs cycle, and (c) the respiratory chain. Each segment of the pathway is responsible for a specific set of actions on various products of glucose (Fig. 7.12). Nutrient processing is extremely varied, especially in bacteria, yet in most cases it is based on three basic catabolic pathways. In previous discussions, microorganisms were categorized according to their requirements of oxygen gas, and this relates directly to their mechanisms of energy release. As we shall see, aerobic respiration is a series of reactions that converts glucose to CO_2 and gives off energy.

It relies on free oxygen as the final acceptor for electrons and hydrogens, and produces a relatively large amount of ATP. Aerobic respiration is characteristic of many bacteria, fungi, protozoa, and animals, and is the system we will emphasize here. Facultative and aerotolerant anaerobes may use only the glycolysis scheme to incompletely oxidize or ferment glucose. In this case, oxygen is not required, organic compounds are the final electron acceptors, and a relatively small amount of ATP is produced. Some strictly anaerobic microorganisms metabolize by means of anaerobic respiration.

Aerobic Respiration

This pathway is the principal energy-yielding scheme for aerobic heterotrophs, and it provides both ATP and metabolic intermediates for many other pathways in the cell,

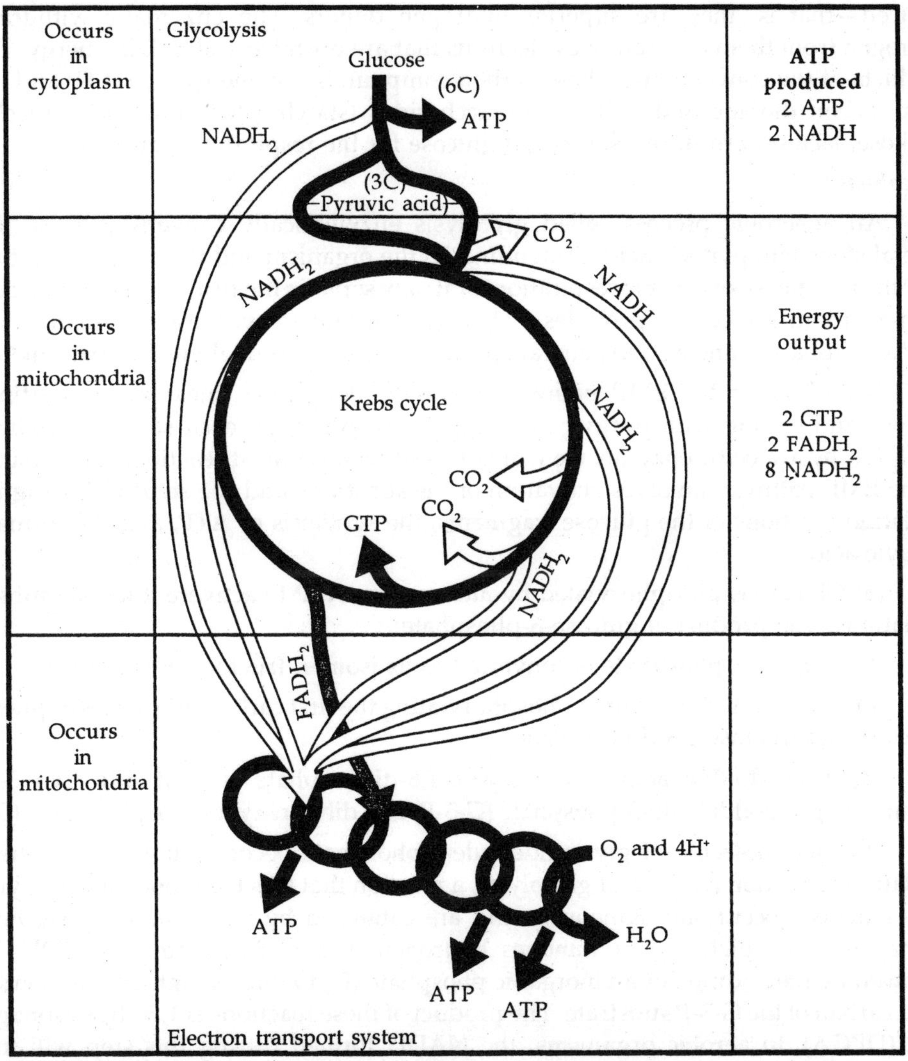

Fig. 7.12 : Aerobic respiration.

including those of protein, lipid, and carbohydrate synthesis (Fig. 12). Aerobic respiration in microorganisms can be summarized by an equation :

$$\text{Glucose} + 6\ O_2 + 38\ \text{ADP} + 38\ P_i \rightarrow 6\ CO_2 + 6\ H_2O + 38\ \text{ATP}$$

The seeming simplicity of the equation for aerobic respiration conceals its complexity. Fortunately, we do not have to present all of the details to address some important concepts concerning its reactants and products. These ideas are : (a) The steps in the oxidation of glucose. (b) The involvement of coenzyme carriers and the final electron acceptor, (c) where and how ATP originates, (d) where carbon dioxide originates, (e) where oxygen is required and (f) where water originates.

Carbohydrates like glucose are good fuels because these compounds are highly reduced—that is, they are superior hydrogen donors. The enzymatic withdrawal of hydrogen from them also removes electrons that are unstable and rich in energy. The end products of the conversion of these carbon compounds are energy-rich ATP and energy-poor carbon dioxide and water. Polysaccharides (starch, glycogen) and disaccharides (maltose, lactose) are stored sources of glucose for the respiratory pathways.

Glycolysis

An anaerobic process called glycolysis enzymatically converts glucose through several steps into pyruvic acid. Depending on the organism and the conditions, it may be only the first phase of aerobic respiration, or it may serve as the primary metabolic pathway (fermentation). Glycolysis provides a significant means to synthesize a small amount of ATP anaerobically and also to generate pyruvic acid, an essential intermediary metabolite.

Glycolysis or the EMP pathway proceeds along nine linear steps, starting with glucose and ending with pyruvic acid (Fig. 7.13). Although each step of metabolism is catalyzed by a specific enzyme, we will not mention it for most reactions. The first portion of the EMP pathway involves activation of the substrate, and the steps following involve oxidation reactions of the glucose fragments, the synthesis of ATP, and the formation of pyruvic acid.

(a) Glucose is phosphorylated by means of an ATP to activate it for the subsequent oxidations. The product is glucose-6-phosphate.

(b) Glucose-6-phosphate is converted to its isomer, fructose -6-phosphate.

(c) Another ATP is spent in phosphorylating the first carbon of fructose-6-phosphate, which yields fructose-1,6-diphosphate.

(d) Now doubly activated, fructose-1,6-diphosphate is split into two 3-carbon fragments : glyceraldehyde-3-phosphate (G-3-P) and dihydroxyacetone phosphate (DHAP).

(e) Each molecule of glyceraldehyde-3-phosphate becomes involved in the single oxidation-reduction reaction of glycolysis, a reaction that sets the scene for ATP synthesis. Two reactions occur simultaneously and are catalyzed by the same enzyme. First, the coenzyme NAD picks up or removes hydrogen from G-3-P, forming NADH. This is followed by the addition of an inorganic phosphate (P_i) to form a high-energy bond on the third carbon of the G-3-P substrate. The product of these reactions is 1, 3-diphosphoglyceric acid (DPGA). In aerobic organisms, the NADH formed during this step will undergo further reactions in the electron transport system, where the final H acceptor will be oxygen and additional ATPs will be generated.

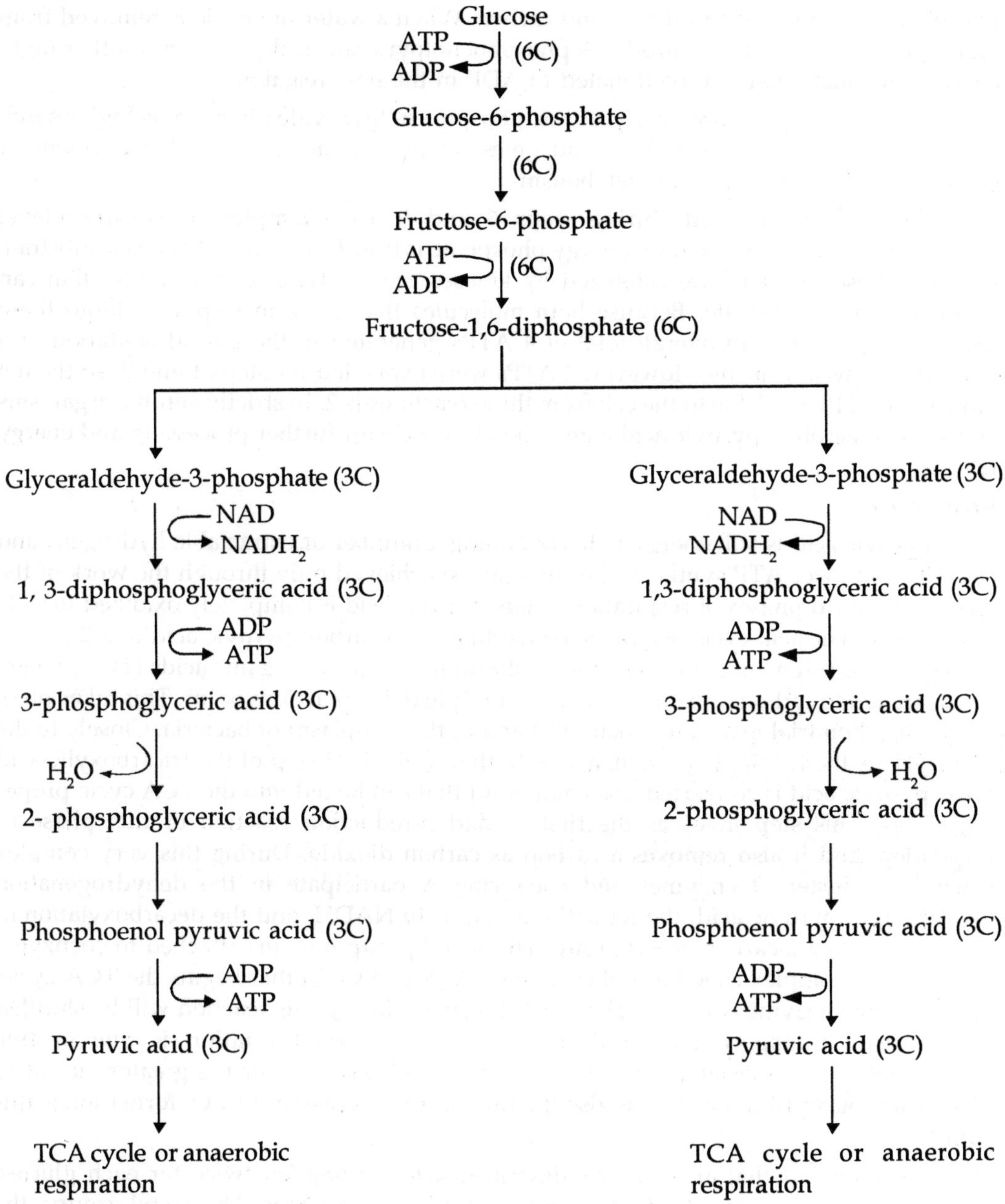

Fig. 7.13 : Glycolysis cycle.

(f) One of the high-energy phosphates of 1, 3-DPGA is donated to ADP, resulting in a molecule of ATP. The product of this reaction is 3-phosphoglyceric acid.

(g, h) A substrate for the synthesis of a second ATP is produced. First, 3-phosphoglyceric acid is converted to 2-phosphoglyceric acid through the shift of a phosphate from the third to the second carbon. When a water molecule is removed from 2-phosphoglyceric acid, this produces phosphoenolpyruvate and generates another high-energy phosphate that will be donated to ADP in the next reaction.

(i) In the final reaction of glycolysis, phosphoenolpyruvate gives up its high-energy phosphate to form a second ATP, and gives off pyruvic acid, also called pyruvate, a compound with many roles in metabolism.

The 2 ATPs formed during steps 6 and 9 are examples of substrate-level phosphorylation, in that the high-energy phosphate is transferred directly from a substrate to ADP. These reactions are catalyzed by kinases, special types of transferase that can phosphorylate a substrate. Because both molecules that arose in step 4 undergo these reactions, this makes an overall total of 4 ATPs generated in the partial oxidation of a glucose to 2 pyruvic acids. However, 2 ATPs were expended for steps 1 and 3, so the net number of ATPs available to the cell from these reactions is 2. In strictly aerobic organisms and some anaerobes, pyruvic acid enters the TCA cycle for further processing and energy release.

Kreb's cycle

Pyruvic acid is still energy-rich, containing a number of extractable hydrogens and electrons to power ATP synthesis, but this can be achieved only through the work of the second and third phases of respiration where pyruvic acid is completely oxidized to CO_2. The tricarboxylic acid cycle begins by converting the 3-carbon pyruvic acid to a 2-carbon fragment. The fragment is then channeled through a circuit of organic acids (TCA) where it is degraded to carbon dioxide and where multiple redox reactions occur. This takes place in the mitochondrial matrix in eukaryotes and in the cytoplasm of bacteria. Closely to the TCA cycle is the respiratory chain, a system that in the first step of the tricarboxylic acid cycle, pyruvic acid is converted to a compound that can be fed into the TCA cycle proper (Fig. 7.14). This step involves the first oxidation-reduction reaction of this phase of respiration, and it also removes a carbon as carbon dioxide. During this very complex reaction, a cluster of enzymes and coenzyme A participate in the dehydrogenation (oxidation) of pyruvic acid, the reduction of NAD to NADH, and the decarboxylation of pyruvic acid to a 2-carbon acetyl group. The acetyl group remains attached to coenzyme A, forming a complex called acetyl coenzyme A (acetyl CoA) that begins the TCA cycle. This is a huge enzyme complex. The NADH formed during this reaction will be shuttled into electron transport and used to generate ATP; its formation is one of five dehydrogenations associated with the TCA cycle and accounts for the greater output of ATP in aerobic respiration. This is also the first of three occasions of CO_2 formation is this pathway.

Be reminded that all reactions described actually happen twice for each glucose because of the two pyruvates that are released during glycolysis. The acetyl groups, the remaining fragments of glucose that started the glycolytic pathway, are subsequently dismantled by first combining with a 4-carbon oxalloacetate molecule to form citric acid, a 6-carbon molecule. Although this seems a cumbersome way to dismantle such a small molecule, it is necessary in biological systems for extracting larger amounts of energy. The tricarboxylic acid cycle has eight steps, beginning with citric acid formation and ending with the organic acid oxalloacetic acid. As we take a single spin around the TCA wheel,

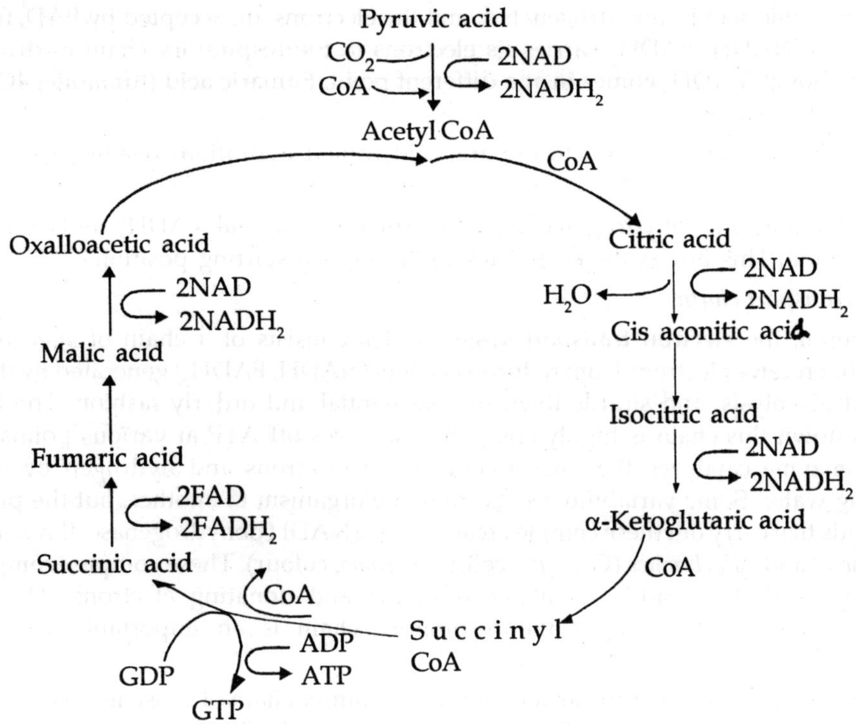

Fig. 7.14 : The TCA cycle.

it will be instructive to keep track of : (a) The numbers of carbons of each substrate and product (#C), (b) reactions where CO_2 is generated, (c) the involvement of the electron carriers NAD and FAD, and (d) the site of ATP synthesis. The reactions in the TCA cycle are :

(a) Oxalloacetic acid (oxalloacetate; 4C) reacts with the acetyl group (2C) on acetyl CoA, thereby forming citric acid (citrate; 6C), and releasing coenzyme A to react with another acetyl.

(b) Citric acid is converted to its isomer, isocitric acid (isocitrate; 6C), to prepare this substrate for the decarboxylation and dehydrogenation of the next step.

(c) Isocitric acid is acted upon by an enzyme complex including NAD or NADP (depending on the organism), in a reaction that generates NADH or NADPH, splits off a carbon dioxide, and leaves α-ketoglutaric acid (α-ketoglutarate; 5C).

(d) Alpha-ketoglutaric acid serves as a substrate for the last decarboxylation reaction and yet another redox reaction involving coenzyme A and yielding NADH. The product is the high-energy compound succinyl CoA (4C).

(e) Succinyl CoA releases its high-energy bond, this time not to an ADP but to its relative and metabolic equivalent, GDP (guanosine diphosphate), forming GTP. GTP will later be converted to ATP. This is the single substrate level phosphorylation in the TCA cycle. This reaction leaves succinic acid (succinate; 4C).

(f) Succinic acid is dehydrogenated, and the electrons are accepted by FAD, forming $FADH_2$. Like NADH, $FADH_2$ carries its electrons to the respiratory chain to drive ATP synthesis, though $FADH_2$ comes in at a different point. Fumaric acid (fumarate; 4C) is the product of this reaction.

(g) The addition of water to fumaric acid (called hydration) results in malic acid (malate; 4C).

(h) Malic acid is dehydrogenated (with formation of a final NADH), and oxalloacetic acid is formed. This brings the cycle back to its original starting position.

Electron transport chain

Overall, the electron transport system (ETS) consists of a chain of special redox carriers that receive electrons from reduced carriers (NADH, $FADH_2$) generated by the TCA cycle and glycolysis, and shuttle them in a sequential and orderly fashion. The flow of electrons down this chain is highly energetic and gives off ATP at various points. At its end, an enzyme catalyzes the final acceptance of electrons and hydrogen by oxygen, producing water. Some variability exists from one organism to another, but the principal compounds that carry out these complex reactions are NADH dehydrogenase, flavoproteins, coenzyme Q and *cytochromes* (Gr. *Cyto*, cell, and *kroma*, colour). These complex compounds contain a metal that readily facilitates receiving and donating electrons. The highly compartmentalized structure of the respiratory chain is an important factor in its function.

The sequence of electron carriers in the respiratory chain of most aerobic organisms is : (a) NADH dehydrogenase, (b) flavin mononucleotide (FMN), (c) coenzyme Q, (d) cytochrome b, (e) cytochrome c_1, (f) cytochrome c, and (g) cytochrome aa_3. Conveyance of the NADHs from the TCA cycle and glycolysis to the first carrier sets in motion the remaining six steps. With each redox exchange, the energy level of the reactants is lessened. The released energy is captured and used by the ATP synthase complex stationed near the ETS. Each NADH that enters electron transport gives rise to 3 ATPs (Fig. 16). This coupling of ATP synthesis to electron transport is termed oxidative phosphorylation. Since $FADH_2$ from the TCA cycle is metabolically equivalent to FMN, it releases only enough energy to synthesize 2 ATPs.

According to a widely accepted concept called the *chemiosmotic hypothesis*, as the electron transport carriers shuttle electrons, they actively pump hydrogen ions (protons) into the outer compartment of the mitochondrion. This process sets up a concentration gradient of hydrogen ions called the proton motive force (PMF). The PMF also generates a difference in charge between the outer membrane compartment (+) and the inner membrane compartment (-). The potential energy inherent in such a separation of charge can be harnessed to produce ATP. Although the exact mechanism is not completely understood, it appears that special pores penetrating the ATP synthase complex transport hydrogen ions (protons) from the outside compartment back into the mitochondrial matrix. This process releases sufficient free energy for the synthesis of ATP from an ADP and P_i. Bacterial ATP synthesis occurs by means of this same overall process. However, bacteria have the ETS stationed in the cell membrane and the direction of the proton movement is from the cytoplasm to the periplasmic space.

Energy output

The total of five NADHs (four from the TCA cycle and one from glycolysis) can be used to synthesize:

15 ATPs for ETS (5×3 per electron pair)

and

$15 \times 2 = 30$ ATPs per glucose

The single $FADH_2$ produced during the TCA cycle results in : 2 ATPs per electron pair

and $2 \times 2 = 4$ ATPs per glucose

Close examination of this view will review several important facets of aerobic respiration : (a) The total yield of ATP is 40:4 from glycolysis, 2 from the TCA cycle, and 34 from electron transport. However, since 2 ATPs were expended in early glycolysis, this leaves a maximum of 38 ATPs. The actual totals may be lower in certain eukaryotic cells and bacteria. (b) Six carbon dioxide molecules are generated during the TCA cycle. (c) Six oxygen molecules are consumed during electron transport. (d) Six water molecules are produced in electron transport and 1 in glycolysis, but because 1 is used in the TCA cycle, this leaves a net of 6. This large enzyme complex is specially adapted to receive electrons from cytochrome c, pick up hydrogens from solution, and react with oxygen to form a molecule of water. This reaction, though in actuality more complex, is summarized as follows :

$$2\,H + 2e^- + \tfrac{1}{2}\,O_2 \rightarrow H_2O$$

Some species lack one or more of the redox steps, while others have several alternative electron transport schemes, since many bacteria lack cytochrome C, this variation can be used to differentiate among certain genera of bacteria. The oxidase test is diagnostic of cytochrome C. It is used to help identity members of the genera *Neisseria* and *Pseudomonas,* and some species of *Bacillus.* Another variation in the cytochrome system is evident in certain bacteria (*Klebsiella, Enterobacter*) that can grow even in the presence of cyanide because they lack cytochrome oxidase. Cyanide would ordinarily cause rapid death in humans and other eukaryotes because it blocks cytochrome oxidase, thereby completely terminating aerobic respiration. A potential side reaction of the respiratory chain in aerobic organisms is the incomplete reduction of oxygen to superoxide ion ($:O_2^-$) and hydrogen peroxide (H_2O_2). Because these toxic oxygen products can be very damaging to cells, aerobes have various neutralizing enzymes (*superoxide dismutase* and *catalase*). One exception is the genus *Streptococcus,* which can grow well in oxygen yet lacks both cytochromes and catalase. The tolerance of these organisms to oxygen can be explained by the neutralizing effects of a special peroxidase.

Anaerobic Respiration

Some bacteria have evolved an anaerobic respiratory system that functions like the aerobic cytochrome system except that it utilizes oxygen containing salts, rather than free oxygen, as the final electron acceptor. The reaction in species such as *Escherichia coli* is represented as :

Nitrite reductase

$$NO_3^- + H_2 \rightarrow NO_2^- + H_2O$$

Reduction here is the removal of oxygen from nitrate. A test for this reaction is one of the battery of physiological tests used in identifying bacteria. Some species of *Pseudomonas* and *Bacillus* possess enzymes that can further reduce nitrite to nitric oxide (NO), nitrous oxide (N_2O), and even nitrogen gas (N_2). This process, called *denitrification,* is a very important step in recycling nitrogen in the biosphere.

Technically speaking, fermentation (L. *Fervere,* to boil or *fermenatum,* yeast) is the incomplete oxidation of glucose or other carbohydrates in the absence of oxygen, a process that uses organic compounds as the terminal electron acceptors and yields a small amount of ATP. Over time, the term *fermentation* has acquired several looser connotations. Originally, Pasteur called the microbial action of yeast during wine production *ferments,* and to this day, biochemists use the term in reference to the production of ethyl alcohol by yeasts acting on glucose and other carbohydrates. Fermentation is also what bacteriologists call the formation of acid, gas, and other products by the action of various bacteria on pyruvic acid. The process is a common metabolic strategy among bacteria.

From one standpoint, fermentation permits independence from molecular oxygen and allows colonization of anaerobic environments. It also enables microorganisms with a versatile metabolism to adapt to variations in the availability of oxygen. For them, fermentation provides a means to grow even when oxygen levels are too low for aerobic respiration. After initially hydrolyzing cellulose to glucose, they ferment the glucose to

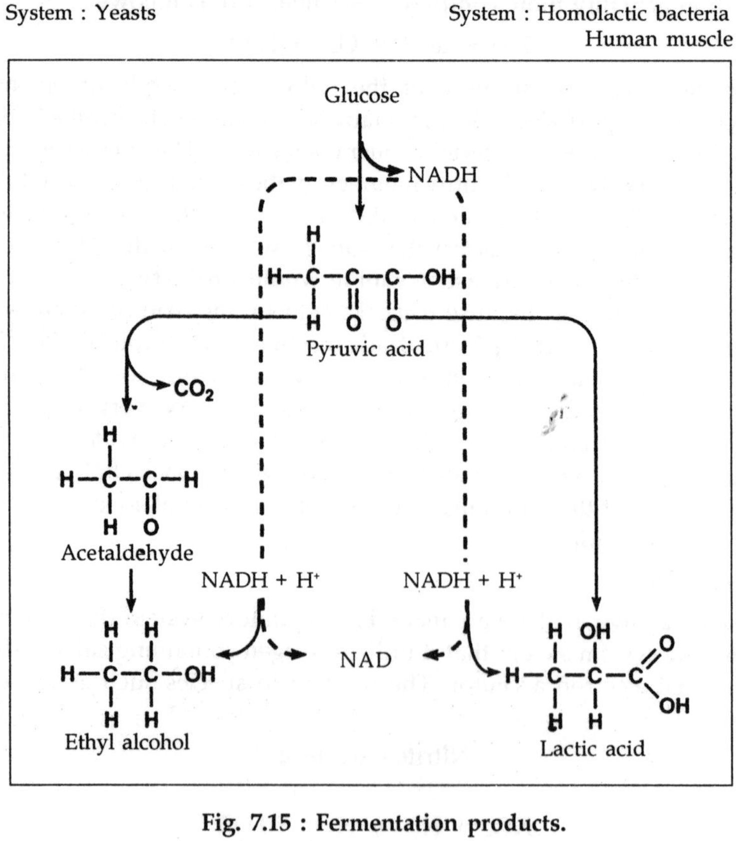

Fig. 7.15 : Fermentation products.

organic acids that are then absorbed as the bovine's principal energy source. Even human muscle cells can undergo a form of fermentation that permit short periods of activity after the oxygen supply in the muscle has been exhausted. Muscle cells convert pyruvic acid into lactic acid, which allows anaerobic production of ATP to proceed for a time.

Fermentation Products

Alcoholic beverages (wine, beer, whiskey) are perhaps most prominent among fermentation products, as are solvents (acetone, butanol), organic acids (lactic, acetic), dairy products, and many other foods. Fermentation products can be grouped into two general categories—alcoholic fermentation products and acidic fermentation products (Fig. 7.15). Alcoholic fermentation occurs in yeasts species that have metabolic pathways for converting pyruvic acid to ethanol. This process involves a decarboxylation of pyruvic acid to acetaldehyde, followed by a reduction of the acetaldehyde to ethanol. In oxidizing the NADH formed during glycolysis, this step regenerates NAD, thereby allowing the glycolytic pathway to continue. These processes are crucial in the production of beer and wine, though the actual techniques for arriving at the desired amount of ethanol and the prevention of unwanted side reactions are important tricks of the brewer's trade. Alcohols other than ethanol can be produced during bacterial fermentation pathways. Certain clostridia produce butanol and isopropanol through a complex series of reactions. Although this process was once an important source of alcohols for industrial use, it has been largely replaced by a nonmicrobial petroleum process.

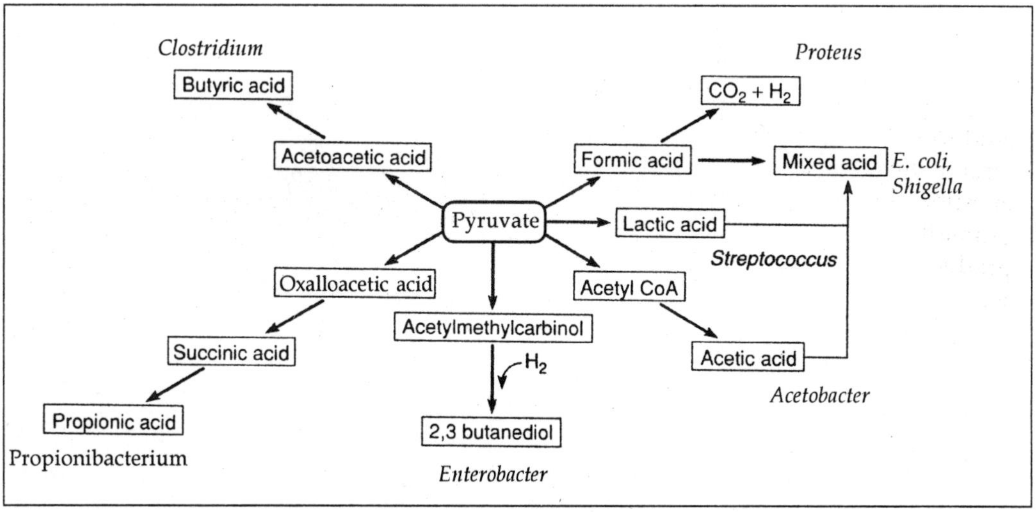

Fig. 7.16 : Products of pyruvate fermentation.

Lactic acid bacteria ferment glucose in the same way that humans do—by reducing it to lactic acid. If the product of this fermentation is *only* lactic acid, as in certain species of *Streptococcus* and *Lactobacillus*, it is termed *homolactic*. When glucose is fermented to a mixture of lactic acid, acetic acid, and carbon dioxide, as is the case with *Leuconostoc* and other species of *lactobacillus*, the process is termed *heterolactic fermentation*. Many members of the family Enterobacteriaceae (*Escherichia, Shigella,* and *Salmonella*) possess enzyme systems for converting pyruvic acid to several acids simultaneously (Fig. 7.16). Mixed acid fermentation produces a combination of acetic, lactic, succinic, and formic acids, and lowers the pH of a medium to about 4.0. *Propionibacterium* produces primarily propionic acid, which gives the characteristics flavour to Swiss cheese while fermentation gas produces the holes. Some members also further decompose formic acid completely to carbon dioxide and hydrogen gases. Because enteric bacteria commonly occupy the intestine, this fermentative activity accounts for the accumulation of some types of gas—primarily CO_2 and H_2—in the intestine. Some bacteria oxidize the organic acids and produces the neutral end product 2,3-butanediol. In fact, so broad is the meaning of the word fermentation that the large-scale industrial syntheses by microorganisms often utilize entirely different mechanisms from those described here, and even occur aerobically. This is particularly true in antibiotic, hormone, vitamin, and amino acid production.

❑❑❑

8

GENETICS OF BACTERIOPHAGE

In bacteriophage genetics, λ phage was integral to studies dealing with the initiation of DNA replication. Interchangeable σ factors were discovered using T4 while λ proved central to our understanding of transcriptional repressors and activators. Another phage we discuss, μ, has taught us much regarding DNA rearrangements. As you will see, bacteriophages remain critically important research tools. For example, the λ family of phages has enjoyed a resurgent interest in the genomic era. The availability of complete genomic sequences has revealed a pivotal role for members of this phage family in the development of *E. coli* virulence. The genomes of many pathogens contain entire sections, called pathogenicity islands, that encode virulence genes with an AT content different from the rest of the chromosome. The DNA flanking these islands often contain remnants of viral genes, like scars marking how the organism may have acquired these genes. In addition to their evolutionary significance, the unique recombination systems of λ have been exploited to simplify prokaryotic and eukaryotic genetic engineering strategies.

CHARACTERS OF BACTERIOPHAGES

Characterization and classification of the bacteriophage viruses poses many problems. From a practical point of view, they can be distinguished on the basis of their natural host, host range, and other similar characteristics. In addition, they can be characterized on the basis of their RNA or DNA content. Each phage generally contains only one kind of nucleic acid. Structural symmetry and susceptibility or resistance to ether and other solvents and to other chemical agents also provide additional criteria for classification (Fig. 8.1).

The nucleic acid may be either double stranded, single stranded, circular, or linear. Morphologically, phages generally display a considerable degree of geometric symmetry. They generally have a head or capsid composed of identical protein subunits called capsomeres. The nucleic acid core is housed within the capsid. Some bacteriophages have a prominent tail structure. The capsid structure may assume an icosahedral form. The

Phages	Morphology of virus		Type and amount of nucleic acid	Growth cycle at 37°C			
	Head (nm)	Tail (nm)		Latent period (min)	Average yield per Cell	Lysogeny	Specificities
T1	Icosahedral 50	10 × 150	DNA, 2.5×10^7	13	150	-	Resistant to drying
T2, T4, T6	Prolated icosahedral (65 × 95)	25 × 110	DNA, 1.2×10^8	21-25	150-400	-	Contain glucosylated HMC
T3, T7	Icosahedral 47	10 × 15	DNA, 2.4×10^7	13	300	-	Give semitemperate mutants
T5	Icosahedral 65	10 × 17	DNA, 7.5×10^7	40	200	-	DNA injection in two steps
λ, ϕ 80	Icosahedral 54	10 × 140	DNA, 3.3×10^7	35	100	+	DNA circularizes, *in vitro* or *in vivo*
P1	Icosahedral 65	12 × 150	DNA, 6×10^7	45	80	+	General transduction
P2	Icosahedral 50	10 × 150	DNA, 2.2×10^7	30	120	+	Multiple chromosomal locations
ϕ X174 S13	Icosahedral 30	None	DNA, 1 strand 1.7×10^6	13	180	-	Circular DNA
f2, MS2 pili	Icosahedral 24	None	RNA, 9×10^5	22	20,000	-	Male specific attachment to F pili
f1, fd	None	6 × 800	DNA, 1 strand 1.3×10^6	30	100-200 (Continuous release)	-	Male specific, circular DNA
χ*	Icosahedral 67.5	12.5 × 230	DNA	60	200	-	Attaches to motile flagella

* *Salmonella* phage

Fig. 8.1 : Characters of *E. coli* phages.

entire infectious unit is generally referred to as a virion. The smallest phage contain nucleic acid (either DNA or RNA) with a molecular weight on the order of 1×10^6 daltons (Fig. 8.2). This relatively short-chain nucleic acid can obviously only code for a small

Corticoviridae
    Icosahedral capsid with lipid layer, circular supercoiled dsDNA
Cystoviridae
    Enveloped, icosahedral capsid, lipids, three molecules of linear dsRNA
Fuselloviridae
    Pleomorphic, envelope, lipids, no capsid, circular supercoiled dsDNA
Inoviridae, genus Inovirus
    Long filaments with helical symmetry, circular ssDNA
Leviviridae
    Quasi-icosahedral capsid, one molecule of linear ssRNA
Lipothrixviridae
    Enveloped filaments, lipids, linear dsDNA
Microviridae
    Icosahedral capsid, circular ssDNA
Myoviridae
    Icasohedral capsid, dsDNA.
Plasmaviridae
    Pleomorphic, envelope, lipids, no capsid, circular supercoiled dsDNA
Podoviridae
    Tail short and noncontractile, head isometric
Rudiviridae
    Helical rods, linear dsDNA
Siphoviridae, morphotype B1
    Tail long and noncontractile, head isometric
Tectiviridae
    Icosahedral capsid with inner lipoprotein vesicle, linear dsDNA, "tail" produced for DNA injection

Fig. 8.2 : Bacteriophage's families.

number of genes. These genes, of necessity, must encode information that governs the formation of the basic viral subunits. The presence of such a small number of genes reflects the degree to which viruses are dependent on the host cell. Each species of microorganism is susceptible to a limited subset of viruses. Put another way, each phage has a specific (and limited) host range of susceptible microorganisms that it can infect. For example, a hypothetical phage A might only infect one strain of *E. coli* while phage B could infect many different strains of *E. coli.* Phage A is viewed as having a narrow host range relative to that of phage B. Host range depends on the presence of specific viral receptors on the host cell surface. These receptors are usually composed of specific carbohydrate groups on lipopolysaccharides (LPS) of cell surface structures.

Some viruses, such as M13, attach to sex pili and are referred to as male-specific phages. Two types of infection cycles may occur following the initial infection. One is a virulent or lytic infection that ends in lysis and death of the host cell. The other type of infection is called lysogenic, which may be quite unapparent since the host cell does not die. In fact, the general appearance and activity of the lysogenized host cell may not be altered in any overt manner. A lytic type of phage only undergoes virulent infection. Bacteriophage infections of bacteria follow a characteristic pattern. Once a population of bacterial cells has been inoculated with a given number of bacteriophages, the number of detectable infectious particles rapidly decreases. This stage is termed the eclipse phase and represents that portion of the time following infection during which bacteriophages cannot be detected either in the culture medium or within the cell. The overall period encompassing adsorption and eclipse is referred to as the latent period. After the latent period has been completed, infectious viruses begin to be released from the cell.

Replication cycle may be completed in following steps :

(a) Prepenetration and penetration of viral nucleic acid into the cell.

(b) Removal of nucleocapsid and intracellular synthesis of viral subunits.

(c) Assembly of subunits into complete virus particles and release of virus particles after completion of virus replication cycle.

Lysogenic infection occurs when the nucleic acid of the virus enters the host cell, becomes aligned with the genome of the cell, and eventually is integrated into the cell genome. In this way, the virus is replicated in unison with replication of the host cell genome. At some later time, information coded within the viral genome may be induced to function. This triggers replication of the virus and entry into the lytic stage. Tumor viruses may initiate lysogenic infections of animal cells in a manner comparable to lysogenic infections of bacteria by bacteriophages.

T4 PHAGE

T4 possesses an oblong head (80×120 nm) and a contractile tail (95×20 nm). Within the capsid is a linear, double-stranded DNA molecule (1.3×10^8 daltons or 1.69×10^5 base pairs). Associated with the DNA are a number of polyamines (putrescine, spermidine, and cadaverine). The base composition of T4 DNA is unusual in that all of the cytosine residues have been replaced by a modified base, hydroxymethylcytosine (HMC). The reason for this replacement will become apparent from subsequent discussion. While virions in general are thought not to possess enzymes, a number of enzymatic activities have been found to be associated with the T4 virion.

The one-step growth experiment described earlier was originally developed with T4. The curve reflects a series of events beginning with the tail fibers of T4 binding to specific *E. coli* cell surface receptors (adsorption). Once the pins of the base plate contact the outer surface of the cell, the tail sheath contracts and the tail core penetrates the cell wall (Fig. 8.3).

Expression of Genes in T4 Phage

Transcription of certain T4 genes begins almost immediately. The immediate-early genes are transcribed at 30 seconds postinfection and the delayed-early or middle genes are transcribed at 2 minutes postinfection (Fig. 8.4). The products of these genes establish the infection. Since transcription of the immediate early genes is directed by σ^{70} promoters, normal host RNA polymerase is used. Among the early gene products are nucleases and other T4 proteins that stop host transcription, unfold the host chromosome, and degrade host DNA to nucleotides. A unique feature of T4 DNA is that its DNA contains the unusual base 5-hydroxymethylcytosine (hm-dCTP) :

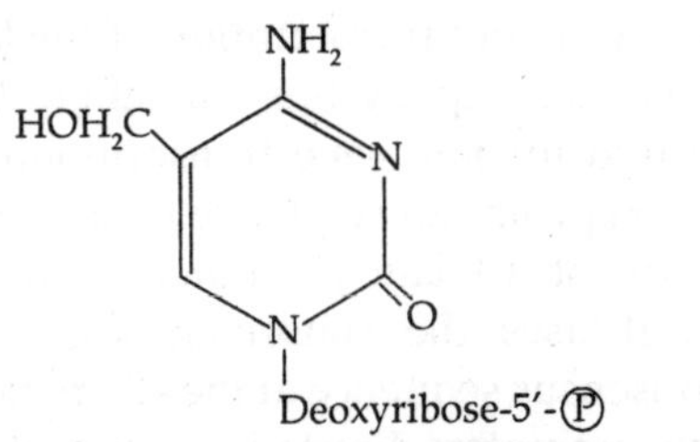

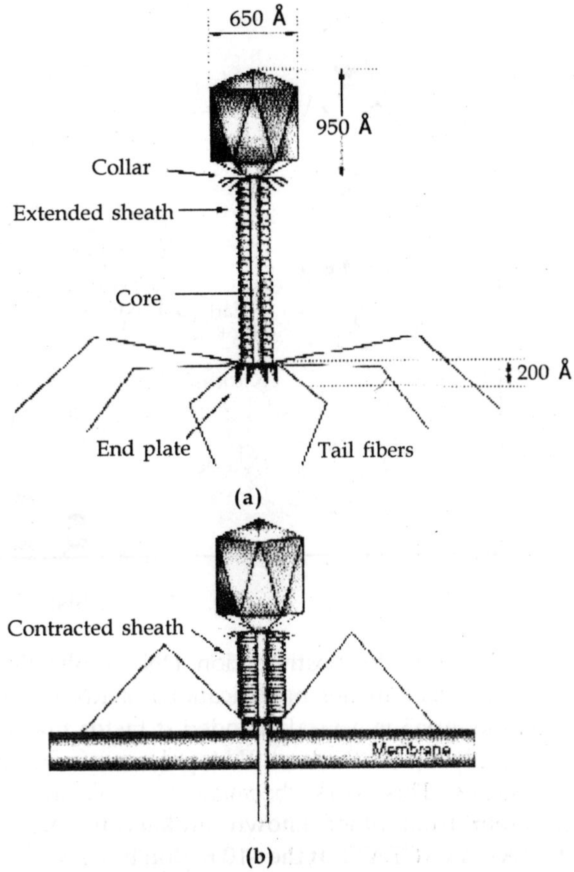

(b)

Fig. 8.3 : Structure of T4 phage.

This base is synthesized as hm-dCMP from dCTP by a T4 enzyme prior to incorporation of the base into T4 DNA. Following incorporation, this nucleotide is glucosylated by a T4-encoded HMC glucosyltransferase. These modifications serve two basic functions : (a) phages with HMC-modified DNA can use cytosine-specific endonucleases to scavenge deoxyribonucleotides from host DNA without damaging their own DNA; and (b) the HMC-modified phages can shuttle back and forth undetected between strains carrying cytosine-specified restriction-modification systems, since the phages do not contain cytosine proper.

Alt and Mod proteins catalyze the ADP ribosylation of Arg265 in the RNA polymerase α subunits. This modification lowers the affinity of core polymerase for the housekeeping σ^{70} protein and is part of a strategy to reduce host RNA synthesis. The Alt protein is actually packaged in the T4 capsid and injected with the DNA during infection. Mod, on the other hand, is synthesized after infection. Transcription of middle genes requires the MotA protein, a positive regulator that binds to a TAT/AGCTT DNA sequence 13 bp upstream of a –10 box, converting the region into an active promoter even though the region lacks a –35 box. T4 also produces a 10 kDa anti-σ-70 protein (AsiA) that prevents transcription of the T4 σ^{70} early gene promoters.

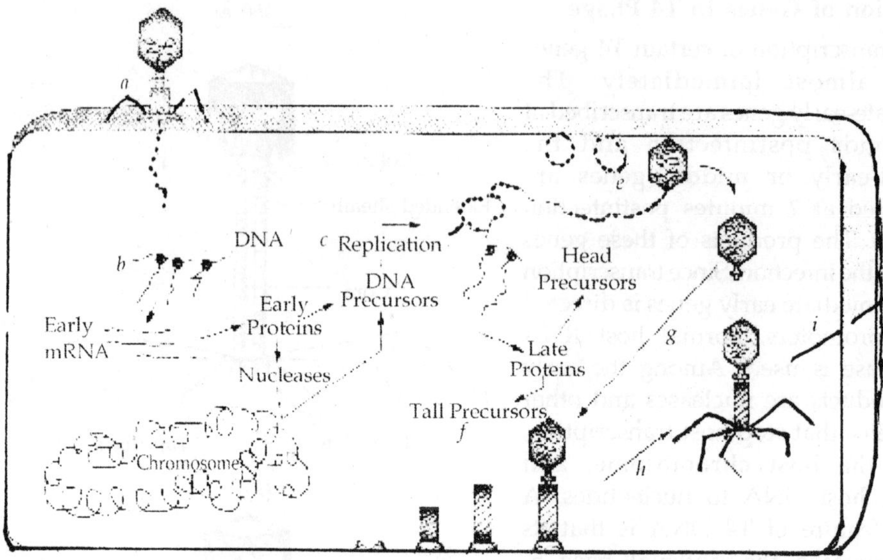

Fig. 8.4 : Life history of T4 phage.

At 5 minutes postinfection, DNA replication has begun and transcription of the late genes begins. Further modifications of RNA polymerase accompany this transition. For example, gp55 is a viral-encoded σ factor that is required for late gene transcription. It essentially hijacks the host RNA polymerase by altering its promoter sequence recognition properties. This works because the molecular structure of T4 late promoters is quite different from other known prokaryotic promoters. It uses the consensus sequence TATAAATACTATT at the –10 region but exhibits no consensus sequence at the –35 region. The alternate σ gp55 and the unusual structure of late gene promoters directs T4 transcription toward the late genes. Another curiosity of T4 transcription is that expression of late genes is closely tied to DNA replication. One of the proteins involved with late gene transcription (gp33) serves as a link between RNA polymerase-σ^{gp55} and components of the T4 replication machinery.

In spite of the complexity of the T4 capsid, the assembly of the various components (e.g., heads, tail fibers, base plate) is remarkably efficient. It is fascinating that final assembly of the various phage component parts occurs spontaneously, without guidance. This feature is referred to as self-assembly and is due to the remarkable affinity each component has for its mate. The entire process results in the production of at least 200 particles per cell by 25 to 30 minutes postinfection (at 37 °C, in rich medium). At the completion of the T4 multiplication cycle the cell will lyse. Lysis is accomplished by a sudden cessation of respiration and requires the phage t gene product plus the phage e product (T4 lysozyme). The t protein is an example of a holin protein that assembles in the cell membrane and allows passage of the e lysozyme through the inner membrane, allowing e lysozyme access to the outer membrane to degrade lipopolysaccharide carbohydrates. Holin proteins in general have to be regulated; otherwise, the cell will prematurely lyse. One piece of the timing puzzle has finally been exposed. It involves lysis inhibition, an intriguing and historically important feature of T4 phage. Over 50 years ago, it was observed that the normal lysis of T4-infected cells could be inhibited by secondary T4 infection occurring at least 3 minutes after the primary infection. Lysis inhibition by superinfecting phages is effective even if the superinfecting DNA does not replicate. T4 mutants in the rI, rII, and rIII genes are not subject to lysis inhibition and exhibit unusual plaque morphology.

Genome of T4 Phage

The resolution of this paradox can be found in the realization that the ends of each T4 DNA molecule are terminally redundant— that is, a duplication of genetic information occurs at both ends of a single molecule. This phenomenon is the result of the headful mechanism utilized for the packaging of T4 DNA into the T4 capsid. After DNA replication, replica molecules can recombine with each other at their terminal redundancies, creating very long molecules (concatemers) many genomes in length. Each mature virion is then made by packaging a "headful" of DNA cut from such a T4 concatemer. One headful, however, is 105% of a single genome. Therefore, the packaging process will generate a population of phages carrying all possible circular permutations of the T4 genome (Fig. 8.5).

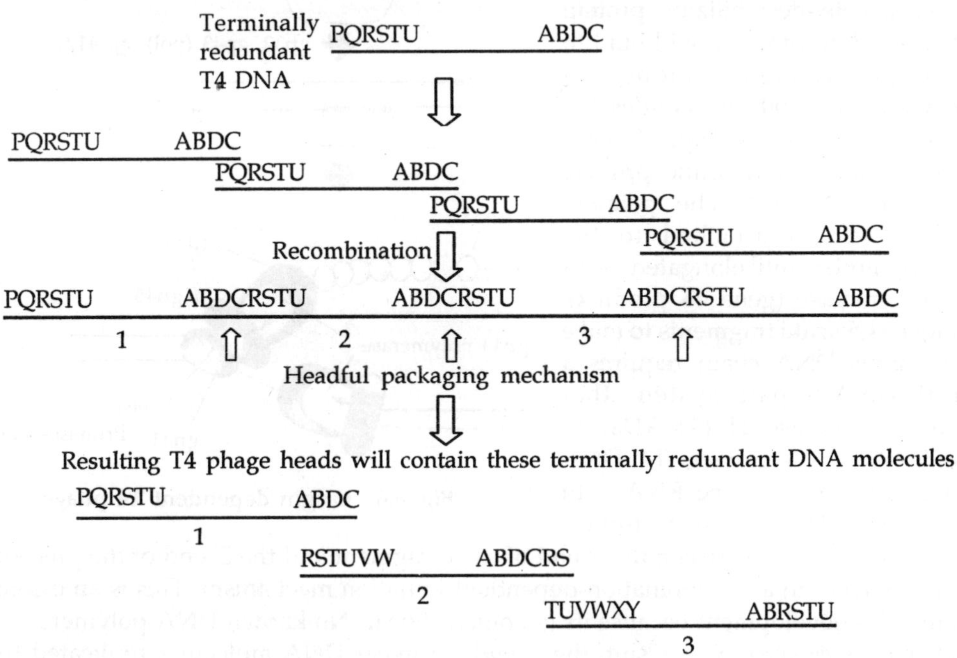

Fig. 8.5 : Redundant T4 DNA.

There is an origin-dependent "early" mode of initiation involving multiple origins spaced around the T4 genome. Transcription in these areas is essential to initiating replication. Transcription at an origin creates a substrate for RNAse H. Following cleavage of the RNA in the RNA:DNA hybrid, the remaining RNA serves as a replication primer. One major initiation site occurs between genes rII and 42, and there are up to four additional sites at other locations. As with host DNA replication, there is a leading or continuously synthesized strand and a lagging strand. Okazaki fragments in T4 are about 2000 nucleotides long (Fig. 8.6). The major protein responsible for nucleotide synthesis is gp43, the T4 DNA polymerase (mw. 110,000). Accessory proteins include g44, g45, and g62, which collectively form a sliding clamp on the DNA much as DnaB protein does on chromosomal DNA. The gene 4l protein is a DNA helicase that will also serve to unwind dsDNA at the replication fork. A complex of gp4l and gp6l is envisioned to move along the lagging strand in the 5' to 3' direction dependent on nucleotide hydrolysis. The gene 32 product, a helix-destabilizing protein (ssDNA-binding protein), will bind to the single-stranded regions. At intervals of about 2000 nucleotides, the gp41/61 complex stops at specific sites to make pentaribonucleotide primers (pppApCpNpNpNp). The primers probably remain attached to the priming complex until elongated by T4 DNA polymerase (gene 43 product). Joining the Okazaki fragments to make a continuous DNA chain requires a special DNA repair system that includes T4 RNase H (44 kDa) to degrade the RNA primer, T4 DNA polymerase to replace the RNA with DNA, and T4 DNA ligase to form a phosphodiester bond between the 3' end of each fragment and the 5' end of the preceding one. T4 switches to a recombination-dependent initiation mechanism. This is an essential feature of T4 development because its genome is linear. No known DNA polymerase can initiate chains *de novo*. As a result, the 3' ends of linear DNA molecules replicated from specific origins will remain unreplicated. The first clue as to how T4 solves this problem was the discovery that its genome is terminally redundant. The redundancy extends 3 to 4 kb (Fig. 8.7). The 3' single-stranded end of another T4 molecule can then invade the homologous region positioned in the middle of the concatemer, forming a new replication fork. Although these four genes (g43, g44, g45, g62) are transcribed by the host RNP, the T4 *mot* product influences promoter selection by RNAP, stimulating transcription of g43 and g45 while inhibiting transcription of g44.

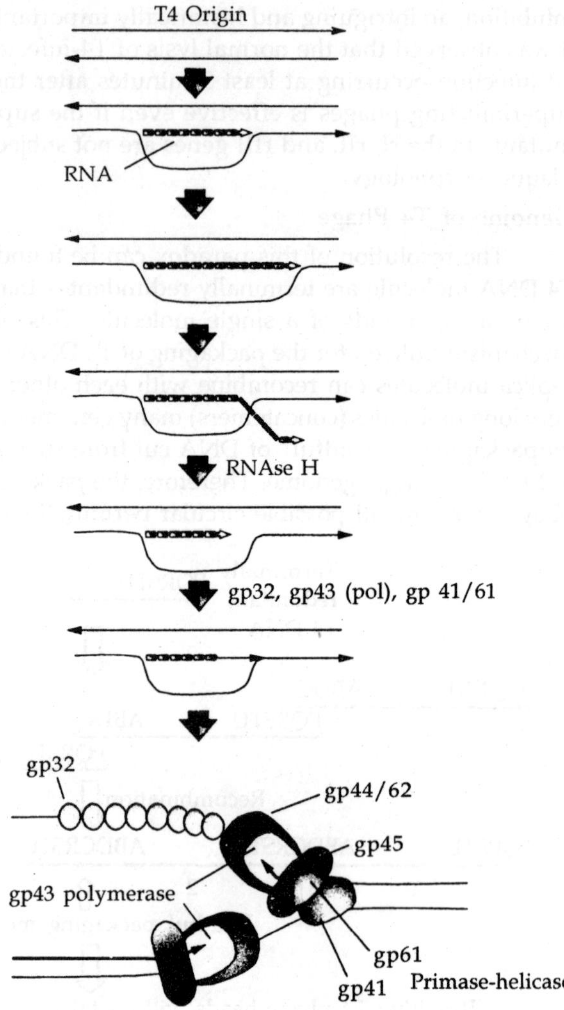

Fig. 8.6 : Origin dependent T4 phage.

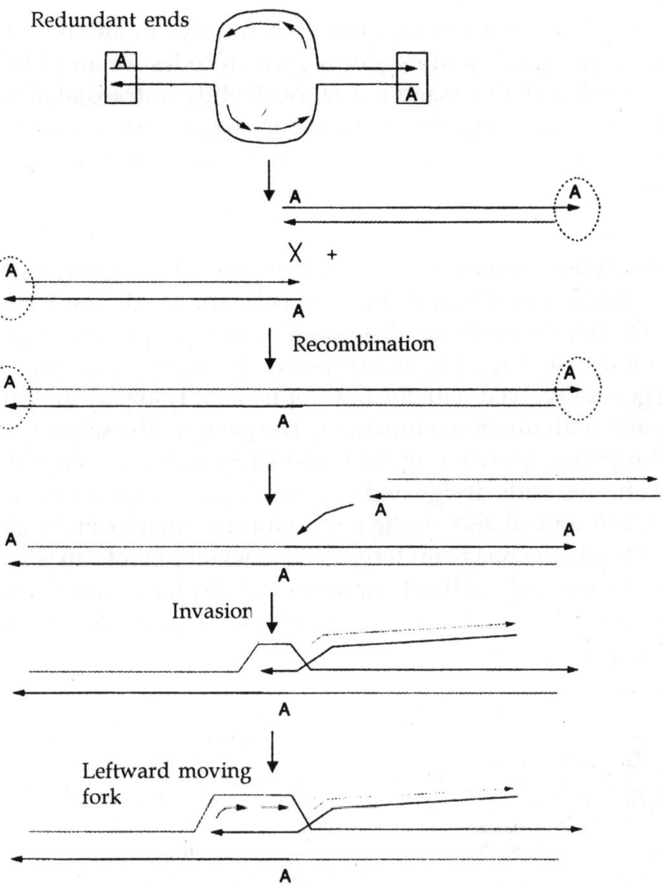

Fig. 8.7 : Initiation of T4 DNA replication depends on recombination.

Regulation of T4 Gene Expression

At the translational level two systems are recognized; autogenous translational repression by the gene 32 product (helix-destabilizing protein) and translational repression of many T4-induced early mRNAs by the phage *regA* product. Each protein repressor was specific to only one species of mRNA. This mechanism appears the same for the T4 gene 32 product. However, translational regulation by RegA (12,000 mw) is unique in that it appears to repress the translation of different mRNA species. The mechanism appears to be the simple recognition of and binding to a consensus mRNA sequence [(AUG)UACAAU-3')] thus blocking ribosome access. Secondary mRNA structures appear to be unnecessary.

The intervening sequences must be excised from the primary transcripts prior to translation. Few examples of introns have been documented in prokaryotes. The first was a 1 kb intron within the thymidylate synthase (*td*) gene of bacteriophage T4. Since then, two more T4 genes, *nrdB* and *sunY*, have been shown to contain introns. The T4 introns are all self-splicing, in which a series of transesterifications, or phosphodiester bond transfers, occurs with the RNA functioning as an enzyme. The introns with *td* and *sunY* are actually mobile introns that can catalyze their transfer to another DNA molecule. A novel form of gene splicing is found in gene 60, which codes for an 18 kDa subunit of the T4 topoisomerase. Within the message is a 50 nucleotide untranslated region. However, this region is not removed as an intron from the message. It seems a pair of 5 nucleotide direct repeats flanking the IVS can base pair, forming a hairpin the brings codons on either side close together.

λ PHAGE

Mutations in certain λ genes (*cI, cII, CIII*) will prevent lysogeny, so the mutant phage will produce clear plaques in which all infected cells are lysed. The λ genome (48.5 kDa) expresses approximately 50 proteins. The genes of this phage can be grouped into four categories based on the their roles in either lysis or lysogeny : (a) Genes involved in lytic development, (b) genes involved with the development of lysogeny, (c) genes participating in both, and (d) genes with unknown functions. The genes in the second and fourth groups are nonessential for phage growth (Fig. 8.8). The DNA within a λ capsid is linear (48,502 bp) but contains cohesive ends designated *cos* sites. Upon phage attachment and injection the *cos* ends of λ DNA anneal and are ligated to form a circular molecule. There are two phases of λ replication as the phage undergoes lytic development : an early phase in which the circular molecules replicate bidirectionally and a late phase, approximately 10 minutes postinfection, consisting of a rolling circle mechanism that generates long concatemeric

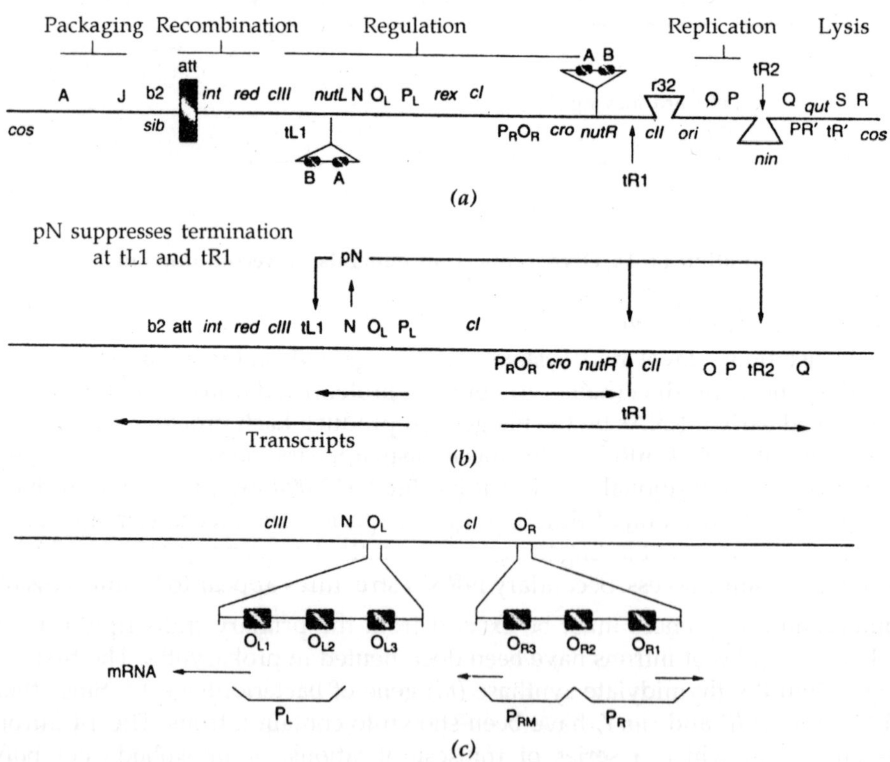

Fig. 8.8 : Genetic organization of λ phage.

molecules containing multiple copies of the λ genome. The concatemers are then acted upon by the λ-packaging system and are used to fill preformed capsids with λ DNA.

Transcription in λ phage

The leftward transcript results in the production of the antiterminator N while the rightward transcript codes for Cro protein. Transcription into genes beyond the terminators requires N protein acting as an antiterminator at t_{L1} and t_{R1}. Six factors produced by *E. coli* are required for effective N utilization : *nusA* (71 min), *nusB* (9 min), *nusC* (*rpoB,* 90 min), *nusD* (*rho,* 84min), *nusE* (*rpsJ,* 74 min), and *nusG* (90 min). The *nus* acronym stands for N-utilization substance. *E. coli nus* mutants fail to support λ growth by blocking the action of pN rather than its synthesis. As noted above, *nusC, nusD,* and *nusE* mutations occurred in the genes for RNA polymerase β subunit, Rho, and ribosomal protein S10 respectively.

The presence of NusA, NusB, and/or NusG on core polymerase enables pN to modify RNA polymerase when a *nut* site is encountered. The modified RNA polymerase can then transcribe past t_{R1} and t_{L1}, resulting in the synthesis of *CII* and *CIII* message, respectively, and past t_{R2}, yielding another antiterminator pQ. The *nut* site in λ mRNA is divided into three parts : BoxA, a site that binds NusB and S10; a spacer region; and stem-loop structure called BoxB that binds N and NusA. NusB binds to the BoxA sequence of nascent message and communicates with RNA polymerase through the S10 collaborator. N-antiterminator protein binds to the *nut* BoxB hairpin loop and interacts with RNA polymerase through NusA, an interaction stabilized by NusB—S10 and NusG. NMR studies have revealed that the N—BoxB interaction involves the aromatic ring of a tryptophan residue in N actually stacking with an unpaired adenine in Box B. Several minutes after λ infection, the synthesis of early (N and Cro) and delayed early (*CII, P, O, Q* and *CIII, red, int*) transcripts diminishes due to the action of Cro repressor on P_L and P_R following its binding to O_L and O_R. However, the Q antiterminator will allow a P'_R transcript to extend through the late genes involved with capsid production and lysis. In contrast to N, which binds mRNA, Q binds to a DNA site called *qut.* RNA polymerase initiating at P'_R pauses at +16/+17, which allows Q to bind and modify RNAP, shortening the pause. The P'_R extended transcript eventually terminates at b (remember the λ genome is circular at this point).

Cro v/s CI Repressor

Although both Cro and CI bind to the same operators (O_L and O_R), these repressor proteins serve distinct physiological functions due to several unique features of the operators. First, within each operator there exist three CI (and Cro)-binding sites. Each binding site (designated O_{L1}, O_{L2}, etc.) is similar in sequence and pcssesses axes of 2-fold hyphenated symmetry. The two repressors have different relative affinities for each of these binding sites. At low concentrations, dimers of CI bind to O_{R1} and O_{R2}, blocking transcription from PR. Actually, CI repressor binds cooperatively to O_{R1} and O_{R2}, binding first at O_{R1}, which increase the affinity of O_{R2} for CI, then at O_{R2}. Binding of CI to O_{R2} will activate P_{rm} (promoter for repressor maintenance), ensuring continued production of CI. In contrast, Cro protein binds more tightly to O_{R3}. The study of both Cro and CI repressors revealed a motif that has proven to be common to many repressor proteins—that is, the helix-turn-helix motif. The DNA - binding domain of CI resides at the amino-terminal ends of this protein. In this region are several alpha helices, two of which participate in DNA binding.

Synthesis of Repressors

Transcription from P_{re} is positively regulated by the CII protein. CII also activates the integrase gene whose product is required for λ DNA to integrate into the host chromosome (prophage state). Thus, CII coordinately regulates the two transcriptional units required for lysogeny. One of the main criteria for whether lysis or lysogeny occurs is the amount of CII activator produced upon infection. This amount is dependent on several factors : (a) The level of *cII* mRNA, (b) the rate of *cII* mRNA translation, and (c) the rate of CII processing. The level of *cII* mRNA is controlled by CI and Cro repressors as well as by pN, which is required for transcription past t_{R1} into *cII*. Translational control of *cII* expressions involves the host HimA protein. Efficient translation of CII mRNA requires HimA, which appears to allow ribosome access to the sequestered Shine-Dalgarno sequence and translation initiation codon in CII mRNA.

The first two amino acids (N-formyl methionine and valine) must be removed in order to form an active CII protein. Also, turnover (or degradation) of CII is very important. The products of the *hflA* and *hflB* loci (94 and 72 centisomes, respectively) are responsible for CII degradation. HflB is actually a membrane-bound protease called FtsH. FtsH is responsible for degrading CII as well as σ32 and the SecY component of the *E. coli* protein secretory system. The *hflA* locus consists of three genes : *hflK, hflC,* and *hflX*. HflK and C are membrane proteins (periplasmic side) that interact with the FtsH protease and regulate its activity. It is the apparent role of the CIII protein to inhibit this protease, thus allowing sufficient accumulation of CII to promote transcription from P_{re} (i.e., CI production). Once again, a high multiplicity of infection will result in a large amount of CIII. This inhibits FtsH (HflB) protease from degrading CII. It is interesting, too, that the overall energy state of the cell affects the process of lysogenization. Under conditions where cAMP levels are low, lysogeny is the preferred developmental route, while the reverse is true where cAMP levels are high. It was discovered that cAMP and CRP appear to inhibit the transcription or activity of HflA. Thus, high cAMP levels lead to lowered HflA protease activity.

Integration and Excision Mechanism

Integration involves site-specific recombination between the phage *att* and bacterial *att* attachment sites. The integration requires the *int* gene product plus integration host factor. Prophage attachment normally occurs at the *att* site located between the *gal* and *bio* operons at 17 minutes. Note the permutation of gene order that occurs during the integration—that is, A and R are distal loci in the linear λ DNA present at infection but end up as proximal loci in the prophage. The *int* and *xis* loci are tightly linked, and their coding regions partially overlap. The *int* locus is controlled by the P_{Int} promoter while transcription of *xis* usually results from P_L.

Retroregulation (Negative) of *int* by *sib*

The *sib* locus within the *b* region is responsible for this. Upon exposure to UV light, activation of the host *recA* coprotease results in the autocleavage of CI protein. Transcription of λ is allowed to occur from P_L and will proceed through *xis* and *int*. Both Int and Xis are needed for prophage excision. At this point, Int is no longer required and could prove detrimental if it causes reintegration of λ in a cell ravaged by UV irradiation. Regulation by *sib* is posttranscriptional and occurs only in *cis*. The basis for *sib* retroregulation (retro because *sib* is located downstream from *int*) is the formation of a stem-loop structure near the 3' end of *int* message (Fig. 8.9). This stem-loop structure is a substrate for RNase III; thus, RNase III cleaves at the stem-loop region and the 3' end is degraded by a 3' exonuclease. Retroregulation, however, only regulates the P_L transcript but not the transcript resulting from P_I. The reason is that transcription from P_L is regulated by N protein, which, upon binding RNA polymerase, renders it resistant to most termination signals. Thus, the P_L transcript will proceed through the *sib* region, bypassing a termination signal. The CII-activated P_I transcript terminates within *sib*.

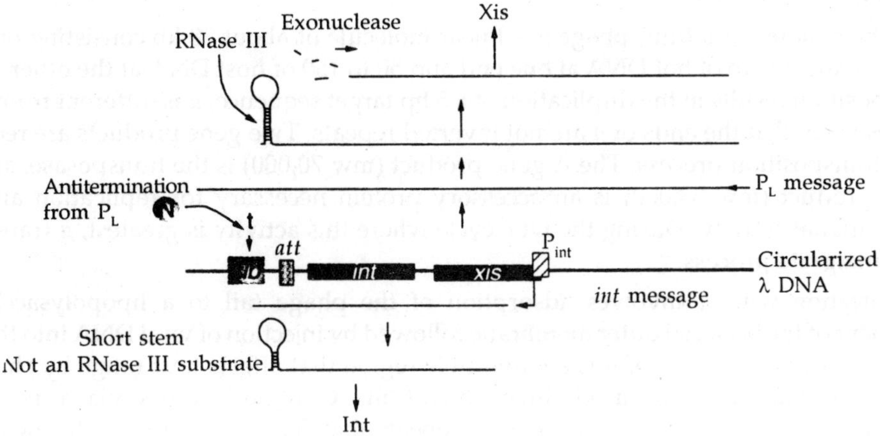

Fig. 8.9 : Retroregulation by the *sib* region.

Replication of λ Phage

Evidence suggests that the amino-terminal end of the O protein (as four dimers) interacts with *ori* at four iteron sequences as the first step in a chain of interactions that control replication. The result is a nucleosomelike structure called the O-some. The λ P protein's job is to compete with DnaC in the host replication machinery for the DnaB helicase. The λ P-DnaB complex then interacts with the O-some, forming a preprimosome complex. The recruitment of DnaB into the λ-replication complex sequesters DnaB away from host replication machinery, with the consequence of decreasing *E. coli* DNA replication. Activation of DnaB within the λ *ori* complex is different from its activation at the chromosomal origin. Three *E. coli* heat shock proteins, DnaK, DnaJ, and GrpE, release λ P from the complex, triggering DnaB activation.

It is thought that DnaB can only unwind DNA in one direction from a preprimosome complex that includes λ O protein. Transcription directed toward the complex on the opposite side dislodges λ O, which is then degraded by the ClpXP protease. Because of this transcriptional requirement, replication is directly inhibited by CI repressor binding to the right operator. As noted previously, early bidirectional replication of λ only occurs for a brief period after which a rolling circle mode of replication, referred to as σ replication, begins to make the long concatemeric DNA required for packaging. Searches for a gene that would mediate the switch from θ- to σ-replication modes have led to the theory that DnaA protein is involved. Transcription from P_R is required for initiation of bidirectional replication, and DnaA protein activates transcription from P_R. Since λ DNA contains many DnaA-binding sites, as the λ copy number increases, so, too, does the number of DnaA-binding sites, leading to the hypothesis that the titration of DnaA by these sites is part of the switch to σ replication. The *gam* and *red* genes are clearly important for the switch to σ replication. The host *recBC* product, exonuclease V, will degrade the tails produced during σ-mode replication. The λ *gam* product inhibits exonuclease V activity, thus allowing concatemers to form. A gam^- mutant, therefore, will only undergo *θ*-mode replication. However, concatemers are required for packaging to occur.

μ PHAGE

The genome of μ (mu) phage is a linear molecule of about 38 kb consisting of 36 kb of μ DNA and 1.5 kb of hot DNA at one end and 50 to 150 of host DNA at the other. While *μ* transposition results in the duplication of a 5 bp target sequence, *μ* is different from other transposons in that the ends of *μ* are not inverted repeats. Two gene products are required for the transposition process. The A gene product (mw 70,000) is the transposase, and the B gene product (mw 33,000) is an accessory protein necessary for replication and full transpositional activity. During the lytic cycle where this activity is greatest, *μ* transposes by a cointegrate process.

Infection with *μ* involves adsorption of the phage tail to a lipopolysaccharide component of the bacterial outer membrane followed by injection of viral DNA into the host cytoplasm. The G segment of μ is an invertible region that affects host range by changing the type of tail fiber produced. Inversion of the G region occurs via a reciprocal recombination between the short inverted repeats that flank this region. This switching occurs during the prophage state and is controlled by the *gin* product (a recombinase). The switch determines which of two alternative tail fibers are produced in a manner similar to what was described for *S. enterica* flagellar phase variation. μG (+) phages, with Sv and U at the left and of G, will infect *E. coli* K-12 or *S. enterica.* μG (-) phages, with S'v and U' at the left end of G, will infect such strains as *E. coli C, Shigella sonnei, Erwinia cloacae,* or other *Erwinia* species. Following infection, the linear *μ* DNA circularizes, but the circular DNA is proteinase sensitive, suggesting that circularization may be the result of proteins covalently attached to the DNA. Ten minutes after infection, much of the infecting *μ* DNA is associated with the chromosome. Transposition to the chromosome from infecting phage involves a nonreplicative simple insertion as opposed to replicative cointegrate formation (Fig. 8.10).

The original prophage copy is not excised prior to the onset of replication. The newly replicated *μ* copies also remain associated with the chromosome. Each new round of replication involves additional convalent attachment of the ends of μ to a distant target site. In effect, *μ* replicates by transposition to numerous sites on the bacterial chromosome. Both the *A* and *B* gene products are required for replication and transposition. In addition, transposition requires both the c and S ends of *μ*. Two *c* ends or two *S* ends are not suitable substrates for the transposition enzymes. The A gene product (MuA) recognizes the ends of the phage genome, assembles as a tetramer, and catalyzes the two basic transposition reactions. The protein introduces single-strand nicks at each end of *μ* and promotes strand transfer where the 3' ends of the nicked DNA attack a new DNA site. The *B* product (MuB) stimulates transposition by biding to target DNA and to MuA, essentially recruiting target DNA to the transpososome complex. In addition, MuB binding to MuA stimulates the strand transfer activity of MuA and prevents premature disassociation of the MuA tetramer

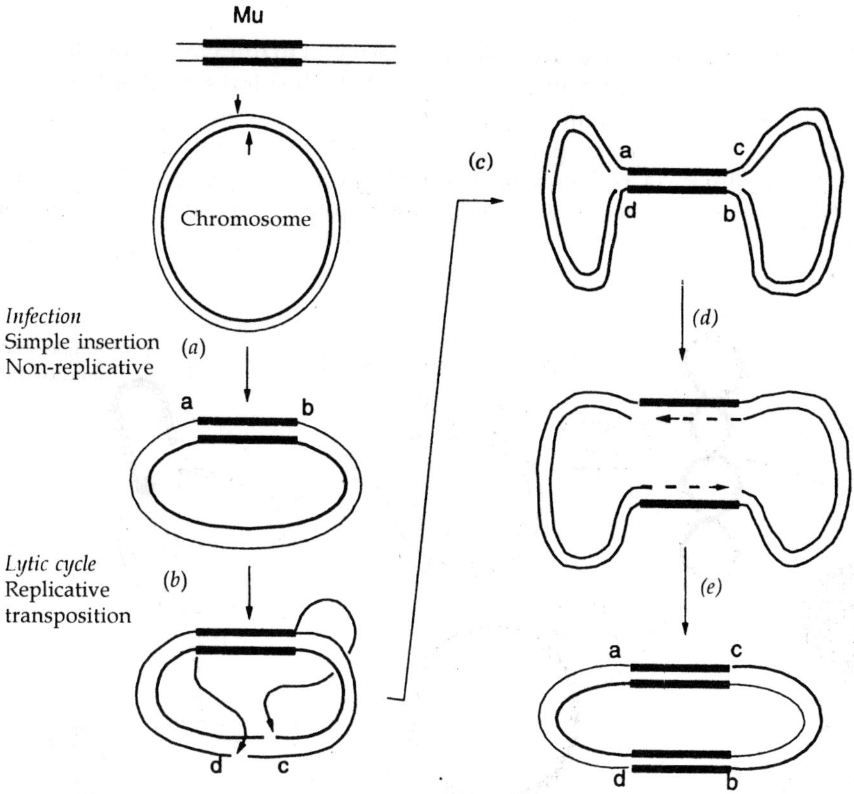

Fig. 8.10 : Replication and integration of μ DNA.

by the ClpX chaperone. Recent evidence indicates that MuB is an ATPase and that ATP hydrolysis acts as a switch, regulating the affinity of MuB for target DNA. Multiple rounds of transposition into the same target molecule are actively discouraged, a phenomenon known as target *immunity*.

Mu repressor binds to an operator region 1 kb from the left end of the phage genome. When bound, μ repressor negatively controls the synthesis of the transposase and other functions involved in transposition, thus maintaining μ in a lysogenic state. Accumulation of a second repressor called Ner can initiate the lytic cycle by binding to another operator that controls production of μ repressor. The μ repressor is also selectively degraded by the ClpXP protease. This will trigger lytic growth by allowing synthesis of the μA and μB proteins. ClpX, therefore, has a dual role in controlling transposition. It limits transposition by disassembling μA and can stimulate transposition by targeting μ repressor for degradation. At the end of the lytic cycle, the various μ-DNA copies scattered around the chromosome are cut and packaged into phage particles. Packaging of μ chromosomes into capsids occurs by a headful mechanism starting from a *pac* site located at the C end. One cut is made 100 nucleotides outside this end, with the second cut made at a fixed length past the *S* end.

ϕX174 PHAGE

ϕX174 has been used extensively in studies designed to unravel the mysteries of how the *E. coli* chromosome, 1000 times its size, is replicated. The viral strand of ϕX174 contains 5386 bases and is packaged within an icosahedral capsid with a knob at each vertex (Fig. 8.11). The viral strand is designated (+) because it is the template strand and can be transcribed directly by RNA polymerase. The strand contains gene for 11 gene products. However, the combined molecular weights of these proteins exceed the limit imposed by the available DNA.

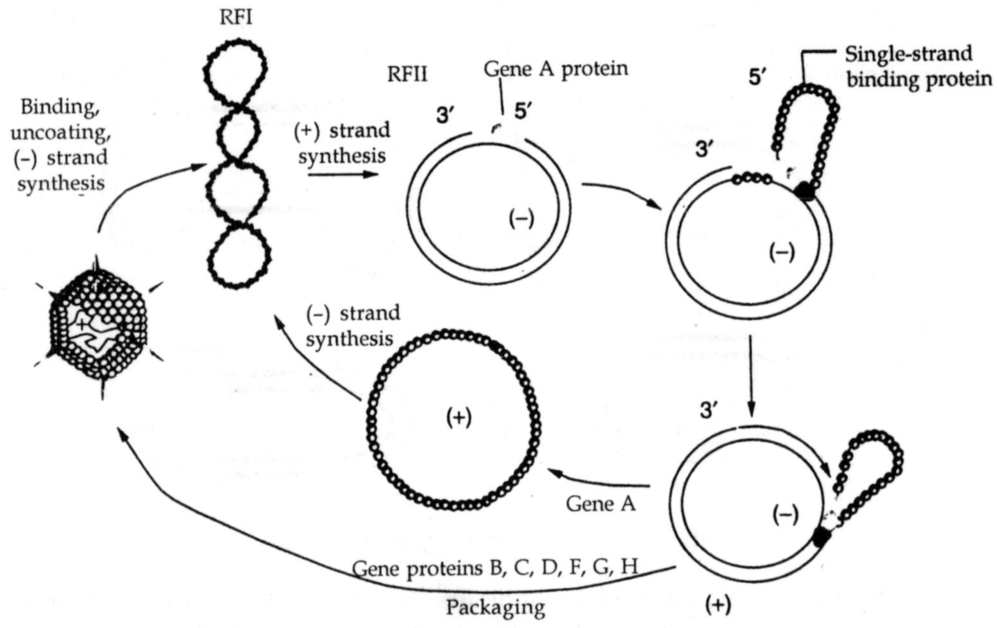

Fig. 8.11 : Life history of ϕX174 phage.

The region originally assigned to gene *A* is now known to also contain all of gene *B*. This phenomenon of overlapping genes is possible because gene *B* is translated in a different reading frame from gene *A*. In a similar fashion, gene *E* is encoded in its entirety within gene *D*. Another translational control machanism further expands the use of gene *A*. The 37 kDa gene *A* protein is formed by reinitiating translation at an internal AUG codon within the gene A message. The same translational phase is used, but the functions of the two proteins are distinct. During infection, the phage adsorbs irreversibly to the lipopolysaccharide in the outer membrane of rough strains of *E. coli* and *S. enterica*. Rough strains lack the O antigen outer chains but retain intact core and lipid A. Attachment is accomplished via the spikes on the capsid. Following attachment, the virus enters the eclipse stage in which the eclipsed virus has lost infectivity for new cells. Uncoating of the virus is coupled to replication of the viral genome, forming a double-stranded replicative form (RFI). The binding of ϕX174 particles and RFI occurs at areas of adhesion between inner and outer membranes and possibly affords direct access to the host chromosomal replicative apparatus fixed at the inner surface. This SS $\Rightarrow$ RFI replication is dependent solely on host replication proteins and includes prepriming, priming, elongation, gap filling, ligation, and supercoiling. The first stage of replication (SS $\Rightarrow$ RF) occurs within 1 minute of infection. The second replication stage involves RF $\Rightarrow$ RF synthesis and occurs up to 20 minutes postinfection. Two additional proteins are required for RF $\Rightarrow$ RF replication. These are the phage-encoded gene *A* protein and *E. coli rep* product. The gene A product nicks supercoiled RFI in the (+) strand between residues 4305 and 4306, forming the open circular form RFII. The gene *A* protein remains attached to the 5' end of the (+) nicked strand by one of two tyrosine residues : Tyr-343 or Tyr-347. By complexing with RepA (a DNA helicase), the gene *A* protein remains indirectly attached to the (−) strand and is dragged around the genome with the helicase. Leading strand replication then occurs from the 3' end of the open (+) strand, regenerating a new duplex, while the old (+) strand is displaced, extended, and coated with single-stranded DNA-binding proteins. After traversing the entire length of the template, the gene *A* protein cleaves the regenerated origin, again attaching to the 5' phosphoryl end by one of the two tyrosines noted above. The now free 3'-OH end of the (+) strand can attack the original gene *A* protein-5'-phosphoryl bond, generating a single-strand circle coated with SSB.

❑❑❑

9

Miscellaneous Bacterial Diseases

SPIROCHETES BACTERIA

Spirochetes have helical form and mode of locomotion, that appear especially striking in live, unstained preparations using the dark-field or phase-contrast microscope. These contain Gram-negative cell wall and a well-developed periplasmic space that encloses the flagella. Although internal flagella are constrained somewhat like limbs in a sleeping bag, their flexing propels the cell by rotation and even crawling motions. The spirochetes are classified in the Order Spirochaetales, which contains two families and five genera. The majority of spirochetes are free-living saprobes or commensals of animals and are not primary pathogens. Treponemes are thin, somewhat regular, coiled cells, are strict parasites with complex growth requirements that necessitate cultivating them in live cells. Their oxygen requirements are minimal, some being strictly anaerobic and others microaerophilic. Diseases caused by *Treponema* are called treponematoses. The subspecies *Treponema pallidum pallidum* is responsible for venereal and congenital syphilis; the subspecies *T.p. endemicum* causes nonvenereal endemic syphilis, or bejel; and *T.p. pertenue* causes yaws. *Treponema carateum* is the cause of pinta. Infection begins in the skin, progresses to other tissues in gradual stages, and is often marked by periods of healing interspersed with relapses.

Syphilis Spirochete

The disease was first recognized at the close of the fifteenth century in Europe, a period coinciding with the return of Columbus from the West Indies, which led some medical scholars to conclude that syphilis was introduced to Europe from the New World. However, a more probable explanation contends that the spirochete evolved from a related subspecies, perhaps an endemic treponeme already present in the Mediterranean basin. The combination of the immunologically naive population of Europe, the European wars, and sexual promiscuity set the stage for world-wide transmission of syphilis that continues to this day. The term syphilis first appeared in a poem entitled "Syphilissive Morbus Gallicus" by Fracastorius (1530), about a mythical shepherd whose name eventually became synonymous with the disease from which he suffered. Another early Latin name for it was *lues venerea*—literally, the plague of love. Attempting to distance themselves from the disease, various peoples at various time have called syphilis the "Italian disease" or the "French disease."

It is an extremely fastidious and sensitive bacterium that cannot survive for long outside the host, being rapidly destroyed by heat, drying, disinfectants, soap, high oxygen tension, and pH changes. It survives a few minutes to hours when protected by body secretions and about 36 hours in stored blood. Research with human subjects has demonstrated that the risk of infection from an infected sexual partner is 12% to 30%. Less common modes of transmission are passage to the fetus in utero and laboratory or medical accidents. Syphilis, like other STDs, has experienced periodic increases during times of social disruption. Epidemics tend to be concentrated in larger metropolitan areas among prostitutes, their contacts, and intravenous drug abusers. In some areas, the case rate has doubled. Because many cases go unreported, the actual incidence is likely to be several times higher than these reports show. Syphilis is also on the rise worldwide, especially in Africa and Asia. Persons with syphilis often suffer concurrent infection with other STDs. Coinfection with the AIDS virus is an especially deadly combination with a rapidly fatal course.

The number of cells required for infection using human volunteers was established at 57 organisms. At the binding site, the spirochete multiplies and penetrates the capillaries by dissolving the hyaluronic acid between endothelial cells. Within a short time, it moves into the circulation, and the body is literally transformed into a large receptacle for incubating the pathogen—virtually any tissue is a potential target. Specific factors that account for the virulence of the syphilis spirochete appear to be outer membrane proteins. It produces no toxins and does not appear to kill cells directly. Studies have shown that although phagocytes seem to act against it and several types of antitreponemal antibodies are formed, cell-mediated immune responses are unable to contain it. The primary lesion occurs when the spirochetes invade the spaces around arteries and stimulate an inflammatory response.

Untreated syphilis is marked by distinct clinical stages designated as primary, secondary, and tertiary syphilis. It also has latent periods of varying duration during which the disease is quescent. The spirochete appears in the lesions and blood during the primary and secondary stages, and thus is communicable at these times. Syphilis can be classified in following types :

Primary syphilis—The earliest indication of syphilis infection is the appearance of a hard chancre (L. *Cancer*) at the site of inoculation, after an incubation period that varies from nine days to three months. The chancre begins as a small, red, hard bump that enlarges and breaks down, leaving a shallow crater with firm margins. The base of the ulcer beneath the encrusted surface swarms with spirochetes. Most chancres appear on the internal and external genitalia, but about 20% occur on the lips, oral cavity, nipples, fingers, or around the rectum. Because genital lesions tend to be painless they may escape notice in some cases. Lymph nodes draining the affected region become enlarged and firm, but systemic symptoms like fever or headache are virtually absent.

Secondary syphilis—About three weeks to six months (average is six weeks) after the chancre heals, the secondary stage appears. By then, many systems of the body have been invaded, and the symptoms are more profuse and intense. Initially there is fever, headache, and sore throat, followed by lymphadenopathy and a peculiar red or brown rash that breaks out on all skin surfaces, including the palms and the soles. Like the chancre, the lesions contain viable spirochetes and disappear spontaneously. The major complications develop in the bones, hair follicles, joints, liver, eyes, brain, and kidneys.

Latency and tertiary syphilis—Latency is divisible into early and late phases, and though antitreponeme antibodies are readily detected, the parasite itself is not. The final stage of disease, late or tertiary syphilis, is quite rare today because of widespread use of antibiotics to treat other infections. By the time a patient reaches this phase, the combined action of the latent infection and the body's response to it produce severe pathological complications. Cardiovascular syphilis results from damage to the small arteries in the aortic wall. As the fibers in the wall weaken, the aorta is subject to distension and fatal rupture. The same pathologic process can damage the aortic valves, resulting in insufficiency and heart failure. In one form of tertiary syphilis, painful swollen syphilitic tumors called gummas (L. *Gummi,* gum) develop in tissues such as the liver, skin, bone, and cartilage. Gummas are usually benign and only occasionally lead to death, but they can impair function. Neurosyphilis can involve any part of the nervous system, but it shows particular affinity for the blood vessels in the brain, cranial nerves, and dorsal roots of the spinal cord. The diverse reactions include severe headaches, convulsions, mental derangement, the Argyll Robertson pupil, atrophy of the optic nerve and blindness.

Congenital syphilis—*Treponema pallidum* can pass from a pregnant woman's circulation into the placenta and can be carried throughout the fetal tissues. An infection leading to congenital syphilis can occur in any of the three trimesters, though it is most common in the second and third. The pathogen inhibits fetal growth and disrupts critical periods of development with varied consequences, ranging from mild to the extremes of spontaneous miscarriage or stillbirth. Early congenital syphilis encompasses the period from birth to two years of age and is usually first detected three to eight weeks after birth. Infants often demonstrate such symptoms as nasal discharge, skin eruptions and loss, bone deformation, and nervous system abnormalities. The late form gives rise to an unusual assortment of stigmata in the bones, eyes, inner ear, and joints, and causes the formation of Hutchinson's teeth.

Clinical and laboratory diagnosis—Not only do the stages mimic other diseases, but their appearance can be so separated in time as to seem unrelated. The chancre and secondary lesions must be differentiated from bacterial, fungal, and parasitic infections, tumors, and even allergic reactions. Overlapping symptoms of concurrent, sexually transmitted infections such as gonorrhea or chlamydiosis can further complicate diagnosis. The clinician must weigh presenting symptoms, patient history, and microscopic and serologic tests in rendering a definitive diagnosis. For specific diagnosis of primary, early congenital, and to a lesser extent, secondary syphilis, no laboratory test is more valuable than dark-field microscopy of a suspected lesion. The lesions are gently squeezed or scraped to extract clear serous fluid. A wet mount prepared from the exudate is then observed for the characteristic size, shape, and motility of *T. pallidum.* A single negative test is insufficient to exclude syphilis, since the patient may have removed the organism by washing, sò two follow-up tests without prior washing are recommended.

Syphilis testing—If dark-field or direct antigen tests are negative, serologic tests provide valuable though indirect diagnostic support. These tests are based upon detection of antibody formed in response to *T. pallidum* infection. Several of tests are variations on the original test developed by Wasserman using cardiolipin, a natural constituent of many cells, as the antigen. Although anticardiolipin antibodies are not specific for syphilis, the test is an effective way to screen the population for people who may be infected. Blood tests given prior to marriage have been used as a screening test for syphilis. Blood tests are also suggested for high risk groups such as homosexuals, male and female prostitutes, people with other STDs, and pregnant women.

In the case of a positive results, it is important that a series of serologic tests be carried out to detect an elevated antibody titer indicative of active infection, since a single positive test may be due to a prior cured infection. Because the most common screening tests (RPR and VDRL) are based on reactions to a substance found normally in human tissue, biological false positives can occur, especially in patients with autoimmune diseases or impaired immunity. Typical of these specific tests is the *T. pallidum* micro-hemagglutination assay (MHA-TP), which employs red blood cells that have been coated with treponemal antigen. Agglutination of the cells by serum indicates antitreponemal antibodies and infection. Another standard test is an indirect immunofluorescent method called the FTA-ABS test. The test serum is first absorbed with treponemal cells and reacted with antihuman globulin antibody labelled with fluorescent dyes. If antibodies to the treponeme are present, the fluorescence on the outside of these cells is highly visible with a fluorescent microscope. The most sensitive and specific of all is the *T. pallidum* immobilization (TPI) test, in which live syphilis spirochetes are mixed with the test serum and observed microscopically for loss of motility.

Syphilis control—Penicillin G retains its status as a wonder drug in the treatment of all stages and forms of syphilis. It is given parenterally in large doses with benzathine or procaine to maintain a blood level lethal to the treponeme for at least seven days. Alternate drugs (tetracycline and erythromycin) are less effective and are indicated only if penicillin allergy has been documented. The core of an effective prevention program depends upon detection and treatment of the sexual contacts of syphilitic patients. Public health departments and physicians are charged with the task of questioning the patients and tracing their contacts. All individuals identified as being at risk, even if they show no signs of infection, are given immediate prophylactic penicillin in a single, long-acting dose. The barrier effect of a condom provides superior protection, but washing after sexual intercourse alone is not an adequate protective measure. Protective immunity apparently does arise in humans and in experimentally infected rabbits, which raises the prospect of an effective immunization program in the future.

Bejel—Bejel is also known as endemic syphilis and nonvenereal childhood syphilis. The pathogen, a subspecies of *T. pallidum,* is harbored by a small reservoir of nomadic and seminomadic people in arid areas of the Middle East and North Africa. It is a chronic, inflammatory childhood disease transmitted by direct contact or shared household utensils and other fomites, and is facilitated by minor abrasions or cracks in the skin or a mucous membrane.

Yaws—It is endemic to warm, humid tropical regions of Africa, Asia, and South America. The microbe is readily spread by direct contact with skin lesions or fomites. Crowded living conditions and poor community or personal hygiene are contributing factors. The earliest sign is a large, abscessed papule called the "mother yaw," usually on the legs or lower trunk. After the initial lesion has healed, a secondary crop of moist nodular tumors appears, which erode the skin, periosteum, and bones, but do not penetrate to the viscera. Late stage yaws can result in mutilating ulcerations of the face and extremities. Yaws can be prevented by correcting predisposing conditions, shielding minor skin injuries from mechanical insect vectors, mass treatment, and surveillance for new cases.

Pinta—This is a chronic skin infection caused by *T. carateum.* Transmission evidently requires several years of close personal contact accompanied by poor hygiene and inadequate health facilities. Even though pinta is not currently widespread, the disease is still found in isolated populations inhabiting the tropical forest and valley regions of Mexico and Central and South America. Infection begins in the skin with a dry, scaly papule reminiscent of psoriasis of leprosy. In time, pigmented secondary macules and blanched tertiary lesions appear.

LEPTOSPIROSIS

There are only two species in the genus : *Leptospira interrogans* (Gr. *Leptos,* slender, and *spiera,* a coil) which causes leptospirosis in humans and animals, and *L. biflexa,* a harmless, free-living saprobe. The two species are serologically, genetically, and physiologically distinct. *L. interrogans* demonstrates nearly 200 serotypes distributed among various animal groups, which accounts for the extreme variations in leptospirosis among humans. Leptospirosis is a zoonosis associated with wild animals such as rodents, skunks, raccoons, foxes, and some domesticated animals, particularly horses, dogs, cattle, and pigs. Although these reservoirs are distributed throughout the world, the disease is concentrated mainly in the tropics. Leptospires shed in the urine of an infected animal can survive for several months in natural or alkaline soil or water. Infection occurs almost entirely through contact of skin abrasions or mucous membranes with animal urine or some environmental source containing urine. Transmission does not seem to occur through ingestion, animal bites, inhalation, or from human to human.

Pathogenesis—Leptospirosis proceeds in two phases, and its principal targets are the kidneys, liver, brain, and eyes. During the early or leptospiremic phase, the pathogen appears in the blood and cerebrospinal fluid. Symptoms are sudden high fever, chills, headache, muscle aches, conjunctivitis, and vomiting. During the second, or immune phase, the blood infection disappears due to the action of phagocytes, complement, and IgM antibodies. This period is marked by milder fever, headache due to leptospiral meningitis, and *Weil's syndrome,* a cluster of symptoms characterized by kidney invasion, hepatic disease, jaundice, anemia, and neurologic disturbances.

Diagnosis and control—Isolation is accomplished by inoculating a specimen into special media or laboratory animals. Because leptospiral infection stimulates a strong humoral response, it is possible to test the patient's serum for its antibody titer. A fast, specific, and effective test called the macroscopic slide agglutination test is most often employed for routine screening. Live or formalinized *L. interrogans* is mixed with the patient's serum and observed for agglutination or lysis with a darkfield microscope. Treatment with penicillin or tetracycline by the fourth day of illness rapidly reduces symptoms and shortens the course of disease, but after this time, therapy is less effective. Strain-specific vaccines made from killed cells are available for humans, dogs, and cattle, but with so many possible serotypes, these can confer protection only in areas where exposure to a specific endemic strain is anticipated. Vaccination is aimed at those with greatest risk such as combat troops training in jungle regions and animal care and livestock workers.

BORRELIA

Relapsing Fever

Borrelia are comparatively larger, ranging from 0.2 to 0.5 µm in width and from 10 to 20 µm in length, and they contain 4 to 30 irregularly spaced and loose coils with an abundance (30 to 40) of periplasmic flagella. The nutritional requirements of *Borrelia* are so complex that the bacterium can be grown in artificial media only with difficulty. Human infections with *Borrelia,* termed borrelioses, are all transmitted by some type of arthropod vector, usually ticks or lice. The two most important human diseases are relapsing fever and Lyme disease.

Epidemiology—*Barrelia hermsii,* the cause of tick-borne relapsing fever, is carried by soft ticks of the genus *Ornithodoros.* The mammalian reservoirs of this zoonosis are squirrels, chipmunks, and other wild rodents, and the human is generally an accidental host. The spirochetes mature and persist in the salivary glands and intestines of the tick, and both the bite itself and the subsequent scratching initiate infection. Tick-borne relapsing fever occurs sporadically in the United States, usually in campers, backpackers, and forestry personnel who frequent the higher elevations of western states. Whenever famine, war, or natural disasters are coupled with poor hygiene, crowding, and inadequate medical attention, epidemics of louse-borne relapsing fever occur. Such conditions favour the survival and spread of the louse vector *Pediculus humanus,* which harbors the spirochete *B. recurrentis* in its body cavity.

Pathogenesis—The pathological manifestations are similar in both tick- and louse-borne relapsing fever. Even though the blood level can reach an imposing 500,000 borrelias per milliliter, there is little sign of disease during the 2- to 15-day incubation period. The incubation period ends abruptly with onset of high fever, shaking chills, headache, and fatigue. Later features of the disease include nausea, vomiting, muscle aches, and abdominal pain. Extensive damage to the liver, spleen, heart, kidneys, and cranial nerves occurs in many cases. Half of the patients hemorrhage profusely into organs, and some develop a rash on the shoulders, trunk, and legs. As the name *relapsing fever* indicates, the fever follows a fluctuating course that is explained by changes in the parasite and the attempts of the immune system to control it. *Borrelia* have adopted a remarkable strategy for evading the immune system and avoiding destruction. They change surface antigens during growth, so that in time, the initial antibodies become useless. These antigenically altered cells survive, multiply, and cause a second wave of symptoms. Eventually, the immune system forms new antibodies, but it is soon faced with yet another antigenic form. A single strain has been known to generate 24 distinct serological types.

Diagnosis and control—A patient's history of exposure, clinical symptoms, and the presence of *Borrelia* in blood smears are very definitive evidence of borreliosis. Except for pregnant women and children under seven years old, tetracycline is the treatment of choice. Chloramphenicol, erythromycin, and doxycycline are also effective antimicrobial agents. Because vaccines are not available, prevention of relapsing fever is dependent upon controlling rodents and avoiding tick bites.

Lyme Disease

Lyme disease is a recently described syndrome that has become the most important borreliosis in the United States. Its spirochetal agent, *Borrelia burgdorferi*, is transmitted primarily by hard ticks of the genus *Ixodes*. In the northeast part of the United States, *Ixodes scapularis* ticks pass through a complex two-year cycle that involves two principal hosts. As a larva or nymph, it feeds on the white-footed mouse, where it picks up the infectious agent. The nymph is relatively nonspecific and will try to feed on nearly any type of vertebrate, thus it is the form most likely to bite humans. The adult tick reproductive phase of the cycle is completed on deer. The greatest concentrations of Lyme disease are in areas having high mouse and deer populations. Most of the cases recorded since 1978 have occurred in New York, Pennsylvania, Connecticut, New Jersey, Rhode Island, and Maryland, though the number in the Midwest and West is growing. Highest risk groups include hikers, backpackers, and people living in newly developed communities near woodlands and forests.

In the 1970s, an enigmatic cluster of arthritis cases appeared in the town and surrounding suburbs of Old Lyme, Connecticut. This phenomenon caught the attention of nonprofessionals and professionals alike, whose persistence and detective work ultimately disclosed the unusual nature and epidemiology of Lyme disease. The story of its unraveling began in the home of Polly Murray, who along with her family, was beset for years by recurrent bouts of stiff neck, swollen joints, malaise, and fatigue that seemed vaguely to follow a rash from tick bites. When Mrs. Murray's son was diagnosed as having juvenile rheumatoid arthritis, she became skeptical. Conducting her own literature research, she began to discover inconsistencies. After similar observations were reported by Judith Mensch, another resident of the same region, the findings came to the attention of Dr. Allen Steere, a rheumatologist with a CDC background. He was able to forge the vital link between Mrs. Murray's case histories, the disease symptoms, and the presence of unique spirochetes in ticks preserved by some of the patients. In the years since Lyme disease was formally characterized, retrospective studies showed that this is not really a new disease. The unique bull's-eye rash was reported in Europe at the turn of the century. Recent PCR analysis of tick museum specimens from 50 years ago documents the presence of *B. burgdorferi*.

An early symptom in 70% of cases is a rash at the site of a larval tick bite. The lesion, called *erythema migrans*, looks something like a bull's-eye, with a raised erythematous ring that gradually spreads outward and a pale central region. Other early symptoms are fever, headache, stiff neck, and dizziness. If not treated or if treated too late, the disease can advance to the second stage, during which cardiac dysrhythmias and neurological symptoms such as facial palsy develop. After several weeks or months, a crippling polyarthritis can attack joints, especially in the European strain of the agent. Diagnosis of Lyme disease is aided by the presence of ringshaped lesions, isolation of spirochetes from the patient, and serological testing with an ELISA method, though false-positive reactions are common. Early treatment with tetracycline and amoxicillin has been effective, and other antibiotics are being tested. Since dogs can also acquire the disease, a vaccine has been marketed to protect them, and a human vaccine for high-risk populations is currently being developed. Avoiding endemic areas in tick season, wearing protective clothing, using insect sprays, and fastidiously removing ticks can prevent Lyme disease and other tick-borne infections as well.

VIBRIO AND CAMPYLOBACTER

Cholera

Vibrios are comma-shaped, curved rods with a single polar flagellum. *Campylobacters* are short spirals that have one or more flagella and a corkscrew type of motility. Spirilla are inflexible coiled rods that move by multiple polar flagella. *Vibrio* is a member of the Family Vibrionaceae, and *Campylobacter* and *Spirillum* are in the Family Spirillaceae. *Vibrio* (L. *Vibrare,* to shake) shares many cultural and physiological characteristics with members of the Enterobacteriaceae, a closely related family. They are fermentative, grow on ordinary or selective media containing bile at 37 °C, and are oxidase positive. They possess unique O (somatic) antigens, H (flagella) antigens, and membrane receptor antigens that provide some basis for classifying members of the family.

Epidemiology—Epidemic cholera, or Asiatic cholera, has been a devastating disease for centuries. Although the human intestinal tract was once thought to be the primary reservoir, it is now known that the parasite is free-living in certain endemic regions. In nonendemic areas such as North America, the microbe is spread by water and food contaminated by asymptomatic carriers. Cholera is relatively uncommon in the United States. However, sporadic outbreaks have occurred along the Gulf of Mexico and the cholera vibrio is sometimes isolated from shellfish in this region. Cholera is a strictly human disease, and there are no demonstrable animal reservoirs. The microbe's natural reservoir appears to be the soil and water of two large Indian Rivers, the Ganges and the Brahmaputra. In these endemic areas, transmission of cholera is associated with pilgrimages to the river and religious practices involving mass bathing. Periodically during the past two centuries, this vibrio has migrated by means of waterways and human carriers to areas of Asia, Africa, Europe, the Middle East, and South America, and in the process, it has caused at least seven pandemics. The most recent epidemics started in 1991 in Peru and spread to several other South American countries. In a few months, several hundred thousand people were affected, and thousands died. Epidemiologists traced many cases to ingestion of contaminated raw seafood. The pattern of cholera transmission and the onset of epidemics and greatly influenced by the season of the year and the climate.

Pathogenesis—This hostile environment influences the size of the infectious dose (10^8 cells), though certain types of food also shelter the pathogen more readily than others. At the mucosa of the duodenum and jejunum, the vibrios penetrate the mucous barrier using their flagella, adhere to the outside of the epithelium, and multiply there. The cells are strictly epipathogens that do not enter the cells or invade the mucosa. The virulence of *V. cholerae* (Gr. *Chole,* bile) is due entirely to an enterotoxin called *cholera toxin* (CT) that disrupts the normal physiology of intestinal cells. When this toxin binds to specific intestinal receptors, a secondary signalling system is activated. Under the influence of this system, the cells shed large amounts of electrolytes into the intestine, an event that is accompanied by profuse water loss. Most cases of cholera are mild or self-limited, but in children and weakened individuals, the disease can explode with tremendous force.

Following an incubation period of a few hours to a few days, symptoms begin abruptly with vomiting, followed by copious watery feces called secretory diarrhea. This voided fluid is odorless and contains flecks of mucus, hence the description "rice-water stool." Fluid losses of nearly one liter per hour have been reported in severe cases, and an untreated patient can lose up to 50% of body weight during the course of the disease. The diarrhea causes loss of blood volume, acidosis from bicarbonate loss, and potassium depletion that predispose the patient to muscle cramps, severe thirst, flaccid skin, and sunken eyes, and in young children, coma, convulsions, and fever. Secondary circulatory consequences can include hypotension, tachycardia, cyanosis, and collapse from shock within 18 to 24 hours. If cholera is left untreated, death can occur in less than 48 hours and the mortality rate approaches 55%.

Diagnosis and control—During epidemics, clinical evidence is usually sufficient to diagnose cholera. But confirmation of the disease is often required for epidemiologic studies and detection of sporadic cases. *V. cholerae* can be readily isolated and identified in the laboratory from stool samples. Direct dark-field microscopic observation reveals characteristic curved cells with brisk, darting motility as confirmatory evidence. Immobilization or fluorescent staining of feces with group-specific antisera is supportive as well. The key to cholera therapy is prompt replacement of water and electrolytes, since their loss accounts for the severe morbidity and mortality. This can be accomplished by various rehydration techniques that replace the lost fluid and electrolytes. Cases in which patient is unconscious or has complications from severe dehydration require intravenous replenishment as well. Only after the patient has been treated for physiologic disruption can antibiotics to control the growth of the microbe come into play.

Oral antibiotics such as tetracycline and drugs such as trimethoprim-sulfa can terminate the diarrhea in 48 hours, and they also promote recovery and diminish the period of vibrio excretion. However, prompt and aggressive detection and treatment alone are insufficient to control cholera. Effective prevention is contingent upon proper sewage disposal and water purification. Detecting and treating carriers with mild or asymptomatic cholera is a serious goal, but one that is frequently difficult to attain because of inadequate medical provisions in those countries where cholera is endemic. To date, results of vaccination have been lackluster.

Gastroenteritis

Vibrio parahaemolyticus is a salt tolerant inhabitant of coastal waters where it associates with zooplankton and colonizes the chitinous exoskeleton of shellfish such as shrimp and crabs. In temperate zones, vibrios survive over the winter by settling into the ocean sediment, and when resuspended by upwelling during the warmer seasons, they become incorporated into the food chain, eventually growing on fish, shellfish, and other edible seafood.

Symptoms—The great majority of cases appear in individuals who have eaten raw, partially cooked, or poorly stored seafood. The vehicles most often implicated are squid, mackerel, sardines, crabs, tuna, shrimp, oysters, and clams. Outbreaks tend to be concentrated along coastal regions during the summer and early fall. The incubation period of nearly 24 hours is followed by explosive, watery diarrhea accompanied by nausea, vomiting, abdominal cramps, and sometimes fever. Symptoms caused by the enterotoxin and hemolysin normally last about 72 hours but can persist for 10 days. Treatment of severe cases can require fluid and electrolyte replacement, and occasionally antibiotics, but because the disease is self-limiting, hospitalization is not usually indicated. Control measures aim to keep the bacterial count in all seafood below the infective dose by continuous refrigeration during transport and storage, sufficient cooking temperatures, and prompt serving.

Intestinal Ulcers

Campylobacters (*Helicobacter*) are slender, curved or spiral bacilli propelled by polar flagella at one or both poles, often appearing in S-shaped or gull-winged pairs. These bacteria tend to be microaerophilic inhabitants of the intestinal tract, genitourinary tract, and oral cavity of humans and animals. Species of *Campylobacter* most significant in medical and veterinary practice are *C. jejuni* and *C. fetus*. A close relative, *Helicobacter pylori*, has recently been implicated in certain types of stomach ulcers.

Pathogenesis—Although the human stomach is usually regarded as a hostile habitat, an unusual vibrio. *Helicobacter pylori*, has found its own special niche there. Not only does it thrive in the acidic environment, but evidence mounting since its discovery in 1983 has also clearly linked it to a variety of gastrointestinal ailments. It is known to cause an inflammatory condition of the stomach lining called gastritis and some types of stomach and duodenal ulcers, and it is an apparent cofactor in the development of a common type of stomach cancer called adenocarcinoma. It occurs in the stomachs of 25% of healthy middle-aged adults and in more than 60% of adults over 60 years of age. *H. pylori* is probably transmitted from person to person by the oral- oral or oral-fecal route. It seems to be acquired early in life and carried asymptomatically until its activities begin to damage the digestive mucosa. Since other animals are also susceptible to *H. pylori* and even develop chronic gastritis, it has been proposed that the disease is a zoonosis transmitted from an animal reservoir.

First, it bores through the outermost mucus that lines the epithelial tissue. Then it attaches to specific binding sites on the cells and entrenches itself. It turns out that one receptor specific for *Helicobacter* is the same receptor as type O blood, which accounts for the higher rate of ulcers in people with this blood type. As the immune system recognizes and attacks the pathogen, infiltrating white blood cells begin to damage the epithelium to some degree, leading to chronic active gastritis. *H. pylori* is isolated primarily from biopsy specimens. Since it gives off large amounts of urease, it can be initially identified by inoculating a rapid urease test or analyzing the amount of ammonia in a patient's breath. Other physiologic tests can be used in confirming its identity. The knowledge of this infection has useful applications in treatment. Gastritis and ulcers have traditionally been treated with drugs (Tagamet, Zantac) that suppress symptoms by slowing the secretion of acid in the stomach and must be taken continuously for indefinite periods.

Campylobacter jejuni is considered one of the most important causes of bacterial gastroenteritis worldwide. Early results of epidemiologic and pathologic studies have shown that this species is a primary pathogen transmitted through contaminated beverages and food, especially water, milk, meat, and chicken. The source of a Vermont outbreak of enteritis was traced to a water storage reservoir that supplied most of the community and eventually caused disease in 20% of the population. When ingested, *C. jejuni* cells reach the mucosa at the last segment of the small intestine (ileum) near its junction with the colon; they adhere, burrow through the mucus, and multiply. Symptoms of headache, fever, abdominal pain, and bloody or watery diarrhea commence after an incubation period of one to seven days. Diagnosis of *C. jejuni* enteritis requires isolation from fecal samples and occasionally blood samples. More rapid presumptive diagnosis can be obtained from direct examination of feces with a dark-field microscope, which accentuates the characteristic curved rods and darting motility. Resolution of infection occurs in most instances with simple, nonspecific rehydration and electrolyte balance therapy. In more severely affected patients, it may be necessary to administer erythromycin, tetracycline, aminoglycosides, or quinolones. Traditionally of interest to the veterinarian, *C. fetus* (subspecies *venerealis)* causes a sexually transmitted disease of sheep, cattle, and goats. Its role as an agent of abortion in these animals has considerable economic impact on the livestock industry. About 40 years ago, the significance of *C. fetus* as a human pathogen was first uncovered, though its exact mode of transmission in humans is yet to be clarified.

Rat Bite Fever

Rat bite fever, is caused by a rigid spiral called *Spirillum minus.* The primary host for the pathogen is the rat, in which it causes septicemia, eye infections, and lung infections. It is transmitted to human hosts almost exclusively by bites from infected wild rodents or their predators. Fleas or other arthropods do not serve as vectors, and direct human-to-human or fomite transmission has not been demonstrated. In humans, the initial wound heals readily, but after an incubation period of two weeks, the site becomes swollen and inflamed, turns purple and ulcerates. The local lymph nodes enlarge and become tender. Fever, malaise, headache, and a rash radiating from the wound site develop but then progressively diminish. Diagnosis is accomplished by dark-field examination of specimens, but culturing is not attempted because *S. minus* cannot be cultivated on artificial media. The infection responds to penicillin, streptomycin, and tetracycline. Another form of rat bite fever caused by *Streptobacillus moniliformis* is more common in the United States, where several cases have occurred in biomedical personnel who handle laboratory rats.

RICKETTSIA

Typhus

Typhus disease caused by *Rickettsia.* It eventually became evident, however, that they are indeed bacteria, possessing a Gram-negative cell wall, binary fission, metabolic pathways for synthesis and growth, and both DNA and RNA. They are among the smallest cells, ranging from 0.3 to 0.6 µm wide and from 0.8 to 2.0 µm long, and are nonmotile pleomorphic rods or coccobacilli. The precise nutritional requirements of the rickettsiae have been difficult to demonstrate because of a close association with host cell metabolism. Their obligate parasitism originates from an inability to metabolize AMP, an important precursor to ADP and ATP, which they must obtain from the host. Rickettsiae are generally sensitive to environmental exposure, although *R. typhi* can survive several years in dried flea droppings, and *Coxiella burnetii* shows extreme resistance to physical and chemical agents because of an endospore-like survival cell.

The life cycle of many rickettsiae depends upon a complex exchange between bloodsucking arthropod hosts and vertebrate hosts, including humans. Humans accidentally enter the zoonotic life cycles through occupational contact with the animals except in the cases of louseborne typhus and trench fever. Because humans are often incidental or "dead-end" hosts, they are not a regular source of infection. Most vectors apparently harbor rickettsiae with no ill effect, but others, like the human body louse, die from typhus infection and do not continuously harbor the pathogen. In certain vectors such as the tick of Rocky Mountain spotted fever, rickettsiae are transferred transovarially, from the infected female to her eggs. This event has profound impact, since the microbe is inherited continuously through multiple generations of ticks, creating a long-standing reservoir. These arthropods feed on the blood or tissue fluids of their mammalian hosts, but not all of them transmit the rickettsial pathogen through direct inoculation with saliva. Ticks and mites directly inoculate the skin lesion as they feed, but fleas and lice harbor the infectious agents in their intestinal tracts. During their stay on the host's body, these latter insects defecate or are smashed, thereby releasing the rickettsias onto the skin or into a wound. Ironically, scratching the bite helps the pathogen invade deeper tissues.

Pathogenesis—The bacteria recognize, enter, and multiply within endothelial cells, causing necrosis of the vascular lining. Among the immediate pathological consequences are vasculitis, perivascular infiltration by inflammatory cells, vascular leakage, and thrombosis. These pathologic effects are manifested by skin rash, edema, hypotension, and gangrene. Intravascular clotting in the brain accounts for the stuporous mental changes and other neurological symptoms that sometimes occur. Isolation of most rickettsiae from clinical specimens requires a suitable live medium and specialized laboratory facilities, including controlled access and safety cabinets. The usual choices for routine growth and maintenance are the yolk sacs of embryonated chicken eggs, chick embryo cell cultures, and to a lesser extent, mice and guinea pigs.

Epidemiology—The extensive investigations of Dr. Howard Ricketts and Stanislas von Prowazek in the early 1900s led to the discovery of the vector and the rickettsial agent, but not without mortal peril, since both men died of the very disease they investigated. Humans are the sole hosts of human body lice and the only reservoirs of *R. prowazekii*. The louse spreads infection by defecating into its bite wound or other breaks in the skin. Infection of the eye or respiratory tract can take place by direct contact or inhalation of dust containing dried louse feces, but this is a rarer mode of transmission. Infected lice die from their infection in one to three weeks without transmitting the rickettsiae to their eggs. The entire metamorphic cycle (egg to adult) can take place upon an infested person's garments or other inanimate objects. Crowding, infrequently changing or sharing clothing greatly favour transmission. The overall incidence of epidemic typhus in the United States is very low, with no epidemics since 1922. Recent sporadic cases involving transmission from the flying squirrel indicate a possible animal reservoir. The first clinical manifestations are sustained high fever, chills, frontal headache, and muscular pain. Within seven days, a generalized rash appears, initially on the trunk and then spreads to the extremities. Personality changes, oliguria (low urine output), hypotension, and gangrene complicate the more severe cases. Untreated disease continues for three weeks, ending in either rapid recovery or death. Recovery usually confers resistance to typhus, but in some cases, the rickettsiae are not completely eradicated by the immune response and enter into latency.

After several years, a milder recurring form of the disease known as *Brill-Zinsser disease* can appear.

Treatment and control—The standard chemotherapy for typhus is tetracycline or chloramphenicol. Despite antibiotic therapy, however, the prognosis can be poor in patients with advanced circulatory or renal complications. Eradication of epidemic typhus is theoretically possible by exterminating the vector. Widespread dusting of human living quarters with insecticides has provided some environmental control, and individual treatment with an anti-louse shampoo or ointment is also effective.

Endemic Typhus

The agent for endemic typhus is *Rickettsia typhi* (*R. mooseri*), which shares many characteristics with *R. prowazekii* except for pronounced virulence. Synonyms for this rickettsiosis are endemic typhus, murine (mouse) typhus, and flea-borne typhus. The disease occurs in certain parts of Central and South America and in the Southeast, Gulf Coast, and Southwest regions of the United States, where *R. typhi* is regularly harbored by mice and rats. Transmission to the human population is chiefly through infected rat fleas that inoculate the skin, though disease is occasionally acquired by inhalation. In the United States, most reported cases arise sporadically among persons working in rat-infested facilities such as shipyards, grain elevators, warehouses, and markets. Many more cases probably occur, but they escape notice because they are scattered or not severe enough to warrant medical attention. The clinical manifestations of endemic typhus included fever, headache, muscle aches, and malaise. After five days, a skin rash, transient in milder cases, begins on the trunk and radiates toward the extremities. Symptoms dissipate in about two weeks. Tetracycline and chloramphenicol are effective therapeutic agents, while various pesticides are available for vector and rodent control.

Scrub Typhus

Scrub typhus, was first described nearly two centuries ago in Japan. The illness, caused by *R. tsutsugamushi*, is endemic to Southeast Asia, Australia, and India. Current incidence of this disease is difficult to trace, but it appears to occur sporadically in endemic areas. Scrub typhus rickettsias live naturally in the bodies of chigger mites. Infected mammalian and avian hosts help perpetuate the reservoir. Humans are attacked by chiggers while passing through forests and parkland. At the site of a bite, a distinctive black scab develops in one to three weeks. Early symptoms of fever, headache, and muscle aches resemble those of endemic typhus, and in about half the cases, a generalized rash originates on the trunk and radiates peripherally. Severe cases are accompanied by mental confusion, delirium, pneumonia, and circulatory collapse. Mortality in untreated disease reaches 50%, but with prompt diagnosis and adequate therapy, fatalities can be greatly reduced.

Rocky Mountain Spotted Fever

The rickettsial disease with greatest impact on people living in North America is Rocky Mountain spotted fever (RMSF). Ricketts identified the etiologic agent *Rickettsia rickettsii* in smears from infected animals and patients, and later discovered that it was transmitted by ticks. Despite its geographic name, this disease occurs infrequently in the western United States. The majority of cases are concentrated in the southeast and eastern seaboard regions. Infections occur most frequently among children living in rural and mountain areas, generally in the spring and summer months when the tick vector is most active. The principal reservoirs and vectors of *R. rickettsii* are hard ticks such as the would tick (*Dermacentor andersoni*), the American dog tick (*D. variabilis*, among others), or the Lone Star tick (*Ambylomma americanum*). The dog tick is probably most responsible for transmission to humans since it is the major vector in the southeastern United States. The tick passes the pathogen to its offspring, to various domestic and wild mammals, and to humans.

Pathogenesis—The distinctive spotted rash usually comes on within two to four days after the prodromium. It starts on the wrist and ankles, moves to the arms and legs, converges toward the chest, and eventually covers the entire body. Early lesions are slightly mottled like measles, but later ones are macular, maculopapular, and even petechial. In the most severe, untreated cases, the enlarged lesions merge and can become necrotic, predisposing to gangrene of the toes or fingertips. Other grave manifestations of disease are cardiovascular disruption, including hypotension, thrombosis, and hemorrhage. Conditions of restlessness, delirium, convulsions, tremor, and coma are signs of the often overwhelming effects on the central nervous system. Fatalities occur in an average of 20% of untreated cases and 5-10% of treated cases. Although death usually occurs late in the disease, there have been instances of mortality within three days of onset of symptoms.

Diagnosis and control—Careful clinical observation and patient history are thus essential. Indications sufficiently suggestive to start antimicrobial therapy are the following : (a) A cluster of symptoms, including sudden fever, headache, and rash; (b) recent contact with ticks or dogs; and (c) possible occupational or recreational exposure in the spring or summer. A recent boon to early diagnosis is a method for staining rickettsias directly in a tissue biopsy using fluorescent antibodies. Isolating rickettsiae from the patient's blood or tissues is desirable, but it is expensive and requires specially qualified personnel and laboratory facilities. The drug of choice for suspected and known cases is tetracycline (doxycycline) administered every day for one week. Chloramphenicol is used if diagnosis has failed to differentiate between spotted fever and meningococcal meningitis, since this drug is effective in both cases. So widespread are endemic areas and so deeply entrenched is *R. rickettsii* in reservoir animals that anyone involved in outdoor activities should wear protective clothing, boots, leggings, and insect repellent containing DEET. Individuals exposed to heavy infestation should routinely inspect their bodies (and their dogs!) for ticks and remove ticks gently without crushing, preferably with forceps or fingers protected with gloves, because it is possible to become infected by tick feces or body fluids.

Transmission Vectors of Typhus

(i) Ticks—Two families of ticks are differentiated—soft (argasid) and hard (ixodid)—based on the absence or presence of a dorsal shield called the scutum. Argasids seek specific, sheltered habitats in caves, barns, burrows, or even bird cages, and they parasitize wildlife during nesting seasons and domesticated animals throughout the year. Endemic relapsing fever is transmitted by this type of tick. In contrast, the ixodids prefer confined habitats, and many have adapted to a wide-ranging life-style, hitchhiking along as their hosts wander through forest, savannah, or desert regions. Depending upon the species, ticks feed during larval, nymph, and adult metamorphic stages. The longevity of ticks is formidable; metamorphosis can extend for two years and adults can survive for four years away from a host without feeding. Many tick-borne agents cause mild infection in their primary wildlife hosts, but serious infection in livestock and humans.

(ii) Mites—Most species are free-living, and only a handful of the parasitic ones are of direct medical importance. The house mouse mite, *Liponyssoides sanguineus*, is a vector for the etiologic agent of rickettsialpox. Although domestic mice and rats are the natural hosts for this mite, it is not specific in its food getting and will readily nibble on humans if the opportunity arises. Larval mites called chiggers develop in the undergrowth of jungles, grasslands, coniferous forests, cultivated fields, gardens, and riverbanks that are frequented by humans. The larvae may escape notice at first, since they are minute and the bite is initially painless. But the bite can become intensely itchy in sensitive individuals. Once the larval stage is passed, the nymph and adult feed on nonvertebrate hosts.

(iii) Fleas—They are known for their extreme longevity and resistance, and many are notorious in their monospecificity, passing with ease from wild or domesticated mammals to humans. In response to mechanical stimulation and warmth, fleas jump onto their targets and crawl about, feeding as they go. A well-known vector is the oriental rat flea *Xenopsylla cheopis* that transmits *Rickettsia typhi,* the cause of murine typhus. The flea harbors the pathogen in its gut, and periodically contaminates the environment with virulent rickettsias by defecating.

(iv) Lice—The lice of humans usually occupy head and body hair (*Pediculus humanus*) or pubic, chest, and axillary hair (*Phthirus pubis*). They feed by gently piercing the skin and sucking blood and tissue fluid. Infection develops when the louse or its feces are inadvertently crushed and rubbed into wounds, skin, eyes, or mucous membranes. Lice are involved in transmitting *R. prowazekii* (epidemic typhus) and *Borrelia recurrentis* (relapsing fever).

Q Fever

Its origin was mysterious for a time, until Harold Cox working in Montana and Frank Burnet in Australia discovered the agent later named *Coxiella burnetii*. This rickettsia is highly resistant, due to its production of an unusual type of spores, and it is apparently harbored by a wide assortment of vertebrates and arthropods, especially ticks. Arthropod vectors play an essential role intransmitting the rickettsiae between wild and domestic animals, but transmission to humans is largely by means of environmental contamination and airborne spread. Sources of infectious material include urine, feces, milk, and airborne particles from infected animals. *Coxiella burnetti* has been isolated from most regions of the world. California and Texas have the highest case rates in the United States, though most cases probably go undetected. People at highest risk are farm workers, meat cutters, wool and leather workers, laboratory technicians, and consumers of raw milk. The clinical manifestations typical of *Coxiella* infection are abrupt onset of fever, chills, head and muscle ache, but only rarely are skin rashes seen. Disease is sometimes complicated by pneumonitis, hepatitis, and delayed endocarditis, which are occasionally fatal.

Trench Fever

Since 1945, cases have occurred infrequently in endemic regions of Europe, Africa, and Asia. The causative agent is *Bartonella (Rochalimaea) quintana,* a species that, like epidemic typhus, cycles between humans and lice, but unlike the typhus rickettsia, does not multiply intracellularly and does not kill the louse vector. Highly variable symptoms can include a five-to-six-day fever (hence five-day, or quintana, fever); leg pains, especially in the tibial region (shinbone fever); headache; chills; and muscle aches. A macular rash can also occur. Another rickettsia, *Bartonella henselae,* is the most common agent of cat scratch disease (CSD), an infection connected with a cat scratch or bite. The symptoms start after 1-2 weeks, with a cluster of small papules at the site of inoculation. In a few weeks, the lymph nodes along the lymphatic drainage swell and can become pus-filled. Most infections remain localized and resolve in a few weeks, but drugs such as tetracycline, erythromycin, and rifampin can be effective therapies.

CHLAMYDIA

Trachoma

Like rickettsias, the chlamydias are obligate parasites that depend on certain metabolic constituents of host cells for growth and maintenance. They show further resemblance to the rickettsias with their small size, Gram-negative cell wall, and pleomorphic morphology, but they are markedly different in several aspects of their life cycle. The species of greatest medical significance is *Chlamydia trachomatis,* a very common pathogen involved in sexually transmitted, neonatal, and ocular disease (trachoma), *Chlamydia pneumoniae,* the cause of one type of atypical pneumonia, and *Chlamydia psittaci* a zoonosis of birds and mammals that causes ornithosis in humans. Chlamydias alternate between two distinct stages : (a) A small, metabolically inactive, infectious form called the elementary body that is released by the infected host cell, and (b) a larger, noninfectious, actively dividing form called the reticulate body that grows within the host cell vacuoles (Fig. 9.1). Elementary bodies are tiny, dense spheres shielded by a rigid, impervious envelope that ensures survival outside the eukaryotic host cell. Reticulate bodies are finely granulated and have thin cell walls. Studies of the reticulate bodies indicate that they are energy parasites, entirely lacking enzyme systems for catabolizing glucose and other substrates and for synthesizing ATP, though they do possess ribosomes and mechanisms for synthesizing proteins, DNA, and RNA.

Pathogenesis—The reservoir of pathogenic strains of *Chlamydia trachomatis* is the human body. The microbe shows astoundingly broad distribution within the population, often being carried with no symptoms. Elementary bodies are transmitted in infectious secretions, and although infection can occur in all age groups, disease is most severe in infants and children. The two human strains are the trachoma strain, which attacks the squamous or columnar cells of mucous membranes in the eyes, genitourinary tract, and lungs, and the lymphogranuloma venereum (LGV) strain, which invades the lymphatic tissues of the genitalia.

The two forms of chlamydial eye disease, ocular trachoma and inclusion conjunctivitis, differ in their patterns of transmission and ecology. Ocular trachoma, an infection of the epithelial cells of the eye, is an ancient disease and a major cause of blindness in certain parts of the world. Although a few cases occur yearly in the United States, several million cases occur endemically in parts of Africa and Asia. Transmission is favoured by contaminated fingers, fomites, flies, and a hot, dry climate. The first signs of infection are a mild conjunctival exudate and slight inflammation of the conjunctiva. This is followed by marked infiltration of lymphocytes and macrophages into the infected area. As these

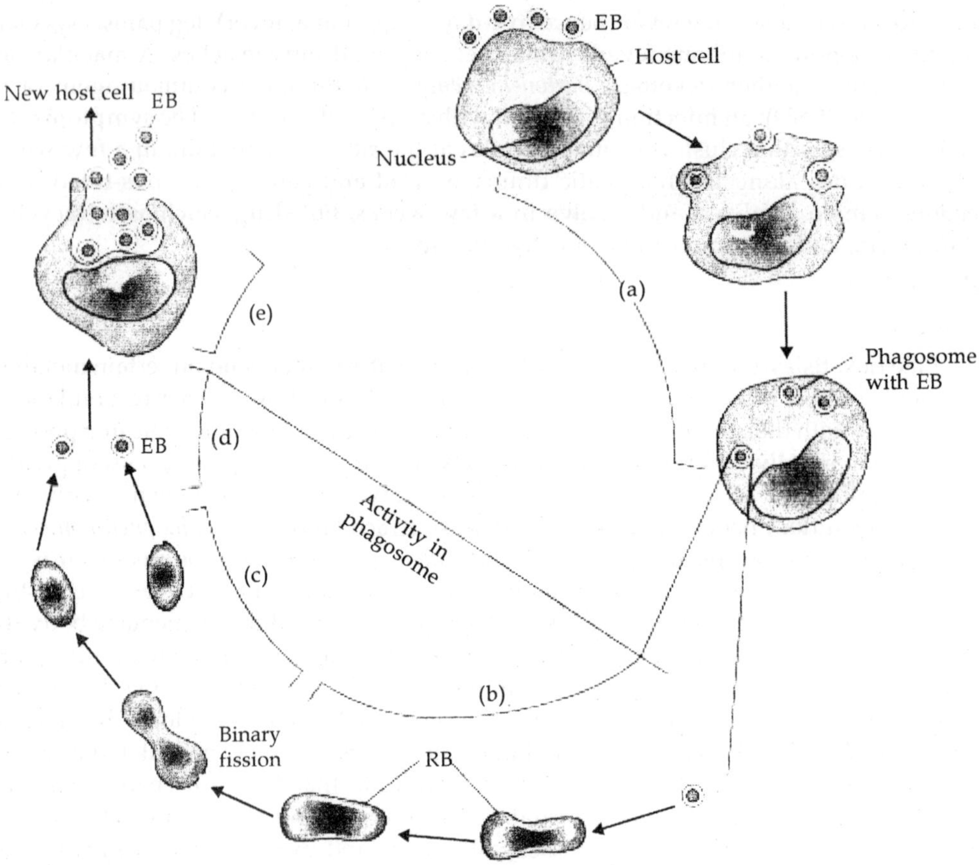

Fig. 9.1 : The life history of *Chlamydia.*

cells build up, they impart a pebbled (rough) appearance to the inner aspect of the upper eyelid. In time, a vascular pseudomembrane of exudate and inflammatory leukocytes forms over the cornea, a condition called *pannus* that lasts a few weeks and usually heals. Complications contributing most to corneal damage and impaired vision are repeated infections, secondary bacterial infection, and deformation of tear ducts and eyelashes, which predispose to dry eyes and corneal abrasion. Early treatment of this disease with tetracycline of sulfa drugs is highly effective and prevents all of the complications.

Inclusion conjunctivitis is usually acquired through contact with secretions of an infected genitourinary tract. Infantile conjunctivitis develops 5 to 12 days after a baby has passed through the birth canal of its infected mother, and is the most prevalent form of conjunctivitis (100,000 cases per year). The initial signs are conjunctival irritation, a profuse adherent exudate, redness, and swelling. Although the disease is generally self-limited and heals spontaneously, trachoma-like scarring occurs often enough to warrant routine prophylaxis of all newborns (as for gonococcal infection). Because the traditional silver nitrate solution is ineffective against *C. trachomatis*, antibiotics such as erythromycin silver nitrate solution is ineffective against *C. trachomatis*, antibiotics such as erythromycin and tetracycline must be instilled into the eyes. Most inclusion conjunctivitis in adults has occurred in people with concurrent genital infection. These cases appear to result from self-inoculation or contamination with secretions. Other cases have been traced to contact with swimming pool water and shared towels.

About 70% of infected women harbor it asymptomatically on the cervix, while 10% of infected males show no signs or symptoms. Depending on the condition of the host, the bacterium is capable of causing a whole spectrum of diseases, many of which resemble gonorrheal infections. A syndrome appearing among males with chlamydial infection is an inflammation of the urethra called nongonococcal urethritis(NGU). This diagnosis is derived from the symptoms that mimic gonorrhea, yet do not involve gonococci. Chlamydia is the most frequent cause of this syndrome. Women with symptomatic chlamydial infection have cervicitis accompanied by a white drainage, endometritis, and salpingitis. As is often the rule with sexually transmitted diseases, chlamydia frequently appears in mixed infections with the gonococcus and other genitourinary pathogens, thereby greatly complicating treatment. When a particularly virulent strain of *Chlamydia* chronically infects the genitourinary tract, the result is a severe, often disfiguring disease called lymphogranuloma venereum. The external genitalia, anus, rectum, and the inguinal lymph nodes are the typical targets of invasion. Chlamydiae enter through tiny nicks or breaks in the perigenital skin or mucous membranes and form a small, painless vesicular lesion that often escapes notice. Other acute symptoms are headache, fever, and muscle aches. As the lymph nodes near the lesion begin to fill with granuloma cells, they enlarge and become firm and tender.

Diagnosis and control—Because chlamydias reside intracellularly, specimen sampling requires enough force to dislodge some of the cells from the mucosal surface. Genital samples are taken with a swab inserted a few centimeters into the urethra or cervix, rotated, and removed. Although the most reliable diagnosis comes from culture in chicken embryos, mice, or tissue culture, this procedure is too costly and time-consuming to be used routinely in STD clinics; however, it is an essential part of diagnosing neonatal infections. The most sensitive and specific tests currently available are a direct assay of specimens using immunofluorescence and a PCR-based probe. Methods useful in diagnosing inclusion conjunctivitis are Giemsa or iodine stains, but they are not recommended for urogenital specimens because of low sensitivity and the possibility of obtaining false- negative results in asymptomatic patients. Urogenital chlamydial infections are most effectively treated with drugs that act intracellularly, such as tetracyclines, erythromycin, and rifampin. Penicillin and the aminoglycosides are not effective and must not be used. Because of the high carrier rate and the difficulty in detection, prevention of chlamydial infections has been and will continue to be very difficult. As a general rule, sexual partners of infected people should be treated with drug therapy to prevent reinfection, and sexually active people can achieve some protection with a condom.

Ornithosis

As outbreaks of this disease appeared in areas of the world having no parrots, it became evident that other birds could carry and transmit the microbe to humans and other animals. In light of this, the generic term ornithosis has been suggested as a replacement.

Ornithosis is a worldwide zoonosis that is carried in a latent state in wild and domesticated birds but becomes active under stressful conditions such as overcrowding. In the United States, poultry have been subject to extensive epidemics that killed as many as 30% of flocks. Infection is communicated to other birds, mammals, and humans by contaminated feces and other discharges that become airborne and are inhaled. The symptoms of ornithosis mimic those of influenza and pneumococcal pneumonia. Early manifestations are fever, chills, frontal headache, and muscle aches, and later ones are coughing and lung consolidation. Unchecked, infection can lead to systemic complications involving the meninges, brain, heart, or liver. Although most patients respond well to tetracycline or erythromycin therapy, recovery is often slow and fraught with relapses. A strict human pathogen, *Chlamydia pneumoniae,* is distinctly different and not closely related to the other chlamydias. It has been linked to a type of respiratory illness that includes pharyngitis, bronchitis, and pneumonitis. It is usually a mild illness in young adults, though it can cause a severe reaction in asthmatic patients that is responsible for increased rates of death in this group.

MYCOPLASMA

Atypical Pneumonia

Animal species of mycoplasmas are sometimes referred to as membrane parasites because they form intimate associations with membranes of the respiratory and urogenital tracts. Since mycoplasmas bind to specific receptor sites on cells and adhere so tenaciously that they are not easily removed by usual defense mechanisms, infections are chronic and difficult to eliminate. Mycoplasmas are inhibited by antibiotics that repress protein synthesis but are resistant to penicillin and cephalosporins that block cell wall synthesis. *Mycoplasma pneumoniae* is a human parasite that is the most common agent of primary atypical pneumonia (PAP). This syndrome is atypical in that its symptoms do not resemble those of pneumococcal pneumonia. PAP can also be caused by rickettsias, chlamydias, respiratory syncytial viruses, and adenoviruses. Mycoplasmal pneumonia is transmitted by aerosol droplets among people confined in close living quarters, especially families, students, and the military. *M. pneumoniae* selectively binds to specific receptors of the respiratory epithelium and inhibits ciliary action. Gradual spread of the bacteria over the next two to three weeks disrupts the cilia and the epithelium. Even though most mycoplasmas are neither invasive nor particularly toxigenic, the sloughing off of the epithelium causes inflammation and other tissue damage. The first symptoms—fever, malaise, sore throat, and headache—are not suggestive of pneumonia. A cough is not a prominent early symptom, and when it does appear, it is mostly unproductive. As the disease progresses, nasal symptoms, chest pain, and earache can develop.

Diagnosis—Because a culture can take two or three weeks, early diagnosis of mycoplasma pneumonia is difficult and relies chiefly on close clinical observation. Stains of sputum appear devoid of bacterial cells, leukocyte counts are within normal limits, and X-ray findings are nonspecific. Some physicians find the appearance of small blister on the tympanic membrane a useful sign. Others use the patient's history to rule out other bacterial and viral agents. Tetracycline and erythromycin inhibit mycoplasmal growth and help to rapidly diminish symptoms, but they do not stop the shedding of viable mycoplasmas.

Mycoplasma hominis and *Ureoplasma urealyticum* are regarded as weak, sexually transmitted pathogens. They are frequently encountered in samples from the urethra, vagina, and cervix of new-borns and adults. These species initially colonize an infant at birth subsequently diminish through early and late childhood. A second period of colonization and persistence is initiated by the onset of sexual intercourse. Evidence linking genital mycoplasmas to human disease is substantial and growing every year. *Uroplasma urealyticum* is implicated in some types of nonspecific or nongonococcal urethritis and prostatitis. There is increasing evidence that this mycoplasma plays a role in opportunistic infections of the fetus and fetal membranes. It appears to cause some cases of miscarriage, stillbirth, premature birth, and respiratory infections of newborns. *Mycoplasma hominis* is more often associated with vaginitis, pelvic inflammatory disease, and kidney inflammation. This intracellular pathogen tentatively named *M. incognitus,* was isolated from the spleen, liver, brain, and blood of AIDS patients. At first it appeared to be a potent cofactor accounting for rapid deterioration and cell death in some AIDS patients; however, researchers have now found that it is capable of causing a fatal infection on its own. This mycoplasma has been associated with aggressive flulike illness in people who are negative for HIV infection, and it appears to suppress the immune system on its own.

DENTAL DISEASES CAUSED BY BACTERIA

Periodontium

The mouth contains a diversity of surfaces for colonization, including the tongue, teeth, gingiva, palate, and cheeks, and it provides numerous aerobic, anaerobic, and microaerophilic microhabitats for the estimated 1,000 different oral species with which humans coexist. The habitat of the oral cavity is warm, moist, and greatly enriched by the periodic infusion of food. In most humans, this association remains in balance with little adverse effect, but in people with poor or nonexistent oral hygiene, it teeters constantly on the brink of disease. Most dental diseases develop on the tooth and its surrounding supportive structures, or periodontium. A tooth is composed of a crown that protrudes above the gum and a root that is inserted into a bony socket. The outer surface of the crown is protected by a dense coating of enamel, an extremely hard, noncellular material composed of tightly packed rods of $CaPO_4$ that cannot be replaced once the tooth has matured. The root is surrounded by a layer of cementum, which anchors the tooth to the periodontal membrane that lines the socket. The major portion of the tooth inside the crown and root is composed of a highly regular calcified material called dentin, and the very core contains a pulp cavity that supplies the living tissues with blood vessels and nerves. The root canal is the portion of the pulp that extends into the roots. Places where the bone protrudes toward the crown are covered by connective tissue and a mucous membrane called the gingiva, or gum.

Dental pathology generally affects both hard and soft tissues (Fig. 9.2). Although both categories of disease are initiated when microbes adhere to the tooth surface and produce dental plaque, their outcomes vary. In the case of dental caries, the gradual breakdown of the enamel leads to invasive disease of the tooth itself, whereas in soft-tissue disease, calcified plaque damages the soft gingival tissues and predisposes them to bacterial invasion.

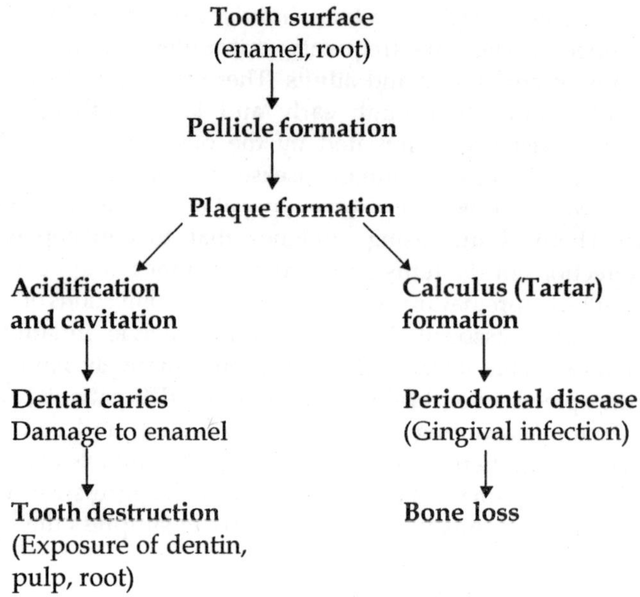

Fig. 9.2 : Events leading to dental diseases.

Dental Caries

It is a complex mixed infection of the dentition that gradually destroys the enamel and often lays the groundwork for the destruction of deeper tissues. It occurs most often on tooth surfaces that are less accessible and harder to clean, and on those that provide pockets or crevices where bacteria can cling. Caries commonly develop on enamel pits and fissures, especially those of the occlusal surfaces, though they can also occur on the smoother crown surfaces and subgingivally on the roots. Over the years, several views have been put forth to explain how dental caries originate. At various times it has been believed that sugar, microbes, or acid cause teeth to rot; however, studies with germfree animals have now shown that no single factor can account for caries. Caries development occurs in many phases and requires multiple interactions involving the anatomy, physiology, diet, and bacterial flora of the host.

Plaque formation—Within a few moments, it develops a thin, mucous coating called the acquired pellicle, which is made up of adhesive salivary proteins. This structure presents a potential substrate upon which certain bacteria first gain a foothold. The most prominent pioneering colonists are the cariogenic genera streptococcus and Actinomyces. These Gram-positive bacteria have adhesive receptors such as fimbriae and slime layers that allow them to cling to the tooth surfaces and to each other, forming a foundation for the dense, whitish mass called plaque (Fr. a patch). Fed by a diet high in sucrose, glucose and certain complex carbohydrates, *Streptococcus mutans* and *S. gordonii* produce a gummy polymer of glucose called dextran that helps them attach to smooth enamel surfaces, holds the plaque together, and adds to its bulk. As these primary invaders continue to grow and plaque begins to build up, the scene is set for the invasion and aggregation of additional species of bacteria. Among secondary invaders are species of *Lactobacillus, Bacteroides, Fusobacterium, Neisseria,* and *Treponema.*

Etching of enamel—The role of plaque in caries development relates directly to streptococci and lactobacilli, which produce acid as they ferment dietary carbohydrates. If this acid is immediately flushed from the plaque and diluted in the mouth, it has little effect. However in the denser regions of plaque, the acid can accumulate indirect contact with the enamel surface. Eventually the pH of the environment adjacent to the enamel is lowered below 5, which is acidic enough to begin to dissolve (decalcify) the calcium phosphate of the enamel in that spot. This initial lesion can remain localized in the enamel (first-degree caries) and be repaired with various inert materials (fillings). Once the deterioration has reached the level of the dentin (second-degree caries), tooth destruction speeds up, and the tooth can be rapidly destroyed.

Periodontal Disease

Most kinds are due to bacterial colonization and varying degrees of inflammation that occur in response to gingival damage. The most common predisposing condition occurs when the plaque becomes mineralized (calcified) with calcium and phosphate crystals. This process produces a hard, porous substance called calculus on both the sub- and supra-gingival tooth surfaces. Supra-gingival calculus is usually not harmful and can even protect the tooth from caries, but its presence in the relatively sheltered gingival sulcus initiates a pathological lesion that leads to periodontal disease. Calculus and plaque accumulating in the gingival sulcus irritates the delicate gingival membrane, and the chronic trauma to this area causes a pronounced inflammatory reaction. This event, in turn, creates an avenue of invasion for a variety of normal flora. These bacteria are primarily anaerobic and Gram-negative, and include *Veillonella* (cocci), *Porphyromonas* and *Bacteriodes* (rods), *Fusobacterium* (spindle-shaped rods), and numerous spirochetes. In response to the mixed infection, the damaged area becomes infiltrated by neutrophils and macrophages, and later by lymphocytes, which cause further inflammation and tissue damage.

The initial signs of gingivitis are swelling, loss of normal contour, patches of redness, and increased bleeding of the gingiva. Spaces or pockets of varying depth also develop between the tooth and the gingiva. If this condition persists, a more serious disease called periodontitis results. This is the natural extension of the disease into the periodontal membrane and cementum. The deeper involvement increases the size of the pockets and can cause bone resorption severe enough to loosen the tooth in its socket. Another serious periodontal disease is acute necrotizing ulcerative gingivitis (ANUG), formerly called trench mouth or Vincent's disease. This disease is a synergistic infection involving *Treponema vincentii* and other spirochetes and fusobacteria that are sufficiently aggressive to invade the gingival tissues and severely damage them. It develops rapidly and is associated with severe pain, bleeding, pseudomembrane formation, and necrosis.

The practice of putting a baby down to nap with a bottle of fruit juice or formula can lead to rampant dental caries. In addition to diet, numerous anatomic, physiologic, and hereditary factors influence oral diseases. The structure of the tooth enamel can be influenced by genetics and by environmental factors such as fluoride, which strengthens the enamel bonds. The amount of calcium available to developing fetuses and neonates can also affect tooth development. Saliva that is more fluid in consistency inhibits oral colonization, and its content of inhibitory factors such as antibodies and lysozyme can also help prevent dental disease. The greatest control of dental diseases is preventive dentistry, including regular brushing and flossing to remove plaque, because stopping plaque build-up automatically reduces caries and calculus production. Mouthwashes are relatively ineffective in controlling plaque formation because of the high bacterial content of saliva and the relatively short acting time of the mouthwash.

10

Physical and Chemical Control Measures of Microorganisms

Salting, smoking, pickling, and drying foods, and exposure of food, clothing, and bedding to sunlight were prevalent practices among early civilizations. The Egyptians showed surprising shophistication and understanding of decomposition by embalming the bodies of their dead with strong salts and pungent oils. They introduced filtration of wine and water as well. The Greeks and Romans burned clothing and corpses during epidemics, and they stored water in copper and silver containers. The armies of Alexander the Great reportedly boiled their drinking water and buried their wastes. Burning sulfur to fumigate houses and applying sulfur as a skin ointment also date from this approximate era. During the great plague pandemic of the Middle Ages, it was commonplace to bury corpses in mass graves, burn the clothing of plague victims, and ignite aromatic woods in the houses of the sick in the belief that fumes would combat the disease. In a desperate search for some sort of protection, survivors wore peculiar garments and anointed their bodies with herbs, strong perfume, and vinegar. These attempts may sound foolish and antiquated, but it now appears that they may possibly have had some benefits. Burning wood releases formaldehyde, which could have acted as a disinfectant; herbs, perfume, and vinegar contain mild antimicrobial substances.

The methods of microbial control belong to the general category of decontamination procedures, in that they destroy or remove contaminants. In microbiology, contaminants are microbes present at a given place and time that are undesirable or unwanted. Most decontamination methods employ either physical agents such as heat or radiation, or chemical agents such as disinfectants, antiseptics, or drugs. This separation is convenient, but the categories overlap in some cases—for instance, radiation can cause damaging chemicals to form, or chemicals can generate heat.

This targeted population is rarely simple or uniform; in fact, it often contains mixtures of microbes with extreme differences in resistance and harmfulness. Contaminants that can have far-reaching effects if not adequately controlled include bacterial vegetative cells and endospores, fungal hyphae and spores, yeasts, protozoan trophozoites and cysts, worms, insects and their eggs, and viruses.

TERMINOLOGY USED IN CONTROL OF MICROBES

To complicate matters, the everyday use of some of these terms can at times be vague and inexact. For example, occasionally one may be directed to sterilize or disinfect a patient's skin, even though this usage does not fit the technical definition of either term.

Sterilization

Sterilization is a process that destroys or removes all viable microorganisms, including viruses. Any material that has been subjected to this process is said to be sterile. (Gr. *Steira,* barren). These terms should be used only in the strictest sense for methods that have been proved to sterilize. An object cannot be slightly sterile or almost sterile – it is either sterile or not sterile. Control methods that sterilize are generally reserved for inanimate objects, because sterilizing parts of the human body would call for such harsh treatment that it would be highly dangerous and impractical. Sterilized products – surgical instruments, syringes, and commercially packaged foods, just to name a few – are frequently essential to human well-being. Although most sterilization is performed with a physical agent such as heat, a few chemicals called *sterilants* can be classified as sterilizing agents because of their ability to destroy spores. At times, sterilization is neither practicable nor necessary, and only certain groups of microbes need to be controlled. Some antimicrobial agents eliminate only the susceptible vegetative states of microorganisms but do not destroy the more resistant endospore and cyst stages.

Antimicrobial Agents

A bactericide is a chemical that destroys bacteria except for those in the endospore stage. It may or may not be effective on other microbial groups. A fungicide is a chemical that can kill fungal spores, hyphae, and yeasts. A virucide is any chemical known to inactivate viruses, especially on living tissue. A sporicide is an agent capable of destroying bacterial endospores. A sporicidal agent can also be a sterilant because it can destroy the most resistant of all microbes.

Bacteriostatic Agents

These can be used in combination with various prefixes to denote a condition in which microbes are temporarily prevented from multiplying but are not killed outright. Although killing or permanently inactivating microorganisms is the usual goal of antimicrobial control, microbistasis does have meaningful applications. Bacteriostatic agents prevent the growth of bacteria on tissues or on objects in the environment, and *fungistatic* chemicals inhibit fungal growth. Materials used to control microorganisms in the body have microbistatic effects because microbicidal compounds would generally by too toxic to human cells.

Germicides

A germicide (L. *Germen,* germ, and *caedeve,* to kill) also called a microbicide, is any chemical agent that kills pathogenic microorganisms. A germicide can be used on inanimate (nonliving) materials or on living tissue, but it ordinarily cannot kill resistant microbial cells. Any physical or chemical agent that kills germs is said to have germicidal properties. The related term, disinfection, (L. *Dis,* apart, and *inficere,* to corrupt) refers to the uses of a physical process or a chemical agent (a disinfectant) to destroy vegetative pathogens but not bacterial endospores. It is important to note that disinfectant are normally used only on inanimate objects because, in the concentrations required to be effective, they can be toxic to human and other animal tissue.

FACTORS AFFECTING MICROBIAL DEATH

Death is a phenomenon that involves the permanent termination of an organism's vital processes. Signs of life in complex organisms such as animals are self-evident, and death is made clear by loss of nervous function, respiration, or heartbeat. In contrast, death in microscopic organisms that are composed of just one or a few cells is often hard to detect, because they reveal no conspicuous vital signs to begin with. Lethal agents (such as radiation and chemicals) do not necessarily alter the overt appearance of microbial cells. Even the loss of movement in a motile microbe cannot be used to indicate death. The destructive effects of chemical or physical agents occur at the level of a single cell. As the cell is continuously exposed to an agent such as intense heat or toxic chemicals, various cell structures become dysfunctional, and the entire cell can sustain irreversible damage. If the microbe is so disturbed metabolically and structurally that it is unable to multiply, then it is no longer viable. The permanent loss of reproductive capability, even under optimum growth conditions, has become the accepted microbiological definition of death.

With this concept, workers in the field of microbial control have defined the conditions required to destroy microorganisms and have pinpointed the manner by which antimicrobial agents kill cells. Standards of sterilization and disinfection in medicine, dentistry, and industry have been established largely by means of rigorous experimentation. Literally hundreds of testing procedures have been developed for evaluating physical and chemical agents. The cells of a pure culture show marked variations in susceptibility to a given microbicidal agent. Death of the whole population is not instantaneous, but occurs in a logarithmic manner and requires a certain time of exposure. The most susceptible cells (younger, actively growing cells) die immediately, whereas less susceptible cells (older, inactive ones) have greater resistance and require longer exposure times.

The effectiveness of a particular agent is governed by several factors besides time. The following additional factors influence the action of antimicrobial agents :

(a) A higher load of contaminants requires more time to destroy.

(b) In most actual circumstances of disinfection and sterilization, the target population is not a single species of microbe but a mixture of bacteria, fungi, spores, and viruses, presenting an even greater spectrum of microbial resistance.

(c) pH and temperature of environment.

(d) The concentration (dosage, intensity) of the agent. For example, UV radiation is most microbicidal at 260 nm; most disinfectants are more active at higher concentrations.

(e) The presence of solvents, interfering organic matter, and inhibitors. Large amount of saliva, blood, and feces can inhibit the actions of disinfectants and even heat.

Permanent cessation of reproduction is the accepted criterion for microbial death; however, differentiating a microbistatic effect from a microbicidal one can present a problem. This is particularly true of some disinfectants and antiseptics that only retard cell growth. Cells kept from multiplying long enough can eventually become unable to do so, and can therefore be dead by definition.

MODE OF ACTION OF ANTIMICROBIAL AGENTS

Determining an agent's precise mode of action can be very difficult, and we do not yet completely understand how agents such as radiation and heat achieve their effects. Current evidence indicates that various parts of the cell accumulate injuries until it ceases to function. Agents can be classified according to the following degrees of selectiveness : (a) Agents that are less selective in their scope of destructiveness inflict severe damage on many cell parts, and are generally very biocidal (heat, radiation, some disinfectants); (b) moderately selective agents with intermediate specificity (certain disinfectants and antiseptics); and (c) more selective agents (drugs) whose target is usually limited to a specific cell structure or function and whose effectiveness is restricted only to certain microbes. The cellular targets of physical and chemical agents fall into four general categories : (1) The cell wall, (2) the cell membrane, (3) cellular synthetic processes (DNA, RNA), and (4) proteins.

Effects on Cell Wall

Several types of chemical agents damage the cell wall by blocking its synthesis, digesting it, or breaking down its surface. A cell deprived of a functioning cell wall becomes fragile and is lysed very easily. Examples of this mode of action include some antimicrobial drugs (penicillins) that interfere with the synthesis of the cell wall in bacteria. The enzymes lysozyme and lysostaphin digest specific bonds in the cell wall of Gram-positive bacteria, eroding and weakening it; in addition, one effect of detergents and alcohol is to disrupt cell walls.

Effects on Cell Membrane

All microorganisms have a cell membrane composed of lipids and proteins, and even some viruses have an outer membranous envelope. If this membrane is disrupted, a cell loses its selective permeability and can neither prevent the loss of vital molecules nor bar the entry of damaging chemicals, which leads to cell death. Detergents called surfactants work as microbicidal agents because they lower the surface tension of cell membranes. Surfactants are polar molecules with hydrophilic and hydrophobic regions that can

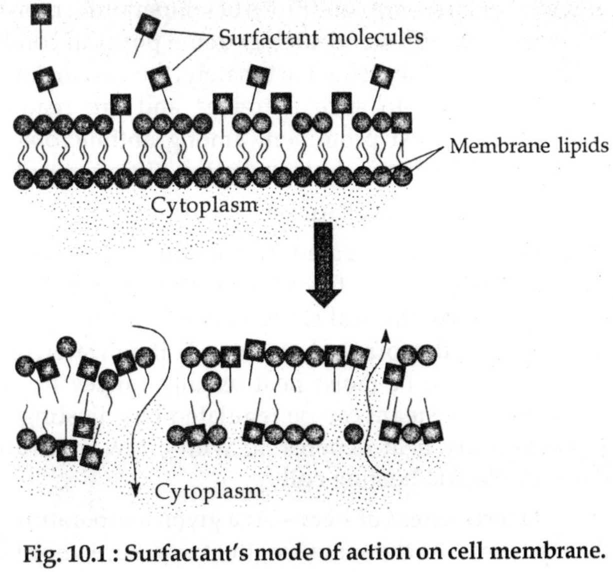

Fig. 10.1 : Surfactant's mode of action on cell membrane.

physically bind to the lipid layer and penetrate the internal hydrophobic region of membranes. In effect, this process opens up the once tight interface, leaving leaky spots that allow injurious chemicals to seep into the cell and important ions to seep out (Fig. 10.1).

Effects on Protein and Nucleic Acids

Any agent that interferes with accurate translation also prevents synthesis of a complete, functioning protein. For instance, the antibiotic chloramphenicol binds to the ribosomes of bacteria in a way that stops peptide bonds from forming. In its presence, many bacterial cells are inhibited from forming proteins required in growth and metabolism, and are thus inhibited from multiplying. Most of the agents that block protein synthesis are drugs used in antimicrobial therapy. The nucleic acids are likewise necessary for the continued functioning of microbes. DNA must be regularly replicated and transcribed in growing cells, and any agent that either impedes these processes or changes the genetic code is potentially antimicrobial. Some agents bind irreversibly to DNA, preventing both transcription (formation of RNA) and translation, while others are mutagenic agents. Gamma, ultraviolet, or X radiation causes mutations that result in permanent inactivation of DNA.

The antimicrobial properties of some agents arise from their capacity to disrupt, or denature, proteins. In general, denaturation occurs when the bonds that maintain the secondary and tertiary structure of the protein are broken. This will cause the protein to unfold or create random, irregular loops and coils. One way that proteins can be denatured is through coagulation by moist heat. Chemicals such as strong organic solvents (alcohols, acids) and phenolics also coagulate proteins. Other antimicrobial agents, such as metallic ions, attach to the active site of the protein and prevent it from interacting with its correct substrate. Regardless of the exact mechanism, such losses in normal protein function can promptly arrest metabolism.

PHYSICAL MEASURES OF CONTROL

Microorganisms have adapted to the tremendous diversity of habitats the earth provides, even severe conditions of temperature, moisture, pressure, and light. For microbes that normally with stand such extreme physical conditions, our attempts at control would probably have little effect. Fortunately, the vast majority of microbes that must be controlled are not adapted to such extremes and are readily controlled by abrupt changes in environment. Most prominent among antimicrobial physical agents is heat. Other less widely used agents include radiation, filtration, ultrasonic waves, and even cold.

Heat Used as Physical Agent

Increasingly elevated temperatures (exceeding the maximum) are microbicidal, whereas lower temperatures (below the minimum) tend to have inhibitory or microbistatic effects. The two physical states of heat used in microbial control are moist and dry. Moist heat occurs in the form of hot water, boiling water, or steam (vaporized water). In practice, the temperature of moist heat usually ranges from 60 to 135 °C. As we shall see, the temperature of steam can be regulated by adjusting its pressure in a closed container. The expression dry heat denotes air with a low moisture content that has been heated by a flame or electric heating coil.

Effectiveness of heat—At a given temperature, moist heat works several times faster than dry heat. At the same length of exposure, moist heat kills cells at a lower temperature than dry heat. The mechanisms by which microorganisms are destroyed by heat are rather complex. Both forms of heat disrupt important cell components, but it appears that their specific modes of action are different. Exposure to moist heat generally coagulates and therefore denatures cell proteins. The greater effectiveness of moist heat has been attributed to this particular mode of action, since protein (enzyme) denaturation occurs more rapidly and at a lower temperature if moisture is present. Dry heat also has several effects on microbes. For instance, it can oxidize cells, and at extremely high temperatures, reduce them to ashes. It dehydrates cell components, thereby concentrating the protoplasm. In time, it also denatures proteins and DNA, but proteins are more stable in dry heat than in moist heat, and higher temperatures are required to inactivate them.

Destruction of spores usually requires temperatures above boiling, although resistance varies widely. In boiling water (100 °C), the spores of *Bacillus anthracis* (the agent of anthrax) can be destroyed in a few minutes, whereas the spores of some thermophilic and anaerobic species can require several hours. Differences in the resistance of spores to moist and dry heat are also evident. The thermal death requirements of vegetative cells also vary, though not to the same extent as for spores. Among bacteria, the death times with moist heat range extending from 50 °C for 3 minutes *(Neisseria gonorrhoeae)* to 60 °C for 60 minutes *(Staphylococcus aureus)*. It is worth noting that vegetative cells of spore formers are just as susceptible as vegetative cells of non-spore formers, and that pathogens are neither more nor less susceptible than nonpathogens. Other microbes, including fungi (yeasts, molds, and some of their spores), protozoa, and worms, are rather similar in their sensitivity to heat. Viruses are surprisingly resistant to heat, with a tolerance range extending from 55 °C for 2-5 minutes (adenoviruses) to 60 °C for 600 minutes (hepatitis virus).

Thermal death measurement—As a general rule, higher temperatures allow shorter exposure times, and lower temperatures require longer exposure times. A combination of these two variables constitutes of thermal death time, or TDT, defined as the shortest length of time required to kill all microbes at a specified temperature. The TDT has been experimentally determined for the microbial species that are common or important contaminants in various heat-treated materials. Another way to compare the susceptibility of microbes to heat is the thermal death point (TDP), defined as the lowest temperature required to kill all microbes in a sample in 10 minutes. Many perishable substances are processed with moist heat. Some of these products are intended to remain on the shelf at room temperature for several months or even years. The chosen heat treatment must render the product free of agents of spoilage or disease. At the same time, the quality of the product and the speed and cost of processing must be considered. For example, in the commercial preparation of canned green beans, one of the cannery's greatest concerns is to prevent growth of the agent of botulism. From several possible TDTs for *Clostridium botulinum* spores, the cannery must choose one that kills all spores but does not turn the beans to mush. Out of these many considerations emerges an optimal TDT for a given processing method. Commercial canneries heat low-acid foods at 121 °C for 30 minutes, a treatment that sterilizes these foods.

Measures of Moist Heat Control

Steam under pressure—This pressure is measured at 15 pounds per square inch (psi), or 1 atmosphere. In order to raise the temperature of steam above this point, it must be pressurized in a closed chamber. This phenomenon is explained by the physical principle that governs the behaviour of gases under pressure. When a gas is compressed, its temperature rises in direct relation to the amount of pressure. So, when the pressure is increased to 5 psi above normal atmospheric pressure, the temperature of steam rises to 109°C. When the pressure is increased to 10 psi above normal, its temperature will be 115°C, and at 15 psi (a total of 2 atmospheres), it will be 121°C. Health and commercial industries use an autoclave for this purpose, and a comparable home appliance is the pressure cooker. The autoclave comes in various forms, all of which have a fundamentally similar plan—a cylindrical metal chamber with an airtight door on one ends, racks to hold materials, and a steam-filled jacket around the sterilizing chamber. Its construction includes a complex network of valves, pressure and temperature gauges, and ducts for regulating and measuring pressure and conducting the steam into the chamber. Sterilization is achieved when the steam condenses against the objects in the chamber and gradually raises their temperature. Experience has shown that the most efficient pressure/temperature combination for achieving sterilization is 15 psi, which yields 121°C.

It is possible to use higher pressure to reach higher temperatures, but this will not significantly reduce the exposure time and can harm the items being sterilized. Because this temperature will develop only if saturated steam is fed into the inner chamber, the modern autoclave is engineered to automatically evacuate the air in the chamber and fill it with steam. It is also important to avoid overpacking or haphazardly loading the chamber, which prevents steam from circulating freely around the contents and impedes the full contact that is necessary. The duration of the process is adjusted according to the bulkiness of the items in the load (thick bundles of material or large flasks of liquid) and how full the chamber is.

Autoclave—The autoclave is an excellent device to sterilize heat-resistant materials such as glassware, cloth (surgical dressings), rubber (gloves), metallic instruments, liquids, paper, some media, and some heat-resistant plastics. If the items are heat-sensitive (plastic petri dishes) but will be discarded, the autoclave is still a good choice.

Intermittent sterilization—Materials that cannot withstand the high temperature of the autoclave can be subjected to intermittent sterilization, also called *tyndallization*. This technique requires a chamber to hold the materials and a reservoir for boiling water. Items in the chamber are exposed to free-flowing steam for 30-60 minutes. This temperature is not sufficient to reliably kill spores, so a single exposure will not suffice. Assuming that surviving spores will germinate into less resistant vegetative cells, the items are incubated at appropriate temperatures for 23-24 hours, and then again subjected to steam treatment. This cycle is repeated for three days in a row. Since the temperature never gets above 100 °C, highly resistant spores that do not germinate may survive even after three days of this treatment. Intermittent sterilization is used most often to process heat-sensitive culture media, such as those containing sera, egg, or carbohydrates and some canned foods.

Boiling water—Because a single processing at 100 °C will not kill all resistant cells, this method is, strictly speaking, disinfection rather than sterilization. When materials are exposed to boiling water for 30 minutes, all non-spore-forming pathogens will be killed, including resistant species such as the tubercle bacillus and staphylococci. Probably the greatest disadvantage with this method is that the items can be easily recontaminated when removed from the water. Boiling is also a recommended method of disinfecting unsafe drinking water. In the home, boiling water is a fairly reliable way to sanitize and disinfect materials for babies, food preparation, and utensils, bedding, and clothing from the sickroom.

Pasteurization—Because microbes have the potential for spoiling these foods or causing illness, heat is frequently used to reduce the microbial load and destroy pathogens. Pasteurization is a technique in which heat is applied to liquids to kill potential agents of infection and spoilage, while at the same time retaining the liquid's flavor and food value. This process is considered a form of moist heat because of the high moisture content of the beverage. Ordinary pasteurization techniques require special vats and heat exchangers that expose the liquid to 71.6 °C for 15 seconds (flash method) or to 63–66 °C for 30 minutes (batch method). The first method is preferable because it is less likely to exchange flavour and nutrient content, and it is more effective against certain resistant pathogens such as *Coxiella* and *Mycobacterium.* Although these treatments inactivate most viruses and destroy the vegetative stages of 97-99% of bacteria and fungi, they do not kill endospores or *thermoduric* species (mostly nonpathogenic lactobacilli, micrococci, and yeasts). Milk is not sterile after regular pasteurization. In fact, it can contain 20,000 microbes per milliliter or more, which explains why even an unopened carton of milk will eventually spoil. In recent times, containers labelled sterile milk have appeared on supermarket shelves.

The primary targets of pasteurization are non-spore forming pathogens : *Salmonella* species (a common cause of food infection), *Campylobacter jejeuni* (acute intestinal infection), *Listeria monocytogenes* (listeriosis), *Brucella* species (undulant fever), *Coxiella burnetii* (Q fever), *Mycobacterium bovis* and *M. tuberculosis* species, and several enteric viruses. In 1985, a California epidemic of listeriosis afflicted hundreds of people and resulted in 40 deaths, mostly among infants and elderly persons. Epidemiologists searching for the infection source were first led to a particular type of fresh, soft cheese that had been eaten by the patients and was contaminated with *Listeria.*

Dry Heat

The temperatures and times employed in dry heat vary according to the particular method, but in general, they are greater than with moist heat. Incineration in a flame or electric heating coil is perhaps the most rigorous of all heat treatments. The flame of a Bunsen burner reaches 1,870°C at its hottest point, and furnance/incinerators operate at temperatures of 800-6,500°C. Direct exposure to such intense heat ignites and reduces the microbes, and sometimes their vehicles, to ashes and gas. Incineration of microbial samples on inoculating loops and needles using a Bunsen burner is a very common practice in the microbiology laboratory. This method is fast and effective, but it is also limited to metals and heat-resistant glass materials. Incinerators are regularly employed in hospitals and research labs for complete distruction and disposal of infectious materials such as syringes, needles, cultural materials, dressings, bandages, bedding.

The so-called dry oven is usually electric (occasionally gas) and has coils that radiate heat within an enclosed compartment. Heated, circulated air contacts the items, transferring its heat in the process. Sterilization is accomplished by exposure of items to 150°- 180°C for 2 to 4 hours, which ensures through heating of the objects and destruction of spores. The dry oven is used in laboratories and clinics for heat-resistant items that do not sterilize well with moist heat. Substances appropriate for dry ovens are glassware, powders, and oils that steam does not penetrate well, and metallic instruments that can be corroded by steam. This method is not suitable for plastics, cotton, and paper, which may burn at the high temperatures, or for solutions, which will dry out. Another limitation is the time required for it to work.

Cold and Dessication use as Physical Agent

It must be emphasized that cold merely retards the activities of most microbes. While it is true that some microbes are killed by cold temperatures, most are not adversely affected by gradual cooling, long-term refrigeration, or deep-freezing. In fact, freezing temperatures, ranging from 70°C to 135°C provide an environment that can preserve cultures of bacteria, viruses, and fungi for long periods. Some microbes (psychrophiles) grow very slowly even at freezing temperatures and can continue to secrete toxic products. Unawareness of these facts is probably responsible for numerous cases of food poisoning from frozen foods that have been defrosted at room temperature and then inadequately cooked. Pathogens able to survive several months in the refrigerator are *Staphylococcus aureus, Clostridium* species (spore formers), *Streptococcus* species, and several types of yeasts, molds, and viruses. Vegetative cells directly exposed to normal room air gradually become dehydrated or *desiccated.* More delicate pathogens such as the pneumococcus , the spirochete of syphilis, and the gonococcus die after a few hours of air drying, but many others are not killed by drying and some are even preserved.

Endospores of *Bacillus* and *Clostridium* are viable for thousands of years under extremely arid conditions. Staphylococci and streptococci, if protected by dried secretions (pus, saliva), and the tubercle bacillus, when surrounded by droplets of sputum, can remain viable in air and dust for lengthy periods. Many viruses (especially nonenveloped) and fungal spores can also withstand long periods of desiccation. Desiccation can be a valuable way to preserve foods such as meats, vegetables, and fruits because it greatly reduces the amount of water available to support microbial growth. It is interesting to note that a combination of freezing and drying—*lyophilization*—is a common method of preserving microorganisms and other cells in a viable state for many years. Pure cultures are frozen instantaneously on dry ice and exposed to a higher vacuum that rapidly removes the water. This avoids the formation of ice crystals that would damage the cells. As a general rule, chilling, freezing, and desiccation should not be construed as methods of disinfection or sterilization because their antimicrobial effects are erratic and uncertain, and one cannot be sure that they kill pathogens.

Radiation used as Physical Agent

For our purposes, radiation is defined as energy emitted from atomic activities and dispensed at high velocity through matter or space. Radiation can behave as waves or particles, depending upon the conditions. Here we will consider the wavelike character of electromagnetic radiation and the particulate character of particle radiation. The wavelengths of electromagnetic radiation range from high-energy, short wavelength gamma rays at one extreme to low-energy, very long radio waves at the other. For the most part, only electromagnetic radiations at the gamma ray, X-ray (also called roentgen ray), and ultraviolet ray (UV) levels are suitable for microbial control. Particle radiation consists of subatomic particles such as electrons, protons, and neutrons that have been freed from the atom. The only particle radiation feasible for antimicrobial applications is the high-speed electron. Visible light does exert some antimicrobial effects, but it is not predictable enough in its effects to control all microbes. Infrared rays can be destructive by virtue of the heat they produce, but they have fewer applications than other heating methods.

Ionization and nonionization radiation's mechanism—To understand the actual physical effects of radiation on microbes, it will be helpful to visualize the process of *irradiation*, or bombardment with radiation, at the cellular level (Fig. 10.2). When a cell is

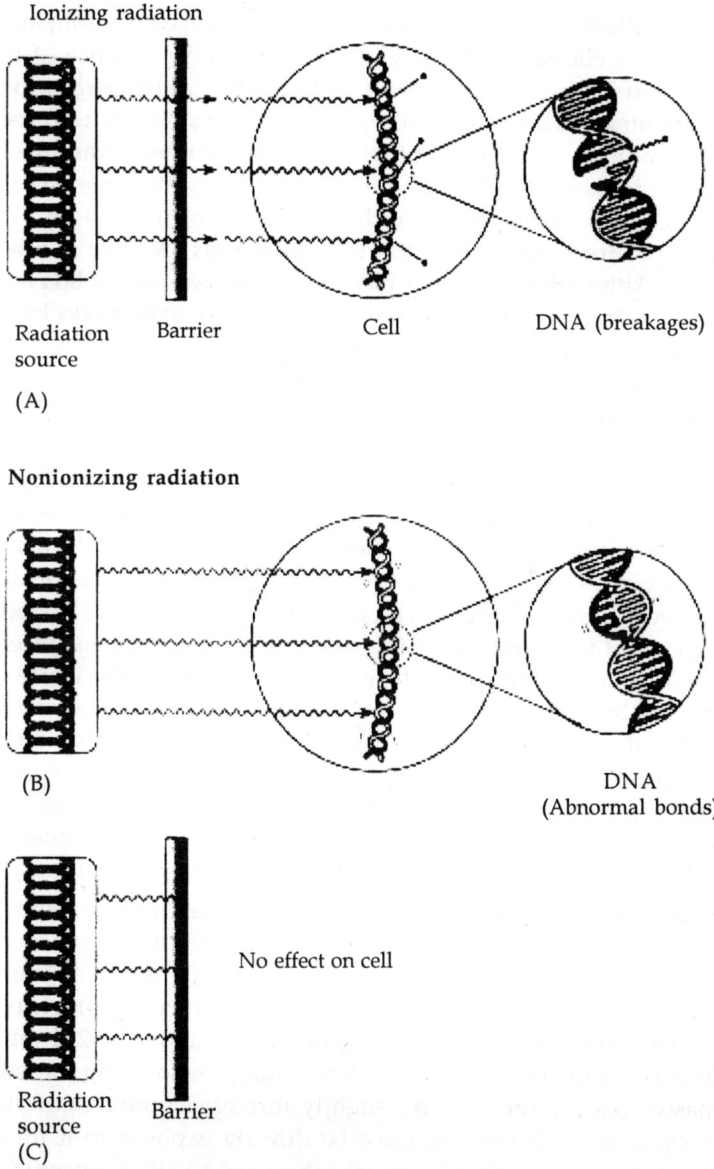

Fig. 10.2 : Irradiation effects on cells.

bombarded by certain waves or particles, its molecules absorb some of the available energy, leading to one of two consequences : (a) If the radiation ejects orbital electrons from an atom, it causes ions to form. This is the effect of *ionizing radiation.* One of the most sensitive targets for ionizing radiation is the DNA molecule, which can sustain mutations on a broad scale. Secondary lethal effects appear to be chemical changes in organelles and the production of toxic substances. Gamma rays, X rays, and highspeed electrons are all ionizing in their effects. (b) *Nonionizing radiation,* best exemplified by UV, excites atoms by raising them to a higher energy state, but it does not ionize them.

Ionizing radiation – It is a highly effective alternative for sterilizing materials that are sensitive to heat or chemicals. Because it sterilizes in the absence of heat, irradiation is sometimes termed *cold sterilization.* Devices that emit ionizing rays include gamma ray machines containing radioactive cobalt, X-ray machines similar to those used in medical diagnosis, and cathode ray machines that operate like the vacuum tube in a television set. Items are placed in these machines and irradiated for a short time with a carefully chosen dosage. The dosage is measured in rads (radiation absorbed dose). Depending on the application, exposure ranges from 0.5 to 5 megarads (Mrads; a megarad is equal to 1,000,000 rads). Although all ionizing radiations can penetrate solids and liquids, gamma rays are most penetrating, X rays are intermediate, and cathode rays least penetrating. The microbial forms with the greatest resistance to radiation are certain Gram-positive cocci *(Deinococcus),* because of extremely efficient DNA repair mechanisms, and bacterial spores, due to their density. Viruses are second in resistance, followed by yeasts and molds.

Consumer organizations are concerned that radiation poses a danger both in potential toxicity and radioactivity that it can then impart to foods. Two known problems that can occur with irradiated food are : (a) Changes in flavour and nutrition, and (b) the possibility of introducing undesirable chemcial reactions. These disadvantages have been reduced by carrying out the process in very cold, oxygen-free chambers with extremely low doses of radiation. Ionizing radiation is currently approved in the United States for cutting down the microbial load and for disinfecting cured meats, some spices, and seasonings. It is also used on fresh pork to destroy *Trichinella* worms (the cause of trichinosis); on chicken carcasses to control *Salmonella;* on wheat to kill insect eggs; and on fresh fruits and vegetables to cut down on surface microbes that can increase the rate of spoilage. Sterilizing medical products with ionizing radiation is a rapidly expanding field. Drugs, vaccines, medical instruments (especially plastics), syringes, sutures, surgical gloves, and tissues such as bone, skin, and heart valves for grafting all lend themselves to this mode of sterilization. Its main advantages include speed, high penetrating power.

Nonionizing radiation – Absorption of most of the sun's UV waves by the atmosphere prevents its full intensity from being felt and thus limits its practical use. Ultraviolet radiation ranges in wavelength from approximately 100 nm to 400 nm. It is most lethal from 240 nm to 280 nm (with a peak at 260 nm). In everyday practice, the source of UV radiation is the germicidal lamp, which generates radiation at 254 nm. Due to its lower energy state, UV radiation is not as penetrating as ionizing radiation. Because UV radiation passes readily through air, slightly through liquids, and only poorly through solids, the object to be disinfected must be directly exposed to it for full effect. As UV radiation passes through a cell, it is initially absorbed by DNA. Specific molecular damage occurs on the pyrimidine bases (thymine and cytosine), which form abnormal linkages with each other called *pyrimidine dimers.* The fact that they occur between adjacent bases on the same DNA strand, and interfere with normal DNA replication and transcription, accounts for the inhibition of growth and cellular death that result, in addition to altering DNA, UV radiation also disrupts cells by generating toxic photochemical products (free radicals). UV rays are a powerful tool for destroying fungal cells and spores, bacterial vegetative cells, protozoa, and viruses.

Germicidal lamps can cut down on the concentration of airborne microbes as much as 99%. They are used in hospital rooms, operating rooms, schools, food preparation areas, nursing homes, and military housing. UV disinfection of air has proved effective in reducing post-operative infections, preventing the transmission of infections by respiratory droplets, and curtailing the growth of microbes in food processing plants and slaughterhouses. UV irradiation of liquids requires special equipment to spread the liquid into a thin, flowing film that is exposes directly to a lamp. This method can be used to treat drinking water (in place of chlorination) and to purify other liquids (milk and fruit juices) as an alternative to heat. UV treatment has proved effective in freeing vaccine antigens and plasma from contaminants. One major disadvantage of UV is its poor powers of penetration. It does not adequately penetrate opaque materials such as glass, metal, cloth, plastic, and even paper.

Sound Waves use as Physical Agent

These frequencies range from 15,000 to more than 200,000 cycles per second (supersonic to ultrasonic). Such vibrations transmitted through a water-filled chamber (sonicator) induce pressure changes and minute, bubblelike cavities in the liquid, a phenomenon called cavitation. The short-lived bubbles swell and collapse, creating intense points of turbulence that can stress and burst cells in the vicinity. Gram-negative rods are most sensitive to ultrasonic vibrations, and Gram-positive cocci, fungal spores, and bacterial spores are most resistant to them. Sonication not only helps destroy some microbes, but it also forcefully dislodges foreign matter from objects. Heat generated by sonic waves (up to 80°C) also appears to contribute to the antimicrobial action. Ultrasonic devices are used in dental and some medical offices to clear debris and saliva from instruments before sterilization and to clean dental restorations. However, most sonic machines are not predictable enough to be used in disinfection or sterilization.

Filtration is Potent Physical Agent

The basis for this technique is simple : A fluid (air or liquid) is strained through a layer of material with openings large enough for the fluid to pass through but too small for microorganisms to pass through. Most modern microbiological filters are thin membranes of cellulose acetate, polycarbonate, and a variety of plastic materials (Teflon, nylon) whose pore size can be carefully controlled and standardized. Ordinary substances like charcoal, diatomaceous earth, or unglazed porcelain are also used in some applications. Viewed microscopically, most filters are perforated by very precise, uniform pores. The pore diameters vary from coarse (8 μm) to ultrafine (0.02 μm), permitting selection of the minimum particle size to be trapped. Those with the smallest pore diameters permit true sterilization by removing viruses, and some will even remove large proteins. A sterile liquid filtrate is typically produced by suctioning the liquid through a sterile filter into a presterilized container.

Filtration sterilization is used to prepare liquids that cannot withstand heat, including serum and other blood products, vaccines, drugs, IV fluids, enzymes, and media. Filtration has been employed as an alternative to sterilize milk and beer without altering their flavour. It is also an important step in water purification. Its usage extends to filtering out particulate impurities (crystals, fibers, etc.) that can cause severe reactions in the body. It has the disadvantage of not removing soluble molecules (toxins) that can cause disease and not removing all viruses. Filtration is also an efficient means of removing airborne contaminants that are a common source of infection and spoilage. High efficiency particulate air (HEPA) filters are widely used to provide a flow of sterile air to hospital rooms and sterile rooms.

Osmotic Pressure uses as Physical Agent

When two solutions with differing concentrations of solute are separated by a semipermeable membrane, there will occur a passage of water, through the membrane, in the direction of the higher concentration. The trend is toward equalizing the concentration of solute on both sides of a membrane. The solute concentration within microbial cells is approximately 0.95 percent. Thus if cells are exposed to solutions with higher solute concentration, water will be drawn out of the cell. The process is called plasmolysis. The reverse process, that is, passage of water from a low solute concentration into the cell, is termed plasmoptysis. The pressure built up within the cell as a result of this water intake is termed osmotic pressure. These phenomena can be observed more conveniently with animal cells since they do not have rigid cell walls. Plasmolysis results in dehydration of the cell, and as a consequence metabolic processes are retarded partially or completely. The antimicrobial effect is similar to that caused by desiccation. Because of the great rigidity of microbial cell walls (except for protozoa), the cell-wall structure does not exhibit distortions as a result of plasmolysis or plasmoptysis.

CHEMICAL MEASURES OF CONTROL

At the present time, approximately 10,000 different antimicrobial chemical agents are manufactured; probably 1,000 of these are utilized routinely in the allied health sciences and the home. Despite the genuine need to avoid infection and spoilage, the abundance of products available on the drugstore shelf advertised to "kill germs disinfect, antisepticize, clean and senitize, deodorize, fight plaque, or purify the air indicates a preoccupation with eliminating microbes from the environment that, seems excessive. Antimicrobial chemicals occur in the liquid, gaseous or even solid state and vary from disinfectants and antiseptics to sterilants and preservatives. Liquid agents generally have a base of water, alcohol, or a mixture of two, into which various solutes are dissolved.

Selection of Antimicrobial Agents

Although actual chemical practices of decontamination vary widely, some desirable quantities in a germicide have been identified including : (a) Rapid action even in low concentrations, (b) solubility in water or alcohol and long-term stability, (c) broad-spectrum microbicidal action without being toxic to human and animal tissues, (d) penetration of inanimate surfaces to sustain a cumulative or persistent action, (e) resistance to becoming inactivated by organic matter, (f) noncorrosive or nonstaining properties, (g) sanitizing and deodorizing properties, and (h) inexpensiveness and ready availability. As yet, no chemical can completely fulfill all of these requirements, but glutaraldehyde and hydrogen peroxide approach this ideal. Germicides are evaluated in terms of their effectiveness in destroying microbes on medical and dental materials. The three levels of chemical decontamination procedures are *high, intermediate,* and *low*. High-level germicides kill endospores, and, if properly used, are sterilants. Materials that necessitate high-level control are medical devices—for example, catheters, heart-lung equipment, and implants—that are not heat-sterilizable and are intended to enter body tissues during medical procedures. Intermediate-level germicides kill fungal spores, resistant pathogens such as the tubercle bacillus, and viruses. They are used to disinfect items that come in intimate contact with the mucous membranes but are noninvasive. Low levels of disinfection eliminate only vegetative bacteria, vegetative fungal cells, and some viruses.

Factors Affecting Germicidal Activity

A chemical's strength or concentration is expressed in various ways, depending upon convention and the method of preparation. The content of many chemical agents can be expressed by more than one notation. In dilutions, a small volume of the liquid chemical (solute) is diluted in a larger volume of solvent to achieve a certain ratio. For example, a common laboratory phenolic disinfectant such as Amphyl is usually diluted 1:200—that is, one part of chemical has been added to 200 parts of water by volume. Solutions such as chlorine that are effective in very high dilutions are expressed in parts per million (ppm). In percent solutions, the solute is added to water by weight or volume to achieve a certain percentage in the solution. Alcohol, for instance, is used in percentages ranging from 50% to 95%.

Most compounds require adequate contact time to allow the chemical to penetrate and to act on the microbes present. The composition of the material being treated must also be considered. Smooth, solid objects are more reliably disinfected than those with pores, pockets, or spaces that can trap soil. An item contaminated with common biological matter such as serum, blood, saliva, pus, tissue fluid, fecal material, or urine presents a problem in disinfection. Large amounts of organic material can hinder the penetration of a disinfectant and, in some cases, form bonds that reduce its activity. Germicides containing chlorine, mercurials, and some detergents are inactivated to some extent in the presence of a high load of organic substances.

Phenols and Alcohols use as Chemical Agent

Lister became aware of Pasteur's studies which incriminated germs as the cause of infection. Accordingly, he instituted the practice of applying a solution of phenol (carbolic acid) to surgical incisions and wounds. The reduction in infections was striking. Later he developed the practice of spraying phenol into the operating room area to control infection. Phenol has the additional distinction of being the standard against which other disinfectants of a similar chemical structure are compared to determine their antimicrobial activity.

Phenol and phenolic compounds are very effective disinfectants. A 5% aqueous solution of phenol rapidly kills the vegetative cells of microorganisms; spores are much more resistant. Many derivatives of phenol have been prepared and evaluated for their antimicrobial activity (Fig. 10.3). Hexylresorcinol, a derivative of phenol, is marketed in a solution of glycerin and water. It is a strong surface-tension reductant, which may account in part for its high bactericidal activity. A commercial product containing hexylresorcinol, S. T. 37, is so named because of its surface-tension value. Hexylresorcinol preparations are employed as general antiseptics.

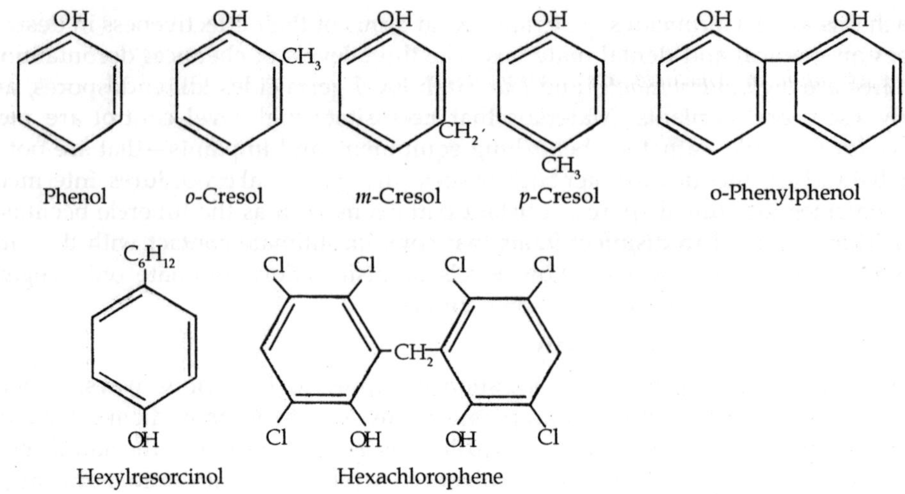

Fig. 10.3 : Phenol and phenolic derivatives.

Phenolic substances may be either bactericidal or bacteriostatic, depending upon the concentration used. Bacterial spores and viruses are more resistant than are vegetative cells. Some phenolics are highly fungicidal. The antimicrobial activity of phenolics is reduced at an alkaline pH and by organic material. Low temperatures and the presence of soap also reduce antimicrobial activity. Pure crystalline phenol is colourless. Aqueous solutions of from 2 to 5% can be employed to disinfect such materials as sputum, urine, feces, and contaminated instruments or utensils. Solutions of pure phenol have limited application. However, derivatives of phenol diluted in detergents or some other carrier find use in many commercial antiseptic and disinfectant preparations. One of the widely used phenolic derivative is o-phenylphenol.

Exposure of microbial cells to phenolic compounds produces a variety of effects. Depending upon the concentration of the phenolic compound to which microbial cells were exposed, researchers have described results such as disruption of cells, precipitation of cell protein, inactivation of enzymes, and leakage of amino acids from the cells. Although the specific mode of action is not clear, there is a consensus that the lethal effect is associated with physical damage to the membrane structures in the cell surface, which initiates further deterioration.

Ethyl alcohol cannot be relied upon to produce a sterile condition. Concentrations which are effective against vegetative cells are practically inert against bacterial spores. Methyl alcohol is less bactericidal than ethyl alcohol; furthermore, it is highly poisonous. Even the fumes of this compound many produce permanent injury to the eyes, and is not generally employed for the destruction of microorganisms. The higher alcohols—propyl, butyl, amyl, and others—are more germicidal than ethyl alcohol. In fact, there is a progressive increase in germicidal power as the molecular weight of alcohols increases. Since alcohols of molecular weight higher than that of propyl alcohol are not miscible in all proportions with water, they are not commonly used in disinfectants.

Halogens used as Chemical Agents

Pure iodine is a bluish-black cyrstalline element having a metallic luster. It is only slightly soluble in water but readily soluble in alcohol and aqueous solutions of potassium or sodium iodide. The element is traditionally used as a germicidal, agent in a form referred to as tincture of iodine. There are several preparations available, such as 2% iodine plus 2% sodium iodide diluted in alcohol, 7% iodine plus 5% potassium iodide in 83% alcohol, and 5% iodine and 10% potassium iodide in aqueous solution. Iodine is also used in the form of substances known as iodophors. Iodophors are mixtures of iodine with surface-active agents which act as carriers and solubilizers for the iodine. One of these agents is polyvinylpyrrolidone (PVP); the complex can be expressed as PVP-I. Iodine is released slowly from this complex.

Iodine is a highly effective bactericidal agent and is unique in that it is effective against all kinds of bacteria. Iodine also possesses sporicidal activity; however, the rate at which the spores are killed is markedly influenced by the conditions under which they are exposed, e.g., amount of organic material and extent of dehydration. In addition, it is highly fungicidal and is to some extent virucidal. Iodine solutions are chiefly used for the disinfection of skin, and for this purpose they rank among the best disinfectants. The effectiveness of iodine preparations for the reduction of the microbial flora of skin is well recognized. Oxidizing agents can irreversibly oxidize and thus inactivate essential metabolic compounds such as proteins with sulfhydryl groups. It has also been suggested that the action may involve the halogenation of tyrosine units of enzymes and other cellular proteins requiring tyrosine for activity.

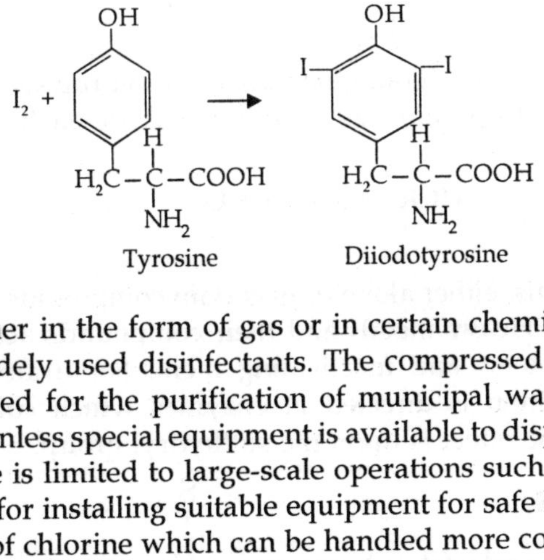

Chlorine, either in the form of gas or in certain chemical combinations, represents one of the most widely used disinfectants. The compressed gas in liquid form is almost universally employed for the purification of municipal water supplies. Chlorine gas is difficult to handle unless special equipment is available to dispense it. Hence, its usefulness in the gaseous state is limited to large-scale operations such as water-purification plants, where it is feasible for installing suitable equipment for safe handling. There are available many compounds of chlorine which can be handled more conveniently than free chlorine and which, under proper conditions of use, are equally effective as disinfectants. One class of compounds in this category is the hypochlorites. Calcium hypochlorite, $Ca(OCl)_2$ and sodium hypochlorite, NaOCl, are popular compounds. The chloramines represent another category of chlorine compounds used as disinfectants, sanitizing agents, or antiseptics. Chemically they are characterized by the fact that one or more of the hydrogen atoms in an amino group of a compound are replaced with chlorine. The simplest of these is monochloramine, NH_2Cl. Chloramine-T and azochloramide, two of several germicidal compounds in this category, have more complex chemical structures.

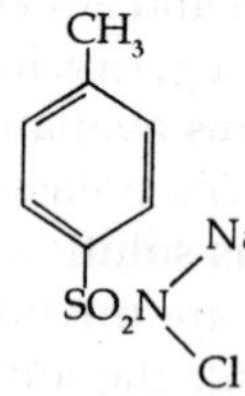
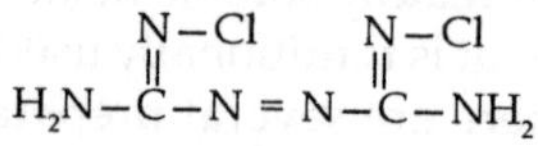

Chloramine-T Azochloramide

Medical students were required to wash their hands and soak them in a hypochlorite solution before examining patients. Chlorine compounds are very widely used to control microorganisms. Major applications are in water treatment, in the food industry, for domestic uses, and in medicine. Products containing calcium hypochlorite are used for sanitizing dairy equipment and eating utensils in restaurants. Solutions of sodium hypochlorite of a 1% concentration are used for personal hygiene and as a household disinfectant; higher concentrations of 5 to 12% are also employed as household bleaches and disinfectants and for use as sanitizing agents in dairy and food-processing establishments. The amount of hypochlorite added should provide a residual concentration of approximately 1 mg per liter of free chlorine. The antimicrobial action of chlorine and its compounds comes through the hypochlorous acid formed when free chlorine is added to water :

$$Cl_2 + H_2O \rightarrow HCl + HClO \text{ (hypochlorous acid)}$$

Similarly, hypochlorites and chloramines undergo hydrolysis, with the formation of hypochlorous acid. The hypochlorous acid formed in each instance is further decomposed :

$$HClO \rightarrow HCl + O$$

Heavy Metals

Most of the heavy metals, either alone or in certain compounds, exert a detrimental effect upon microorganisms. Heavy metals and their compounds act antimicrobially by combining with cellular proteins and inactivating them. For example, in the case of mercuric chloride the inhibition is directed at enzymes which contain the *sulfhydryl* grouping. Note below that the effect is upon the sulfhydryl group.

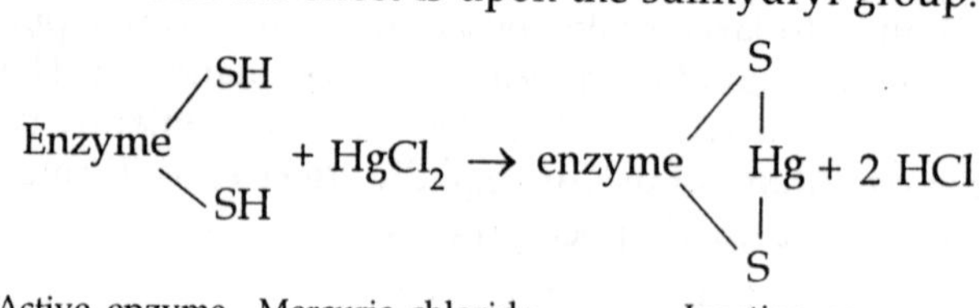

Active enzyme Mercuric chloride Inactive enzyme

Dyes use as Chemical Agents

As a rule Gram-positive organisms are more susceptible to lower concentrations of these compounds than are Gram-negative ones. Crystal violet will inhibit Gram-positive cocci at a dilution of 1 : 200,000 to 1 : 300,000; 10 times this concentration is required to inhibit *Escherichia coli. Staphylococcus aureus* is inhibited by malachite green at a concentration of 1 : 1,000,000; a concentration of about 1 : 30,000 is required to inhibit *E. coli.* Certain media can be made selective by the incorporation of low concentrations of the dyes crystal violet, brilliant green, or malachite green. Gram-positive bacteria will be inhibited. Media of this kind are used extensively in public health microbiology, where detection of *E. coli* is important. Susceptibility to various dyes can also be used for identification of bacteria. Three species of *Brucella* can be distinguished by their patterns of resistance to several dyes. Crystal violet has also been used as a fungicide. Two examples of dyes derived from acridine are acriflavine and tryptoflavine. These compounds exhibit selective inhibition against bacteria, particularly staphylococci and gonococci. Gonococci are inhibited by tryptoflavine in dilutions of 1 : 10,000,000 to 1 : 50,000,000. They possess little, if any, antifungal activity. Presently, they have less application than before the advent of antibiotics and other chemotherapeutic agents. They are used to some extent for the treatment of burns and wounds and for ophthalmic application and bladder irrigation.

Detergents use as Chemical Agents

Soap is a poor detergent in hard water. For this reason many new more efficient cleaning agents have been developed, called *surfactants* or synthetic detergents, many of which are superior to soap. They do not form precipitates in alkaline or acid water, nor do they produce deposits with minerals found in hard water. They are extensively used in laundry dishwashing powders, shampoos, and other washing preparations. Some are also highly bactericidal.

(a) Those which ionize with the detergent property resident in the anion are referred to as anionic detergents. For example,

$[C_9H_{19}COO)^- Na^+$	$[C_{12}H_{25}OSO_3]^- Na^+$
A soap	Sodium lauryl sulfate

(b) A third category of detergents is nonionic; i.e., they do not ionize. However, these substances do not possess significant antimicrobial activity.

Insofar as reduction of the microbial flora from surfaces such as skin and clothing is concerned, the real value of ordinary soaps lies in the mechanical removal of microorganisms. Soaps reduce surface tension and thereby increase the wetting power of the water in which they are dissolved. Soapy water has the ability to emulsify and disperse oils and dirt. The microorganisms become enmeshed in the soap lather and are removed by the rinse water.

Ammonium Compounds use as Chemical Agents

The R_1, R_2, R_3, and R_4 groups are carbon groups linked to the nitrogen atom. The R groups may be any one of a large number of different alkyl groups. Accordingly, a very large number of different quaternary ammonium compounds have been synthesized and evaluated for their antimicrobial activity. Several are available commercially as effective antimicrobial agents for a variety of uses. The bactericidal power of the quaternaries is exceptionally high against Gram-positive bacteria, and they are also quite active against Gram-negative organisms. Another of their characteristics is the ability to manifest bacteriostatic action far beyond their bactericidal concentration. For example, the limit of bactericidal action for a given compound may be at a dilution of 1 : 30,000; yet it may be bacteriostatic in dilutions as high as 1 : 200,000. The action of these compounds demonstrates the need to distinguish between static and lethal activity in test procedures for the evaluation of disinfectants.

The combined properties of germicidal activity and detergent action, plus such other features as low toxicity, high solubility in water, stability in solution, and noncorrosiveness, have resulted in many applications of quaternaries as disinfectants and sanitizing agents. They are used as skin disinfectants, as a preservative in ophthalmic solutions, and in cosmetic preparations. Quaternaries are widely used for control of microorganisms on floors, walls, and other surfaces in hospitals, nursing homes, and other public places. They are used to sanitize food and beverage utensils in restaurants as well as surfaces and certain equipment in food-processing plants. A variety of damaging effects of quaternaries upon microorganisms have been observed. These include denaturation of proteins, interference with glycolysis, and membrane damage. Experimental evidence suggests that the most likely site of the damage to the cell is the cytoplasmic membrane; the quaternaries alter the vital permeability features of this cell structure.

Aldehydes use as Chemical Agents

Among the class of chemicals with the general formula RCHO (aldehydes), several of the low-molecular-weight compounds are antimicrobial. Two of the most effective are formaldehyde and glutaraldehyde. Formaldehyde (HCHO) is the simplest compound in the aldehyde series. It is a gas that is stable only in high concentrations and at elevated temperatures. At room temperature it polymerizes, forming a solid substance. The important polymer is paraformaldehyde, a colourless substance which rapidly yields formadehyde upon heating. Formaldehyde is also marketed in aqueous solution as formalin, which contains 37 to 40% formaldehyde. Formaldehyde in solution is useful for sterilization of certain instruments. Formaldehyde in gaseous form can be used for disinfection and sterilization of enclosed areas. Formalin and paraformaldehyde are two principal sources of formaldehyde when it is used for gaseous disinfection. Vaporization of formaldehyde from either of these sources into an enclosed area for an adequate time will cause sterilization, vegetative cells being killed more quickly than spores. Humidity and temperature have a pronounced effect on the microbicidal action of formaldehyde; in order to sterilize an enclosure the temperature must be about room temperature (22°C) and the relative humidity between 60 to 80 percent.

Formaldehyde is an extremely reactive chemical. It combines readily with vital organic nitrogen compounds such as proteins and nucleic acids. It is likely that interaction of formaldehyde with these cellular substances accounts for its antimicrobial action. Certain kinds of medical devices that need to be available in a sterile condition are made of materials that are damaged by heat. Examples are plastic syringes, blood transfusion apparatus, and catheterization equipment. The same is true for routinely used laboratory ware, such as plastic pipettes, Petri dishes, and other equipment, that is packaged and sterilized ready for use. On occasion there is need to disinfect or sterilize an enclosed area. Sterilization by means of gaseous agents is effective and practical for such situations.

Gaseous Compounds use as Chemical Agents

Ethylene oxide is a liquid at temperatures below 10.8 °C (51.4°F). Above this temperature it vaporizes rapidly. Vapors of this compound in air are highly flammable even in low concentrations. In this respect it is very much like diethyl ether. This objectionable feature was overcome by preparing mixtures of ethylene oxide in carbon dioxide or Freon, which are now available commercially. The carbon dioxide-ethylene oxide or Freon-ethylene oxide mixtures are nonflammable, and there is no alteration of the microbicidal activity of the ethylene oxide. Ethylene oxide is a unique and powerful sterilizing agent. Its use for sterilizing heat-or moisture-sensitive materials in hospitals, industry, and laboratories has become universal. Bacterial spores, which are many times more resistant than vegetative cells as measured by other antimicrobial agents, show little resistance to destruction by this agent. It must be used with caution, although devices are available for its safe, routine laboratory use. The commercially available apparatus for this purpose is essentially an autoclave modified to allow the chamber to be filled with the gas under controlled conditions. The concentration of ethylene oxide, as well as the temperature and humidity, are critical factors which together determine the time required to achieve sterilization. Modern autoclaves are equipped with controls to maintain the desired concentration of ethylene oxide and the proper temperature and humidity.

Ethylene oxide has been established as an effective sterilizing agent for heat-and moisture-sensitive materials. Effective usage requires careful control of three parameters : ethylene oxide concentration, temperature, and moisture. The varieties of materials on which it is used include spices, biological preparations, soil, plastics, certain medical preparations, and contaminated laboratory equipment. It has been used in the space program by both the Americans and the Russians for decontaminating spacecraft components. Among the advantages already mentioned for this agent as a sterilant is its remarkable penetration and its broad spectrum of activity against microorganisms, including spores.

The mode of action of ethylene oxide is believed to be alkylation reactions with organic compounds such as enzymes and other proteins. Alkylation consists in the replacement of an active hydrogen atom in an organic compound, e.g., the hydrogen atom in a free carboxyl, amino, or sulfhydryl group, with an alkyl group. In this reaction the ring in the ethylene oxide molecule splits and attaches itself where the hydrogen was originally. It is not flammable like ethylene oxide but is a vesicant and lachrymator and consequently must be handled with care. It lacks the penetrating power of ethylene oxide but is considerably more active against microorganisms; it is sporicidal, fungicidal, and virucidal. Whereas the usual concentration of ethylene oxide for sterilization purposes is 400 to 800 mg/liter, only 2 to 5 mg of β-propiolactone is required. β-propiolactone is very effective in destroying microorganisms on surfaces.

LABORATORY TECHNIQUES : ANTIMICROBIAL ASSAYS

Dilution Tube Method

In each the chemical agent is tested against a specified microorganism referred to as the test organism.

(a) Liquid water-soluble substances appropriately diluted are dispensed into sterile test tubes, to which are added a measured amount of the test organism. At specified intervals, a transfer is made from this tube into tubes of sterile media that are then incubated and observed for the appearance of growth. It is necessary in this type of procedure to ascertain whether the inhibitory action is bactericidal and not bacteriostatic.

(b) The chemical agent is incorporated into an agar medium or broth, inoculated with the test organism, incubated, and then observed for (a) decrease in the amount of growth or (b) complete absence of growth.

(c) A plate of agar medium is inoculated with the test organism, and the chemical agent is placed on this medium. Following incubation, the plate is observed for a zone of inhibition (no growth) around the chemical agent. This is particularly suitable for semisolid preparations. It may also be used for liquid solutions, in which case the solution is first impregnated in absorbent paper or confined by a hollow cylinder placed on the agar surface.

Phenol-Coefficient Method

This procedure is suitable for testing disinfectants miscible with water and exerting their antimicrobial action in a manner similar to that of phenol. The test organism employed in this procedure is a specific strain of either *Salmonella typhi* or *Staphylococcus aureus*. The temperature at which the test is performed, the manner of making subcultures, the composition of the subculture medium. The size of the test tubes, and other details of the test are spelled out in the official procedure. Briefly, the test is performed as follows :

To a series of dilutions of the disinfectant being tested (5 ml per tube), 0.5 ml of 24-hour broth culture of the test organism is added. At the same time, similar additions, in the same amounts, are made to a series of dilutions of phenol. All tubes (disinfectant + organisms and phenol + organisms) are placed in a 20°C water bath. At intervals of 5, 10, and 15 min, subcultures are made with a loop transfer needle into sterile tubes of medium. The inoculated subculture tubes are incubated and subsequently examined for growth. The greatest dilution of the disinfectant killing the test organism in 10 min but not in 5 min is divided by the greatest dilution of phenol showing the same result. The number obtained by this division is the phenol coefficient of the substance tested. It should be emphasized that no single microbiological test method is suitable for the evaluation of all germicidal chemicals for all applications recommended. Therefore, one must exercise care in selecting a test method for a specific chemical agent, so the results obtained will be meaningful and reproducible and lend themselves to some degree of practical interpretation. The ultimate criterion for the effectiveness of a germicidal agent is its performance under practical conditions.

□□□

11

MICROORGANISMS AND HUMAN INTERACTIONS

The first step in host – parasite interactions is the encounter between the microbe and the host. This can occur through a variety of methods ranging from direct contact with an infected animal or contaminated fomite, inhalation of contaminated aerosolized droplets or particles, ingestion of a contaminated food or beverage, intimate sexual contact, arthropod vectors, or contaminated needles and syringes. Following contact with the microbe, one of two possible situations develops : the microbe will be lost/shed without any significant interaction with the host or the microbe may establish a relationship with the host. This relationship can be a coexistence that is harmless to the microbe and host (commensalism), a mutual benefit to both the host and the microbe (symbiosis), or a benefit to the microbe at the expense of the host (parasitism). There are other types of relationships that can be established, but we will not consider them at this time. The microbe (parasite) in commensalism is referred to as a commensal while the benefiting parasite in parasitism is referred to as a pathogen. The host in or on which the parasite lives (plants, animals, or humans) is unharmed/benefits in commensalism/symbiosis while in parasitism the host is harmed or damaged in some manner. In parasitism, damage to the host may result from pathogen-encoded structures or products and/or the host's response to the infection. This damage results in the production of a set of signs and symptoms characteristic of a disease state. Commensal organisms typically inhabit or infect host body surfaces, and by the nature of their relationship they have evolved so as to occupy and effectively compete within a specific host niche without causing damage to the host under normal circumstances.

As such, commensal organisms have become an important asset to the host by supplying important nutrients, breaking down material making nutrients available to the host, and/or occupying space, making it difficult for potential pathogenic microbes to establish themselves within the host. This population of commensal microbes can produce a biofilm that covers many of the host surfaces and is generally referred to as normal or resident flora for the host. Commensals are proposed to block the degradation of I*k*B, which normally inhibits the function of NF-*k*B. NF-*k*B is a key transcription factor in host cells; in particular, it activates the transcription of tumor necrosis factor (TNF) and interleukin (IL) -8, which in turn promote an inflammatory response. Thus, commensals are proposed to actively suppress the host inflammatory response. For a pathogen, the next step after colonization is the entry of the microbe or a microbial product(s) across an epidermal or epithelial cell surface to enter deeper well-vascularized (both lymph and blood circulation) tissue.

With the exception of microbes transmitted by blood- sucking arthropod vectors, or contaminated needles or syringes, entry into the host is for many microbes the difference between life as a commensal organism, confined to the host's body surface, and life as a pathogen, able to infect and elicit damage to host tissue. Entry into the host may or may not involve the physical entry of the microbe into deeper tissue but instead may be limited to the entry of a microbe-derived product such as an exotoxin. The physical entry of the microbe involves microbial products, which mediate the direct invasion of epithelial cells or specialized epithelial cells called M cells, such as invasins. In order to multiply and disseminate, the pathogen must be able to acquire nutrients to support its growth.

HUMAN USED AS HABITAT

Some microbes become implanted there as colonists (normal flora), some are rapidly lost (transients), and others invade the tissues. Such intimate contact with microbes inevitably leads to infection, a condition in which pathogenic microorganisms penetrate the host defenses, enter the tissues, and multiply. When the cumulative effects of the infection damage or disrupt tissues and organs, a disease results. A disease is defined as any deviation from health. There are hundreds of different diseases caused by such factors as infections, diet, genetics, and aging. The pattern of the host-parasite relationship can be viewed as a series of stages that begins with contact, progresses to infection, and ends in disease. Because of numerous factors relating to host resistance and degree of pathogenicity, not all contacts lead to infection and not all infection lead to disease. With its constant source of nourishment and moisture, relatively stable pH and temperature, and extensive surfaces upon which to settle, the human body provides a favourable habitat for an abundance of microorganisms. In fact, it is so favourable that, cell-for-cell, microbes outnumber human cells ten to one. The large and mixed collection of microbes adapted to the body has been variously called the normal resident *flora, indigenous* flora, though some microbiologists prefer to use the terms *microflora* and *commensals.*

Development of the flora proceeds in a relatively orderly and characteristic way. Of the seemingly limitless influx of microbes, only some are capable of persisting in and on the body, and the vast majority are destroyed by host defenses or removed. Those that remain survive by adapting to a particular microhabitat—one that fills requirements for food, moisture, presence or lack of oxygen, layers of dead cells, and even other microorganisms. In general, most anatomical surfaces directly exposed to the environment can and do harbor microbes. Moist areas through which food passes (the alimentary tract) tend to have the largest and most diverse populations. By contrast, organs and fluids inside the body cavity and the central nervous system remain free of normal flora during life because host defense mechanisms maintain their sterility. Interactions with indigenous flora have devolved over millions of years into complex and dynamic relationships that directly and indirectly influence many aspects of body function.

For the most part, effects of normal flora are protective or nutritional. For instance, the flora can modify its microhabitat by altering pH and oxygen tension; excreting chemicals such as fatty acids, gases, alcohol, and antibiotics; or creating barriers. Such reactions can favour colonization by more beneficial types of microbes. They can also inhibit potential pathogens by restricting their population size and activity through antagonism and competition for nutrients. Although relatively stable, the flora fluctuates to a limited extent with general health, age, variations in diet, hygiene, hormones, and drug therapy. Most species are not serious pathogens and remain harmless as long as they do not penetrate superficial skin and mucosal barriers. However, a few species have the potential to become medically important if the balance of the flora shifts. This is seen in the overgrowth of *Clostridium difficile* or *Candida albicans* in patients undergoing antibiotic therapy. After being temporarily disturbed, the normal balance of flora is restored by surviving residents harbored in protected areas and by contact with the environment.

The event that first exposes the infant to microbes is the breaking of the fetal membranes, at which time microbes from the mother's vagina can enter the womb. Comprehensive exposure occurs during the birth process itself, when the baby unavoidably comes in intimate contact with the birth canal. Within 8 to 12 hours after delivery, the newborn typically has been colonized by bacteria such as streptococci, staphylococci, and lactobacilli, acquired primarily from its mother. The nature of the flora initially colonizing the large intestine depends upon whether the baby is bottle-or breast-fed. Bottle-fed infants tend to acquire a mixed population of coliforms, lactobacilli, enteric streptococci, and staphylococci. In contrast, the intestinal flora of breast-fed infants consists primarily of *Bifidobacterium* species whose growth is favoured by a growth factor from the milk. Milestones that contribute to development of the adult pattern of flora are eruption of teeth, weaning, and introduction of the first solid food. Although exposure to microbes is unavoidable and even necessary for the maturation of the infant's flora, contact with pathogens is dangerous, because the neonate is not yet protected by a full complement of flora, and due to its immature immune defenses, is extremely susceptible to infection.

Specific Flora

Studies of the flora have shown that most people harbor certain specially adapted bacteria, fungi, and protozoa.

Flora of the human skin—Its major layers are the epidermis, an outer layer of dead cells continually being sloughed off and replaced, and the dermis, which lies atop the subcutaneous layer of tissue. Depending on its location, skin also contains hair follicles and several types of glands, and the outermost surface is covered with a protective, waxy cuticle that can help microbes adhere. The normal flora resides only in or on the dead cell layers, and except for follicles and glands, it does not extend into the dermis or subcutaneous levels. The nature of the population varies according to site. Oily, moist skin supports a more prolific flora than dry skin. Humidity, occupational exposure, and clothing also influence its character. Ordinarily, there are two cutaneous populations. The transient population, or exposed flora, clings to the skin surface, but does not ordinarily grow there. It is acquired by routine contact, and it varies markedly from person to person and over time. The transient flora includes any microbe a person has picked up, often species that do not ordinarily live on the body, and it is greatly influenced by the hygiene of the individual. The resident population lives and multiplies in deeper layers of the epidermis and in glands and follicles. The composition of the resident flora is more stable, predictable, and less influenced by hygiene. The normal skin residents consist primarily of bacteria (notably *Staphylococcus, Corynebacterium* and *Propionibacterium*) and yeasts. Moist skin folds, especially between the toes, tend to harbor fungi, while lipophilic mycobacteria and staphylococci are prominent in sebaceous secretions of the axilla, external genitalia, and external ear canal.

Flora of the alimentary tract—It encompasses the oral cavity, esophagus, stomach, small intestine, large intestine, rectum, and anus. Stating that the alimentary canal harbors flora may seem to contradict our earlier statement that internal organs are sterile, but it is not an exception to the rule. How can this be true? In reality, the alimentary tract is a long, hollow tube (with numerous pockets and curves), bounded by the mucous membranes of the oral cavity on one extreme and those of the anus on the other. Since the innermost surface of this tube is exposed to the environment, it is topographically outside the body, so to speak. The shifting conditions of pH, oxygen tension, and differences in microscopic anatomy of the alimentary tract are reflected by the variations in or distribution of the flora. Some microbes remain attached to the mucous epithelium or its associated structures, and others dwell in the *lumen*. Although the abundance of nutrients invites microbial growth, the only areas that harbor appreciable permanent flora are the oral cavity, large intestine, and rectum. The esophagus undergoes wavelike contractions (peristalsis), a process that constantly flushes microorganisms; the stomach acid inhibits most microbes; and peristalsis and digestive enzymes help exclude flora from all but the terminal segment of the small intestine.

Mouth flora—Microhabitats, including the cheek epithelium, gingiva, tongue, floor of the mouth, and tooth enamel, provide numerous adaptive niches for hundreds of different species to colonize. The most common residents are aerobic *Streptococcus* species—*S. sanguis, S. salivarius, S. mitis*—that colonize the smooth superficial epithelial surfaces. Two species, *S. mutans* and *S. sanguis,* make a major contribution to dental caries by forming sticky dextran slime layers in the presence of simple sugars. The adherence of dextrans to the tooth surface establishes a medium that attracts other bacteria. Once the teeth have erupted, an anaerobic habitat is established in the gingival crevice that favours the colonization of anaerobic bacteria that can be involved in dental caries and periodontal infections. The instant that saliva is secreted from ducts into the oral cavity, it becomes laden with resident and transient flora. Saliva normally has high bacterial count, a fact that tends to make mouthwashes rather ineffective and a human bite very dangerous.

Large intestine flora—The large intestine (cecum and colon) and the rectum harbor a huge population of microbes (10^8-10^{11} per gram of feces). So abundant and prolific are these microbes that they comprise 10-30% of the fecal volume. Even an individual on a long-term fast passes feces consisting primarily of bacteria. Due to the state of the intestinal environment, the predominant fecal flora are strictly anaerobic bacteria (*Bacteroides, Bifidobacterium*). Coliforms (L. *colum,* a sieve) such as *Escherichia coli, Enterobacter,* and *Citrobacter* are present in smaller numbers. Many species ferment waste materials in the feces, generating vitamins (B_{12}, vitamin K, pyridoxine, riboflavin, and thiamine) and acids

(acetic, butyric, and propionic acids) of potential value to the host. Intestinal bacteria contributes to intestinal odor by producing *skatole* (Gr. *Skatos*, dung), amines, and gases (CO_2, H_2, CH_4 and H_2S). Intestinal gas is known in polite circles as *flatus*, and the expulsion of it as *flatulence*. Some of the gas arises through the action of bacteria on dietary carbohydrate residues from vegetables such as cabbage, corn, and beans. The bacteria produce an average of 8.5 liters of gas daily, but only a small amount is ejected in flatus. Late in childhood, members of certain ethnic groups lose the ability to secrete the enzyme lactase. When they ingest milk or other lactose-containing dairy products, lactose is acted upon instead by intestinal bacteria, and severe intestinal distress can result.

Flora of the respiratory tract—Inhaled air regularly contains microbes that are filtered out, destroyed, or expelled, although some can adapt to specific regions of this habitat. *Staphylococcus aureus* preferentially resides in the nasal entrance, nasal vestibule, and anterior nasopharynx, and *Neisseria* species take up residence in the mucous membranes of the nasopharynx behind the soft palate. Lower still, in the vicinity of the tonsils and lower pharynx, are assorted streptococci and species of *Haemophilus* that colonize the tonsils and lower pharynx.

Genitourinary tract flora—The internal reproductive organs are kept sterile through physical barriers such as the cervical plug and other host defenses. The kidney, ureter, bladder, and upper urethra are presumably kept sterile by urine flow and regular bladder emptying. Since the urethra in women is so short (about 3.5 cm long), it can form a passage for bacteria to the bladder and lead to frequent urinary tract infections. The principal residents of the urethra are nonhemolytic streptococci, staphylococci, corynebacteria, and occasionally, coliforms. The vagina presents a notable example of how changes in physiology can greatly influence the composition of the normal flora. An important factor influencing these changes in women is the hormone estrogen. Estrogen normally stimulates the vaginal mucosa to secrete the carbohydrate glycogen, which certain bacteria (primarily *Lactobacillus* sp.) ferment, thus lowering the pH to about 4.5. Before puberty, a girl produces little estrogen, little glycogen, and has a vaginal pH of about 7. These conditions favour the establishment of diphtheroids, staphylococci, streptococci, and some coliforms. As hormone levels rise at puberty, the vagina begins to deposit glycogen and the flora shifts to the acid-producing lactobacilli. It is thought that the acidic pH of the vagina during this time prevents the establishment and invasion of microbes with potential to harm a developing fetus. The estrogen-glycogen effect continues, with minor disruptions, throughout the childbearing years until menopause, when the flora returns to a mixed population similar to that of prepuberty.

ORGANELLES INVOLVED IN HUMAN-PARASITE INTERACTIONS

The first step in host—parasite interactions is adherence to host surfaces. This is necessary for the colonization of the host and the ability to cause disease. Because it is a critical step in host—parasite interactions, microbial pathogens have evolved a variety of structures and mechanisms to promote their adherence to host cells.

Fimbriae (Pili)—Recently, researchers have proposed that the term *fimbriae* be used to describe structures involved in bacterial-host cell adherence and that the term *pili* be used to describe structures involved in bacteria-to-bacteria adherence such as sex pili involved in conjugation. A complex multiprotein apparatus carries out pilus biogenesis.

A common scheme for pilus formation involves a chaperone/usher pathway. In this scheme, pilus subunits are exported to the periplasm where they interact with chaperone proteins that stabilize the pilus subunits, preventing them from prematurely interacting with each other during their time in the periplasm. The pilus subunit is then transferred to the usher protein complex, which forms an oligomeric channel in the outer membrane and translocates the subunits across the outer membrane. In this way, the subunits are assembled to form a relatively long threadlike structure possessing a tip protein complex that is adhesive and directly interacts with the target receptor.

Type I pili were first identified for their ability to agglutinate red blood cells. Interestingly, this binding is blocked by the addition of mannose, so it is referred to as mannose-sensitive hemagglutination (MSHA). Mannose-sensitive type I pili are produced by a large variety of Gram-negative bacteria—for example, *E. coli, Salmonella enterica, Klebsiella* spp., and *Vibrio* spp. It was later found that some type I—like pili are not inhibited from binding by mannose and therefore mediate mannose-resistant hemagglutination (MRHA) .

Type II pili are thought to be mutants of type I pili, since they are similar to type I pili but are nonadhesive. **Type III** pili mediate MRHA but only when red blood cells are pretreated with tannic acid. **Type IV** pili were originally described for their ability to mediate MRHA and as being immunologically different from type I pili. However, it was later shown that they are assembled through a different pathway than other pili. Many of the pili genes have been sequenced and therefore they can be compared/aligned to one another and grouped based on similarities at the DNA or deduced amino acid sequence level. More recently, pili have been named based on specific characteristics owned by them-for example, pyelonephritis-associated pili (Pap or P pili), bundle-forming pili (BFP), or toxin coregulated pili (TCP).

P pili of uropathogenic *E. coli* (UPEC) are among the best-studied pili, particularly in terms of their biogenesis. Thus, they have become a paradigm for the process of pilus formation. P pili are essential adhesin factors for UPEC strains to colonize the urinary tract and cause pyelonephritis. P pili are similar to type I pili in terms of gene organization and analogous sequences, but type I pili are flexible structures whereas P pili are rigid rodlike structures. The P pili are encoded for by the *pap* operon. The major subunit of P pili is encoded by the *papA* gene. The PapD protein acts as a chaperonin required for transport of the pilus subunits from the inner membrane to the outer membrane, where they are accepted by the PapC protein, which is proposed to function as an usher protein. Two additional proteins, PapF and PapK, play a role in tip fibrillum synthesis. The P pili attach to specific host cell receptors in the upper urinary tract via the tip adhesin protein PapG. PapG binds to α–D-galactopyranosyl (1-4)-β-D-galactopyranoside [Galα-(1-4)-Gal] moieties of glycolipids on cells lining the upper urinary tract.

Afimbrial adhesins—*Bordetella pertussis,* the causative agent of whooping cough in children, expresses a number of adhesin factors including pili and several afimbrial adhesins. All these adhesin factors contribute to the difficult task of adherence to the ciliated respiratory epithelium and colonization of the upper respiratory mucosa. Examples of the latter include filamentous hemagglutination (FHA), pertactin, pertussis toxin and the BrkA protein. FHA is a high-molecular-weight protein sharing homology with the high-molecular-weight (HMW) afimbrial adhesins of *Haemophilus influenzae* and many eukaryotic cell-to-cell adhesins. FHA possesses the RGD (arginine-glycine-aspartate) motif characteristic of the recognition site of adhesion molecules that bind to eukaryotic surface integrin proteins. FHA binds to receptors on the ciliated respiratory epithelium and the leukocyte integrin CR3; the latter triggers bacterial uptake into macrophages without activating NADPH oxidase and the respiratory burst. Furthermore, the S2 and S3 subunits of pertussis toxin share similarities with eukaryotic selectin proteins-that is, cell adhesion molecules present on leukocytes (L-selectins) and endothelium (E- and P-selectins) that bind to mucin-like cell adhesion molecules (CAMs). Both pertactin and BrkA also possess RGD motifs and appear to be involved in host cell adherence. Another example of an afimbrial adhesin molecule is the YadA protein of *Yersinia enterocolitica,* which binds to cell- associated fibronectin. Similarly, *Streptococcus pyogenes* also produces at least two afimbrial adhesin structures that bind to cellular fibronectin : Protein F and M protein/ lipoteichoic acid (LTA) complexes.

Host cell specificity—In enteric bacteria the production of different pili or fimbriae determines the regions of the intestinal tract that the pathogen will adhere to and colonize. The extreme acidic pH achieved in the stomach normally kills most ingested bacteria. However, *Helicobacter pylori* is able to colonize the stomach. It binds preferentially to the gastric mucosa by binding to the Lewis blood group antigen expressed on gastric mucosal epithelium; however, the adhesin has not been definitively identified. This close interaction with the gastric mucosal epithelium is beneficial, since it is believed that the immediate environment around the gastric mucosa is neutralized to prevent damage to the epithelial cells. Passage out of the stomach into the intestine provides several new challenges for the pathogen. Strains of *Vibrio cholerae,* enteropathogenic *E. coli* (EPEC), and enterotoxigenic *E. coli* (ETEC) adhere to and colonize the mucosal epithelium of the duodenum and proximal jejunum. EPEC strains produce damage to the jejunal epithelium (attaching and effacing or A/E lesion) by forming microcolonies referred to as localized adherence (LA). LA requires the production of bundle-forming pili (BFP) encoded by genes located on the EAF plasmid. In addition, the adherence of *V. cholerae* to epithelium of the proximal small intestine requires the chromosomally encoded toxin coregulated pilus (TCP). Interestingly, the colonization of ETEC strains of the proximal small intestine in humans requires plasmid-encoded fimbriae referred to as colonization factor antigens (CFAs); and, in piglets and calves, pili called K88 pili.

Salmonella enterica serovar *Typhimurium* adherence to epithelial cells in the distal small intestine involves at least two pili systems : long polar fimbriae (LPF) and plasmid-encoded fimbriae (PEF). LPF appear to bind to specialized epithelial cells, associated with Peyer's patches, called M cells, and PEF appear to bind to receptors on intestinal epithelium. LPF appear to be necessary for adherence associated with *Salmonella* invasion while PEF appear to be involved in colonization of the mucosal surface without invasion.

Colonization of the large intestine or colon requires that the incoming pathogen bind to colonic epithelium while competing with the normal resident flora for available space. Strains of enterohemorrhagic *E. coli (EHEC)* colonize the large intestine, causing A/E lesions via fimbriae encoded on a 60 MDa plasmid. Strains of enteroaggregative *E. coli* (EAEC) adhere to colonic mucosa in a distinct arrangement called aggregative adherence mediated by pili known as aggregative adherence fimbriae (AAF).

Similar roles for afimbrial adhesins in binding to different host cells/tissues have been observed. FHA, pertactin, pertussis toxin, and BrkA of *B. pertussis* function in binding to receptors on ciliated respiratory epithelium, allowing this bacterium to cause disease in the upper respiratory tract; and, at least, FHA binds to the $\alpha\beta_2$ integrin CR3 on macrophages, which is believed to stimulate uptake of the bacterium into the macrophage without inducing the respiratory burst. This allows *B. pertussis* to evade an important host antimicrobial mechanism.

SECRETION SYSTEMS OF VIRULENCE FACTORS

Interestingly, many of the secretion systems are conserved and highly homologous between species. In addition, many unrelated virulence factors can share basic transport systems. Furthermore, protein secretion in Gram-negative bacteria is more complex than that of Gram-positive bacteria due to the presence of an outer membrane.

Secretion systems of Gram- negative bacteria—Type I secretion systems represent a family of structurally and functionally related protein complexes involved in the export of proteins, lacking the classic N-terminal signal sequence, through the inner and outer membrane to the exterior without a periplasmic intermediate. Thus, these are *sec*-independent pathways. Type I secretion pathways utilize three components to accomplish their task: an inner membrane ABC transporter protein, an outer membrane protein, and a periplasmic spanning protein that is anchored into the inner membrane and associated with the ABC transporter component. The prototype type I secretion system is involved in *E. coli* α-hemolysin secretion. However, several other virulence factors have been shown to exhibit type I secretion including : invasive adenylyl cyclase (CyaA) of *Bordetella pertussis,* leukotoxins (LktA) of *Pasteurella haemolytica,* alkaline protease (AprA) of *Pseudomonas aeruginosa,* LipA lipase and PtrA protease of *Serratia marcescens,* and PtrBC protease of *Erwinia chrysanthemi.*

Type II secretion systems are widely distributed among Gram-negative bacteria. This type of secretion involves two genetically and biochemically distinct translocation steps. This step translocates the protein from the cytoplasm to the periplasm, where the signal peptide is cleaved and the protein folds into its secondary shape and perhaps tertiary assembly prior to being translocated across the outer membrane by the type II secretion system. Type II secretion systems are composed of at least 12 different gene products. Many plant and animal pathogens use this type of secretion for the export of exotoxins and hydrolytic enzymes important for their pathogenesis. The best-studied type II secretion system is the Pul system of *Klebsiella oxytoca* involved in pullulanase secretion; others include the Xcp system of *Pseudomonas aeruginosa,* which secretes elastase, exotoxin A, and phospholipase C; the Esp pathway involved in cholera toxin secretion in *Vibrio cholerae;* the Out system, which secretes pectic enzymes and cellulases in *Erwinia chrysanthemi;* and the Exe pathway implicated in amylase and protease secretion in *Aeromonas hydrophila.*

Type III secretion systems (TTSS) have relatively recently been described as major routes for export of virulence factors in human and animal pathogens. Export via TTSS occurs independently of the Sec system. However, many of the components of the TTSS apparatus appear to require the Sec pathway for their export. Unlike other secretion systems, which can export proteins that are active in the extracellular milieu, TTSS appear to have specifically evolved for the direct translocation of proteins from the bacterial cytoplasm into the host cell cytoplasm. Thus, secretion via TTSS is believed to be regulated animal, and human pathogens have been shown to encode TTSS apparati. The chromosomal loci that encode the TTSS and its secreted proteins are referred to as pathogenicity islands. Examples of pathogens that produce TTSS include (a) *Yersinia* species (*Y. pestis, Y. enterocolitica,* and *Y. pseudotuberculosis*), which secrete Yops proteins that inhibit host phagocyte activity among other functions; (b) *Salmonella enterica* serovars, which encode at least two different TTSS on two separate *Salmonella* pathogenicity islands (SPI-1 and SPI-2) and secrete proteins that promote invasion of nonphagocytic cells within the small intestine (SPI-1) and systemic infection and disease (SPI-2); (c) enteropathogenic (EPEC) and enterohemorrhagic (EHEC) *Escherichia coli,* rabbit-diarrheagenic (RDEC-1) *Escherichia coli,* and the rodent pathogen *Citrobacter rodentium,* which produce related TTSS apparati encoded on a pathogenicity island referred to as the locus of enterocyte effacement (LEE) and secrete proteins required for the generation of the attaching and effacement (A/E) lesion; (d) *Shigella* species, which secrete Ipa proteins required for invasion of nonphagocytic cells and induction of apoptosis in host cells.

Type IV secretion systems are another mechanism of macromolecular export in Gram-negative bacteria. Moreover, the cells targeted for secreted substrate delivery are also diverse, ranging from bacteria to fungi to plants to animals. The three defining type IV secretion systems are the VirB/D4 transfer system of *Agrobacterium tumefaciens* that exports T-DNA into susceptible plant cells, leading to plant tumors; the conjugal transfer system of conjugative IncN plasmid pKM101; and the pertussis toxin exporter Ptl of *Bordetella pertussis.* Additional type IV secretion systems identified to play a role in virulence include the Cag system of *Helicobacter pylori* required for export of the CagA protein, which causes rearrangements of the host cell cytoskeleton, and the Dot/Icm system of *Legionella pneumophila* which secretes an unidentified effector, allowing survival and growth in macrophages possibly by blocking phagosome-lysosome fusion.

Autotransporters are represented by a family of proteins that appear to mediate their own export through the outer membranes. This type of secretion utilizes the Sec pathway for export across the inner membrane, but all the information for translocation across the outer membrane is located in the protein itself. Thus, this represents an additional mechanism of export utilized by Gram-negative bacteria. The paradigm for this form of secretion is the IgA protease of *Neisseria gonorrhoeae.* Other members of this family include IgA protease, putative adherence and invasion proteins of *Haemophilus influenza;* BrkA, pertactin, and other outer membrane proteins in *Bordetella pertussis;* a serine protease of *Serratia marcescens;* the vacuolating cytotoxin of *Helicobacter pylori;* SepA protein secreted by *Shigella flexneri;* and EspC protein secreted by EPEC strains. An additional secretion mechanism has recently been described in Gram-negative bacteria. The two-partner secretion (TPS) pathways are composed of a secretion domain-possessing exoprotein of the TpsA family and a channel-forming outer membrane transporter protein of the TpsB family. Interestingly, each TpsB transporter appears to be specific for the export of its cognate exoprotein and tends to be encoded by genes organized into an operon. Based on analysis of genome sequences, TPS pathways appear to be widespread among Gram-negative bacteria. Three TPS systems have been relatively well characterized : secretion of the filamentous *hemagglutinin* (FHA) of *Bordetella pertussis,* the ShlA Ca^{+2} – independent cytolysin (hemolysin) of *Serratia marcescens.* and the high-molecular-weight adhesins (HMW1 and HMW2) of *Haemophilus influenza.*

HMW2) of *Haemophilus influenza.*

Secretion system of Gram-positive bacteria—Surface proteins or secreted proteins all possess a classic signal sequence peptide. In addition, such proteins have a region to tandem repeats and a C-terminal region rich in proline and glycine residues and a conserved hexapeptide (LPXTGE). Examples of surface-associated or secreted proteins in Gram-positive bacteria include F, M and G proteins of *Streptococcus pyogenes,* the P1 adhesin of *Streptococcus mutans,* internalin of *Listeria monocytogenes.*

EXOTOXINS

The host and its normal resident flora are established and geared to efficiently obtain nutrients. The incoming pathogen must either outcompete the host/resident flora or develop way to increase the availability of nutrients to support its growth. Thus, many pathogens either secrete, or put on their surface, enzymes that can degrade host constituents and/or lyse cells to release nutrients. For example, pathogens may secrete proteases/ peptidases that degrade host proteins (e.g., elastase, collagenase, IgA protease) to smaller peptides or amino acids or secrete lipases that degrade lipids or phospholipids to their fatty acid and sugar moieties. The latter may result in cell lysis, since cell membranes are rich in (phospho) lipids. Therefore, the bacterium attempting to obtain nutrients damages the host through the breakdown of its constituents.

A-B toxins—These possess one or more binding or B subunits that lack enzyme activity but bind to host receptors. These subunits are required for entry of the toxin into the target cell. An activity or A subunit possesses the enzyme activity or toxic activity that acts on the target cell. A-B toxins are synthesized by both Gram-negative and Gram-positive bacteria and target a variety or different host cell proteins and functions. One group of A-B toxins possesses a common enzyme activity but differs in the protein or function it targets. This group utilizes the redox cofactor nicotinamide adenine dinucleotide (NAD) as a substrate, cleaving the bond between the nicotinamide (NAm) group and the ribose molecule to yield NAm and adenosine diphosphate ribose (ADP-ribose or ADP-R). This activity is referred to ADP ribosylation activity, and the toxins possessing this activity are known as ADP ribosyltransferases or ADP-ribosylating toxins. Examples of ADP-ribosylating toxins are diverse in terms of the bacteria that produce them and the protein targeted. The classic example of an ADP ribosylating toxin is the diphtheria toxin (DT) produced by *Corynebacterium diphtheriae.* DT and the identically acting exotoxin A (ExoA), from *Pseudomonas aeruginosa,* both ADP- ribosylate the eukaryotic elongation factor (EF)-2, inhibiting protein synthesis and ultimately leading to cell death. DT binds to the heparin-binding epidermal growth factor (EGF), entering the cell through an early endosome while ExoA binds to α_2-macroglobulin, entering the cell following retrograde transport to the endoplasmic reticulum (ER).

Cholera toxin—*Vibrio cholerae* and heat-labile enterotoxin (LT) of enterotoxigenic *E. coli* (ETEC) are both ADP-ribosylating toxins that target the Gsα regulatory protein of adenylyl cyclase (AC), resulting in increased AC activity and elevated levels of cAMP. Elevated cAMP levels in the enterocytes affected by CTX and LT cause increased Cl^- ion secretion and inhibition of Na^+ ion absorption, ultimately resulting in watery diarrhea in the patient. CTX is encoded by a bacteriophage (CTXϕ) while LT is plasmid-encoded. Both bind to surface gangliosides such as GM1 and enter the cell via retrograde transport to the

Pertussis toxin (PT)—*Bordetella pertussis* is an ADP- ribosylating toxin that targets the $G_{i,o}$ regulatory protein of AC, preventing its inhibition and resulting in elevated levels of cAMP. In contrast, to CTX and LT, PT targets respiratory epithelium and results in many symptoms associated with whooping cough. PT is secreted via a type IV secretion system and binds to a surface receptor on respiratory epithelium via its B subunits, delivering the A subunit into the cell cytoplasm. The C2 toxin from *Clostridium botulinum* types C and D, toxin A (Tox-A) and toxin B (Tox-B) of *C. difficile,* and iota toxin of *C. perfringens* are ADP-ribosylating toxins that also possess binding/translocation activity. Other A-B toxins do not possess ADP-ribosylating activity but instead act through a variety of different mechanisms. The Shiga toxin (STX) of *Shigella* spp. and the Shiga-like toxins (SLT) of enterohemorrhagic *E. coli* (EHEC) strains both possess N-glycosidase activity that targets the 28S rRNA, damaging the ribosome, blocking protein synthesis, and eventually killing the cell. Following binding via its B subunits, the A subunit enters the cell through retrograde transport to the ER. Invasive adenylyl cyclase toxin (ACT) of *Bordetella pertussis* enters the cell directly, where in the cytoplasm it acts as a calmodulin dependent adenylyl cyclase that raises cAMP levels in the cell, generally disrupting intracellular signaling pathways.

Anthrax toxin—Anthrax toxin of *Bacillus anthracis* is composed of an edema factor (EF), lethal factor (LF), and protective antigen (PA). The PA subunit acts as the binding or B subunit for both EF and LF, which act independent of each other, The EF, like ACT, is a calmodulin-dependent adenylyl cyclase that produces similar effects on the cell. The LF is Zn^{+2}-metalloprotease, which cleaves host cell mitogen-activated protein kinase kinases (MAPKK), inhibiting cell proliferation and likely causing cell death. The cytolethal distending toxins (CDT) of *Haemophilus ducreyi, Shigella* spp., *Campylobacter* spp., and several *E. coli* strains are A-B toxins that indirectly inhibit $p34^{cdc2}$ dephosphorylation, arresting the cell cycle in G2 and resulting in cytoplasmic distention and cell enlargement; the CDT of *C. jejuni* possesses deoxyribonuclease I-like activity that appears to be required for its action.

Tetanus toxin—*Clostridium tetani* and botulism neurotoxin (BoNT) of *Clostridium botulinum* both are A-B toxin composed of a heavy chain (B subunit) and light chain (A subunit). The light chain possesses metalloprotease activity that cleave synaptosomal-associated proteins. Both TeNT and BoNT inhibit neuroexocytosis. However, TeNT acts on central inhibitory interneurons, preventing inhibitory neurotransmitter release, causing constant muscle stimulation and a spastic paralysis; BoNT acts on peripheral motor neurons, blocking acetylcholine (ACh) release and muscle stimulation, leading to a flaccid paralysis.

Surface-acting toxins—Strains of both *Staphylococcus aureus* and *Streptococcus pyogenes* produce toxins that act as superantigens. They nonspecifically cross-bridge class II MHC molecules on antigen presenting cells and the T cell (antigen) receptor (TCR), which results in the polyclonal activation of helper T cells and copious cytokine secretion. Superantigen toxins include TSST-1, SPE-A, SPE-C, SSA, SEA-SHE, and SpeB. The heat-stable enterotoxins (ST) produced by strains of ETEC, *Yersinia enterocolitica, Vibrio cholerae,* and *Citrobacter freundii* bind to receptors on the cell surface to activate guanylyl cyclase activity, elevating intracellular cGMP levels and activating cGMP kinase II. The net effect is very similar to the effect of LT or CTX on enterocytes, inducing Cl^- ion secretion and blocking Na^+ ion absorption. Several bacteria produce exotoxins that form pores in the cell membranes of host cells, resulting in cell lysis. Examples of these lysins are streptolysin O (SLO) of *S. pyogenes*, which binds to cholesterol in the membrane and polymerizes to form large pores, disrupting cell membrane integrity; similarly, perfringolysin of *C. perfringens*, pneumolysin of *Streptococcus pneumoniae*, listeriolysin O (LLO) of *Listeria monocytogenes*, and tetanolysin of *C. tetani* also form large pores in the cell membrane, causing enormous electrolyte, metabolite, and protein loss with eventual cell death. Other pore-forming toxins generate smaller pores and include the α–toxin and leukocidin of *S. aureus* and aerolysin of *Aeromonas hydrophila*.

Injected toxins—The YpkA and YopO of *Y. pseudotuberculosis* and *Y. enterocolitica* both are protein kinases, but their target proteins are unknown. In addition, YopH of *Y. pseudotuberculosis* is a protein tyrosine phosphatase that dephosphorylates the Crk-associated tyrosine kinase substrate Cas and inhibits its phagocytosis by host macrophages. The YopE protein [a GTPase-activating protein (GAP) for Rho GTPases] and the YopT protein of *Y. pseudotuberculosis* and *Y. enterocolitica* cause disruption of the actin cytoskeleton of the host cell. Exoenzyme S or ExoS of *Pseudomonas aeruginosa* has two functional domains : the N-terminal domain acts as a GAP on Rho GTPases, causing disruption of the actin cytoskeleton, while the C-terminal domain acts to ADP- ribosylate the Ras protein. The SopE protein of *Salmonella enterica* sv. *Typhimurium* has guanine-nucleotide exchange factor (GEF) activity and activates Rho GTPases, causing actin cytoskeleton reorganization that results in membrane ruffling.

INFECTION

The microbe enters the body, attaches itself, invades (crosses host barriers), multiplies in a target tissue, and finally is released to the exterior by various pathways. The type and severity of an infection depend on numerous factors, most of which relate to the pathogenicity of the microbe and the condition of the host. Pathogenicity, we learned earlier, is a broad concept that describes a microorganism's potential to cause an infection or disease. Pathogenic microbes have been traditionally divided into two categories, depending upon the nature of the microbe-host relationship. True pathogens (primary pathogens) are capable of causing infection and disease in healthy persons with normal immune defenses. Examples include the influenza virus, plague bacillus, rabies virus, malarial protozoan, and other microorganisms with well developed qualities of virulence (L. *Virulentia*, Virus, poison). Primary pathogens are generally associated with a distinct, recognizable disease. The degree of pathogenicity of true pathogens ranges from weak to potent—that is, infection with the cold virus is mild and nonfatal, while infection with the rabies virus is nearly 100% fatal.

Opportunists are not generally considered pathogenic to the normal, healthy person and do not usually have well-developed virulence properties. It could be argued that most microbes have some capacity for pathogenicity. For example, a severely compromised host may be challenged by a large inoculum in a vulnerable site. Another side of this idea is that anyone who becomes infected may be temporarily compromised to an extent. Recognizing that classifying pathogens as either true or opportunistic may be an oversimplification, the Centers for Disease Control and Prevention has adopted a system of biosafety categories for pathogens based on their degree of pathogenicity and the relative danger in handling them. With this system, presented in appendix D, pathogens are assigned a level or class. Microbes not known to cause disease in humans (such as

Micrococcus luteus) are given class 1 status; moderate risk agents such as *Staphylococcus aureus* are assigned to class 2; readily transmitted and virulent agents such as *Mycobacterium tuberculosis* are in class 3; and the highest risk microbes are in class 4. Differences in pathogenicity can be accounted for on the basis of virulence. Virulence takes into account the ability of microbes to invade a host and produce toxins (toxigenicity). These properties are called virulence factors. Virulence can be due to single or multiple factors. In some microbes, the causes of virulence are clearly established, but in others, they are not.

The Portals of Penetration

The source of the infectious agent can be exogenous, originating from a source outside the body (the environment or another person or animal), or endogenous, already existing on or in the body. For the most part, the portals of entry are the same anatomical regions that also support normal flora : the skin, alimentary tract, respiratory tract, and urogenital tract. The majority of pathogens have adapted to a specific portal of entry, one that provides a habitat for further growth and spread. This adaptation can be so restrictive that if certain pathogens enter the wrong portal, they will not be infectious. For instance, inoculation of the nasal mucosa with the influenza virus invariably gives rise to the flu, but if this virus contacts only the skin, no infection will result. Likewise, contact with athlete's foot fungi in small cracks in the toe webs can induce an infection, but inhaling the fungus spores will not infect a healthy individual. Occasionally, an infective agent can enter by more than one portal.

Skin portal of penetration—The actual sites of entry are usually nicks, abrasions, and punctures (many of which are tiny and inapparent) rather than smooth, unbroken skin. *Staphylococcus aureus* (the causes of boils), *Streptococcus pyogenes* (an agent of impetigo), the fungal dermatophytes, and agents of gangrene and tetanus gain access through damaged skin. The viral agent of cold sores (*Herpes simplex*, Type 1) enters through the mucous membranes near the lips. Some infectious agents create their own passageways into the skin using digestive enzymes. For example, certain helminth worms burrow through the skin directly to gain access to the tissues. Other infectious agents enter through bites. The bites of insects, ticks, and other animals offer an avenue to a variety of viruses, rickettsias, and protozoa. Although the conjunctiva, the outer protective covering of the eye, is ordinarily a relatively good barrier to infection, bacteria such as *Haemophilus aegyptius* (pinkeye), *Chlamydia trachomatis* (trachoma), and *Neisseria gonorrhoeae* have special affinity for this membrane. Foreign bodies in the eye or minor injuries can serve as a portal for *Herpes simplex* virus and *Acanthamoeba.*

Gastrointestinal tract as portal of penetration—Most enteric pathogens possess specialized mechanisms for entering and localizing in the mucosa of the small or large intestine. The best known enteric agents of disease are Gram-negative rods in the genera *Salmonella, Shigella, Vibrio,* and certain strains of *Escherichia coli*. Viruses that enter through the gut are poliovirus, hepatitis A virus, echovirus, and rotavirus. Important enteric protozoans are *Entamoeba histolytica* (amebiasis) and *Giardia lamblia* (giardiasis).

Respiratory portal of penetration—The extent to which an agent is carried into the respiratory tree is based primarily on its size. In general, small cells and particles are inhaled more deeply than larger ones. Infectious agents with this portal of entry include the bacteria of streptococcal sore throat, meningitis, diphtheria, and whooping cough; and the viruses of influenza, measles, mumps, rubella, chickenpox, and the common cold. Pathogens that are inhaled into the lower regions of the respiratory tract (bronchioles and lungs) can cause pneumonia, an inflammatory condition of the lung. Bacteria (the pneumococcus, *Klebsiella, Mycoplasma*) and fungi (*Cryptococcus* and *Pneumocystis*) are a few of the agents involved in pneumonias. All types of pneumonia are on the increase due to the greater susceptibility of AIDS patients to them.

Urogenital portals of penetration—The older term, venereal disease, has been replaced by the more descriptive term sexually transmitted disease. The microbes of STDs enter the skin or mucosa of the penis, external genitalia, vagina, cervix, and urethra. Some can penetrate an unbroken surface, while others require a cut or abrasion. The once predominant sexual diseases syphilis and gonorrhea have been supplanted by a large and growing list of STDs headed by genital warts, chlamydia, and herpes. Evolving sex practices have increased the incidence of STDs that were once uncommon, and diseases that were not originally considered STDs are now so classified.

Penetration of pathogens during pregnancy and birth—The placenta is ordinarily an effective barrier against microorganisms in the maternal circulation. However, a few microbes such as the syphilis spirochete can cross the placenta, enter the umbilical vein, and spread by the fetal circulation into the fetal tissues. Other infections such as *Herpes simplex* occur perinatally when the child is contaminated by the birth canal. The common infections of fetus and neonate are grouped together in a unified cluster known by the acronym STORCH that medical personnel must monitor. STORCH stands for syphilis, toxoplasmosis, other diseases (hepatitis B, AIDS, and chlamydial), rubella, cytomegalovirus, and *Herpes simplex* virus. The most serious complications of STORCH infections are spontaneous abortion, congenital abnormalities, brain damage, prematurity, and stillbirths.

Inoculum Concentration

For most agents, infection will proceed only if a minimum number, called the *infectious dose* (ID), is present. This number has been determined experimentally for many microbes. In general, microorganisms with smaller infectious dose have greater virulence. The ID varies from the astonishing number of 1 rickettsial cell in Q fever to about 10 infectious cells in tuberculosis, giardiasis, and coccidioidomycosis. The ID is 1,000 cells for gonorrhea and 10,000 cells for typhoid fever in contrast to 1,000,000,000 cells in cholera. Lack of an infectious dose will generally not result in an infection. Yet if the quantity is far in excess of the ID, the onset of disease can be more rapid.

Mechanism of Pathogen Establishment

Since this often involves a specific interaction between molecules on the microbial surface and receptors on the host cell, adhesion can determine the specificity of a pathogen for its host organism and in some cases the specificity of a pathogen for a particular cell type. Once attached, the pathogen is poised advantageously to invade the sterile body compartments. Bacterial pathogens attach most often by mechanisms such as fimbriae (pili), flagella, and adhesive slimes or capsules, while viruses attach by means of specialized receptors. Protozoa can infiltrate by means of their organelle of locomotion; for example, *Balantidium coli* is said to penetrate by the boring action of its cilia. In addition, parasitic worms are mechanically fastened to the portal of entry by suckers, hooks, and barbs.

Virulence factors from a microbial perspective are simply adaptations it uses to invade and establish itself in the host. These same factors determine the degree of tissue damage that occurs. Effects vary from pathogens such as cold viruses that invade the tissues but multiply with relatively minor effects to very harmful pathogens that severely damage and even kill the host. For convenience, we will divide the virulence factors into exoenzymes, toxins, and antiphagocytic factors.

Extracellular enzymes—Many pathogenic bacteria, fungi, protozoa, and worms secrete exoenzymes that break down and inflict damage on tissues. Other enzymes dissolve the host's defense barriers and promote the spread of microbes to deeper tissues. Examples of enzymes are : (a) Mucinase, which digests the protective coating on mucous membranes and is a factor in amebic dysentery. (b) Keratinase, which digests the principal component of skin and hair, is secreted by fungi that cause ringworm. (c) Collagenase digests the principal fiber of connective tissue and is an invasive factor of *Clostridium* species and certain worms. (d) Hyaluronidase digests hyaluronic acid, the ground substance that cements animal cells together. Some enzymes react with components of the blood. Coagulase, an enzyme produced by pathogenic staphylococci, causes clotting of blood or plasma. By contrast, the bacterial kinases (streptokinase, staphylokinase) do just the opposite, dissolving fibrin clots and expediting the invasion of damaged tissues.

Bacterial toxins—Toxigenicity, the power to produce toxins, is a genetically controlled characteristic of many species and is responsible for the adverse effects of a variety of diseases generally called toxinoses. A type of toxinosis in which the toxin is spread by the blood from the site of infection is a toxemia (tetanus and diphtheria, for example), while one is caused by ingestion of toxins is an intoxication (botulism). A toxin is named according to its specific target of action : Neurotoxins act on the nervous system; enterotoxins act on the intestine; hemotoxins lyse red blood cells; and nephrotoxins damage the kidneys. An unbound toxin molecule secreted by a living bacterial cell into the infected tissues is an exotoxin. A toxin that is not secreted but released only after the cell is damaged or lysed is an endotoxin. Exotoxins are proteins with a strong specificity for a target cell and extremely powerful, sometimes deadly, effects. They generally affect cells by damaging the cell membrane and initiating lysis or by disrupting intracellular function. Hemolysins (Gr. *Haima,* blood, and *lysis,* dissolution) are a class of bacterial exotoxin that disrupts the cell membrane of red blood cells (and some other cells too). This damages causes the red blood cells to hemolyze—to burst and release hemoglobin pigment. Hemolysins that increase pathogenicity include the streptolysins of *Streptococcus pyogenes* and the alpha (α) and beta (β) toxins of *Staphylococcus aureus.*

VARIOUS SYMPTOMS OF DISEASES

A sign is any objective evidence of disease as noted by an observer; a symptom is the subjective evidence of disease as sensed by patient. In general, signs are more precise than symptoms, though both can have the same underlying cause. For example, an infection of the brain might present with the sign of bacteria in the spinal fluid and symptom of headache. Or a streptococcal infection might produce a sore throat (symptom) and inflamed pharynx (sign). Disease indicators that can be sensed and observed can qualify as either a sign or a symptom. When a disease can be identified or defined by a certain complex of signs and symptoms, it is termed a *syndrome.*

Inflammation Symptoms

The earliest symptoms of disease result from the activation of the body defense process called inflammation (L. *Inflammatio,* to set on fire). The inflammatory response includes cells and chemicals that respond non-specifically to disruptions in the tissue. Some common symptoms of inflammation include fever, pain, soreness, and swelling. Signs of inflammation include edema, (Gr. *Oidema,* swelling) the accumulation of fluid in an afflicted tissue; *granulomas* and *abscesses,* which are walled-off collections of inflammatory cells and microbes in the tissues; and *lymphadenitis,* swollen lymph nodes.

Symptoms of Blood Infection

Leukocytosis (a white blood cell) is an increase in the level of white blood cells, whereas *leukopenia* (lack of white blood cells) is a decrease. Other signs of infection revolve around the occurrence of a microbe or its products in the blood. The clinical term for blood infection, septicemia, refers to a general state in which microorganisms are multiplying in the blood and are present in large numbers. When small numbers of bacteria or viruses are found in the blood, the correct terminology is bacteremia or viremia, which means that these microbes are present in the blood but are not necessarily multiplying. During infection, a normal host will invariably show signs of an immune response in the form of antibodies in the serum or some type of sensitivity to the microbe.

EXIT PORTALS : EMPTYING THE HOST

A parasite is equally unsuccessful if it does not have a provision for leaving its host and moving to other susceptible hosts. With few exceptions, pathogens depart by a specific avenue called the portal of exit. In most cases, the pathogen is shed or released from the body through secretion, excretion, discharge, or sloughed tissue. The number of infectious agents in these materials is usually very high, which increases both virulence and the likelihood that the pathogen will reach other hosts. In many cases, the portal of exit is the same as the portal of entry, but a few pathogens use a different route.

Respiratory and salivary portals—The most effective means of releasing these secretions is coughing and sneezing, although they can also be released during talking and laughing. Tiny particles of liquid released into the air form aerosols or droplets that can spread the infectious agent to other people. The agents of tuberculosis, influenza, measles, and chickenpox most often leave the host through airborne droplets. Droplets of saliva are the exit route for several viruses, including those of mumps, rabies, and infectious mononucleosis.

Scales of skin—A large proportion of household dust is actually composed of skin scales. A single person can shed several billion skin cells per day, and some persons, called shedders, disseminate massive numbers of bacteria into their immediate surroundings. Skin lesions and their exudates can serve as portals of exit in warts, fungal infections, boils, *Herpes simplex*, smallpox, and syphilis.

Exit of feces—Some intestinal pathogens grow in the intestinal mucosa and create an inflammation that increases the motility of the bowel. This speeds up peristalsis, resulting in diarrhea, and the more fluid stool provides a more rapid exit for the pathogen. A number of metazoan parasites release cysts and eggs through the feces. Feces containing pathogens are a public health problem when allowed to contaminate drinking water or when used to fertilize crops.

Urogenital tract as exit portal—This is also the source of neonatal infections such as *Herpes simplex, Chlamydia* and *Candida albicans,* which infect the infant as it passes through the birth canal. Less commonly, certain pathogens that infect the kidney are discharged in the urine—for instance, the agents of leptospirosis, typhoid fever, tuberculosis, and schistosomiasis.

Bleeding—Although the blood does not have a direct route to the outside, it can serve as a portal of exit when it is removed or released through a vascular puncture made by natural or artificial means. Blood-feeding insects such as mosquitos and fleas are common transmitters of pathogens. The AIDS and hepatitis viruses are transmitted by shared needles or through small gashes in a mucous membrane caused by sexual intercourse. Blood donation is also a means for certain microbes to leave the host, though this is now unusual because of close monitoring of the donor population and blood used for transfusions.

Microorganism's Persistence

Following the initial symptoms in certain chronic infectious diseases, the infectious agent retreats into a dormant state called latency. Throughout this latent state, the microbe can periodically become active and produce a recurrent disease. The viral agents of *Herpes simplex, Herpes zoster,* hepatitis B, AIDS, and Epstein-Barr can persist in the host for long periods. The agents of syphilis, typhoid fever, tuberculosis, and malaria also enter into latent stages. The person harboring a persistent infectious agent may or may not shed it during the latent stage. If it is shed, such persons are chronic carriers who serve as sources of infection for the rest of the population. Some diseases leave sequelae (L. *Sequi,* to follow) in the form of long-term or permanent damage to tissues or organs.

EPIDEMIOLOGY OF DISEASES

Let us now turn our attention to the effects of diseases on the community—the realm of epidemiology (Gr. *Epidemios,* prevalent). By definition, this term involves the study of the frequency and distribution of disease and other health-related factors in defined human populations. It involves many disciplines—not only microbiology, but anatomy, physiology, immunology, medicine, psychology, sociology, ecology, and statistics—and it considers many diseases other than infectious ones, including heart disease, cancer, drug addiction, and mental illness. The epidemiologist is a medical sleuth who collects clues on the causative agent, pathology, sources and modes of transmission, and tracks the numbers and distribution of cases of disease in the community. In fulfilling these demands, the epidemiologist asks, who, when, where, how, what, and why ? about diseases. Epidemiologists are concerned with all of the factors covered earlier in this chapter : virulence, portals of entry and exit, and the course of disease. But they are also interested in surveillance—that is, collecting, analyzing, and reporting data on the rates of occurrence, mortality, morbidity, and transmission of infections. This involves keeping data for a large number of diseases seen by the medical community and reported to public health authorities.

Physicians and hospitals report all notifiable diseases that are brought to their attention. Case reporting can focus on a single individual or collectively on group data. Local public health agencies first receive the case data and determine how they will be handled. In most cases, health officers investigate the history and movements of patients to trace their prior contacts and to control the further spread of the infection as soon as possible through drug therapy, immunization, and education. In sexually transmitted diseases, patients are asked to name their partners so that these persons can be notified, examined, and treated. It is very important to maintain the confidentiality of the persons in these reports. The principal government agency responsible for keeping track of infectious diseases nationwide is the Centers for Disease Control and Prevention (CDC) in Atlanta, Georgia, which is a part of the United States Public Health Service. The CDC publishes a weekly notice of diseases (the *Morbidity and Mortality Report*) that provides weekly and cumulative summaries of the case rates and deaths for about 37 notifiable diseases, highlights important and unusual diseases, and presents data concerning disease occurrence in the major regions of the United States.

Epidemiological statistics—It is a cumulative statistic, usually represented as the percentage of the population having a particular disease at any given time. Disease incidence measures the number of new cases over a certain time period, as compared to the general healthy population. This statistic, also called the case or morbidity rate, indicates both the rate and the risk of infection. The equations used to figure these rates are :

$$\text{Prevalence} = \frac{\text{Total number of cases in population}}{\text{Total number of persons in population}} \times 100 = \%$$

$$\text{Incidence} = \frac{\text{Number of new cases}}{\text{Number of healthy persons}} = \text{Ratio}$$

Prior to exposure, the prevalence and incidence in this population are both zero (0/50). If in one week, 5 out of the 50 people contract the disease, the prevalence is 5/50 = 10%, and the incidence is 1 in 9 (5 cases compared to 45 healthy persons). If after two weeks, 5 more students contract the flu, the prevalence becomes 5 + 5 = 10/50 = 20% , and the incidence becomes 1 in 8 (5/40). When dealing with large populations, the incidence is usually given in numbers of cases per 1,000 or 100,000 population. Statistics of concern to the epidemiologist are the rates of disease with regard to sex, race, or geographical region. Also of importance is the mortality rate, which measures the total number of deaths in a population due to a certain disease. Over the past century, the overall death rate from infectious diseases has dropped, although the number of persons afflicted with infectious diseases has remained relatively high. An infectious disease that exhibits a relatively steady frequency over a one time period in a particular geographic local is endemic. For example, Lyme disease is endemic to certain areas of the United States where the tick vector is found. A certain number of new cases are expected in these areas every year. When a disease is sporadic, occasional cases are reported at irregular intervals in unpredictable locales. Rabies and typhoid fever are reported sporadically in the United States. When statistics indicate that the prevalence of an endemic or sporadic disease is increasing beyond what is expected for that population, the pattern is described as an epidemic.

A recent example is tuberculosis. From 1975 until 1985, the case rate declined from 34,000 to a historic low of 22000. Beginning in 1985, a resurgence in TB cases occurred that continued until 1993. Many of these new cases occurred in AIDS patients and many were due to multidrug-resistant strains. The spread of an epidemic across continents is a pandemic, as exemplified by AIDS and influenza. One important epidemiological truism might be called the "iceberg effect." Regardless of case reporting and public health screening, a large number of cases of infection in the community go undiagnosed and unreported. In the instance of salmonellosis, approximately 40,000 cases are reported per year. Epidemiologists estimate that the actual number is more likely somewhere between 400,000 and 4,000,000.

Investigative strategies of epidemiology—A few sporadic cases are seen by physicians and are eventually reported to authorities. However, it can take several reports before any alarm is registered. Epidemiologists and public health departments must piece together odds and ends of data from a series of apparently unrelated cases and work backwards to reconstruct the epidemic pattern. A completely new disease requires even greater preliminary investigation, because the infectious agent must be isolated and linked directly to the disease. All factors possibly impinging on the disease are scrutinized. Investigators search for *clusters* of cases indicating spread between persons or a public (common) source of infection; they also look at possible contact with animals, contaminated food, water, and public facilities; and at human interrelationships or changes in community structure.

TRANSMISSION OF INFECTIOUS AGENTS

A disease is communicable when an infected host can transmit the infectious agent to another host and establish infection in the host. The transmission of the agent can be direct or indirect, and the ease with which the disease is transmitted varies considerably from one agent to another. If the agent is highly transmissible, especially through direct contact, the disease is contagious. Influenza and measles move readily from host to host and thus are contagious, whereas leprosy is only weakly communicable. Because they can be spread through the population, communicable diseases will be our main focus in the following sections. In contrast, a noncommunicable infectious disease does not arise through transmission of the infecious agent from host to host. The infection and disease are acquired through some other, special circumstance. Noncommunicable infections occur primarily when a compromised person is invaded by his own microflora (as with certain pneumonias, for example) or when an individual has accidental contact with a facultative parasite that exists in a nonliving reservoir such as soil. Some examples are certain mycoses acquired through inhalation of fungal spores and tetanus, in which *Clostridium tetani* spores from a soiled object enter a cut or wound.

Transmission of Communicable Diseases

The spread of diseases is by direct or indirect contact with animate or inanimate objects, and can be horizontal or vertical. The term *horizontal* means the disease is spread through a population from one infected individual to another; *vertical* signifies transmission from parent to offspring via the ovum, sperm, placenta, or milk. The extreme complexity of transmission patterns among microorganisms makes it very difficult to generalize.

Direct transmission—In order for microbes to be directly transferred, some type of contact must occur between the skin or mucous membranes of the infected person and that of the infectee. It may help to think of this route as the portal of exit meeting the portal of entry without the involvement of an intermediate object or substance. Included in this category are fine droplets sprayed directly upon a person during sneezing or coughing.

Most sexually transmitted diseases are spread directly. Additionally, infections that result from kissing, nursing, placental transfer, or bites by biological vectors are direct.

Indirect transmission—This form of communication is especially pronounced when the infected individual contaminates inanimate objects, food, or air through his activities. The transmitter of the infectious agent can be either openly infected or a carrier.

Transmission by vectors—A *common vehicle* is a single material that serve as the source of infection for many individuals. Some specific types of vehicles are food, water, various biological products, and fomites. A fomite (L. *Fomes,* tinder) is an inanimate object that harbors and transmits pathogens. The list of possible fomites is as long as your imagination allows. Probably highest on the list would be objects commonly in contact with the public such as doorknobs, telephones, push buttons, and faucet handles that are readily contaminated by touching. Shared bed linens, handkerchiefs, toilet seats, toys, eating utensils, clothing, personal articles, and syringes are other examples. Although paper money is impregnated with a disinfectant to inhibit microbes, pathogens are still isolated from a fairly large proportion of bills as well as coins. Outbreaks of food poisoning often result from the role of food as a common vehicle. The source of the agent can be soil, the handler, or a mechanical vector. In the type of transmission termed the *oral-fecal route,* a fecal carrier with inadequate personal hygiene contaminates food during handling, and an unsuspecting person ingests it. Hepatitis A, amebic dysentery, shigellosis, and typhoid fever are often transmitted this way. Because milk provides a rich growth medium for microbes, it is a significant means of transmitting pathogens from diseased animals, infected milk handlers, and environmental sources of contamination.

Transmission by air—Indoor air (especially in a closed space) can serve as an important medium for the suspension and dispersal of certain respiratory pathogens via droplet nuclei and aerosols. Droplet nuclei are dried microscopic residues created when microscopic pellets of mucus and saliva are ejected from the mouth and nose. They are generated forcefully in an unstifled sneeze or cough or mildly during other vocalizations. Although the larger beads of moisture settle rapidly, smaller particles evaporate and remain suspended for longer periods. Droplet nuclei are implicated in the spread of hardier pathogens such as the tubercle bacillus and the influenza virus. Aerosols are suspensions of fine dust or moisture particles in the air that contain live pathogens. Q fever is spread by dust from animal quarters, and psittacosis by aerosols from infected birds. An unusual outbreak of coccidioidomycosis (a lung infection) occurred during the 1994 southern California earthquake.

Transmission by hospitals—Infectious diseases acquired as a result of a hospital stay are known as nosocomial (Gr. *Nosos,* disease, and *komeion,* to take care of) infections. This concept seems, incongruous at first thought, because a hospital is regarded as a place to get treatment for disease, not a place to acquire a disease. Yet it is not uncommon for a surgical patient's incision to become infected or a burn patient to develop a case of pneumonia in the clinical setting. The rate of nosocomial infections can be as low as 0.1% or as high as 20% of all admitted patients at any one time, but the average is about 5%. In light of the number of admissions, this amounts to from 2 to 4 million cases per year, which result in 20,000 to 40,000 deaths. Nosocomial infections cost time and money. After all, the hospital both attracts and creates compromised patients, and it serves as a collection point for pathogens. Some patients become infected when surgical procedures or lowered defenses permit resident flora to invade their bodies. Other patients acquire infections directly or indirectly from fomites, medical equipment, other patients, medical personnel, visitors, air, and water.

The health care process itself increases the likelihood that infectious agents will be transferred from one patient to another. Treatments using reusable instruments such as respirators and thermometers constitute a possible source of infectious agents. Indwelling devices such as catheters, prosthetic heart valves, grafts, drainage tubes, and tracheostomy tubes form a ready portal of entry and habitat for infectious agents. An additional problem is the tendency for drug-resistant strains of microorganisms to develop in hospitals, thereby further complicating treatment. Gram-negative intestinal flora (*Escherichia coli, Klebsiella, Pseudomonas*) are cultured in slightly more than half of patients with nosocomial infections. Gram-positive bacteria (staphylococci and streptococci) and yeasts make up most of the remainder. True pathogens such as *Mycobacterium tuberculosis, Salmonella,* hepatitis B, and influenza virus can be transmitted in the clinical setting as well. The potential seriousness and impact of nosocomial infections have required hospitals to develop committees that monitor infectious outbreaks and develop guidelines for infection control and aseptic procedures.

Surgical asepsis involves a high level of disinfection, antisepsis, and sterilization as would be required in surgery. Instruments, dressings, sponges, and all other supplies coming in contact with the patient are sterilized. Personnel must be fully covered in sterile garments, and room surfaces and air must be thoroughly disinfected. *Medical asepsis* includes any practice that lowers the load of infectious microbes in patients, personnel, and the hospital environment. This includes handwashing, decontamination procedures, and isolation of patients. In isolation, various barriers (gloves, mask, gown) and other techniques are used in the patient's room to prevent the entry or exit of infectious agents on health care workers, visitors, and into the surroundings. The aim of four categories of isolation is to contain the spread of infectious agents from infected patients or carriers, and one category (reverse isolation) prevents highly compromised patients from coming in contact with pathogens from the outside environment.

12

TOOLS AND TECHNIQUES OF GENETIC ENGINEERING

Scientists from diverse disciplines now have at their disposal a variety of powerful molecular tools to manipulate, alter, and analyze the genetic material of microbes, plants, animals, and viruses. This relatively young science has ushered in a new era in the history of humankind and it has spawned an explosion of medical, agricultural, and industrial applications. The genetic revolution has forced every person to become DNA literate. The media keep us constantly updated on genetic news events ranging from provocative to peculiar to profound : cloning dinosaur DNA, inserting methuselah genes into houseflies to prolong their lives, creating plants that can manufacture human antibodies, and engineering bacteria that can fabricate spider silk. We have been challenged to consider the accuracy of DNA fingerprinting evidence in court and the safety of eating genetically-altered fruits and animal products. The ability to change genetic blueprints and create totally new types of organisms is so alarming that a group of scientists and religious leaders lobbied Congress in 1988, and again in 1995, to legislate against genetic alteration of human gametes and embryos. Some environmental activists have even gene so far as to attempt to block the release of engineered microbes and plants into natural habitats.

Genetic studies over the last 25 years have provided a number of insights into certain measurable and predictable activities of nucleic acids. Unravelling this information also had a practical benefit, which was the development of highly sophisticated laboratory tools and techniques. Traditional methods of genetic analysis have involved observing phenotypic inheritance in a test subject. But with these newer techniques, discovery is more likely to involve seeing actual DNA patterns on electrophoresis gels, locating the exact gene on a chromosome, and producing computer-generated gene sequences.

DNA : THE MOLECULAR TOOL

DNA structure is like a child's building blocks and its language is a special code that imparts information on life processes. It can be wound up and untwisted like a rubber band, opened into two strands like a zipper, diced up into shorter pieces, and spliced back together. One useful property of DNA is that it readily anneals (O.E. *Anaelan*, a process of heating and cooling) or changes, its binding properties in response to heating and cooling. Exposure to temperatures just below boiling (90°-95°C) causes DNA to become temporarily denatured (Fig. 12.1). When heat breaks the hydrogen bonds holding the double helix together, it separates longitudinally into two strands. Each strand displays its nucleotide code so that the DNA in this form can be subjected to tests or replicated. When heating is followed by slow cooling, two single DNA strands rejoin (renature) by hydrogen bonds at complementary sites.

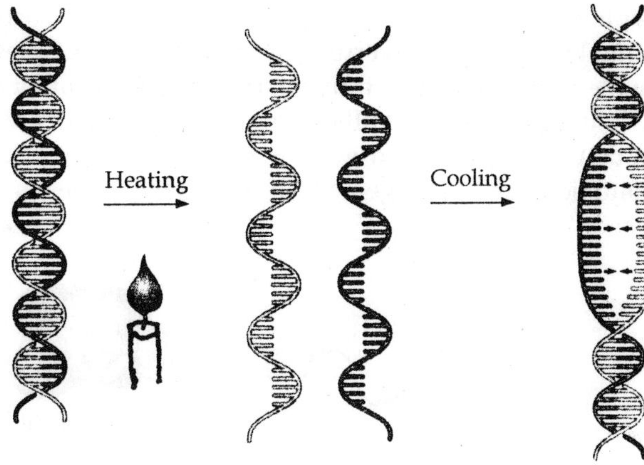

Fig. 12.1 : Annealing of DNA.

RESTRICTION DIGESTION ENZYMES

So far, hundreds of restriction endonucleases have been discovered in bacteria. Each type has a known sequence of 4 to 10 base pairs as its target, so that sites of cutting can be finely controlled. These enzymes have the unique property of recognizing and snipping inverted repeats called palindromes (Fig. 12.2). Examples of three common endonucleases with their sites of cutting are :

Endonuclease	EcoRI	HindIII	HaeIII
Cutting pattern	G|A A T T C	A|A G C T T	G G|C C
	C T T A A|G	T T C G A|A	C C|G G

Endonucleases are named by combining the first letter of the bacterial genus, the first two letters of the species, and the endonuclease number. Thus, EcoRI is the first endonuclease found in *E. coli*, and HindIII is the third endonuclease discovered in *Haemophilus influenza* type d. Those such as HaeIII make straight, blunt cuts on DNA. But more often (EcoRI and HindIII), the enzymes make staggered symmetrical cuts that leave short tails called 'sticky ends'. Such adhesive tails will base pair with complementary tails on other chromosomes or open plasmids. This effect makes it possible to splice genes into specific sites. Fragments of DNA produced by restriction endonucleases are termed *restriction fragment length polymorphisms* (RFLPs) because of the way of endonuclease clips the DNA into pieces of various lengths. They are polymorphic, meaning they vary considerably from one individual to another and they are inherited in predictable patterns. Hundreds of cleavage sites that produce RFLPs are sprinkled throughout genomes. Another enzyme called a ligase is necessary to seal the sticky ends together by rejoining the phosphate-sugar bonds cut by endonucleases. Its main application is in final splicing of genes into plasmids and chromosomes. An enzyme called reverse transcriptase is best known for its role in the replication of the AIDS virus and other retroviruses. It also provides geneticists with a valuable tool for converting RNA into DNA. Copies called copy DNA, or cDNA, can be made from messenger, transfer, ribosomal, and other forms of RNA. The technique provides a valuable means of synthesizing eukaryotic genes from mRNA transcripts. The advantage is that the synthesized gene will be free of the intervening sequences (introns) that can complicate the management of eukaryotic genes in genetic engineering.

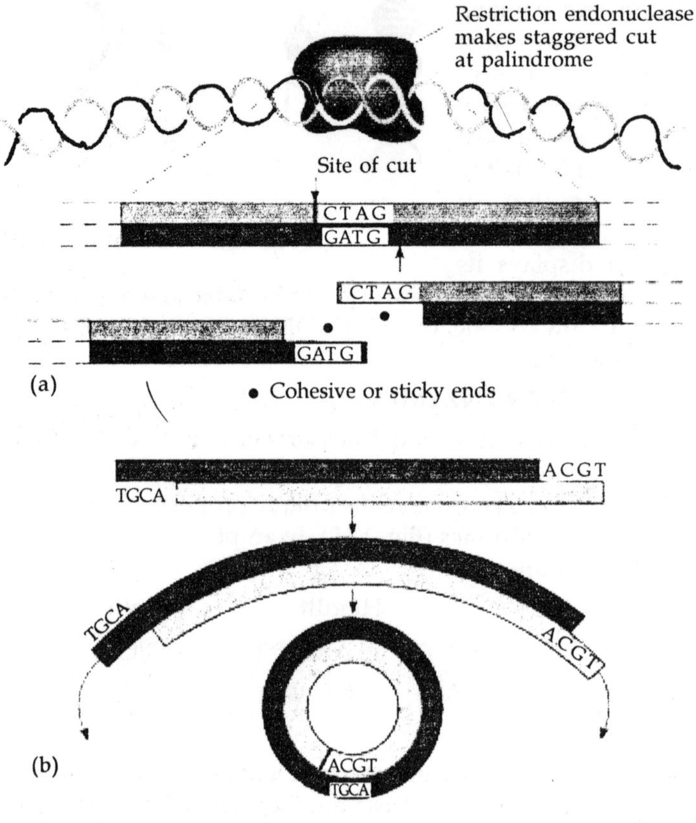

Fig. 12.2 : Restriction digestion of DNA.

DNA Visualization by Naked Eye

One way to produce a readable pattern of DNA fragments is through gel electrophoresis. In this technique, samples are placed in compartments on a special agar gel and subjected to an electric current. Flow of electricity from the negative pole to the positive pole causes the DNA pieces to migrate in the gel substrate toward the positive pole. The rate of movement is based primarily on the size of the fragments. The larger fragments move more slowly and remain nearer the top of the gel, whereas the smaller fragments migrate faster and are positioned closer to the bottom. To determine the positions of DNA fragments, the gel is stained or developed with probes and the pattern is compared against a known standard.

SYNTHESIS AND SEQUENCING OF DNA

The palindromic sequences recognized by endonucleases are usually 4 to 10 bp in length, and an average gene in *Escherichia coli* is approximately 1,300 bp or 1.3 kilobases (kb) and its entire genome is approximately 4.7 million base pairs (Mb). The DNA of the human mitochondrion contains 16 kb and the Epstein-Barr virus (a cause of infectious mononucleosis) has 172 kb. Humans have approximately 3 billion base pairs arrayed along 46 chromosomes. Since large segments of DNA or RNA are difficult to handle and analyze, this problem can be solved by using far shorter pieces of DNA or RNA called oligonucleotides (Gr. *Oligo,* few or scant). Oligonucleotides vary in length from 2 to 200 bps, although the most common ones are about 20 to 30 bps. Although they can be isolated from cells, for most applications oligonucleotides are tailor-made on a gene machine. DNA synthesizers add the desired nucleotides in a controlled series of steps, but are limited to a length of about 200 nucleotides. An enormously useful type of information about a given gene or genome is the precise sequence of the base pairs it contains. This information was very slow in coming until machines called sequencers were developed.

NUCLEIC ACID HYBRIDIZATION

Two different nucleic acids can also hybridize by uniting at their complementary sites. All different combinations are possible : single-stranded DNA can unite with other single-stranded DNA or RNA, and RNA can hybridize with other RNA. This property has been the inspiration for specially formulated oligonucleotide tracers called gene probes. Hybridization probes have practical value because they can detect specific nucleotide sequences in unknown samples. So that areas of hybridization can be visualized, the probes carry radioactive labels, which are isotopes that emit radiation, or luminescent labels, which give off visible light. Reactions can be revealed by placing photographic film in contact with the test reaction. When probes hybridize with an unknown sample of DNA or RNA, they tag the precise area and degree of hybridization and help determine the nature of nucleic acid present in a sample. In a method called the Southern blot. DNA fragments are first separated by electrophoresis and then denatured and immobilized on a special filter.

A DNA probe is then incubated with the sample, and wherever this probe encounters the segment for which it is complementary, it will attach and form a hybrid. Development of the hybridization pattern will show up as one or more bands. Probes are commonly used for diagnosing the cause of an infection from a patient's specimen and identifying a culture of an unknown bacterium or virus. A simple and rapid method called a dot blot does not require electrophoresis. DNA from a test sample is isolated, denatured, placed on an absorbant filter, and combined with a microbe-specific probe. The blot is then developed and observed for areas of hybridization. Commercially available diagnostic kits are now on the market for identifying intestinal pathogens such as *Escherichia coli, Salmonella, Campylobacter, Shigella, Clostridium difficile,* rotaviruses, and adenoviruses. Other bacterial probes exist for *Mycobacterium, Legionella, Mycoplasma,* and *Chlamydia;* viral probes are available for *Herpes simplex* and zoster, papilloma (genital warts), hepatitis A and B, and AIDS. With another method, called *in situ* hybridization, probes are applied to intact cells to visualize the presence and location of specific nucleic acids. It is a very effective way to locate genes on chromosomes.

POLYMERASE CHAIN REACTION

This technique rapidly increases the amount of DNA in a sample without the need for making cultures or using a highly purified form. It is so sensitive that it holds the potential to detect cancer from a single cell or to diagnose an infection from a single gene copy. It is comparable to being able to pluck a single DNA "needle" out of a "haystack" of other molecules and make unlimited copies of the DNA. The rapid rate of PCR makes it possible to replicate a target DNA from a few copies to billions of copies in a few hours. It can amplify DNA fragments that consist of a few base pairs to whole genes containing several thousand base pairs. Primers are synthetic oligonucleotides of a known sequence of 15-30 bases. They serve as landmarks to indicate where DNA amplification will begin. To ensure that the DNA templates do not recombine and interfere with DNA synthesis, processing must be carried out at a relatively high temperature. This necessitates the use of special DNA polymerases isolated from thermophilic bacteria. Examples of these unique enzymes are Taq polymerase obtained from *Thermus aquaticus* and Vent polymerase from *Thermococcus litoralis*.

The first step involves heating target DNA to 94°C to separate it into two strands. Next, the system is cooled to 65°C and primers are added in a concentration that favours binding to their complementary strand of test DNA. This reaction prepares the two DNA strands, now called amplicons, for synthesis. In the third phase, which proceeds at 72°C, DNA polymerase and raw materials in the form of nucleotides are added. Beginning at the free end of the primers on both strands, the polymerases extend the molecule by adding appropriate nucleotides, and produce two complete strands of DNA. It is through repetition of these same steps that DNA becomes amplified. When the DNAs formed in the first cycle are denatured, they become amplicons to be primed and extended in the second cycle. Each subsequent cycle converts the new DNAs to amplicons and doubles the number of copies. The number of cycles required to produce a million molecules is 20, but it can be carried out to 40 cycles. One significant advantage of this technique has been its natural adaptability to automation.

PCR can be adapted to analyze RNA by initially converting an RNA sample to DNA with reverse transcriptase. This cDNA can then be amplified by PCR in the usual manner. It is by such means that ribosomal RNA and messenger RNA are readied for sequencing. The polymerase chain reaction has found prominence as a powerful workhorse of molecular biology, medicine, and biotechnology. It often plays an essential role in gene mapping, the study of genetic defects and cancer, forensics, taxonomy, and evolutionary studies. For all of its advantages, PCR has some problems. A serious concern is the introduction and amplification of nontarget DNA from the surrounding environment. Such contamination can be minimized by using equipment and rooms dedicated for DNA analysis maintained with the utmost degree of cleanliness.

rDNA TECHNOLOGY

Its origins can be traced to 1970, when microbiologists first began to duplicate the clever tricks bacteria do naturally with bits of extra DNA such as plasmids, transposons, and proviruses. The discovery that bacteria can readily accept, replicate, and express foreign DNA made them powerful agents for studying the genes of other organisms in isolation. The practical applications of this work were soon realized by biotechnologists.

An important objective of this technique is to form genetic clones, or duplicates. Cloning involves the removal of a selected gene from an animal, plant, or microorganism (the genetic donor) followed by its propagation in a different host organism. Cloning requires that the desired donor gene first be selected, excised by restriction endonucleases, and isolated. The gene is next inserted into a vector (usually a plasmid or virus) that will insert the DNA into a cloning host. The cloning host is usually a bacterium or yeast that can replicate the gene and translate it into the protein product for which it codes.

Gene Cloning

Some of the first isolated genes came from viruses that have much smaller and more manageable genomes, and later genes were isolated from bacteria and yeasts. At first, the complexity of the human genome made locating specific genes very difficult. However, this process has been greatly improved by tools for excising, mapping, and identifying genes. Common strategies of gene isolation are as follows :

(a) The DNA isolated from cells, separated into fragments by endonucleases, and each fragment is inserted into a vector and cloned. The cloned fragments are Southern blotted and probed to identify desired sequences. This is a long and tedious process, since each fragment of DNA must be examined for the cloned gene.

(b) A gene can be synthesized from isolated mRNA transcripts using reverse transcriptase.

(c) DNA can be synthesized artificially on a machine.

Plasmids are excellent vectors because they are small, well characterized, easy to manipulate, and can be transferred into appropriate host cells through transformation. Bacteriophages also serve well because they have the natural ability to inject DNA into bacterial hosts through transduction. A common vector in early work was an *E. coli* plasmid that carries genetic markers for resistance to antibiotics, although it is restricted by the relatively small amount of foreign DNA it can accept. A modified phage vector, the Charon phage is missing large sections of its genome, so that it can carry a fairly large segment of foreign DNA. In addition, hybrid vectors have been developed by splicing two different vectors together. One that combines both a plasmid and a phage, called a cosmid, is capable of carrying relatively large genomic sequences. For a vector to function as intended, it must carry various specialized genes. Shuttle vectors are chosen expressly for their ability to be accepted by the cloning host. Vectors also contain expression genes for directing the start of replication, transcription, and translation, established sites for the action of restriction endonucleases, and some means of identifying them in culture. The latter property is usually fulfilled by genes for resisting antibiotics such as ampicillin and tetracycline.

One complication of working with large DNA is that bacteria may be unable to clone the larger segments of DNA. An answer to this problem came from basic research in the common brewer's yeast *Saccharomyces cerevisiae*. It was determined that yeast plasmids were able to copy much larger stretches of foreign DNA. These giant cloning vectors, called yeast artificial chromosomes (YACs) carry 10 times the amount of DNA of ordinary plasmids and can enhance the ability to replicate more complex genomes outside of their natural host.

Cloning Hosts

Since this bacterium (*E. coli*) was the original recombinant host, the protocols using it are well established, relatively easy, and reliable. Literally hundreds of specialized cloning vectors have been developed for it. The main disadvantage with this species is its lack of versatility in correctly expressing eukaryotic genes. One alternate host for certain industrial processes and research is the yeast *Saccharomyces cerevisiae,* which is more adapted to processing and modifying eukaryotic gene products. Certain techniques may also employ different bacteria (*Bacillus subtilis*), animal cell culture, and even live animals and plants to serve as cloning hosts.

Chimeric DNA Cloning

The human alpha interferon gene is a linear molecule of approximately 500 bp that codes for a polypeptide of 166 amino acids. It was originally isolated and identified from human blood cells and prepared from mRNA transcripts to be free of introns. This is a necessary step because the bacterial cloning host has none of the machinery needed to excise this nontranslated part of a gene. The first step in cloning is to prepare the isolated interferon gene for splicing into an *Escherichia coli* plasmid. One way this is accomplished is to attach terminal nucleotides (sticky ends) that can mate with the plasmid after it has been digested with an endonuclease. In the case of the HindIII endonuclease, for example, the gene will require one sticky end that reads AA and another that reads TT. In the presence of the endonuclease, the plasmid's circular molecule is nicked open and its sticky ends are exposed. When the gene and plasmid are placed together, their free ends base pair, and a ligase makes the final covalent seal. The resultant gene and plasmid combination is called a *chimera* (Gr. *Chimaira,* a monster with loin's head, a goat's body, and a serpent's tail).

Since the recombinant plasmid enters only some of the cloning host cells, it is necessary to perform a screening test to determine which clones carry the interferon gene. The key to locating these recombinant clones is the fact that the plasmid with the desired gene also carries a gene for ampicillin resistance. Clones are plated out on medium containing ampicillin, and only those clones that are resistant to the drug will form colonies. Recombinant colonies are selected from the plates and grown in culture.

To ensure that the bacteria can express the eukaryotic gene, the plasmid has been modified with the necessary transcription and translation leader sequences. As the *E. coli* culture grows in a special medium under optimum conditions, it transcribes and translates the interferon gene, synthesized the peptide, and secrets it into the growth medium. At the end of the process, the cloning cells and other chemical and microbial impurities are removed from the medium. Final processing to excise a terminal amino acid from the peptide yields the interferon product in a relatively pure form.

USES OF GENETIC ENGINEERING

Our discussion of these applications is divided into the following sections : Bioengineered products; transgenic (genetically recombined) microbes, plants, and animals; gene therapy and genetic medicines; and genomic mapping and analysis.

Biochemical Products

Diseases such as diabetes and dwarfism, caused by the lack of an essential hormone, are treated by replacing the missing hormone. Porcine and bovine insulin were once the only forms available to treat diabetes, even though such animal products can cause allergic reactions in sensitive individuals. In contrast, dwarfism cannot be treated with animal growth hormones, so the only source of human growth hormone (HGH) was originally from the pituitaries of cadavers. At one time, not enough HGH was available to treat the thousands of children in need. Recombinant technology changed the outlook of these and many other conditions by enabling large-scale manufacture of life-saving hormones and enzymes of human origin. Recombinant human insulin, humulin, can now be prescribed for diabeties, and recombinant HGH, or Protropin, can now be administered to children with dwarfism, a premature aging disease called progeria, or failure to grow. In addition to the obvious medical applications, numerous agricultural and industrial products are being developed at a rapid rate. Nucleic acid products also have a number of medical applications.

RECOMBINANT MICROORGANISMS

Foreign genes have been inserted into a variety of microbes, plants, and animals through recombinant DNA techniques developed especially for them. Transgenic "designer" organisms are available for a variety of biotechnological applications. Since they are unique life forms that would never have otherwise occurred, they can be patented. One of the first practical applications of rDNA in agriculture was to create a genetically altered strain of the bacterium *Pseudomonas syringae*. The wild strain ordinarily contains an ice nucleation gene that promotes ice or frost formation on moist plant surfaces. Genetic alteration of the frost gene using recombinant plasmids created a different strain that could prevent ice crystals from forming. A commercial product called Frostban has been successfully applied to stop frost damage in strawberry and potato crops. A strain of *Pseudomonas fluorescens* has been engineered with the gene from a bacterium (*Bacillus thuringiensis*) that codes for an insecticide. These recombinant bacteria are released to colonize plant roots and help destroy invading insects. All releases of recombinant microbes must be approved by the Environmental Protection Agency (EPA) and are closely monitored.

So far, extensive studies have shown that such microbes do not proliferate in the environment and probably pose little harm to humans. But they may give rise to other concerns. A genetically engineered bacterium called *Rhizobium* that increases the ability of plants to fix nitrogen and improves crop yields was due to be released in 1995 over 7 million acres of crops. Although the EPA has originally approved the project, it was blocked when concerns surfaced that the recombinant microbe would spread the genes to other plants, with the possibility of creating 'superweeds'. Microbiologists have invented an unusual means of assaying drug resistance in the pathogen *Mycobacterium tuberculosis*. They engineered a bacteriophage with the genes for luciferase, an enzyme from fireflies that causes light emission. Live *M. tuberculosis* infected with these viruses light up, while dead or dying cells do not. The test is conducted by incubating cultures with the usual drugs used in treatment and observing the viability of the culture. Dying cells indicate that the drug is working, and a healthy culture indicates drug resistance. Viruses can be genetically engineered for transfection purposes. Several viral vectors have been developed for gene therapy and for an experimental AIDS vaccine. In another agricultural application, a strain of virus pathogenic to cabbage looper insects was released into a cabbage patch.

TRANSGENIC PLANTS

Agrobacterium pathogens live in soil and can invade injured plant tissues. Inside the wounded tissue, the bacteria transfer a discrete DNA fragment called T-DNA into the host cell. The DNA is integrated into the plant cell's chromosome, where it directs the synthesis of nutrients to feed the bacteria and hormones to stimulate plant growth. In the case of *Agrobacterium tumefaciens*, increased growth causes development of a tumor called *crown gall disease*, a mass of undifferentiated tissue on the stem. The other species, *A. rhizogenes*, attacks the roots and transforms them into abnormally overgrown "hairy roots". The capacity of these bacteria, especially *A. tumefaciens*, to transfect host cells can be attributed to a large plasmid termed Ti. This plasmid inserts into the genomes of the infected plant cells and transforms them. Even after the bacteria in a tumor are dead, the plasmid genes remain in the cell nucleus and the tumor continues to grow. This plasmid is a perfect vector for inserting foreign genes into plant genomes. The procedure involves removing the Ti plasmid, inserting a previously isolated gene into it, and returning it to *Agrobacterium*. The insertion of genes with the Ti plasmid works primarily to engineer important dicot plants such as potatoes, tomatoes, cotton, and grapes. Examples of agriculturally important transgenic plants are herbicide-resistant cotton, soyabean, and tobacco plants; plants with a built-in insecticide; potatoes and tomatoes with resistance to bacterial pathogens; squash and melons that resist viral pathogens; pea seeds that can resist the destructive effects of weevils during storage.

For plants that cannot be transformed with *Agrobacterium*, seed embryos have been successfully implanted with additional genes using gene guns-small "shotguns" that shoot the genes into plant embryos with minute gold bullets. This technique was used to create wheat that resists pesticides, corn that resists a blight fungus, and viral resistant rice. Another strategy for engineering plants is to insert a strand of antisense RNA that turns off the expression of a target gene. This tomato, called the FLAVR SAVR, does not soften as rapidly as conventional tomatoes, therefore, allowing extra time for the natural flavour to develop on the vine and for transportation and storage.

TRANSGENIC ANIMALS

In fact, animals are so amenable to transfection that several hundred new strains of transgenic animals have been introduced by research and industry. The reason for this movement towards animals is that, unlike bacteria and yeasts, they can express human genes in organs and organ systems that are very similar to humans. One outgrowth has been the design of animal models of human genetic diseases to study disease mechanisms, and then to use these natural systems to test new genetic therapies prior to their use in humans. A common way of delivering foreign genes into an egg is to pulse it with high voltage to enable the gene to enter through small pores in the cell membrane. The technique is so intricate that only a small number of embryos survive; however it only takes a few successes, because a transgenic animal will usually pass the genes on to its offspring. The dominant animals in the genetic engineer's laboratory are transgenic forms of mice. In some early experiments, fertilized mouse embryos were injected with human genes for growth hormone, producing super mice twice the size of normal mice. Mouse zygotes afflicted with a genetic muscle defect were corrected by having the normal gene added. The so-called 'knock out mouse' has become a standard way of producing animals with tailor-made genetic defects.

The technique involves transfecting a mouse embryo with a defective gene and cross breeding the progeny through several generations. Eventually mice will be born with two defective genes and will express the disease. One of the first successful attempts was to develop an animal model of cystic fibrosis. Animal husbandry has been quick to adopt some of these techniques. Pig embryos have been implanted with the bovine growth hormone gene in an attempt to increase the pig's growth rate and decrease fat formation in the meat. The transgenic pigs did indeed have less fat, but unfortunately, the foreign gene appeared to contribute to severe health problems such as arthritis and heart and kidney disease. The pharmaceutical industry has already engineered a number of animals to manufacture commercial quantities of human proteins. A strain of transgenic pig can synthesize certain clotting factors and human hemoglobin as a blood substitute. Also, several companies have successfully harnessed the synthetic abilities of mammary glands.

GENE THERAPY

The inherent benefit of this therapy is to permanently cure the physiologic dysfunction by repairing the genetic defect. There are two strategies for this therapy. In *ex vivo* therapy, the normal gene is cloned in vectors such as retroviruses or adenoviruses that are infectious but relatively harmless. Tissues removed from the patient are incubated with these genetically modified viruses to transfect them with the normal gene. The transfected cells are then reintroduced into the patient's body by transfusion. In contrast, the *in vivo* type of therapy skips the intermediate step of incubating excised patient tissue, and instead the gene is introduced by virus vectors directly into the patient's body. So far, therapy has been hampered by a number of difficulties. First, the systems for delivering the gene into target cells are not completely successful. Second, the rate of productive transfection by the viruses is very low. Third, safety is also a concern. Some experts fear that the virus could insert the DNA into a site that turns on an oncogene or could activate other dormant viruses. In order to deliver the product more precisely, alternate delivery methods are being tested. These include gene guns similar to those previously described for plants, liposomes, and dendrimers. Liposomes are tiny spheres with an exterior lipid bilayer and aqueous center that can carry genes across cell membranes.

In some cells, the DNA appears to be picked up and held free in the nucleus. The transfected cells can even translate the gene for a limited time, but they may not duplicate or pass it on during cell division, and these cells often have a limited life span. One solution tailored for diseases of blood cells is to genetically engineer the long-lived stem cells in bone marrow. If enough stem cells take up and retain the replacement gene, it will be inherited by new blood cells for the life of the patient. Early results hint that transfecting the umbilical blood of new borns is an effective technique to increase uptake of the gene, but whether it can affect a long-term cure awaits further testing. The first gene therapy experiment in humans was initiated in 1990 by researchers at the National Institutes of Health. The subject was a four-year-old girl suffering from a severe immunodeficiency disease caused by the lack of the enzyme adenosine deaminase (ADA). She was transfused with her own blood cells into which the functional ADA gene had been inserted by viruses. Later, another child was also given the same type of therapy. Both children showed remarkable improvement and continue to be healthy but are not yet cured. In trials for cystic fibrosis (CF) therapy, adenoviruses carrying a normal CF gene are delivered directly into the nose and lungs. Gene treatment in both ADA deficiency and cystic fibrosis is impermanent and must be performed every year or so. Even so, if it can reduce the severity of the disease, it will be a boon for people with a number of single-gene defects such as hemophilia and sickle cell anemia. It also shows promise for acquired or non-genetic diseases (cancer, AIDS).

ANTISENSE DNA TECHNOLOGY

With this approach, oligonucleotides are delivered into cells to block transcription or translation of a specific DNA site, thereby bypassing the problem gene and protein altogether. Note that the term *antisense* as used in molecular genetics does not mean "no sense or nonsense". It is used to describe any nucleic acid strand with a base sequence that is complementary to the sense or translatable strand. For example, DNA contains a sense strand that is transcribed and a matching antisense strand that is not usually transcribed. We can also apply this terminology to RNA. Messenger RNA is considered the translatable sense strand, and a strand with the matching sequence of nucleotides would be the antisense strand. To illustrate : if an mRNA sequences reads AUGCGAGAC, then an antisense RNA strand for it would reads UACGCUCUG. In practice, most antisense agents are made of DNA because the technology for large-scale RNA synthesis is less feasible. The therapy takes advantage of the readiness with which an antisense DNA can hybridize with RNA. It is worth noting that the concept of antisense nucleic acids did not originate with genetic engineers. To increase their therapeutic effectiveness, antisense molecules are chemically modified by inserting a sulfur atom in place of one oxygen on the phosphate. This change makes the agents more soluble and improves their uptake by cells. When an adequate dose of antisense DNA is delivered across the cell membrane, it binds to specific sites on any mRNAs that are the targets of therapy. As a result, the reading of that mRNA transcript on ribosomes will be blocked and the gene product will not be synthesized. From the beginning of its development, antisense technology was hailed as a revolution in clinical pharmacology. On the positive side, animal experiments show that antisense drugs are relatively nontoxic and safe to use, and cell culture tests indicate that they can retard synthesis of an undesirable protein. But, as with most new therapies, there are also some stumbling blocks. One is the drugs' high cost and the need to administer them throughout the course of a person's life.

Another drawback has been the inability to get a significant dose of the agent into cells. Whether the DNA could become a permanent part of the genome of cells and lead to cancer and other diseases is another concern. It is also clear that this therapy will not work for many genetic-based diseases such as cystic fibrosis and sickle-cell anemia that are caused by the lack of a functioning gene. However, it promises to be most effective in preventing diseases caused by undesirable gene expression and by viruses. Diseases that might be amenable to this therapy are cancers, autoimmune diseases such as multiple sclerosis, infections such as hepatitis B and AIDS, and Alzheimer's disease. Human clinical trials are underway for blocking the replication of human papillomavirus. And the AIDS virus tribals of acute and chronic myelogenous leukemias, caused by the abnormal expression of a cancer gene, are also being carried out.

One of the unusual variations, triplex DNA, is a triple helix formed when a third strand of DNA inserts into the major groove of the molecule. The inserted strand forms hydrogen bonds with the purine bases on one of the adjacent strands. It seems logical that if the DNA has an extra strand wedged into its structure, this can make its template relatively inaccessible to normal transcription. Indeed, it is theorized that this is yet another natural means for cells to control their genetic expression. The potential therapeutic applications of triplex DNA are still in their infancy. So far, oligonucleotides have been synthesized to form triplex DNA that interact with regulatory sequences in genes. They can interrupt the action of transcription factors and prevent RNA polymerase from forming mRNA transcripts. Studies performed on cell cultures revealed that the genes coding for oncogenes, viruses, and the receptor for interleukin-2 (an immune stimulant) could be blocked with triplex DNA.

GENETIC MAPS AND DNA FINGER PRINTING

DNA technology will in turn lead to understanding of diverse biological phenomena, including development, virulence, disease, and evolution. Its influence is being felt in numerous other fields such as forensics (medical law), genetic pedigreeing, anthropology, neurobiology, and archeology. Gene mapping is the process of delineating the relative order and positions of genes on the chromosomes of living things. Less detailed location or linkage maps show the approximate sites of the genes on chromosomes. Sequence maps, which are much more difficult to arrive at, indicate the exact order of base pairs making up a gene or other chromosomal site. Up until recently only certain viral genomes and the bacterium *Haemophilus influenza* have been completely mapped and sequenced. The gene maps of *Escherichia coli* and the yeast *Saccharomyces cerevisiae* are nearing completion, and other microbes are being rapidly sequenced.

Within the first 10 years, a large number of genes were already mapped to a given chromosome, approximately 7,000 genes of the 100,000 genes in the human library were correlated with a specific location, and a smaller number have been completely sequenced. Space limitations here restrict showing more than an overview of the chromosomal map of humans. The total genome consists of about three billion nucleotides. If this DNA library were printed in book form, it would fill a million pages of text. It has been likened to a "human gene dictionary" or "encyclopedia of humans" composed of 23 volumes, one for each chromosome pair. This text would be tedious reading for all but the most powerful computers. Interestingly, only a small fraction of the genome actually codes for real phenotypic genes, and the rest is something quite remarkable. The outcome of the genome project promises to forever change the human condition. With a map of the human genome, the genetic background of the 5,000 inherited diseases could eventually be disclosed, clearing the way for extensive genetic screening. It would not only enable families to know in advance their risks for certain diseases, but also would guide decisions on correcting or alleviating them through gene and other forms of therapy. Map sites have been discovered for approximately 50 of these diseases.

It is now possible to apply DNA technology in a manner that emphasizes these differences and arrays the entire genome in a pattern for comparison. DNA fingerprinting is best known as a tool of forensic science. But it can also identify hereditary relationships, the risk for hereditary diseases, and the patterns of inheritance in conditions such as Huntington's disease, cystic fibrosis, and Alzheimer's disease. One application of this technique is to identify the risk for genetic disease as represented by Alzheimer's disease. Test DNA from various family members is clipped by restriction endonucleases, electrophoresed, and reacted with gene specific probes that will show exactly which restriction sites (genetic markers) the person possesses.

The genetic markers are not the actual gene that causes the disease, but they are located nearby and are inherited in exactly the same pattern. By comparing fingerprint patterns in a family that is known to carry and express the gene, the genetic family tree for many diseases can be worked out. Futurists predict that a DNA fingerprint will be made for every child at birth and kept on file, much as footprints and regular fingerprints have been used to until now. This same method of analysis can be used for clarifying human paternity and maternity, the pedigree (genetic ancestry) of domestic animals, and the genetic diversity of animals bred in zoos. In conjunction with PCR, it enables analysis of even small fragments of DNA and from this has sprung highly novel applications. In California, it played a role in determining if some hunters had poached deer from a private ranch and in linking a specific mountain lion to a dead woman by examining its stomach contents. The U.S. military has begun keeping DNA profiles on their personnel. When the remains of several "unknown soldiers" who had died in the Persian Gulf War were too mutilated and mixed to be properly identified, the DNA content was analyzed. A comparison to known DNA profiles of soldiers missing in action allowed the remains to be identified and separated. Since even fossilized DNA can remain partially intact, ancient DNA is being studied for comparative and evolutionary studies. Anthropologists have traced the migrations of ancient people by analyzing the mitochondrial DNA found in bone fragments. It has been used as a way to analyze and clone the ancient DNA of mummies, a 120 million-year-old insect, a 40,000-year-old mammoth, and dinosaurs.

13

SOIL AND AQUATIC MICROBIOLOGY

Amoebic dysentery, thermophilic spoilage, actinomycosis, and botulism are new terms, but the need of freedom of earth in wounds, in food, and in clothing is no recent discovery. The demonstration that soil, instead of being all dead, harbors millions of organisms releases that flight of imagination which pictures the soil as a sort of Lilliputian Zoo in which some magic hand has eliminated all barriers and set free every grade of minute but rapacious monster to go roaring after the next lesser grade as its lawful prey. Directly or indirectly the wastes of humans and other animals, their bodies, and the tissues of plants are dumped onto or buried in the soil. Somehow they all disappear, transformed into the substances that make up the soil. It is the microbes that make these changes—the conversion of organic matter into simple inorganic substances that provide the nutrient material for the plant world.

Aquatic microbiology is the study of microorganisms and their activities in fresh, estuarine, and marine waters, including springs, lakes, rivers, bays, and seas. It is the study of the microorganisms—viruses, bacteria, algae, protozoa, and microscopic fungi—which inhabit these natural waters. Some of these microorganisms are indigenous to natural bodies of water; others are transient, entering the water from air or soil or from industrial or domestic wastes. For example, wastewater can be pumped into rivers and coastal waters, or it can be disposed of in deep ocean dump sites. Wastewater usually contains microorganisms which will influence the activities of microorganisms already present in the receiving waters. Aquatic microbiology is the study of microorganisms and their activities are of great importance in many ways. They may affect the health of humans and other animal life; they occupy a key position in the food chain by providing rich nourishment for the next higher level of aquatic life; they are instrumental in the chain of biochemical reactions which accomplish recycling of elements, e.g., in mineralization. Aquatic microbiology, which in previous decades was studied by a relatively few microbiologists, has emerged as one of the more important areas of applied microbiology. Urbanization and consequently the growing demand for water by communities, the importance of natural water as a major food source, the off-shore exploration for oil and minerals, and other developments have resulted in the establishment of federal agencies which exercise jurisdiction over many aspects of natural bodies of water.

PHYSICAL CHARACTERS OF SOIL

From a functional viewpoint, the soil may be considered as the land surface of the earth which provides the substratum for plant and animal life. The characteristics of the soil environment vary with locale and climate. Soils differ in depth, physical properties, chemical composition, and origin (Fig. 13.1). The dominant mineral particles in most soils are compounds of silicon, aluminium, and iron, and lesser amounts of other minerals; including calcium, magnesium, potassium, titanium, manganese, sodium, nitrogen, phosphorus, and sulfur. The mineral constituents of soil range in size from small clay particles (0.002 mm or less) to large pebbles and gravel. The physical structure, aeration, water-holding capacity, and availability of nutrients are determined by the proportion of these particles, which are formed by the weathering of rock and the degradative metabolic activities of microorganisms. Soil can be classified as mineral soils, which have solid matter that is largely inorganic, and organic soils. Which have very little inorganic material. The latter are typically found in bogs and marshes.

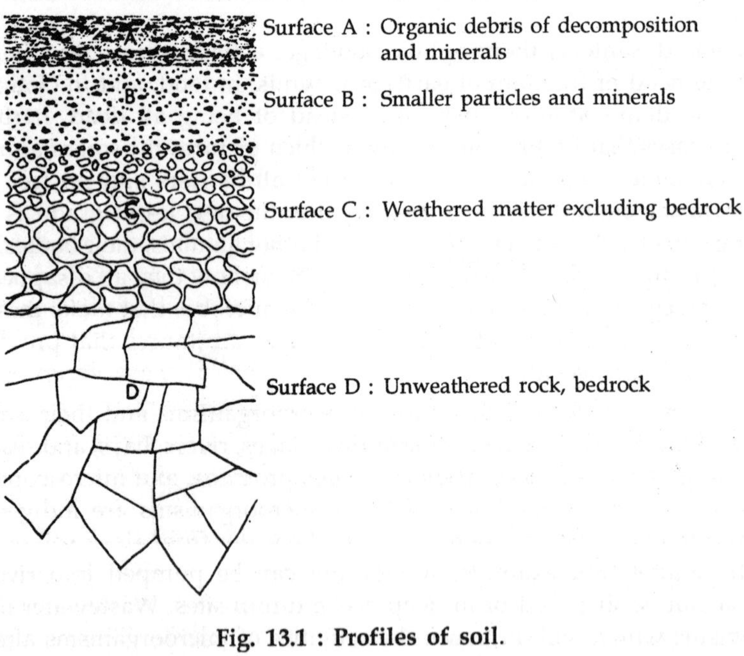

Fig. 13.1 : Profiles of soil.

ORGANIC RESIDUES OF SOIL

In the last stages of decomposition, such material is referred to as humus, a dark-coloured, amorphous substance composed of residual organic matter not readily decomposed by microorganisms. Indeed, the microbial population, both dead cells and living cells, is of such a large magnitude that it contributes significantly to the organic matter of soil. Certain agriculturally important properties are contributed to the soil by humus, which improves the texture and structure of the soil, contributes to its buffering capacity, and increases its water-holding capacity. The amount of water in soil depends on the amount of precipitation and other climatic conditions, drainage, soil composition, and the living population of the soil. Various organic and inorganic components of soil are dissolved in soil water and thus are made available as nutrients for soil inhabitants. The soil atmosphere is derived from air but differs in composition from it because of the biological processes occurring in soil. The gaseous phase of soil consists mainly of carbon dioxide, oxygen, and nitrogen. These gases exist primarily in the spaces between soil particles which are not filled with water, although a small amount of gas, especially carbon dioxide, is dissolved in water.

SOIL MICROFLORA

The vast differences in the composition of soils, together with differences in their physical characteristics and the agricultural practices by which they are cultivated, result in corresponding large differences in the microbial population both in total numbers and in kinds. The conditions described earlier as influencing the growth of organisms under laboratory cultivation are equally applicable to the soil. With specific reference to soil, these conditions can be summarized as follows : (a) Amount and type of nutrients, (b) available moisture, (c) degree of aeration (d) temperature, (e) pH, (f) practices and occurrences which contribute large numbers of organisms to the soil, e.g., floods or addition of manure. The existence of roots and the extensiveness of the root system in soil also influence the numbers and kinds of microorganisms present. Variations of climatic conditions may selectively favour certain physiological types. Interactions between and among microbial species no doubt has an important effect on the members of the population. This is an extremely complex situation. Predatory protozoa and antibiotic-producing actinomycetes may eliminate certain groups of microorganisms. Cellulolytic and proteolytic organisms, on the other hand, may provide nutrients for less versatile biochemical species.

Bacteria, fungi, algae, protozoa, and viruses make up this microscopic menagerie, which may reach a total of billions of organisms per gram. The great diversity of the microbial flora makes it extremely difficult to determine accurately the total number of microorganisms present. Cultural methods will reveal only those physiological and nutritional types compatible with the cultural environment. Direct microscopic counts theoretically should permit enumeration of all except the viruses, but this technique also has its limitations, especially in distinguishing living from dead microorganisms. Very often the microbiological analysis of soil is concerned with the isolation and identification of specific physiological types of microorganisms. Direct microscopic counts as high as several billions per gram have been reported; plate counts from the same samples yield only a fraction of this number (millions). The reason for this discrepancy is that there is such a great variety of nutritional and physiological types of bacteria in soil that no single laboratory environment supports the growth of every viable cell in the inoculum. The following are all likely to be found in soil : Autotrophs and heterotrophs; mesophiles, thermophiles, and psychrophiles; aerobes and anaerobes; cellulose digesters and sulfur oxidizers; nitrogen fixers and protein digesters; and other kinds of bacteria. It is generally agreed that there are many species of bacteria in soil yet to be discovered. Large numbers of actinomycetes, as many as millions per gram, are present in dry warm soils. The most predominant genera of this group are *Nocardia, Streptomyces,* and *Micromonospora.* These organisms are responsible for the characteristic musty or earthy odor of a freshly plowed field. They are capable of degrading many complex organic substances and consequently play an important role in building soil fertility. The actinomycetes are also noted for their ability to synthesize and excrete antibiotics.

Fungi inhabit the soil, and are most abundant near the surface, where an aerobic condition is likely to prevail. They exist in both the mycelial and spore stage. Since growth can take place from either a spore or a fragment of a mycelium, it is difficult to estimate their numbers; however, counts ranging from thousands to hundreds of thousands per gram of soil have been reported. Fungi are active in decomposing the major constituents of plant tissues, namely, cellulose, lignin, and pectin. The physical structure of soil is improved by the accumulation of mold mycelium within it. The binding together of fine soil particles to form water-stable aggregates. This is accomplished by the penetration of mycelium through the soil, forming a network which entangles the small particles.

The population of algae in soil is generally smaller than that of either bacteria or fungi. The major types present are the green algae and diatoms. Their photosynthetic nature accounts for their predominance on the surface of just below the surface layer of soil. In a rich fertile soil, the biochemical activities of algae are dwarfed by those of bacteria and fungi. In some situations, however, algae perform prominent and beneficial changes. Cyanobacteria, the oxygenic photosynthetic bacteria, are known to grow on the surfaces of freshly exposed rocks where the accumulation of their cells results in simultaneous deposition of organic matter. This establishes a nutrient base that will support growth of other bacterial species. The growth and activities of the initial algae and bacteria pave the way for the growth of other bacteria and fungi. The mineral nutrients of the rock are slowly dissolved by acids resulting from microbial metabolism. This process continues with a gradual accumulation of organic matter and dissolved minerals until a condition results that supports growth of lichens, then mosses, then higher plants.

Most soil protozoa are flagellates or amebas; the number per gram of soil ranges from a few hundred to several hundred thousand in moist soils rich in organic matter. From a microbiological standpoint they are of significance since their dominant mode of nutrition involves ingestion of bacteria. Of academic interest is the fact that they demonstrate a preference for certain microbial species.

The microbial population on and around roots is considerably higher than that of root-free soil; the differences are both quantitative and qualitative. Bacteria predominate, and their growth is enhanced by nutritional substances released from the plant tissue, e.g., amino acids, vitamins, and other nutrients; the growth of the plant is influenced by the products of microbial metabolism that are released into the soil. It has been reported that amino acid—requiring bacteria exist in the rhizosphere in larger numbers than in the root-free soil. It has been demonstrated that the microbiota of the rhizosphere is more active physiologically than that of nonrhizosphere soil.

SOIL MICROBE INTERACTIONS

Some of the associations are indifferent or neutral; some are beneficial or positive; others are detrimental or negative. As each different type of association or interaction is discovered, it has been given a specific descriptive label. As you might presume, many of these associations do not fall neatly into discrete categories. Furthermore, and likewise not unexpected, there is the existence of some confusion and contradiction in the use of terms. The term symbiosis, for example, as first proposed, referred to the "living together of dissimilarly named organisms"; it was used as a general term. Later it took on a more specific meaning, namely, an association between bacteria and plants referred to as symbiotic nitrogen fixation which is described later in this chapter. Currently the trend is to use the term *symbiosis* as originally intended, that is, merely as a condition in which the individuals of a species live in close association with individuals of another species. It is conceivable that two different species of microorganisms occupy the same environment without affecting each other. For example, each could utilize different nutrients without producing metabolic end products that are inhibitory.

Mutualism

Mutualism is an example of a symbiotic relationship in which each organism benefits from the association. The manner in which benefit is derived varies. One type of mutualistic association is that involving the exchange of nutrients between two species, a phenomenon called syntrophism. Many microorganisms synthesize vitamins and amino acids in excess of their nutritional requirements. Others have a requirement for one or more of these nutrients. Still others synthesize a particular essential nutrient in suboptimal amounts. Another mutalistic association is characterized by different metabolic products from the association as compared with the sum of the products of the separate species. The growth of the two species in association and the resulting effect on the rate and extent of leaching copper from an ore is shown. Leaching is one of the processes for recovering metals from ore; microorganisms play the important role of oxidizing insoluble metal sulfides to soluble sulfates.

Commensalism

The phenomenon of commensalism refers to a relationship between organisms in which one species of a pair benefits; the other is not affected. This occurs commonly in soil with respect to degradation of complex molecules like cellulose and lignin. For example, many fungi are able to dissimilate cellulose to glucose and beyond (they are cellulolytic). Many bacteria are unable to utilize cellulose, but they can and do utilize the fungal breakdown products of cellulose, e.g., glucose and organic acids. The existence of many nutritionally fastidious bacteria in soil suggests that their growth and survival is dependent upon the synthesis and excretion of vitamins and amino acids by less fastidious species. Another example of commensalism is that of a change in the substrate produced by a combination of species and not by individual species. For example, lignin, a major constituent of wood, is generally resistant to degradation by pure cultures of microorganisms under laboratory conditions.

Antagonistic Associations

When one species adversely affects the environment for another species, it is said to be antagonistic. Such organisms may be of great practical importance, since they often produce antibiotics or other inhibitory substances which affect the normal growth processes or survival of other organisms. Antagonistic relationships are quite common in nature. For instance, both *Staphylococcus aureus* and *Pseudomonas aeruginosa* are antagonistic toward

Aspergillus terreus. Certain *Pseudomonas* pigments inhibit germination of *Aspergillus* spores. *S. aureus* produces a diffusable antifungal material that causes distortions and hyphal swellings in *A. terreous*. Although microorganisms from a variety of natural habitats produce antibiotics, soil microorganisms are the most common producers. It is not unusual for one organism to produce five or six different antimicrobial agents. There is some question about the role of antibiotics in nature; production in the laboratory is under conditions quite different from those in nature. Production of antibiotics in soil may enable the antibiotic-producing organism to thrive successfully in a competitive environment. For example, large populations of actinomycetes have been found in the chitinaceous shells of dead crustaceans in the sea. Organisms that elaborate antibiotics represent the classic example of this phenomenon; however, antibiosis may result from a variety of other conditions operative in mixed populations. Cyanide is produced by certain fungi in concentrations toxic to other microorganisms, and the algae elaborate fatty acids which exhibit a marked antibacterial activity. Other metabolic products that may result from microbial activity in the soil which are likely to be inhibitory to other species are methane, sulfides, and other volatile sulfur compounds. Many soil microorganisms—important examples are the myxobacteria (slime bacteria) and streptomycetes—are antagonistic because they secrete potent lytic enzymes which destroy other cells by digesting their cell wall or other protective surface layers. Presumably the degraded cellular material, as well as the released protoplasmic material, servés as nutrients. Although it might be assumed that organisms producing lytic substances have a selective advantage over sensitive microbes, microbial interactions of this type are difficult to interpret. It appears that in the natural environment producers of lytic substances are often found in close proximity with sensitive organisms and do not predominate over them. A negative association may result from competition among species for essential nutrients. In such situations the best adapted microbial species will predominate or, in fact, eliminate other species which are dependent upon the same limited nutrient substance.

Parasitism

Parasitism is defined as a relationship between organisms in which one organism lives in or on another organism. The parasite feeds on the cells, tissues, or fluids of another organism, the host, which is commonly harmed in the process. The parasite is dependent upon the host and lives in intimate physical and metabolic contact with the host.

NITROGEN CYCLE

Soil microorganisms serve as biogeochemical agents for the conversion of complex organic compounds into simple inorganic compounds or into their constituent elements. The overall process is called mineralization. This conversion of complex organic compounds into inorganic compounds or elements provides for the continuity of elements (or their compounds) as nutrients for plants and animals including people. In this respect we can view the planet earth as a closed system dependent upon the process of recycling for maintenance of life as we know it. In the following paragraphs we shall discuss the role of soil microorganisms with respect to the transformations they bring about on nitrogen, carbon, sulfur, phosphorus, and their compounds.

The sequence of changes from free atmospheric nitrogen to fixed inorganic nitrogen, to simple organic compounds, to complex organic compounds in the tissues of plants, animals, and microorganisms, and the eventual release of this nitrogen back to atmospheric nitrogen (Fig. 13.2). Proteins, nucleic acids, purine and pyrimidine bases, and amino sugars (glucosamine and galactosamine) represent the complex organic nitrogenous substances which are deposited in soil in the form of animal and plant wastes or their tissues. Synthetic processes of microorganisms also contribute some amount of complex organic nitrogen compounds.

The simplest form of nitrogen involved in biological transformations is gaseous elementary nitrogen. The overall transformations in which microorganisms are involved

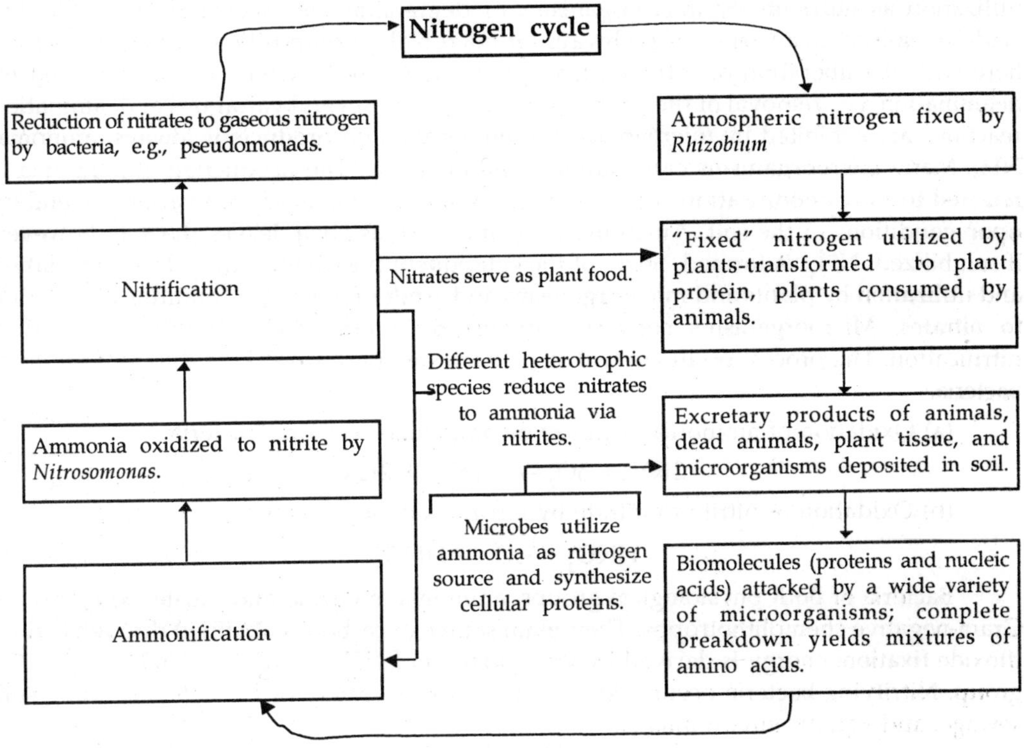

Fig. 13.2 : Nitrogen cycle.

range from nitrogen gas to protein. A great many intermediate products and a corresponding large number of intricate enzymatic reactions are involved in bringing about these changes. The nitrogen in proteins (as well as in nucleic acids) may be regarded as the end of the line as far as synthesis of nitrogenous compounds is concerned. The nitrogen in proteins is "locked" and is not available as a nutrient to plants. In order to set this organically bound nitrogen free for reuse, the first process that must take place is the enzymatic hydrolysis of proteins (proteolysis). This is accomplished by microorganisms capable of elaborating extracellular proteinases that convert the protein to smaller units (peptides). The peptides are then attacked by peptidases, resulting ultimately in the release of individual amino acids. Some bacterial species elaborate large amounts of proteolytic enzymes. Among the most active in this respect are some of the clostridia, e.g., *Clostridium histolyticum* and *C. sporogenes*; a lesser degree of activity is found in species of the genera *Proteus, Pseudomonas,* and *Bacillus*. Many fungi and soil actinomycetes are extremely proteolytic. Peptidases, however, occur widely in microorganisms as demonstrated by the fact that peptones are a common constituent of bacteriological media and provide a readily available source of nitrogen.

Proteolysis

The end products of proteolysis are amino acids. Their fate in the soil may be utilization as nutrients by microorganisms or degradation by microbial attack. Amino acids are subject to a variety of pathways for microbial decomposition. We are concerned here with the liberation of nitrogen from these compounds, which is accomplished by deamination, i.e., removal of the amino group. Although several variations of deamination reactions are exhibited by microorganisms, one of the end products is always ammonia, NH_3. Many microorganisms can deaminate amino acids. The production of ammonia is referred to as ammonification. The fate of the ammonia thus produced varies, depending upon conditions in the soil. Ammonia is volatile and, as such, leaves the soil; however, if solubilized, NH_4^+ is formed. Some of the subsequent possibilities include accumulation and utilization by plants and microorganisms and, under favourable conditions, oxidation to nitrates. Microorganisms convert ammonia to nitrate, and the process is called nitrification. The process occurs in two steps, each step performed by a different group of bacteria.

(a) Oxidation of ammonia to nitrite by ammonia-oxidizing bacteria

$$2NH_3 + 3O_2 \rightarrow 2HNO_2 + 2H_2O$$

(b) Oxidation of nitrite to nitrate by nitrite-oxidizing bacteria

$$HNO_2 + \tfrac{1}{2}O_2 \rightarrow HNO_3$$

Bacteria of both physiological groups, ammonia oxidizers and nitrite oxidizers, are Gram-negative chemolithotrophs. Their main source of carbon is obtained through carbon dioxide fixation; energy is derived by the oxidation of NH_3 or NO_2^- depending upon the group. Nitrifying bacteria occur widely in nature in a variety of habitats, including soil, sewage, and aquatic environments.

Nitrifying bacteria cannot be isolated directly by the usual techniques employed to isolate heterotrophic bacteria. Some of the reasons for this are : they are slow-growing compared with heterotrophs; and they may be present in very small numbers compared with other physiological types. Species of ammonia-oxidizing bacteria vary in morphology (rod, spherical, spiral, or lobular) and usually have an extensive membrane system within their cytoplasm. They frequently form cysts and zooglea. Species of nitrite-oxidizing bacteria exhibit some of the same morphological characteristics as the ammonia oxidizers. Only a few species have been isolated and described. These include *Nitrobacter winogradskyi* and *Nitrospina gracilis*. An interesting historical event involving nitrification and production of gunpowder may be cited. During the Napoleonic wars, France was unable to import nitrate, which was needed for the manufacture of gunpowder. To solve this dilemma, artificial niter beds were made, consisting of soil mixed with animal waste and vegetable materials, ashes, etc. Aeration was performed by turning the heap over from time to time.

After a long period of incubation, crude saltpeter (mineral nitrates) was extracted with hot water. This occurred, of course, long before the specific activities of microorganisms were known.

Nitrate Reduction to Ammonia

Several heterotrophic bacteria are capable of converting nitrates into nitrites or ammonia. This normally occurs under anaerobic conditions, e.g., in waterlogged soil. The oxygen of the nitrate serves as an acceptor for electrons and hydrogen. The process involves several reactions, and the overall result is—

$$HNO_3 + 4H_2 \rightarrow NH_3 + 3H_2O$$

This reaction is not of major significance in well-cultivated agricultural soil.

Denitrification

The transformation of nitrates to gaseous nitrogen is accomplished by microorganisms in a series of biochemical reactions. The process is known as denitrification. From the standpoint of agriculture, this is an undesirable process in that it results in loss of nitrogen from the soil and hence a decline in nutrients for plant growth. Species of several genera of bacteria are capable of transforming NO_3^- to N_2, e.g., *Achromobacter, Agrobacterium, Alcaligenes, Bacillus, Chromobacterium, Flavobacterium, Hyphomicrobium, Pseudomonas, Thiobacillus,* and *Vibrio.* The overall biochemical reaction which expresses the process of denitrification is—

$$2NO_3^- \rightarrow 2NO_2^- \rightarrow 2NO \rightarrow N_2O \rightarrow N_2$$

Environmental conditions in a soil have a significant effect on the level of denitrification. For example, the process is enhanced in soils (a) by an abundance of organic matter, (b) by elevated temperatures (25 to 60°C), and (c) by neutral or alkaline pH. Availability of oxygen has a dual effect. Denitrification proceeds only when the oxygen supply is limited.

Nitrogen Fixation

The conversion of molecular nitrogen into ammonia is known as nitrogen fixation. Two groups of microorganisms are involved in this process : (a) Nonsymbiotic microorganisms—those living freely and independently in the soil; and (b) symbiotic microorganisms—those living in roots of plants. Several types of experiments are used to detect nitrogen fixation by microorganisms. One approach is to demonstrate growth in a nitrogen-free medium. More specific evidence of fixation can be obtained by cultivating the microorganism in the presence of nitrogen labeled with isotopic nitrogen. $^{15}N_2$ can be measured by using a mass spectrometer. In essence, after the organism is grown in the mixture of atmospheric nitrogen and $^{15}N_2$, the culture is examined for evidence of $^{15}N_2$ incorporated in any compounds. Its presence is positive proof that nitrogen has been fixed. The capability of the nitrogen-fixing enzyme to act upon acetylene, discovered in the mid-1960s, has led to the development of a simple, rapid, relatively inexpensive technique now widely used to measure nitrogen fixation. The test is based on the observation that the nitrogen-fixing enzyme (*nitrogenase*) interacts with triple-bonded compounds, e.g., acetylene, to form ethylene as follows :

$$HC \equiv CH \xrightarrow[\text{Nitrogenase}]{2H} H_2C = CH_2$$

The comparable reaction with nitrogen is–

$$N{\equiv}N \xrightarrow[\text{Nitrogenase}]{6H} 2NH_3$$

The amount of ethylene produced is a measure of nitrogenase activity.

The essential reactants in the bacterial nitrogen fixation process are as follows :

(a) This has been characterized as two components, and neither is active without the other. Component I is nitrogenase and component II is nitrogenase reductase. Component I is known as the MoFe protein. Component II, which is a smaller molecule is designated the Fe protein. Both molecules contain sulfur; (b) a strong reducing agent such as ferredoxin or flavodoxin; (c) ATP; (d) a regulating system for NH_3 production and utilization (e) a system that protects the nitrogen-fixing system from inhibition by molecular oxygen.

Nonsymbiotic nitrogen fixation has been studied extensively with *Clostridium pasteurianum* and species of *Azotobacter*. For many years, these bacteria were the only ones known to be capable of this activity. The former is an anaerobic bacillus, and the latter are aerobic oval to spherical cells; both are widely distributed in soils. The nitrogen-fixing capacity of the *Azotobacter* species is greater than that of *C. pasteurianum*. It has been estimated that the amount of nitrogen fixed by the nonsymbiotic process ranges between 20 and 50 Ib/acre annually. This estimate is no doubt subject to much variation depending upon the conditions peculiar to a particular soil.

Symbiotic nitrogen fixation is accomplished by bacteria of the genus *Rhizobium* in association with legumes. Before these bacteria can fix nitrogen, they must establish themselves in the cells of root tissue of the host plant. Infection of the root system by the rhizobia bacteria is closely associated with the formation of an "infection thread" that develops in certain root hairs. The nitrogen-fixing bacteria invade the host plant cells via this infection thread. Some of the cells of the plant are thus infected, causing cell enlargement and an increased rate of cell division, leading to the formation of abnormal growths (nodules) on the root system. It is a process where both the bacteria and the plant benefit by the association. The bacteria convert atmospheric nitrogen to fixed nitrogen which is available to the plant, and in turn, the bacteria derive nutrients from the tissues of the plant. Not all species of *Rhizobium* produce nodulation and nitrogen fixation with any legume. There is a degree of specificity between the bacteria and legumes. For purposes of inoculation with commercial preparation of these bacteria, legumes are divided into seven major categories as follows : Alfalfa, clover, peas and vetch, cowpeas, beans, lupines, and soybeans. *Rhizobium* species or strains effective for one group are less effective or ineffective for other groups. One area of research is directed toward introducing the "package" of nitrogen-fixing genes from bacteria into plant cells. If this were achieved, plants might be capable of directly fixing nitrogen from the atmosphere. This would be a tremendous advance not only for agriculture but for the world at large in terms of producing food more economically and abundantly. Obviously, considerably more research is necessary before this kind of genetic engineering can be attempted at a practical level.

For instance, nitrogenase is easily destroyed by oxygen, and some means of protection of this enzyme complex from oxygen would have to be provided in order for a plant cell to be able to fix nitrogen. Alternatively, it may be possible to modify certain bacteria in a manner so that they would develop a relationship with the root system of other plants, as the *Rhizobium* species grow with legumes. For example, a symbiotic bacterial nitrogen-fixing system with cereal grains would have a tremendous effect on grain production both in yield and cost.

CARBON CYCLE

The ultimate source of organic carbon compounds in nature is the carbon dioxide present in the atmosphere. Although green plants and algae are the most important agents of carbon dioxide fixation, bacteria are also capable of synthesizing organic matter from inorganic carbon. The occurrence of photosynthesis among microorganisms has already been described. Other examples of carbon dioxide transformation or incorporation into organic compounds by bacteria are : (a) Utilization of carbon dioxide by autotrophic bacteria; the carbon dioxide represents the sole source of carbon for these organisms and is transformed by a reduction reaction to carbohydrates. (b) Carbon dioxide fixation by heterotrophic microorganisms is common among bacteria. A specific example of this type of reaction is—

$$CH_3COCOOH + CO_2 \rightarrow HOOCCH_2COCOOH$$

The end product, carbon dioxide, is released into the air and soil. Fresh air contains approximately 0.03 percent carbon dioxide by volume. Bacteria and fungi are the principal microorganisms that degrade organic carbon compounds. Under most natural systems of vegetation, e.g., forests, the amount of organic material in the soil remains approximately the same from year to year. This results from a balance established between the annual litter fall and death of the plants and the capacity of microorganisms to degrade these tissues. The most abundant organic material in plants is cellulose. It is readily attacked by many species of bacteria and fungi. The initial enzymatic attack is by cellulase which splits this long-chain polymer of glucose to cellobiose, which contains two glucose units. In turn, the cellobiose is split to glucose by the enzyme β-glucosidase; glucose is metabolized readily by many microorganisms. Complete oxidation yields CO_2 and H_2O.

SULFUR CYCLE

Some species oxidize and others reduce various sulfur compounds. The microbial transformations of sulfur have counterparts in the microbial transformation of nitrogen. For example, sulfide and ammonia are reduction products of the dissimilation of some organic compounds; both may be oxidized by various bacterial species. Sulfur in its elemental form cannot be utilized by plants or animals. Certain bacteria, however, are capable of oxidizing sulfur to sulfates. Sulfate is assimilated by plants and is incorporated into sulfur-containing amino acids and then into proteins. Degradation of proteins [proteolysis] liberates amino acids, some of which contain sulfur. This sulfur is released from the amino acids by enzymatic activity of many heterotrophic bacteria.

Sulfates may also be reduced to hydrogen sulfide by soil microorganisms. Hydrogen sulfide resulting from sulfate reduction and amino acid decomposition is oxidized to elemental sulfur. This reaction is characteristic of certain pigmented [photosynthetic] sulfur bacteria and is expressed as—

$$CO_2 + 2H_2S \xrightarrow{\text{Light}} [CH_2O]_x + H_2O + 2S$$

The column contains mud, $CaSO_4$, plant tissue (a source of carbohydrate-cellulose), and water. It is exposed to daylight and incubated at room temperature. A variety of heterotrophic microorganisms oxidizes various substrates, depleting the oxygen supply and creating anaerobic conditions. Organic acids serve as the electron donors for the reduction of sulfates and sulfites to hydrogen sulfide by anaerobic sulfate-reducing bacteria, e.g., *Desulfotomaculum*. Photosynthetic microorganisms such as the purple and green sulfur bacteria [*Chromatium* and *Chlorobium*] use hydrogen sulfide as the electron donor to reduce CO_2 –

$$CO_2 + H_2S \xrightarrow{\text{Sunlight}} [CH_2O]_x + S$$

The aerobic sulfur-metabolizing bacteria, *Thiobacillus* spp., develop in the upper portion of the column and oxidize reduced sulfur compounds. The nonsulfur purple bacteria [*Rhodospirillum, Rhodopseudomonas,* and *Rhodomicrobium*] are facultative phototrophs; they grow aerobically in the dark and anaerobically in the light and can utilize sulfide at low levels. The preceding discussion was concerned with transformation of nitrogen, carbon, and sulfur and their compounds. But this represents only a part of the elements and their compounds that are subject to assimilation and dissimilation by microorganisms. The metabolic activity of microorganisms [production of acids] solubilizes phosphate from insoluble calcium, iron, and aluminium phosphates. Phosphates are released from organic compounds such as nucleic acids by microbial degradation. Bacteria change insoluble oxides of iron and manganese to soluble manganous and ferrous salts.

Herbicides are chemical substances that kill plants, especially weeds; pesticides, as the term denotes, are chemical substances that destroy pests. In the context of soil, we think of those pests which adversely affect economic crops—weeds, insects, and pathogenic microorganisms. Thus a more specific nomenclature for substances classified as pesticides would be herbicides, insecticides, fungicides, and nematocides. The wide- scale application of herbicides and pesticides, while improving the crop yield, raises questions as to the short- and long-range effects as they are deposited in the soil. Are they degraded by soil microorganisms, and if so how rapidly? Do they have a temporary [or permanent] effect upon the soil microbiota? Do they constitute a form of runoff pollution to streams and rivers and as such affect aquatic plant life? These are some of the questions that concern the soil microbiologist as well as other soil scientists, biologists, and environmentalists. Naturally, a major research effort is directed toward answering the questions asked above, as well as others.

TYPES OF WATER

According to nature, the waters can be classified into following types :

(a) Atmospheric Water

It has been estimated that about 80,000 cubic miles of water from oceans and 15,000 cubic miles from lakes and land surfaces evaporate annually. The total evaporation is equaled by the total precipitation, of which about 24,000 cubic miles fall on land surfaces. Microorganisms of various kinds are present at different stages of this cyclic process—in atmospheric water, surface water, and groundwater. Because the kinds of aquatic environments are so different, it is not surprising that different species of microbes are considered to be indigenous to specific habitats. The moisture contained in clouds and precipitated as snow, sleet, hail, and rain constitutes atmospheric water. The microbial flora of this water is contributed by the air. In effect, the air is "washed" by atmospheric water, which carries with it the particles of dust to which microorganisms are attached. Most of the microorganisms are thus removed from air during the early stages of precipitation.

(b) Surface Water

Bodies of water such as lakes, streams, rivers, and oceans represent surface water. To a greater or lesser degree, these waters are susceptible to contamination with microorganisms from atmospheric water [precipitation], the surface runoff from soil, and any wastes deliberately dumped into them. Microbial populations vary in both number and kind with the source of water, with composition of the water in terms of microbial nutrients, and with geographical, biological, and climatic conditions.

(c) Ground Water

Ground water is subterranean water that occurs where all pores in the soil or rock-containing materials are saturated. Bacteria as well as suspended particles are removed by filtration, in varying degrees, depending on the permeability characteristics of the soil and the depth to which the water penetrates. Springs consist of ground water that reaches the surface through a rock fissure or exposed porous soil. Wells are made by sinking a shaft into the ground to penetrate the ground water level. Wells less than 100 ft deep are considered to be shallow . Bacteriologically speaking, wells and springs that are properly located produce water of very good quality.

THE AQUATIC ENVIRONMENT

The microbial population in a body of natural water is, to a large extent, determined by the physical and chemical conditions which prevail in that habitat. This generalization applies to all habitats. It is readily apparent that these conditions vary over wide extremes when one compares streams, estuaries, and the open sea.

More than 90 percent of the marine environment [by far the major aquatic habitat] is below 5°C, a condition favourable for the growth of psychrophilic microorganisms. Microorganisms do occur in natural hot springs where temperatures as high as 75 to 80°C prevail. Recently, microbiologists have reported extreme thermophilic microorganisms associated with geothermal vents in the Pacific Ocean floor. These unusual microbes are said to be capable of growing at 250°C and 265 atm of pressure. If these reports can be confirmed, the existence of such organisms will cause all biologists to reevaluate their concept of the maximum temperature that can be tolerated by life on earth. Aside from this extreme, the temperature in lakes, streams, and estuaries is influenced by the seasons, and there are corresponding shifts in the microbial flora. There are striking differences in the hydrostatic pressure of surface waters and of water in oceanic depths. Hydrostatic pressure affects chemical equilibrium, which, in turn, results in lowering the pH of seawater, resulting in a change in the solubility of nutrients such as bicarbonate, HCO_3^- Hydrostatic pressure also increases the boiling point of water, thereby maintaining water in its liquid state at high temperatures and pressures. By definition hydrostatic pressure increases with depth at the rate of 1 atm per 10 m. Barophilic microorganisms, organisms which cannot grow at normal atmospheric pressures, have been isolated from Pacific trenches (depth 1000 to 10,000 m), where enormous hydrostatic pressures exist [>100 atm]. Hydrostatic pressure of the deep sea is an important factor in the occurrence and growth of marine microorganisms in this environment. Deep sea bacteria have now been isolated from sediment, water, and deep sea animals by using special pressure-retaining sampling devices.

In most aquatic habitats, these primary producers are algae, and their growth is restricted to the upper layers of waters through which light can penetrate. The depth of the photic zone varies depending on such local conditions as latitude, season, and particularly the turbidity of the water. Generally, the photosynthetic activity is confined to the upper 50 to 125 m. Carbon dioxide is available largely from HCO_3^-, although some gaseous CO_2 is available. A distinctive characteristic of sea water is its high salt content, which is remarkably constant. The concentration of dissolved salts varies between 33 and 37 g/kg water. The principal salts are the chlorides, sulphates and carbonates of sodium, potassium, calcium, and magnesium. The concentration of salts is usually less in shallow offshore regions and near river mouths. Most marine microorganisms are halophilic; they grow best at salt concentrations of 2.5 to 4.0 percent, whereas those from lakes and rivers are salt sensitive and do not grow at a salt concentration of more than 1 percent.

The Adriatic Sea is sparkling clear at great depths, whereas some near-shore rivers are often turbid. The suspended material responsible for the turbidity includes (a) particles of mineral material which originate from land; (b) detritus, predominantly particulate organic material, such as cellulose, hemicellulose, and chitin fragments; and (c) suspended microorganisms. As previously mentioned, turbidity of the water influences the penetration of light, which in turn affects the photosynthetic zone. Particulate matter also serves as a substrate to which microorganisms adhere or as substrates that are metabolized.

The pH of the sea is 7.5 to 8.5. Optimum growth of most marine species is obtained on media adjusted to pH 7.2 to 7.6. Lakes and rivers may show a wider range in pH depending upon local conditions. Nitrates and phosphates are important inorganic constituents, particularly for the growth of algae. Organic compounds are required for the growth of saprophytic bacteria and fungi. Near-shore waters, which receive domestic wastewater, are subject to intermittent variations in their nutrient load, whereas the nutrient load of the open sea is very low and stable. Industrial wastes may contribute antimicrobial substances to estuaries and coastal waters. Mercury and other heavy metals in small concentrations may inhibit growth of some microorganisms while simultaneously permitting the growth of resistant forms. Resistance is usually coded for by genes associated with R [resistance] plasmids. For example, many pseudomonads and staphylococci are capable of volatilizing mercury, thereby removing its toxic effects from their immediate environment.

The top "layers," especially the surface film, and the bottom sediments harbor the higher concentrations of microorganisms, particularly in deep waters. The aggregation of floating and drifting microbial life in the surface region of the aquatic ecosystem is called plankton. Plankton may be composed primarily of algae [phytoplankton], or it may be predominantly protozoa and other minute animal life [zooplankton]. Phototrophic microorganisms are regarded as the most important plankton since they are the primary producers of organic matter via photosynthesis. Most phytoplanktonic organisms are motile, possess some structural feature, or contain oil droplets which give them buoyancy; all these features aid the organisms in maintaining their location in the photosynthetic zone.

Microbial inhabitants of the bottom region of a body of water are referred to as the benthic organisms. The richest region of an aquatic system in terms of numbers and kinds of organisms is the benthic region. Many aquatic microorganisms inhabit the gut of marine animals, an even richer habitat. The movement of water by wind, tide, or currents accomplishes some redistribution of the microbial flora. A phenomenon called upwelling occurs in an ocean when water rises from a deeper to a shallower depth, usually as a result of divergence of offshore currents or winds. In this process the bottom water carries with it a rich supply of nutrients that are delivered to the surface region. Upwelling occurs off the coasts of California and Peru and is responsible for the high productivity of these regions. Geothermal vents also contribute to the total nutrient budget of the ocean. It has been calculated that vents such as the one near the Galapagos Islands account for most of the nutrients dissolved in the oceans of the world. Prior to discovery of these vents, oceanographers were unable to account for nutrients on the basis of precipitation, input from rivers and streams, and other obvious sources.

TECHNIQUES USED TO STUDY OF AQUATIC MICROBES

Numerous problems are associated with attempts to characterize the microbial flora of aquatic environments. Among these problems are the following :

(a) Many aquatic microorganisms will not grow on the usual laboratory media such as nutrients agar or nutrient broth and consequently cannot be isolated. It is generally acknowledged that the estuaries and oceans contain a large number of microbial species that await discovery.

(b) A high percentage of aquatic bacteria have a natural affinity to grow attached to solid surfaces, either on particulate material or on large organisms.

(c) During the time which elapses between sample collection and transport back to the home-based laboratory, there is a loss of viability of many organisms.

(d) The collection of samples from the depth of the estuary or ocean requires specialized sampling equipment.

The choice of method is determined by the purpose of the examination, e.g.–

a. Microscopic examination for identification and enumeration of algae, bacteria, protozoa, and many fungi.

(i) Direct viable counting of physiologically responsive bacteria.

(ii) Detection of specific bacteria by means of epifluorescent microscopy and fluorescent antibody techniques.

b. Isolation of certain groups of bacteria, e.g., *Escherichia, Pseudomonas*. Many species of bacteria can be cultured by the plate technique on the usual bacteriological media. The membrane-filter technique is applicable for the examination and cultivation of many bacteria from the aquatic environments.

c. Enrichment-culture technique for isolation of specific physiological or metabolic types of microorganisms.

d. Measurement of total mass or biochemical activity.

Methods of examining water bacteriologically are set forth in the book *Standard Methods for the Examination of Water and Wastewater*, prepared and published jointly by the American Public Health Association, the American Water Works Association, and the Water Pollution Control Federation. The Environmental Protection Agency [EPA] has also developed a manual for this purpose under the title *Microbiological Methods for Monitoring the Environment –Water and Wastewater*. It is essential that strict attention be given to the following details when water samples are submitted for bacteriological analysis. The routine bacteriological procedures consist of (a) a plate count to determine the number of bacteria present and (b) tests to reveal the presence of coliform bacteria.

The interpretation of the results of the standard plate count must take into account the fact that the presence of a few pathogenic microorganisms is more significant than water containing many saprophytic bacteria. However, water of good quality is expected to give a low count, less than 100 per milliliter. Plate counts are useful in determining the efficiency of operations for removing of destroying organisms, e.g., sedimentation, filtration, and chlorination. A count can be made before and after specific treatment. The results indicate the extent to which the microbial population has been reduced.

Several selective and differential media greatly expedite the process of examining water for coliform organisms. The standard microbiological technique involves three successive steps : (a) The presumptive test, (b) the confirmed test, and (c) the completed test.

The membrane-filter technique for the bacteriological examination of water :

a. A sterile disk is placed in a filtration unit.

b. A volume of the water to be tested is drawn through this filter disk; the bacteria are retained on the surface of the membrane.

c. The filter disk is removed and placed upon an absorbant pad that has previously been saturated with the appropriate medium. Alternatively, the disk can be placed on the surface of an agar medium in a Petri dish.

d. Upon incubation, colonies will develop upon the filter disk wherever bacteria were entrapped during the filtration process.

AQUATIC MICROBES

It includes the microbiology of lakes, ponds, streams, estuaries, and the sea. It is an all-inclusive term for the study of microorganisms in natural waters. The microbiology of fresh waters constitutes a part of the science of limnology, which is the study of the flora and conditions for life in lakes, ponds, and streams. There is usually a fairly large littoral zone along the shore, which has considerable rooted vegetation and includes regions where light penetrates to the bottom. In open areas, the limnetic zone is determined by the light-compensation level. Photosynthetic activity decreases progressively in the deeper regions of the open water. The benthic zone is composed of soft mud or ooze at the bottom. Together with the profundal zone, the benthic region is largely populated by heterotrophic organisms. When the benthic zone is composed primarily of organic material, the majority of organisms will be anaerobic decomposers. The greatest variety of physiological types is found in the limnetic and littoral zones, and in addition they constitute the most productive regions. Productivity, of course, is affected by the chemical nature of the basin and the nature of imported materials from streams and rivers. Such stratification acts as a barrier to nutrient and oxygen exchange, especially in still water. In the summer, the top layers tend to be warmer than the lower regions, but in the winter, ice, which is less dense, collects on the top ; therefore, a reversal of temperature and mixing occurs in the spring and in the fall, often resulting in massive growth of algae (bloom).

Streams obtain a majority of nutrients from the flow of inorganic and organic materials from the surrounding terrestrial system or lakes or ponds. To a major extent, the microbial flora reflects the immediate terrestrial conditions, including the effects of agricultural and industrial practices. The drastic environmental changes in streams and rivers created by rapidly expanding urbanization on the one hand and changes in farming practices on the other make it impossible to generalize upon typical or characteristic microbial flora. An estuary is a semienclosed coastal body of water which has a free connection with the open sea. Stated differently, it is the coastal adjunct of the marine ecosystem. Compared with ocean waters the estuary lacks constancy in many characteristics. Estuaries are complex systems which receive inputs from a variety of sources. Temperature, salinity, turbidity, nutrient load, and other conditions fluctuate over a wide gradient in space and time.

From these observations it is apparent that the microbial flora of the estuary is subject to considerable fluctuation. Some species are indigenous to specific ecological niches of the estuary; others are transient, having been added from domestic, industrial, agricultural, or atmospheric sources. In areas receiving domestic pollution rich with organic nutrients, predominant bacteria include coliforms, fecal streptococci, and species of *Bacillus, Proteus, Clostridium, Sphaerotilus, Beggiatoa, Thiothrix, Thiobacillus*, and many others. Viruses of the enteric group are also likely to be found. In regions of the estuary that are nutritionally poor, one is likely to find the budding and/or the appendaged bacteria, e.g., *Hyphomicrobium, Caulobacter*, and *Gallionella*, in addition to pseudomonads.

MARINE PLANKTONS

This group of microorganisms is chiefly responsible for the conversion of radiant energy to chemical energy—energy stored in chemical substances that accumulate in the sea. The magnitude of this accomplishment is revealed by a calculation that suggests a requirement of 50 billion metric tons of phytoplankton to support the growth of the potential world fish catch, estimated at 50 million metric tons. Planktonic algae, under certain environmental conditions, may grow into enormous populations with resultant discolouration of the water—a condition referred to as bloom. The characteristic colour of the Red Sea is associated with heavy blooms of a cyanobacterium, *Oscillatoria erythraea*, which contains the pigments phycoerythrin and phycocyanin. "Red tides" are likewise due to the explosive growth of certain planktonic species. Brown, amber, or greenish-yellow discolouration of extensive areas of water occur as result of blooms by other microorganisms. The bacterial population throughout the photosynthetic zone is closely related to the distribution of phytoplankton. The beneficial effect of the plankton may be attributed to both the organic substances they elaborate and the solid surfaces provided for bacterial aggregation. Bacterial populations differ widely with prevailing conditions. The temperature of the marine environment and the degree of salinity would be suitable for the growth of psychrophilic and halophilic physiological type. Most marine bacteria, however, are halotolerant or slightly to moderately halophilic. Among the psychrophilic forms are certain luminous bacteria which can produce light in the presence of oxygen. In general, aquatic bacteria tend to be Gram-negative. It is thought that the Gram-negative envelope provides a structure better suited to support life in nutritionally dilute aquatic environments than the Gram-positive cell wall. Important hydrolytic enzymes are retained in the periplasmic space, rather than being excreted and lost to the aquatic environments, as would be the case for Gram-positive bacteria. In addition lipopolysaccharide (LPS) of the outer membrane affords Gram-negative bacteria protection from certain toxic molecules, e.g., fatty acids and antibiotics, and it may serve to sequester important nutrients from the water. Species of the following genera are commonly found in freshwater : *Pseudomonas, Flavobacterium, Aeromonas*, and *Alcaligenes*. Common to marine and estuarine waters are *Vibrio, Acinetobacter, Pseudomonas, Flavobacterium, Alteromonas*, and *Staphylococcus*. Bacteria in the surface region of the marine environment are often pigmented, a characteristic which may afford protection from the lethal portion of solar radiation. Mold spores and mycelium fragments are present in seawater throughout the photosynthetic zone. These zooplankton animals feed ("graze") upon phytoplankton organisms, bacteria, or detritus. Observations indicate that many zooplankton avoid light, exhibiting diurnal migrations. At night the animals graze on phytoplankton at the surface, and during the day they sink below the photic zone. The microbial population is sparse near the surface of the sea because the intensity of illumination is inhibitory. Beneath the region of photic activity there exists another region inhabited during the day by vertically migrating zooplankton. The bacterial population is distributed more or less uniformly throughout and below these layers, feeding on descending organic material and other nutrients.

PURIFICATION OF WATER

Water contaminated with either domestic or industrial wastes is called nonpotable or polluted water. The objectives of primary concern in providing potable water are freedom from harmful microorganisms and freedom from undesirable or harmful chemicals. These standards apply both to wells or springs serving single families and to water systems serving hundreds of thousands of persons. However, the processes of purifying water are quite different in these two extreme situations. Underground sources – wells and springs – provide most of the water for individual homes in rural areas. Rainwater caught and stored in cisterns is also used to a limited extent; surface water, however, should not be used for drinking purposes unless it is subjected to purification, since there is constant danger of contamination and consequent transmission of disease. Since it is generally impractical to treat surface waters so as to ensure a continuous safe supply for rural domestic purposes, our discussion is limited to water obtained from wells and springs. As it penetrates through the layers of soil, water from wells and springs undergoes filtration, which removes suspended particles, including microorganisms.

Two-thirds of the nation's population is now served by public water work systems. That service, which involves well-supervised production and distribution of safe water, must be given some of the credit for all but wiping out the scourge of typhoid fever of 50 years ago and for virtually eliminating epidemics of water-borne dysentery. The principal operations employed in a municipal water-purification plant to produce water of a quality safe for human consumption are sedimentation, filtration, and disinfection. Sedimentation occurs in large reservoirs, where the water remains for a holding period; large particulate matter settles to the bottom. Sedimentation is enhanced by the addition of alum (aluminum sulphate) at the treatment plant, which produces a sticky flocculent precipitate. Many microorganisms, as well as finely suspended matter, are removed as this precipitate descends through the water in the settling basins. The water is next passed through sand filter beds, a process which removes 99 percent of the bacteria. Subsequently, the water is disinfected to ensure its potability. The majority of municipal water-treatment facilities employ chlorination for disinfection. The chlorine dosage must be sufficient to leave a residual of 0.2 to 2.0 mg/liter free chlorine. The purification process may include procedures for removing minerals that cause the water to be hard, adjusting the pH if the water is too acid or alkaline, removing undesirable colours or tastes, and adding fluoride for the prevention of dental caries.

SANITARY QUALITY DETERMINATION

Conditions in a community or municipality that may influence the quality of water are not static. There may be changes in population, types of industry, and the quantity of sewage and the manner in which it is disposed; consequently, periodic and comprehensive sanitary surveys are necessary. Data obtained from these surveys are of considerable value, both from the standpoint of indicating any changes in operation that may be necessary to prevent complications from developing and of finding the sources of difficulties that may appear unexpectedly. Sanitary surveys reveal whether water is being produced under conditions in which potable water would ordinarily be produced. However, potability can be determined only by chemical and bacteriological laboratory tests. Chemical analysis indicates whether water is polluted and provides other useful information as well; however, it is not sensitive or specific enough to detect minor degrees of sewage contamination.

It is known that the pathogens that gain entrance into bodies of water arrive there via intestinal discharges of humans or other animals. Furthermore, certain bacterial species, particularly *Escherichia coli* and related organisms designated as coliforms, fecal streptococci and *Clostridium perfringens*, are normal inhabitants of the large intestine of humans and other animals and are consequently present in feces. Thus the presence of any of these bacterial species in water is evidence of fecal pollution of human or animal origin. If these organisms are present in water, the way is also open for intestinal pathogens to gain entrance, since they, too, occur in feces. Since the laboratory examination of water for pathogens is beset with the disadvantages already enumerated, techniques have been developed for the demonstration of bacterial species of known excretal origin, particularly organisms of the coliform group. The coliform group of bacteria includes all the aerobic and facultatively anaerobic, Gram-negative, nonsporulating bacilli that produce acid and gas from the fermentation of lactose. The classical species of this group are *Escherichia coli* and *Enterobacter aerogenes*. The relationship of these organisms to others of the enteric group—*Salmonella, Shigella, Klebsiella, Proteus, Serratia,* and other genera—all of which are Gram-negative, nonsporulating bacilli. *E. coli*, as we have already pointed out, is a normal inhabitant of the intestinal tract of humans and other animals. *E. aerogenes* is most frequently found on grains and plants but may occur in human and animal feces. These species bear a very close resemblance to each other in their morphological and cultural characteristics. Consequently, it is necessary to resort to biochemical tests for differentiation. For convenience, these tests collectively are designated as the IMViC (I, indole; M, methyl red; Vi, voges-Proskaner reaction; and C, citrate) reactions.

MISCELLANEOUS MICROBES OF WATER

Some microorganisms besides coliform bacteria are of intestinal origin and could also be used as indicators of fecal contamination of water, such as the fecal streptococci. Other intestinal microorganisms, such as intestinal viruses, are frank pathogens and can cause serious disease. Still other microorganisms are regarded mainly as nuisance organisms, because they create problems of odor, colour, and taste, or cause obstruction of water flow. Fecal streptococci are enteric bacteria found in the intestines of warm-blooded animals, including humans. *Streptococcus faecalis* is representative of this group; other species are *S. faecium*, *S. bovis*, and *S. equinus*.

Many bacteria are capable of elaborating gummy or mucilaginous materials, either as capsular structures or as extracellular excretion products. The organic and inorganic constituents of the water, which provide nutrients for the bacteria, help to determine whether slime is produced and what organisms are responsible for its production. The iron bacteria are one of the most important types of nuisance organisms in water. They transform soluble compounds of iron to insoluble compounds of iron (ferric hydroxide) which may be deposited in a sheath around the organism or secreted so as to form stalks or ribbons attached to the cell (*Gallionella*). Some of the sulfur bacteria are capable of producing and tolerating extreme acidity. Organisms of the genus *Thiobacillus* oxidize elemental sulfur to sulfuric acid and can produce an acidity in the range of pH 1.0; thus they may be responsible for the corrosion of pipes.

Algae are present in all natural aquatic environments. Their nuisance characteristics involve production of turbidity, discolouration, odor, and taste in water. Algae are frequently the primary cause for the clogging of filters during water purification. The diatoms are the most important in this respect, although green and yellow algae are also involved. Aside from these nuisance characteristics, some algae are capable of producing substances toxic to humans and animals.

The enteroviruses are the ones most commonly found in sewage; they include the polio, coxsackie, and echo viruses. The virus that causes infectious hepatitis has been isolated from polluted water and shellfish; occurrence of this disease has been traced to these sources. Rotaviruses are also of major importance. The possibility that virus diseases, particularly the enteric virus diseases, may be waterborne indicates that methods for evaluating the potability of a water supply from a virological standpoint should be developed. Considerable research is underway for the development of a routine test method for the detection of viruses in water and wastewater. At the same time more attention is being given to the assessment of the effectiveness of water treatment processes for the removal and/or inactivation of viruses.

TREATMENT OF WASTEWATER

Treatment and disposal of wastewater and sewage from individual dwellings or other unit structures can be accomplished by anaerobic digestion and/or by aerobic metabolic processes. One of the more common installations used to accomplish this is the septic tank, an anaerobic digesting system. A septic tank is a sewage-settling tank designed to retain the solids of the sewage entering the tank long enough to permit adequate decomposition of the sludge. As sewage enters this type of tank, sedimentation occurs from the upper portion, permitting a liquid with fewer suspended solids to be discharged from the tank. The sedimented solids are subject to degradation by anaerobic bacteria; hence the end products are still very unstable, i.e., high in BOD and odorous. Septic tanks are the most satisfactory method for disposing of sewage from small installations, especially individual dwellings and isolated rural buildings where public sewers are not available. They cannot, however, be relied upon to eliminate pathogenic microorganisms carried in the sewage. Consequently, it is imperative that the drainage from the tank be prevented from contaminating the drinking-water supply.

Wastewater as it arrives at a wastewater-disposal plant is first treated to remove coarse solid materials by a variety of mechanical techniques including screening, grinding, and grit chambers. Subsequent to this it is treated to remove settleable solids. Sedimentation units provide the means for concentrating and collecting the particulate material referred to as sludge. Following sedimentation, the sludge and the liquid effluent are processed separately.

The trickling filter consists of a bed of crushed stone, gravel, slag, or synthetic material with drains at the bottom of the tank. Trickling filters have been described as "a pile of rocks over which sewage or organic wastes slowly trickle." The liquid sewage is sprayed over the surface of the bed either by a rotating arm or through nozzles. The spraying saturates the liquid with oxygen. Intermittent application of the sewage permits maintenance of aerobic conditions in the bed. The filtering medium of the tank becomes coated with a microbial flora, the zoogloeal film, which consists of bacteria, fungi, protozoa, and algae. As the sewage seeps over these surfaces, the microorganisms adsorb and metabolize the organic constituents to more stable end products. This operation may be regarded as a stationary microbial culture (on the stones) fed by a continuous supply of nutrients.

Many species of protozoa and fungi occur throughout the filter; their numbers are influenced in part by the availability of oxygen and nutrients. It is apparent that the microbial activities and interactions are extremely complex in such a heterogeneous environment. Vigorous aeration of sewage results in the formation of a floc; the finely suspended and colloidal matter of sewage forms aggregates designated as floccules. If this floc is permitted to settle and then added to fresh sewage that is again vigorously aerated, flocculation occurs in a shorter time than before. By repetition of this process, i.e., addition of sedimented floc to fresh sewage, aeration, sedimentation, addition of sediment to fresh sewage, aeration, etc., a stage is reached where complete flocculation of the fresh sewage occurs very quickly e.g., a few hours. These particles of floc, i.e., "activated sludge," contain large numbers of very actively metabolizing bacteria, together with yeasts, molds, and protozoa. This combination of microbial growth is very effective in the oxidation of organic compounds. A poor settlement of activated-sludge flocs adversely affects performance of a sewage-treatment plant. The sludge becomes more voluminous and is difficult to control. The principal reason for poor settling [bulking] of activated sludge is the growth of filamentous microorganisms. The use of activated sludge is of great importance in wastewater treatment. This process usually employs an aeration period of 4 to 8 h, after which the mixture is piped to a sedimentation tank.

The effluent from these tanks represents wastewater treated to secondary levels; there is a considerable reduction of suspended solids, and the BOD is reduced. Some of the advantages of this method of treatment are that relatively little land space is needed and the quality of final effluent is such that it does not require high dilution for disposal. Oxidation ponds are shallow ponds designed to allow algal growth on the wastewater effluent. Use of oxidation ponds should be preceded by primary treatment. Oxygen for biochemical oxidation of nutrients is supplied from the air, but the release of O_2 during photosynthesis by the alga *Chlorella pyrenoidosa* provides an additional important source of oxygen.

Advanced wastewater treatment is required when removal of substances beyond the limits normally achieved by conventional primary and secondary processes are necessary. Unit processes have been developed to remove nutrients, simple organic substances, and complex synthetic organic compounds. Processes include biological treatment; however, physical-chemical methods predominate. Common unit processes include biological nitrification-denitrification, filtration, reverse osmosis, carbon adsorption, chemical addition, and ion exchange. Disinfection of wastewaters is necessary to protect public health when the receiving waters are used for purposes such as downstream water supply, recreation, irrigation, or shellfish harvesting. Most facilities use chlorination for disinfection. Current research has proved the serious impact chlorinated waters have on the aquatic life of the receiving water. This has led to the development of several disinfection alternatives. The use of ozone and ultraviolet light is becoming more prevalent.

This process is termed post aeration and is accomplished by mechanical means or a cascading slow technique. Post aeration minimizes the decrease in receiving water-dissolved oxygen that normally occurs where treated wastewater effluent is discharged. Although disposal into surface waters is widely practiced, land application has been and continues to be a feasible alternative to surface-water discharge. Land application of wastewater was given substantial recognition in the Federal Water Pollution Control Amendment of 1972 to implement of the "national goal that discharge of pollutants into navigable waters be eliminated by 1985." It is a fact that municipal and industrial wastewaters have been used nationwide for crop irrigation.

Solids are removed from the primary, secondary, and the advanced stages of the treatment process. Thickening is generally employed to further concentrate the solids or sludge prior to stabilization. Thickening may be accomplished by gravity thickening, similar to sedimentation, or by dissolved air flotation. Many stabilization processes are employed, including aerobic and anaerobic digestion, composting, chemical addition, and heat treatment. The most common process for modern municipal treatment facilities is the anaerobic digestion system. Dewatering if accomplished by physical methods and is often enhanced by the addition of polymer or other chemical coagulant aids. Equipment used for dewatering includes vacuum filters, belt filter presses, plate and frame presses, and centrifuges. The microbial action on the constituents of sludge is termed sludge digestion. An anaerobic condition prevails in these tanks; and the anaerobic and facultative types of bacteria are active. These microorganisms bring about a decrease in organic solids by degrading them to soluble substances and gaseous products. Large amounts of gases consisting mainly of methane (60 to 70 percent) and carbon dioxide (20 to 30 percent), with smaller amounts of hydrogen and nitrogen, are produced during sludge digestion. Conditions affecting microbial growth and metabolism will be reflected in the degree of digestion of the sludge, e.g., inoculum, pH, and temperature. Fresh sludge entering the tanks is inoculated, or seeded. With a portion of ripe sludge (sludge which has already undergone digestion). The ripe sludge may be regarded as an actively growing culture of the types of microorganisms required for rapid digestion of sludge. The proportion of inoculum used and the thoroughness of mixing with the fresh sludge are important. Most rapid digestion of sludge occurs at temperatures which favour growth of thermophilic bacteria (50 to 60°C). A neutral pH (7.0) is optimum for sludge digestion. The temperature is maintained around 30°C, which is below the optimum stated above for most rapid digestion, because it has not been found practical to operate at thermophilic temperatures. The sludge dries into "cakes" which are carried by conveyor belts to incinerators. The incinerator reduces the sludge cake to ash for final disposal in a landfill. An alternative is to spread stabilized dewatered sludge on land. Several municipalities are engaged in land-utilization techniques for disposal of treated sludge. However, such practice requires inexpensive and available land.

Composting is a process where dewatered sludge undergoes decomposition, usually within the thermophilic temperature range. Dewatered sludge is mixed with a bulking agent such as wood chips. The bulking material is added to enhance circulation of air throughout the sludge to improve the stabilization process. The mixture of sludge and bulking material is placed in aerated piles. Oxygen is furnished by forced aeration. The mixture is allowed to cause or biologically decompose for a period of time.

❑❑❑

14

FOOD AND INDUSTRIAL MICROBIOLOGY

Food can serve as a medium for the growth of microorganisms, and this growth may cause the food to undergo decomposition and spoilage. Foods may also carry pathogenic microorganisms and as a result transmit disease. Other microorganisms, if allowed to grow in certain food products, produce toxic substances that result in food poisoning when the food is ingested. Still other microorganisms are used in the preparation and preservation of food products, such as yogurt and sauerkraut. From these general remarks it is clearly evident that the major concern in food microbiology is the control of microorganisms.

The cheap raw materials available in nature as a waste, may be converted into useful commercial products by the activity of microbes. Microbes thus serve a dual purpose. Firstly, they are good agents of disposal of these wastes, and secondly the resultant end products of their breakdown are useful commercial products. Industrial microbiology is an important area of applied microbiology. It refers to the use of microorganisms in commercial enterprises. Cheap raw materials are converted to valuable products through the metabolism of microbes. Microbes for this purpose could be exploited in different ways. For instance this includes (a) synthesis of fermentation products as acids, alcohols or other organic compounds, (b) transformation of one compound into another desired type, (c) the production of enzymes, antibiotics, and insecticides, or (d) the use of microbes themselves as food. The first phase, relatively ancient, involves fermentations for production of alcoholic beverages and traditional dairy products and bread. These processes are non-septic and rely either on a large inoculum or on good housekeeping to prevent microbial spoilage of the product. They all also reflect the synthesis of a primary product of metabolism by the organism, e.g., the enzyme rennin in cheese-making, lactic acid in curd, CO_2 in bread, and alcohol in beers and wines. The second phase was the development of large scale relatively sterile culture facilities in which an inoculum is added. This included processes as large scale cultivation of yeast cells for brewing and baking industries. Efforts were made to develop means of sterilizing large quantities of media, fermentation vessels and pipelines. Steam was often used for this purpose. The growth media were often by-products of other industries, e.g., complex mixture of molasses, and yeast extracts (distiller's by-products). Most of the products come from anaerobic microbial fermentations; e.g., acetone and butanol, and hence the industry that grew up was known as the fermentation industry. The third phase involved the batch culture of microbes under aerobic conditions and aseptic state. Strains of the microbes that are more efficient in antibiotic production were selected.

FRESH FOOD'S MICROORGANISMS

The magnitude of this microbial contamination reflects one or more of the following : The microbial population of the environment from which the food was taken, the condition of the raw product, the method of handling, the time and condition of storage. It is desirable to maintain a very low microbial level of contamination on raw foods; the presence of extremely large numbers of microorganisms suggests that some undesirable event have occurred and that the food is indeed susceptible to further deterioration. The carcass of a healthy animal slaughtered for meat and held in a refrigerated room is likely to have only nominal surface contamination while the inner tissues are sterile. Fresh meat cut from the chilled carcass has its surface contaminated with microorganisms characteristic of the environment and the implements used to cut the meat. Each new surface of meat, resulting from a new cut, adds more microorganisms to the exposes tissues. To improve the microbiological quality of meats, particularly ground beef (in addition to cold cuts and frankfurters), most states have adopted standards , or are considering establishing regulations, to require microbiological standards for these products at the time of purchases. Among the more common species of bacteria occurring on fresh meats are pseudomonads, staphylococci, micrococci, enterococci, and coliforms. The low temperature at which fresh meats are held favours the growth of psychrophilic microorganisms. Freshly dressed eviscerated poultry have a bacterial flora on their surface (skin) that originates from the bacteria normally present on the live birds and from the manipulations during killing, defeathering, and evisceration. Under good sanitary conditions the bacterial count has been reported to be from 100 to 1,000 bacteria per square centimeter of skin surface, whereas under less sanitary conditions the count may increase 100-fold or more. The interior of a freshly laid egg is usually free of microorganisms; its subsequent microbial content is determined by the sanitary conditions under which it is held, as well as the conditions of storage, i.e., temperature and humidity. Microorganisms, particularly bacteria and molds, may enter the egg through cracks in the shells or penetrate the shells when the "bloom" (thin protein coat) covering the shell deteriorates.

Microbial invasion of plant tissue can occur during various stages of fruit and vegetable development, and hence, to the extent that the tissues are invaded, the likelihood of spoilage is increased. A second factor contributing to the microbial contamination of fruits and vegetables pertains to their post-harvest handing. Mechanical handling is likely to produce breaks in the tissue which facilitates invasion by microorganisms. The pH of fruits is relatively acid ranging from 2.3 for lemons to 5.0 for bananas. This restricts bacterial growth but does not retard fungal growth. The pH range for vegetables is slightly higher, pH 5.0 to 7.0, and hence they are more susceptible than fruits to attack by bacteria.

Of particular significance is whether the water is sewage-polluted, in which case the seafood is potentially capable of transmitting various pathogenic microorganisms. The marine bacterium *Vibrio parahaemolyticus* has been responsible for a number of gastroenteritis epidemics in the United States due to consumption of raw of inadequately cooked seafood. This organism occurs widely in the Atlantic, Pacific, and Gulf Coast waters and has been isolated from seafood samples including fish, shellfish, and crustaceans. Shellfish that grow in contaminated water can concentrate viruses and may be sources of hepatitis infection. For example, raw oysters and clams from polluted waters have caused numerous epidemics in various parts of the world. At the time it is drawn from the udder of a healthy animal, milk contains organisms that have entered the teat canal through the teat opening. They are mechanically flushed out during milking. The number present at the time of milking has been reported to range between several hundred and several thousand per milliliter. The counts vary among cows and among the quarters of the same cow and are highest during the initial stages of milking. From the time the milk leaves the udder until it is dispensed into containers, everything with which it comes into contact is a potential source of more microorganisms. This includes the air in the environment, the milking equipment, and the personnel. Disregard of sanitary practices will result in heavily contaminated milk that spoils rapidly.

This change is brought about mainly by *Streptococcus lactis* and *S. cremoris* and certain lactobacilli. The principal change is lactose fermentation to lactic acid; evidence of proteolysis or lipolysis is not detectable by taste or smell. This type of change is sometimes referred to as the normal fermentation of milk.

Bacteria that gain entrance into milk may be classified according to their optimum temperature for growth and heat resistance. Temperature is a very practical consideration, since low temperatures are used to prevent changes due to microorganisms and high temperatures are used to reduce the microbial population, destroy pathogens, and in general improve the keeping quality of the milk. Since certain psychrophiles grow at temperatures just above freezing and some thermophiles grow at temperatures in excess of 65°C, it follows that the temperature at which milk is held will determine which species grow and predominate. Pasteurized milk stored in a refrigerator may be satisfactorily preserved for a week or even longer. But eventually microbial deterioration, manifested by "off" flavour or odor, will become evident because of the accumulation of metabolic products of psychrophilic bacteria. Thermophiles present a problem at the other extreme of the temperature scale. The holding method of pasteurization exposes milk to a temperature of 62.8°C for 30 min; the thermophile *Bacillus stearothermophilus*, however, grows at 65°C. In the dairy industry, thermoduric bacteria are regarded as those which survive pasteurization in considerable numbers but do not grow at pasteurization temperatures. Microorganisms of this category are extremely troublesome from the standpoint of producing raw milk with a low bacterial count. Because they are not killed by pasteurization, the microorganisms may contaminate equipment and accumulate as a result of faulty cleaning.

Milk and dairy products can now be considered model foods from the standpoint of regulations and surveillance of production, processing, and distribution. Furthermore, there are companion standardized methods for analysis of dairy products. No other food can claim the degree of standardized surveillance and analysis that is practiced for milk and milk products. A variety of diseases are potentially transmissible through milk. The source of a pathogenic agent occurring in milk may be either a cow or a human, and it may be transmitted to either. It is also possible for humans to infect cows. For example, mastitis may be caused by a variety of organisms, including *Staphylococcus aureus*. The infecting organism, in some cases, has been traced to humans.

FOOD SPOILAGE BY MICROORGANISMS

The type of food substance and the method by which it is processed and preserved may favour contamination by certain groups of microorganisms. Most foodstuffs serve as good media for the growth of many different microorganisms. Given a chance to grow, the organisms will produce changes in appearance, flavour, odor, and other qualities of foods. The changes that microbes cause in foods are not limited to the results of degradation; they may also be caused by products of microbial synthesis. Some microorganisms discolour foods as a result of pigment production. Slimes may be developed in or on foods by microorganisms capable of synthesizing certain polysaccharides. Milk is an excellent bacteriological medium. In fact, sterile skimmed milk is used routinely as a culture medium. Fresh whole milk contains protein [casein], carbohydrate [lactose], and fat. All of these substrates can be degraded enzymatically by microorganisms. If the degradation of these substrates is extensive, the accumulation of end products will impart undesirable characteristics to the milk. Some microorganisms can synthesize compounds like pigments and slimes which also give undesirable characteristics to the milk.

MICROBIOLOGICAL TECHNIQUES USED FOR FOOD EXAMINATION

In the case of spoiled foods, it is possible to identify the agent responsible for the spoilage; having discovered the agent, it may be possible to trace the source of contamination and the conditions which permitted spoilage to occur. Corrective measures can then be instituted to prevent further spoilage. Microbiological food examination takes advantage of special microscopic techniques and cultural procedures. Extensive use is made of selective and differential media to facilitate the enumeration and isolation of certain types of microorganisms. The particular procedure used is determined by the type of food product being examined as well as by the specific purpose of the examination. For example, a food sample being investigated for possible contamination by *Clostridium botulinum* would be subject to different laboratory tests than one being examined for coliform organisms. The increasing significance of salmonellas in foodborne disease has made it mandatory to develop more rapid, reliable, and reproducible methods for the detection of salmonellas in foods. The procedures selected for examination of a particular sample are, of course, determined by the facts relating to that sample and the purpose of the examination.

Standard microscopic techniques are available for the examination of some food products. For example, a procedure known as the Bread smear is used to make a direct microscopic count of microorganisms in milk. The essential procedures of this technique are : (a) Spreading a measured amount of milk over a known area on a glass slide, (b) staining the film of milk with methylene blue, (c) making a microscopic count of organisms or clumps of organisms in several microscopic fields, and (d) calculating the total number of bacteria per unit volume. A slide designed with a chamber, known as the Howard mold slide, is used, as its name suggests, to enumerate mold filaments in food products such as fruits, juices, and vegetables. When the mold counts obtained by this procedure exceed certain limits, it indicates raw material of poor quality or unsatisfactory sanitary processing. Protozoa can be identified and enumerated by direct microscopic examination.

Plate culture techniques are available for the enumeration of the "total" microbial population or some particular group of microorganisms. The word *total,* of course, needs qualification; the microorganisms enumerated by a cultural techniques are only that portion of the total population which will grow into colonies under the conditions provided, namely, the composition of the medium and the physical conditions of incubation. For example, the standard procedure for counting microorganisms in milk is designed to enumerate bacteria by the standard plate count (SPC). The conditions for the procedure are very specifically articulated in a volume entitled Standard Methods for the Examination of Dairy Products. It is mandatory that the procedure be carried out precisely as specified in the publication. Other culture procedures are available for particular physiological or biochemical types of microorganisms. Prior concentration of the food specimen suspected to be contaminated with viruses may be necessary. Additional provisions are necessary to inhibit bacterial growth in the tissue culture.

FOOD PRESERVATION

The ancient Egyptians and Romans were aware of the preservative effects of salting, drying, and smoking. It has been suggested that the first salt preservation was accomplished by burying the food along the shore, where seawater effected the cure. The American Indians placed strips of fresh bison and venison at the top of a teepee or over a campfire, where preservation was accomplished through drying and smoking. Dried salt cod was a common food for colonial Americans. Perishable foods were stored in caves and springs, where the low temperature prolonged the preservation. Modern methods of food preservation employ elaborate refinements of the primitive processes plus additional new techniques.

All methods of food preservation are based upon one or more of the following principles : (a) Prevention or removal of contamination; (b) inhibition of microbial growth and metabolism (microbistatic action); and (c) killing of microorganisms.

Each step in the preparation of a food for its final treatment is a potential source of contamination. For example, the shell of an egg provides a protective covering which normally excludes microorganisms. However, when the eggs are cracked open in the process of preparing dehydrated egg powder it is likely that the interior of the egg may become contaminated. The extent of the contamination will depend upon the cleanliness of the eggs and the level of aseptic precautions observed in the process. One can recognize more vividly the importance of aseptic technique in the processing of more perishable foods like oysters and crabmeat, each of which requires considerable handling by people.

Heat is widely used to destroy organisms in food products in cans, jars, or other types of containers that restrict the entrance of microorganisms after processing. Steam under pressure, such as in a pressure cooker, is the most effective method of high-temperature food preservation since it can kill all vegetative cells and spores. Food preservation by heat requires knowledge of the heat resistance of microorganisms, particularly spores. In addition, one must consider the rate at which heat penetrates through foods of different consistencies as well as the size of the containers in which they are packed. From such information it is possible to establish satisfactory heat-processing conditions. Much research has been done on this subject, and this accounts for the highly successful results achieved in food preservation by canning. Special laboratory equipment has been designed to determine with precision the heat resistance of various bacterial species, particularly the sporeformers.

In 1810 Nicholas Appert, a Frenchman, published L'Art de Conserver, which described his successful researches in food preservation; and in the same year Peter Durand was granted and English patent describing the use of tin containers for food preservation. The temperatures used for canning foods ranges from 100°C for high-acid foods to 121°C for low-acid foods. The canning process does not guarantee a sterile product. For example, spores of some bacterial species may survive these temperatures. The most important organism to be eliminated in canned foods is the sporeforming anaerobe *C. botulinum*, which is capable of producing a very potent lethal toxin.

The terms pasteurization, pasteurized, and similar terms shall mean the process of heating every particle of milk or milk product to at least 145°F, and holding it continuously at or above this temperature for at least 30 minutes, or to at least 161°F, and holding it continuously at or above this temperature for at least 15 seconds, in equipment which is properly operated and approved by the health authority. The original time-temperature relationships for pasteurization were worked out with *Mycobacterium tuberculosis* since this was regarded as the most heat-resistant pathogen likely to occur in milk. This organism is destroyed when exposed to a temperature of 140°F for 10 min. The pasteurization temperature was set at 143°F for 30 min. Later it was discovered that *Coxiella burnetii*, the causative agent of Q fever which can be transmitted by milk, can survive in milk heated to 143°F for 30 min.

Methods of pasteurization of milk used commercially include a low-temperature holding (LTH) method and a high-temperature short-time (HTST) method. The holding method, or vat pasteurization, exposes milk to 145°F (62.8°C) for 30 min in appropriately designed equipment. The HTST process employs equipment capable of exposing milk to a temperature of 161°F (71.7°C) for 15 seconds. In either method of pasteurization it is essential that the equipment be designed and operated so that every particle of milk is heated to the required temperature and held for the specified time. Precautions must be taken to prevent recontamination after pasteurization.

Phosphatase is an enzyme, present in raw milk and in many tissues, which is destroyed by adequate pasteurization. Thus one can determine whether milk has been properly pasteurized by testing for the absence of this enzyme. The principle of the test is illustrated by the following reaction. Milk, which in its raw condition contains the enzyme phosphatase, is added to a substrate upon which the enzyme will react :

Disodiumphenyl phosphate +	Phosphatase →	Phenol* + Sodium phosphate
Substrate	Enzyme from raw milk	Products

Colour standards are used to interpret the results of this test. This is a very simple testing procedure, yet it provides valuable information about the heat treatment milk has received. Commercial milk-sterilization techniques have been developed which expose milk to ultrahigh temperatures for very short periods of time, for example, 300°F [148.9°C] for 1 to 2 second. In addition, the sterilization process includes steps that eliminate any traces of cooked flavour. The final product is comparable in flavour and nutritional quality to pasteurized milk.

Modern refrigeration and freezing equipment has made it possible to transport and store perishable foods for long periods of time. Refrigerated trucks and railway cars, ships' storage vaults, and the home refrigerator and freezer have improved the quality of the human diet and increased the variety of foods available. Frozen-food production in the United States almost doubled from 11 billion pounds in 1965 to 20 billion in 1975 and is expected to more than double to 48 billion pounds by 1985. Much of this increases will be in prepared frozen foods, whose quantity tripled over the last 10 years and is expected to approach 50 percent of all frozen foods by 1985. Before freezing, the fresh product is steamed [blanched] to inactivate enzymes that would alter the product even at low temperatures. Quick-freeze methods, using temperatures of - 32°C or lower, are considered most satisfactory, smaller crystals of ice are formed, and cell structures in the food are not disrupted. It should be emphasized that freezing foods, no matter how low the temperature, cannot be relied upon to kill all microorganisms. The number and types of viable and nonviable microorganisms present in frozen foods reflect the degree of contamination of the raw product, the sanitation in the processing plant, and the speed and care with which the product was processed. The microbial count of most frozen foods decreases during storage; but many organisms, including pathogens, e.g., species of *Salmonella*, survive for long periods of time at -9 and -17°C.

The increased use of precooked ready-to-serve foods and the prevalence of automatic vending machines for dispensing perishable foods have made it necessary to obtain more data on microbial growth and survival at low temperatures. Note that the type of food product has considerable influence on the rate of bacterial growth at the lower temperatures. The removal of water by drying in the sun and air or with applied heat causes dehydration. The preservative effect of dehydration is due mainly to microbistasis; the microorganisms are not necessarily killed. Growth of all microorganisms can be prevented by reducing the moisture content of their environment below a critical level. The critical level is determined by the characteristics of the particular organism and the capacity of the food item to bind water so that it is not available as free moisture. The cells are plasmolyzed, and metabolism is arrested. Thus the antimicrobial condition imposed by increased osmotic pressure is similar in principle to inhibition by dehydration. Although yeasts and molds are relatively resistant to osmotic changes, processes of food preservation based on this principle are, nevertheless, very useful. Jellies and jams are rarely affected by bacterial action because of high sugar content. However, it is not uncommon to find mold growth on the surface of jelly which has been exposed to air. Condensed milk is preserved in part by the increased concentration of lactose and supplemental sucrose. Similar results are obtained by curing meats and other foods in brines. According to this act, a food is adulterated if any poisonous or deleterious substance has been added which may render it injurious to health. Only a few chemicals are legally acceptable for food preservation. Among the most effective are benzoic, sorbic, acetic, lactic, and propionic acids, all of which are organic acids. Sorbic and propionic acids are used to inhibit mold growth in bread. Nitrates and nitrites used in curing meats, primarily for the preservation of colour, are inhibitory to some anaerobic bacteria. This practice has been the subject of considerable controversy because of the potential of nitrates and nitrites as mutagenic agents and the subsequent relationship to carcinogenesis. Foods prepared by fermentation processes, e.g., sauerkraut, pickles, and silage for animals, are preserved mainly by acetic, lactic, and propionic acids produced during the microbial fermentation.

Because ultraviolet light has very limited penetration power, microorganisms that are embedded or covered are unlikely to be affected. Thus, ultraviolet irradiation is limited to control of microorganisms on surfaces or thin, clear layers of liquid. Examples of applications in the food industry include meat-processing plants, control of surface growth on bakery products, sanitation of equipment, and treatment of water used for the depuration [cleansing] of shellfish. Ionizing radiations are lethal to microorganisms. The fact that they are microbicidal at room temperature and have the ability to penetrate are characteristics that make them attractive candidates for control of microorganisms in foods. Gamma rays and electron beams [beta and cathode rays] have been experimented with extensively for use in the food industry. Canned and packaged foods can be sterilized by an appropriate radiation dosage. This "cold sterilization" produces a rise in temperature of the product of only a few degrees. Radiation pasteurization is a term describing the killing of over 98- but not 100-percent of the organisms by intermediate doses of ionizing radiation. The ionizing radiation resistance of microorganisms does not correspond to their thermal resistance. *Clostridium botulinum* appears to be the most radioresistant organism of importance to the food technologist. Note that the survival of spores is influenced by the material in which they are suspended and that time is not a factor. In the case of radiation,unlike temperature, the radiation death dose rather than radiation death time is determined. Ionizing radiation sterilization provides the possibility of an entirely new approach to food preservation; it could bring about a radical change in industrial methods of food processing. However, despite the extensive research and documentation on the effectiveness of ionization radiation for the preservation of foods, this method of preservation has not been approved in the United States. This is due in part to economic factors as well as to some lingering uncertainties about the effect of the radiation on the food material. In addition, the United States already has well-developed systems for food preservation. This is not the case for all countries.

FERMENTED FOOD

There are many useful applications of microorganisms in the food industry. A variety of important products in our diet are produced with the aid of microbial activity. In the dairy industry, fermented milks are produced by inoculating pasteurized milk with a known culture of microorganisms, sometimes referred to as a starter culture, which can be relied on to produce the desired fermentation, thus assuring a uniformly good product. Several hundred varieties of cheese are manufactured, and with few exceptions, most of them can be made from the same batch of milk. Microorganisms—bacteria or molds—convert the curd of the milk into the desired cheese. For the manufacture of some cheeses, such as blue cheese or Roquefort, it is necessary to inoculate the curd with the microorganism which brings about the changes.

Lactic acid bacteria are chiefly responsible for the desirable type of fermentation required for the production of each of these products. The microorganisms that produce the changes may be the natural flora on the material to be fermented or may be something added as a starter culture. Most commercial sour, sweet, mustard, and mixed pickles are made from fermented salt-stock pickles. The other major type of pickled cucumber is the fermented dill pickle.

IDEAL ORGANISMS : THE MICROORGANISMS

Microorganisms possess a broad variety of enzymes making an array of chemical conversions possible. They have a relatively high metabolic activity which allows conversion to take place rapidly. They possess a large surface area for quick absorption of nutrients and release of end products. Moreover, they have a high rate of multiplication, as a 20-minute generation time for *E. coli* under ideal conditions. In an industrial process, a microorganisms in fact represents a mini-chemical factory. The efficiency of this factory depends upon the liberation of large amount of a single product that may be efficiently isolated and purified. The microbe should be easily cultivated and maintained and should have genetic stability with infrequent mutations. Its value enhances if it can grow on inexpensive, readily available substrate that may be the by-product of other industrial processes. For example, a large amount of whey is produced in cheese production.

We may use an enrichment medium to encourage growth of a particular organism in a mixed culture, or a selective medium which inhibits other microbes in the mixture, selecting only the desired one. They may also be obtained by induced mutations. For a large-scale cultivation of microorganisms, two general processes are used in industry. These are, the batch technique and the continuous flow technique. A large tank with a volume of up to 100,000 gallons of medium is used in this technique. The tank is known as a fermentor, since industrial microbiologists use the term fermentation for any aerobic or anaerobic process catalysed by microorganisms. The medium is sterilised with steam or with a gas as SO_2. Microorganisms, about 2 to 4% of the total volume are added to the fermentor. Oxygen is bubbled if the process is aerobic but omitted if anaerobic conditions are required. The microbial growth is allowed to take place for days, weeks or months. Microbes are removed after fermentation is complete.

In this method (continuous flow technique), the medium is added continuously to the tank to replace that which has been fermented. The isolation and purification of product from the medium is thus an on-going process. An instrument called a chemostat is used to provide a constant flow, keeping microorganisms in the logarithmic phase of growth. In some cases a turbidostat is used. This instrument measures the turbidity of the microbial population to indicate the level of growth. When a certain level is reached, it adds new medium to the tank while withdrawing the spent material.

INDUSTRIAL ORGANIC COMPOUNDS

Though microorganisms are able to produce an enormous kind of organic compounds, we shall here consider only some of the most commonly used chemicals, which include acids, vitamins, enzymes, steroids and gibberellins.

(a) Organic acids—The most commonly used are citric acid, lactic acid, acetic acid and gluconic acid. Citric acid is used in soft drinks, candies, inks and dyes, engraving materials and some foods. Traditionally citric acid was produced by chemical extraction from citrus fruits. However modern technology utilises a mold fungus. The most widely used mold is *Aspergillus niger*, an ascomycetous fungus. The mold is inoculated to a medium of corn meal, molasses, salts and inorganic nitrogen in huge shallow pans or in fermentation tanks. In latter case, sterile air is bubbled through a pipe to provide aerobic conditions, and the culture is known as submerged in the sense that the fungus grows within the medium.

Gluconic acid is used in medicines. This is produced from carbohydrates by *Aspergillus niger* as well as by *Gluconobacter* spp. Glucose is converted to gluconic acid which is then converted to calcium gluconate by adding calcium hydroxide.

Lactic acid is produced by action of bacteria on the whey portion of milk. The commonly used organism is *Lactobacillus bulgaricus,* which produces only lactic acid from lactose. This organism is thus hornofermentative. The bacterium is introduced into whey tanks at 43°C.

Vinegar is produced by action of bacterium on alcohol derived from grapes or other fruits as apples. The fruit juice if fermented by yeasts till the alcohol concentration is about 10-12%. The juice is then sprayed into a tank containing *Acetobacter aceti* growing on the surface of wood shavings, gravel or other substrates. As the alcohol percolates through, the bacterial enzymes convert it first to acetaldehyde and then to acetic acid. The product is circulated many times before collection at the bottom of tank. Residual alcohol is evaporated in the heat.

(b) Amino acids and vitamins—Some bacteria produce amino acids in excess of their requirements. Glutamic acid in this way is produced by some species of *Micrococcus, Arthrobacter* and *Brevibacterium.* Glutamic acid is used as a food supplement and its sodium salt, monosodium glutamate is utilised in food preparations. For lysine synthesis, two organisms are needed, *E. coli* is first cultivated in a medium of ammonium phosphate, glycerol, corn steep liquor and other ingredients. Diaminopimelic acid (DAP) accumulates, some days later, *Enterobacter aerogenes* is added to the mixture. This removes carboxyl group from DAP to yield lysine. This is used in breads and other foods, and is an essential amino acid in man.

Riboflavin (vitamin B_2) and cyanocobalamin (vitamin B_{12}) are also metabolic products of microbes. Riboflavin is obtained from *Ashbya gossypii,* an yeast which produces about 20,000 times the amount of the vitamin that it needs for its own metabolism. Cyanocobalamin is produced by some species of *Streptomyces, Pseudomonas* and *Propionibacterium* grown in cobalt-supplemented medium of malt extract or corn steep liquor.

(c) Enzymes—Amylase is produced by a mold, *Aspergillus oryzae,* and is used as spot-remover in laundry, as an adhesive and in baking (to digest starch to glucose). Pectinase, a product of *Clostridium* spp. is used in retting of flax, and to clarify fruit juices. Proteases, a group of protein-digesting enzymes is produced by *Bacillus subtilis, Aspergillus oryzae* and others. They are used as liquid glues, meat tenderizers, spot-removers, beer clarifiers etc.

(d) Transformations of steroids—Out of several steroids, cortisone, cholesterol and the sex hormones are biologically important steroids. Synthetic manufacture of steroids is expensive and tedious. For instance in a laboratory conversion of bile acids to cortisone requires about 37 steps. However, with *Rhizopus nigricans,* the process could be shortened in which an intermediate– the progesterone is converted to another intermediary closer to cortisone by adding hydroxyl groups to the molecule (Fig. 14.1). At present, the mold *Aspergillus ochraceus* is used for this conversion. Other steroid transformation from progesterone are utilised in the production of contraceptive hormones and adrenal hormones. The organisms are cultivated for some time in a fermentation tank, after which the steroid is added. The desired steroid is isolated at the end of additional incubation period. Species of *Streptomyces, Mucor* and *Penicillium* are utilised for various transformations.

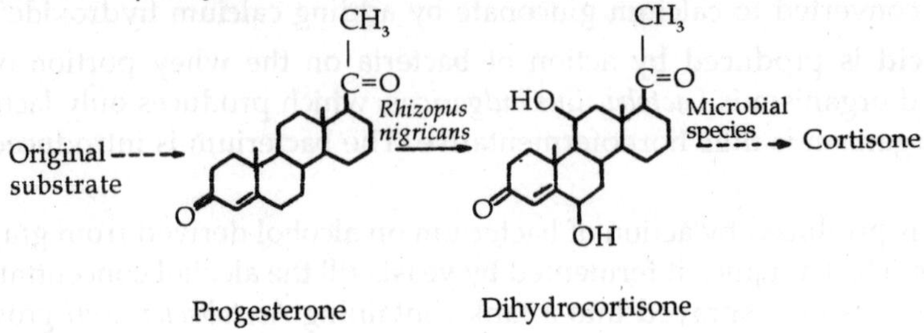

Fig. 14.1 : Steroids produced by *Rhizopus nigricans.*

(e) Gibberellins—These are a series of plant hormones which promote growth by stimulating cell elongation in the stem. They also hasten seed germination and flowering. These are produced during the metabolism of a fungus, *Gibberella fujikuroi,* and may be extracted from it for industrial use.

ALCOHOLIC BEVERAGES

Alcoholic fermentations have been well known since ancient time, though little was known of the nature of the process. Louis Pasteur could show that yeasts were the organisms responsible for such processes.

(a) Beer—Barley seeds are allowed to germinate. During germination the naturally occurring amylases convert the grain starch to sugars, most of which being maltose. The process is called malting and the digested grain as malt. The next step is to mash the grain with water and remove the fluid portion called the wort. Dried petals of a vine, *Humulus lupulus,* called hops are then added to the work to give it flavour, colour and stability. Hops also prevent contamination of the wort, due to presence of two antimicrobial substances in petals. Generally, the yeast is taken from a pervious batch of beer. Some yeasts give a uniform cloudiness to the beer and are carried to the top of the fermentation vat by foaming carbon dioxide. Such yeasts are called top yeasts and the product as ale. Other yeasts ferment the beer more slowly and produce a light beer with less alcohol than ale. These yeasts are known as botton yeasts and their product as lager beer.

After a week, the young beer is transferred to vats for primary and secondary ageing. It may take about six months more. Some yeasts is left in the beer that is to become keg beer and the product is refrigerated to preserve it. The thick wall of the keg traps CO_2 produced during continual fermentation. For canning or bottling the beer is to be pasteurised at 140°F for 13 min to kill the yeast or may be filtered to remove the yeasts. Some yeast is used to seed new wort, the rest for animal feed or pressed to tablets for human consumption.

Sake, though commonly known as rice wine, is somewhat like a rice beer. It is obtained from steamed rice. The rice starch is converted first to sugar by *Aspergillus oryzae* or *Rhizopus* spp. Later *Saccharomyces* spp. ferment the sugar until the alcohol level is approximately 14 percent, when sake is ready for consumption.

(b) Wine—Wine fermentations are carried by controlled cultures of *Saccharomyces ellipsoideus,* a variety of *Saccharomyces cerevisiae* or with the strains of yeast naturally occurring on grapes. Wine is made from ripe fruit, fruit juice, or plant extract such as dandelions. Fermentation usually begins with crushing of the fruit to produce must. Sulphur dioxide may be added to control the process (Fig. 14.2). In natural fermentations, SO_2 is not used, and the yeasts begin to digest sugar. Oxygen may be supplied to promote aerobic growth of the yeast. However, anaerobic conditions are established later. Alcohol production occurs within few days, though ageing may take months or years. During this period, secondary fermentations develop the flavour, aroma and bouquet of the wine. Red wine becomes red as alcohol extracts the colour of grape skin. Additional carbon dioxide production yields Champagne and other sparking wines which are naturally carbonated. Sherry wines result from inoculation with special yeasts to have unique flavours. In dry wines, most or all of the sugar is metabolised, whereas in sweet wines, the fermentation is stopped before entire sugar is consumed. The strongest natural wines have about 16 per cent alcohol. Most table wines average about 10-12 percent alcohol, with fortified wines reaching 22 percent alcohol.

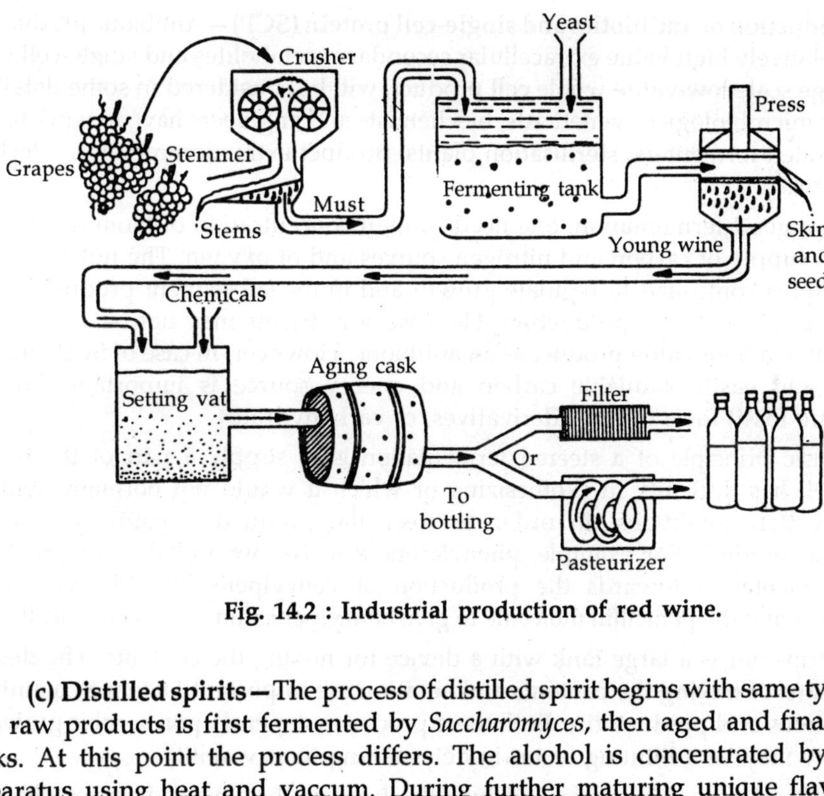

Fig. 14.2 : Industrial production of red wine.

(c) Distilled spirits – The process of distilled spirit begins with same type as for wine. The raw products is first fermented by *Saccharomyces*, then aged and finally matured is casks. At this point the process differs. The alcohol is concentrated by a distillation apparatus using heat and vaccum. During further maturing unique flavour from the chemicals as aldehydes, ethers and volatile acids are added. The alcohol content is then standardised by diluting it with water before bottling. There are four basic types of spirits : Brandy is made from fruit juice, and rum from molasses, whisky from malted cereal grains as for instance, scotch from barley, rye from rye grain and bourbon from corn. The neutral spirits, as vodka is made from potato starch and left unflavored, and gin flavored with juniper oil.

MISCELLANEOUS PRODUCTS

Besides the abovesaid products, some other are also obtained from microorganisms.

(a) Antibiotics – Most antibiotics are produced by *Streptomyces* spp. but a significant number are the products of *Penicillium* or *Bacillus*. About two-thirds of the total world production of antibiotics are the penicillins, which include a major number of synthetic and semisynthetic derivatives of the original penicillin. Although, the microorganisms used to produce antibiotics are all aerobes and are grown under conditions of vigorous aeration, the production process is generally referred to in the technical literature as a fermentation. Antibiotic production is carried out in large, aerated tanks of stainless steel holding up to 30,000 gallons of the medium. In classic technology, huge mats of fungus or actinomycetes, on the surface of tanks were used. New technology, however, employs small fragements of submerged hyphae rotated and agitated in the medium with constant bubbling of oxygen. After several weeks, the microbe is removed and the antibiotic extracted from the medium for further conversion to the desired product.

(b) Production of antibiotics and single-cell protein (SCP) – Antibiotic production as a group of relatively high value extracellular secondary metabolites and single-cell protein as a very large scale low value whole cell product, will be considered in some detail here. In both cases microbiologists, geneticists, biochemists and engineers have to work together to design aerated fermenters, sterilisation plants, product extraction regimes, selection of ideal strain etc.

To get a good fermentation, one needs a high concentration of biomass and hence a continuous supply of carbon and nitrogen sources and of oxygen. The nutrient balance of the medium is controlled to regulate growth and in the subsequent production phase it is optimised for penicillin production. The cost of nutrients may not be a major factor in production of a high value product as an antibiotic. However, in case of SCP, the choice of a cheap and easily available carbon and energy source is important. The main substances are fossil fuels or their derivatives, or carbohydrates.

The basic principle of a steered fermentation is to supply a part of the molecule which the cell has difficulty in synthesizing or which it would not normally synthesize under production conditions, in order to steer the production pathway towards a particular end product. For example, phenylacetic acid is now added to the medium to steer the fermentation towards the production of benzylpenicillin. Phenylacetic acid is incorporated into the penicillin molecule to give benzylpenicillin, that is chemically better.

The fermenter is a large tank with a device for mixing the contents. The design of the tank, means of mixing the contents and aeration are important factors in maintaining sterile conditions arid productivity. Both these processes depend upon aerobic growth and the O_2 supply may be a limiting factor in determining their viability.

I. *Aeratedtower* – Air is sparked into the base of the vessel. It gives cheap, but relatively a poor aeration.

II. *Airlift fermenter* – Air is sparked through only part of the fermenter to cause an efficient circulation within the whole vessel.

III. *Stirred fermenter* – Aeration is done by a mechanical stirrer to provide most efficient, though most expensive aeration.

As large and more productive fermenters are designed, the surface area to volume ratio decreases and hence the heat produced as a by-product of microbial growth cannot be easily dissipated. Instead of the need for the supply of heat in a laboratory-scale fermenter, a cooling system has to be designed. Two processes are used – the batch culture and the continuous culture. We have already explained the main features of these culture processes. For SCP, the continuous culture is more useful. Batch culture on the other hand is used for secondary metabolites such as penicillin, for traditional processes and for small-scale systems.

The scale-up of the culture from the original inoculum to a 50000 litre fermenter requires great care. This is achieved in several consecutive stages of increasing volume leading to an inoculum suitable for the production fermenter. At each stage, the potential productivity of the inoculum is checked and only the best inocula are used.

(c) Insecticides—The bacterium was named *Bacillus thuringiensis*. It is now well known that this microbe produces toxic crystals in older cells. The substance is an alkaline protein. When deposited on leaves it is consumed by caterpillars that are the larval stages of butterflies, moths and other insects. The gut cells are dissolved causing paralysis of larvae. *Bacillus thuringiensis*, at the sporulation stage is harvested and dried as powder which is available in markets.

MICROBIAL CELLS AND ENZYMES

Enzymes have been used in industry for over 70 years, initially in detergents that is still perhaps one of the largest bulk uses of proteases and lipases. In conventional biological industries also microbial enzymes have been used, e.g., proteases and amylases in malting and rennet in cheese manufacture. However, during last ten years or so, immobilised cells and enzymes have been used as production systems.

(a) Enzymes carry out stereospecific reactions with high accuracy, whereas chemical technology results into many side-products from which the desired product is to be purified.

(b) Enzymes are cheap and carry out reactions at low temperature and at atmospheric pressure, whereas chemical catalysts require special and expensive conditions.

(c) Since there is much diversity in microbial world, an enzyme for a desired reaction can be easily found out by merely screening a range of microbes.

(d) Mutants with altered enzyme function can be isolated with appropriate genetic methods. Thus enzymes having different substrate specificities or with different physical properties can be isolated. More recently, whole microbial cells have been used for specific chemical transformations. This should not be confused with an anaerobic fermentation or conventional secondary metabolite production. Here only part of the cell's metabolism is now being utilised, usually a single pathway, and sometimes only a single enzyme. The advantage of using cells is that the expense of purifying the enzyme is avoided and, in some cases, the enzyme is more stable in its natural environment than after purification.

Immobilisation affords a simple way of separating the enzyme or cell from the products when the reaction is complete and is likely to prove invaluable in the development of biological sensors (biosensors)—special electrodes based upon the selectivity and high affinity of enzymes for their substrates.

(a) Sweeteners—Some bacteria and fungi produce an enzyme, glucose isomerase, which converts glucose to an equimolar mixture of glucose and fructose (high fructose syrup) that is sweeter than the equivalent sucrose solution. Glucose is readily and cheaply available in large quantities from the breakdown of starch by amylases.

$$\text{Starch} \xrightarrow{\text{Amylases}} \text{Glucose} \xrightarrow{\text{Glucose isomerase}} \text{High fructose syrup}$$

(b) Amino acids—Some amino acids as phenylalanine can be produced in this way. This is used in production of aspartate a modified dipeptide of aspartic acid and phenylalanine.

(c) Antibiotics and pharmaceuticals—Among antibiotics is the modification of penicillin to change the structure of molecule. Among pharmaceuticals the best example is hydroxylation of steroids to generate more potent drugs.

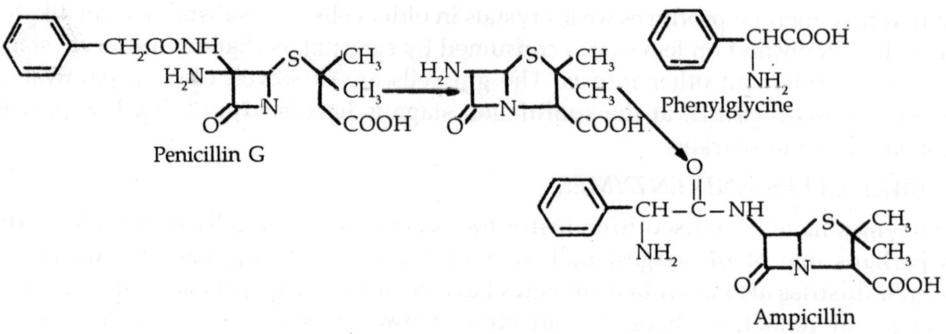

Fig. 14.3 : Penicillin acylase catalyzes the pencillin de-acylation followed by chemical synthesis of ampicillin.

Penicillin acylase catalyses the deacylation of the side chain of the natural penicillin-G produced by industry to yield 6-aminopenicillanic acid (6-APA). Semisynthetic penicillin molecules can then be made by chemical routes (by substituting the removed side chain with other molecule) from the 6-APA, e.g., ampicillin from phenylglycine (Fig. 14.3). Penicillin acylase is produced by *Escherichia coli* and both immobilised cells and the purified enzyme can be used. Current processes use the immobilised enzyme.

❑❑❑

15

Nitrogen Fixation by Microorganisms

The major source of nitrogen to all organisms is that present as the gas dinitrogen in the atmosphere (N_2). The industrial fixation of nitrogen is carried out by the Haber-Bosch process, which takes place at high temperatures (400-600°C) and pressures (100-200 atmospheres) :

$$N_2 + 3H_2 \rightarrow 2NH_3$$

The majority of this fixed nitrogen is used as a fertilizer for crop plants. The release of nitrate into rivers, lakes and oceans following the application of fertilizer has frequently been a cause of public concern, although the precise origin of the nitrate is still a topic of considerable argument. Nitrogen may also be fixed biologically by a range of prokaryotic organisms, and N_2 may be released back to the atmosphere following the action of denitrifying bacteria in the soil. N_2O (dinitrogen oxide) may also be released into the atmosphere where it could play a major role in the breakdown of the ozone layer. Although nitrogen may only comprise 0.5-5% of the dry weight of plant tissues, it is a constituent of vitally important compounds, e.g., amino acids, proteins and nucleic acids. Humans are not able to use any forms of inorganic nitrogen as a nitrogen source, so organic nitrogen mainly as proteins must be an important constituent of the diet. Humans are also unable to synthesize nine amino acids, which are termed essential (Fig. 15.1).

A range of prokaryotic organisms have the capacity to fix nitrogen. The ability to fix nitrogen can be carried out at different concentrations of ambient oxygen. The additional complexity of carrying out nitrogen fixation in the cyanobacteria that are producing oxygen from photosynthesis. Although strictly speaking *Azospirillum* species are free living, they are frequently found in association with the roots of grasses, where they fix nitrogen microaerobically. Probably the most important form of agricultural nitrogen fixation is carried out by the rhizobium-legume symbiosis in the root nodule. However, some rhizobia, e.g., *Azorhizobium, Azotobacter vinelandii, Anabaena, Oscillatoria, Caulinodans* and *Photorhizobium thompsonia* form stem nodules. A number of non-leguminous plants also produce nitrogen-fixing nodules, but the symbiotic prokaryote is an actinomycete *Frankia.* Cyanobacteria can also form nitrogen-fixing symbiotic relationships with a variety of organisms including the coral leaf sponge *Siphonochalina.*

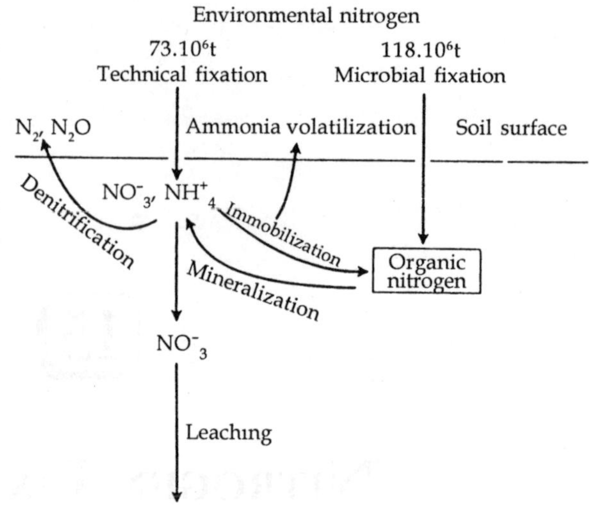

Fig. 15.1 : Processes of nitrogen fixation.

NITROGENASE

Nitrogenase carries out the following reaction—

$$\text{Nitrogen} + 8H^+ + 8e^- \rightarrow 2NH_3 + H_2$$

Each electron transfer is accompanied by the hydrolysis of two ATP molecules, and therefore the reaction also includes :

$$16MgATP \rightarrow 16MgADP + 16ip$$

The majority of studies on nitrogenase were originally carried out on the enzyme isolated from *Klebsiella pneumoniae,* which was shown to contain molybdenum (Mo). The soil bacterium *Azotobacter chroococcum* normally has a Mo-containing nitrogenase but also has the capacity to synthesis a vanadium (V)-containing enzyme if Mo is not available. The closely related species *Azotobacter vinelandii* has, in addition to both Mo and V-nitrogenases, a third iron-containing nitrogenase which functions only when the Mo and V levels are very low. All nitrogenase enzymes contain an Fe protein that is made up of two oxygen sensitive subunits with molecular weights of approximately 62kDa and contain 4Fe and $4S^{2-}$ atoms per dimer. The iron-sulfur center has a midpoint potential of -350mV. When MgATP is bound, a conformational change of the Fe protein takes place and the midpoint potential becomes 100 mV more negative. The molybdenum-containing part of the nitrogenase enzyme is termed the MoFe protein, iron-molybdenum cofactor (FeMoCo) which is derived from homocitrate is bound to the α-subunit of the MoFe protein at a cleft formed from two regions of the polypeptide at Cys-275, Gln-191 and His-195. The most comprehensive description of the mechanism of nitrogenase has been developed. First the reduced Fe-protein-MgATP complex binds to the MoFe protein. Then the electron is transferred with concomitant hydrolysis of MgATP. For simplicity the Mo-Fe protein is shown as E, with a subscript denoting the number of electrons that have been transferred to it. Following the evolution of hydrogen after the transfer of three electrons, nitrogen is then reduced by the sequential addition of electrons and protons until two molecules of ammonia are released. A probable intermediate in this reduction is the hydrazido complex $E = N - NH_2$; it is likely that it is this intermediate that releases hydrazine on acid or base quenching of the reaction. In addition to the reduction of nitrogen, nitrogenase can also reduce a variety of other substrates. These substrates have in common, their small size and in many cases the possession of a triple bond. Notable among these substrates is acetylene (ethyne), the conversion of which to ethylene, is the basis of the most commonly used assay for nitrogen fixation. The ethylene formed by nitrogenase can be readily assayed using a simple gas chromatographic method. The more accurate measurements of nitrogen fixation following the incorporation of $^{15}N_2$ into organic material using a mass spectrometer is expensive and very time consuming. Both the Fe protein and the MoFe protein of nitrogenase are inactivated by O_2, with typical half decay times in air of 45 seconds and 10 min, respectively. For this reason all manipulations during the isolation and assay of the enzyme *in vitro,* have to be carried out under anaerobic conditions. In the case of the MoFe protein, O_2 causes a change in the oxidation state of the Fe centers, which after prolonged exposure results in the loss of Fe, Mo and S^{2-} from the protein. The components of the Mo-independent nitrogenases have been reported to be even more sensitive to O_2, with the half decay time of the VFe protein being only 45 seconds.

GENETICS OF NITROGEN FIXING BACTERIA

The genetics of nitrogen fixation has been extensively studies in *K. pneumoniae.* There are 20 genes (*nif*) involved in the assembly of nitrogenase (Fig. 15.2). The synthesis of nitrogenase is repressed by the presence of combined nitrogen (detected by the intracellular ratio of glutamine : 2-oxoglutarate) and by O_2. In *K. pneumoniae* the general nitrogen regulation system (*ntr*) is also involved. The product of the *ntrA* gene is a sigma factor (σ^{54}) that is required for the recognition of the unusual promoters of the *ntr* and *nif* genes. The

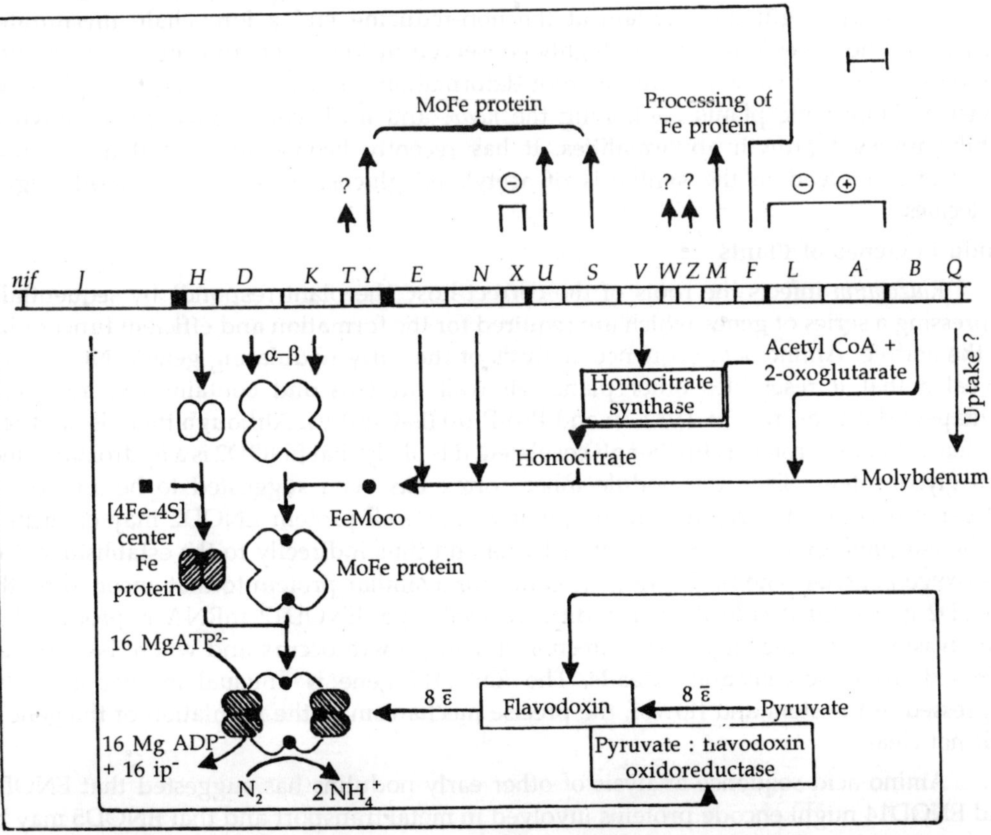

Fig. 15.2 : Roles of gene products in nitrogen fixation by *Klebsiella pneumoniae.*

additional genes required in *Rhizobium* for the formation of the legume root nodule are described below :

Rhizobium- Legume Interaction

The first recognizable reaction of the plant is the curling of the root hair. A cell wall is then broken down and the bacterium is able to enter. An infection thread develops that penetrates the inner cortex of the root. The infected cortex cells increase in size and divide to form a sphere surrounded by uninfected cells and an outer fibrous layer. Within the infected cells the rhizobia divide and enlarge to form bacteroids, which are separated from the plant cytoplasm by the peribacteroid membrane.

The gene *nodD,* appears to be directly under the influence of flavones or isoflavones and to regulate the other genes of the *nod* complex, including the host specific *nodE* and *nod F* genes. Some rhizobia, e.g., *R. meliloti,* have three copies of *nod D* which may vary in their flavonoid recognition and in this way may regulate the host range, or other factors such as the number of nodules formed, or the time between inoculation and nodule formation. Each operon of the *nod* genes is preceded by a conserved DNA sequence the 'nod box', which is presumably regulated by the product of the *nod D* gene. The development of nodules is elicited by specific lipo-oligosaccharide signal molecules released by the *Rhizobium* bacteria. These molecules consist of a chito-oligosaccharide backbone of different chain length that contain at the non-reducing end a long-chain unsaturated fatty acid. The genes *nod ABC* are highly conserved in *Rhizobium* and the corresponding proteins have been shown to induce root deformation, branching and cortical plant cell division. Transgenic plants containing the *nodA* and *nodB* genes have been shown to exhibit unusual growth abnormalities. It has recently been confirmed that the three genes are involved in the synthesis of acetylated glucosamine oligosaccharide signal molecules.

Nodulin Genes of Plants

Rhizobium infects the roots of the correct host, the plant responds by sequentially expressing a series of genes which are required for the formation and efficient functioning of the nodule. Amino acid sequence analysis of the early nodulating gene (ENOD2) has revealed that it resembles other plant cell wall proteins and contains two repeating pentapeptides Pro-Pro-Glu-Tyr-Glu and Pro-Pro-His-Glu-Lys. Although there is no direct evidence that the protein is further glycosylated, it is likely that ENOD2 is a hydroxyproline-rich glycoprotein. Since the nodule inner cortex has been suggested to be an oxygen diffusion barrier, the presence of the putative cell wall protein ENOD2 may contribute to the morphology of the nodule parenchyma and thus indirectly to the establishment of the oxygen barrier. The ENOD12 gene codes for a similar protein to that encoded by the ENOD2 gene, but it is located in a different cell type. ENOD12 mRNA is produced in the invasion zone, the site where infection thread growth occurs and where bacteria are released from the infection threads. The ENOD12 gene is unusual in that it is also expressed in the stem and flower, the precise mechanism of the regulation of the gene is still not clear.

Amino acid sequence analysis of other early nodulins has suggested that ENOD3 and ENOD14 might encode proteins involved in metal transport and that ENOD5 may be a protein in the peribacteroid membrane. Plant cells play significant roles—

(a) To provide a rapid flux of oxygen to support rhizobial respiration at a concentration low enough to prevent the deactivation of nitrogenase.

(b) To supply carbohydrates, usually in the form of C_4 acids to support the rapid rates of rhizobial respiration.

(c) To rapidly remove the ammonia synthesized by the rhizobial bacteroids to prevent the inhibition of nitrogenase.

Leghemoglobin is the most abundant (up to 20%) protein in root nodules. Leghemoglobin is localized in the cytoplasm of the infected plant cells and not inside the peribacteroid membrane. This evidence suggests that leghemoglobin controls the oxygen flux in the plant cytoplasm but not inside the *Rhizobium* bacteroids. The globin is encoded by the host plant, whereas the heme group is primarily synthesized by the bacterium and transported into the host cytoplasm. There is now evidence that the heme precursor δ-amino-levulinic acid is synthesized in the plant cell, whereas the rest of the heme group is assembled by the bacteria. The globin protein is encoded by a family of genes in soybean, where at least four genes have been identified, *Lba, Lbc1, Lbc2* and *Lbc3*. The ammonia excreted by the bacteroids following the action of nitrogenase is assimilated in the plant cytoplasm by the enzyme glutamine synthetase. Following the synthesis of glutamine, the fixed nitrogen is transported to the leaves as amides or ureides. Nodulin-35, which is the second most abundant soluble protein in the nodules of soybean has been shown to be a subunit of uricase II which catalyzes the conversion of uric acid into allantoin. Sucrose is the main source of fixed carbon transported from the leaves to the roots and root nodules. The first enzymes involved in sucrose breakdown is the inappropriately named sucrose synthase, which converts sucrose into UDP- glucose and fructose. The subunit of sucrose synthase have been identified as nodulin 100 of soybean. Free heme causes the sucrose synthase enzyme to dissociate and this inhibition may be involved in the regulation of carbohydrate metabolism during nodule senescence.

BLUE GREEN ALGAE

Cyanobacteria solve the problem of oxygen inactivation of nitrogenase by separating photosynthetic oxygen evolution from nitrogen fixation by space or time. When growing in a medium containing ammonia, the cyanobacteria grow as filaments of undifferentiated vegetative cells. Upon transfer to a nitrogen free medium, 5-10% of the cells undergo differentiation to form thick-walled heterocysts, that are distributed evenly along the cyanobacterial filament. After differentiation, the heterocysts are not able to divide. The heterocyst cell wall consists of three layers : An outer polysaccharide fibrous layer, a homogeneous central layer and an inner glycolipid layer. Mutants deficient in heterocyst glycolipids or polysaccharides lack the capacity to fix nitrogen aerobically, but maintain the capacity to do so in the absence of oxygen. The permeability of heterocysts is such that nitrogen may enter at a rate sufficient to allow nitrogenase to operate but at the same time limit the entry of oxygen. It is possible that the major regulation of gaseous and metabolite exchange takes place through the polar body or 'plug' that links the heterocyst to the vegetative cells.

Metabolism in Heterocyst

The presence of photosystem I ensures that sufficient ATP can be generated by cyclic photophosphorylation. The heterocysts also lack Rubisco and the Calvin cycle enzymes so that CO_2 fixation is limited to the vegetative cells. Thus carbohydrate has to be transported into the heterocysts, probably in the form of maltose. Glucose is oxidized via the pentose phosphate pathway to yield NADPH in the reaction catalyzed by glucose-6-phosphate dehydrogenase (Fig. 15.3).

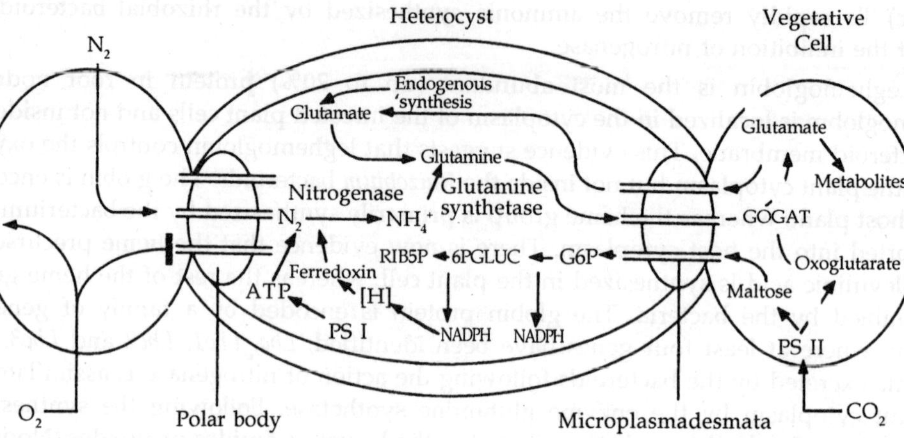

Fig. 15.3 : Carbon and nitrogen exchange between the heterocyst and vegetative cells.

The ammonia formed by nitrogenase is immediately assimilated by glutamine synthetase, and the majority is exported to the vegetative cells in the form of glutamine. Glutamate which is formed by glutamate synthase is returned to the heterocyst to assimilate another molecule to ammonia. In the unicellular cyanobacterium *Gloeothece,* nitrogen fixation takes place at night and photosynthetic CO_2 assimilation and O_2 evolution occur during the day. In *Gloeothece,* nitrogenase synthesis commences two hours before entering the dark period. After the onset of darkness, both synthesis and activity of nitrogenase are supported by ATP, reductant and carbon skeletons formed from the metabolism of stored glucan reserves, accumulated during the light period. The newly fixed ammonia is rapidly assimilated into amino acids.

The molecular nature of the inhibitor and co-inhibitor are unknown, but vegetative cells contain a factor (VFI) that binds to *nifH* and also to *xisA*, a gene involved in heterocyst differentiation. In addition, a gene designated *hetR* is involved in the early stages of heterocyst differentiation in *Anabaena* PCC 7120. Cyanobacteria carrying mutations in this gene do not produce heterocysts, but when present in multicopies, *hetR* allows heterocyst formation even in the presence of fixed nitrogen. In *Anabaena* PCC 7120, a major rearrangement of the *nif* gene region occurs during heterocyst differentiation . During this rearrangement, two regions of DNA are excised from the chromosome, thus producing a functional *nif* regulon. Although transcription of the newly formed operon *nif HDK* closely follows the rearrangement of the *nif* region, the two events are independent of each other. No rearrangement of the *nif* genes have been reported in non-heterocystous cyanobacteria, as *nifHDK* are already clustered.

UPTAKE AND REDUCTION OF NITRATE

In barley, root nitrate uptake is carried out by three different systems. At high external nitrate concentrations, a low affinity system operates that appears to be constitutive and unregulated. At low external concentrations, two high-affinity systems appear to operate. One (K_m for nitrate, 7μM) is constitutive and the second (K_m for nitrate, 15-34μM) is induced by nitrate. Alternatively the nitrate may be temporarily stored in the root vacuole or transported to the leaves for subsequent reduction. The proportion of nitrate reduced in the root or leaves is variable and depends on the concentration of external nitrate, developmental age and plant species. If nitrate reduction and assimilation takes place in the root, higher rates of root respiration are required. The level of expression of the inducible high affinity nitrate transport system is subject to negative feedback regulation dependent on the nitrogen status of the plant. It has been suggested that the feedback regulation is mediated by a product of ammonia assimilation. King *et. al.* (1993) concluded that nitrate itself was able to regulate its own uptake without the requirement of further metabolism. Nitrite may also be taken up by barley roots and is able to induce the nitrate uptake system, suggesting that nitrate and nitrite are transported by the same process.

Nitrate Reductase

The reducing power for nitrate reduction is normally supplied as NADH; however, in some plants, a bispecific enzyme that is able to use either NADH or NADPH as a substrate is also present. In soybean three isoenzymes of nitrate reductase have been isolated; constitutive NADH and NAD(P)H forms and the more extensively studied inducible NADH-dependent form. Nitrate reduction can take place in the root depending on the plant species, developmental age or nitrate supply . In general, as the external nitrate concentration increases, the proportion that is transported to the shoot for reduction increases. NADH-dependent nitrate reductase comprises two identical subunits of 100-115kDa. The redox cofactors FAD, heme (cytochrome b_5) and a molybdenum cofactor (MoCo) are bound to each monomer in a 1:1:1 ratio. Electrons are passed from NADH to nitrate as shown–

NADH $\rightarrow$ FAD $\rightarrow$ Cytochrome $b_5 \rightarrow$ MoCo $\rightarrow$ Nitrate

The monomer is not active by itself and the two subunits are linked by a disulphide bond, which can be reduced without loss of activity. Three catalytic domains have been identified by comparison of the sequence with other proteins of known function. The N-terminal half of the enzyme of 430-460 amino acids is the MoCo domain, determined by homology with sulfite oxidase. The central 81-83 amino acids contain the heme domain, which is a member of the cytochrome b_5 superfamily. The 270-275 amino acids in the C-terminal sequence, comprise the FAD domain. The three domains are linked to one another by two hydrophilic hinge regions, which are very susceptible to proteolytic cleavage and probably account for the early problems in isolating the enzyme in an intact and active state. The individual domains are able to carry out a number of partial reactions, e.g., the heme domain can act as a NADH-dependent cytochrome *c* reductase. Although these partial reactions have been studied in detail it is unlikely that they have any physiological role. In addition nitrate reductase can also convert chlorate into the toxic metabolite chlorite. This reaction is the basis of herbicidal activity and has been used extensively for the isolation of mutants that lack nitrate reductase activity.

Nitrate Reductase Genetics

A range of mutants that lack both nitrate reductase and xanthine dehydrogenase activity are probably defective in the ability to synthesize the MoCo factor. Five loci have been characterized in barley (*Nar2-Nar6*), three in *A. thaliana* (B25, B73 and G1) and six in *Nicotiana plumbaginifolia* (*CnxA-CnxF*). In *A. thaliana,* the analysis of chlorate-resistant mutants has shown that two genes *Ch12* and *Ch13* are involved in the biosynthesis of the nitrate reductase apoenzyme. In barley, the *Nar1* locus has been identified as the NADH-specific nitrate reductase structural gene. None of the *Nar1* mutants are completely deficient in nitrate reductase activity. This is due to the expression of a second gene *Nar7* that is presumed to code of the NAD(P)H nitrate reductase apoenzyme. Double mutants of barley, *Nar1, Nar7,* totally lack the ability to use nitrate as a nitrogen source. In *N. plumbaginifolia* only one functional *Nia* gene is present that codes for the nitrate reductase apoenzyme. Southern blot analysis has shown that in tobacco there is one *Nia* locus per haploid genome, whereas in *A. thaliana* two sequences have been detected. A more complex system exists in barley where one to three copies have been detected and in rice where six to eight copies per haploid genome have been identified, depending on the subspecies. In general three introns of variable size have been detected, which are located at strictly conserved positions.

Nitrite Reductase

The enzyme is present in the leaves and roots where it is located totally in the chloroplast or plastid. The reducing power is obtained from reduced ferredoxin directly from the light reactions in the leaf or from carbohydrate metabolism in the root. Ferredoxin-dependent nitrite reductase comprises on subunit of molecular weight 60-64kDa. The polypeptide contains a siroheme prosthetic group and a 4Fe-4S cluster at the active site .

Nitrite Reductase Genetics

A mutant of barley has been isolated that lacks nitrite reductase activity in the leaves and roots and can only be maintained by growth on ammonia or glutamine. The mutation is in a single nuclear gene. *Nir1* and the plants have been shown to contain elevated levels of nitrate reductase activity. Vaucheret (1992) have expressed antisense nitrite reductase mRNA in tobacco plants. One transformed plant was shown to lack nitrite reductase activity and accumulated nitrite. The plant had a greatly reduced level of ammonia, glutamine and protein in the leaf tissue.

REGULATION MECHANISM OF NITRATE ASSIMILATION

The induction of nitrate reductase was one of the first enzymes in plants shown to be caused by *de novo* protein synthesis. With the onset of molecular biological techniques, it is now possible to follow the induction of nitrate reductase activity at the gene level. Nitrate supply leads to an increase in the steady-state level of mRNA encoding nitrate reductase in a variety of plant species and tissues. This in turn, results in nitrate reductase protein synthesis and increased activity, shortly after nitrate supply to the roots and after a lag in the leaves. The 5′ promoter sequence of the spinach nitrite reductase gene has been shown to confer nitrate-inducible glucuronidase (GUS) activity in transgenic tobacco, confirming that nitrate controls transcription of the nitrite reductase gene. Although nitrate is able to co-regulate the expression of the nitrate and nitrite reductase, there is no evidence at the present time, of the existence of a specific nitrate regulatory gene, using a similar mechanism to that described in fungi. In light-grown plants treated with nitrate, the levels of nitrate and nitrite reductase mRNAs, proteins and enzyme activities slowly decrease after two days in the dark. After return to light there is a rapid increase in the levels of nitrate and nitrite reductase mRNAs and enzyme activities. Diurnal variations in nitrate reductase activity have been known for some time.

More recently strong oscillations in the levels of nitrate reductase mRNA have been observed, with a major peak being detected at the end of the dark period. The circadian rhythm in the level of nitrate reductase mRNA is maintained in continuous light, but disappears slowly in the dark. Oscillations in the levels of nitrite reductase mRNA in phase with nitrate reductase have also been detected. It has been shown that the addition of fructose, glucose and sucrose to dark-adapted tobacco leaves resulted in the induction of nitrate reductase expression in the dark. On the other hand glutamate or glutamine (but not asparagine) down-regulate the expression of nitrate reductase in low light. It had previously been shown that glutamine concentrations correlated in the opposite phase with the expression of nitrate reductase mRNA. The formation of nitrite reductase mRNA was not stimulated by soluble carbohydrates in the dark, but required light for the stimulation of transcription. Repression of nitrite reductase mRNA synthesis was, however, detected to a lesser extent in the presence of glutamine, glutamate or asparagine. It would, therefore, appear that the induction of the transcription of the nitrate and nitrite reductase genes in tobacco leaves is regulated by the balance of soluble sugars and amino acids. The activity of the enzyme was shown to decline under conditions when the rate of photosynthetic CO_2 assimilation was low. Similarly a rapid fall in the activity of nitrate reductase was determined following the transfer from light to dark. It has now been suggested that the nitrate reductase activity in leaves can be regulated *in vivo* in a similar manner to sucrose phosphate synthase.

ASSIMILATION OF AMMONIA

Ammonia is generated via nitrogen fixation and nitrate reduction. A small amount may also be taken up from the soil under acidic conditions, particularly by trees. These three processes are termed primary assimilation. The several secondary metabolic reaction produce ammonia are as follows—

(a) The conversion of glycine into serine in the photorespiratory carbon and nitrogen cycle.

(b) The catabolism of the transport compounds, asparagine, arginine and ureides.

(c) During normal amino acid metabolism, e.g., conversion of cystathionine into homocysteine in methionine biosynthesis and threonine into 2-oxobutyrate in isoleucine biosynthesis. The enzyme phenylalanine ammonia lyase produces ammonia in the conversion of phenylalanine into cinnamate, the first reaction involved in the synthesis of lignin a major constituent of secondary cell walls.

(d) Proteins are frequently hydrolyzed during the germination of seedling or following leaf senescence. Prior to the synthesis of transport compounds, ammonia is liberated through the operation of glutamate dehydrogenase—

$$\text{Glutamate} + NAD^+ + H_2O \rightarrow \text{2-Oxoglutarate} + NH_3 + NADH + H^+$$

Glutamine Synthetase of *Phaseolus*

The enzyme is an octameric protein with a native molecular weight of 350-400 kDa and has a very high affinity for ammonia (K_m = 3-5 μM). Early investigations suggested that there were two isoenzymes of GS present in plants. One located in the cytoplasm and one in the chloroplast or plastid. More recent molecular studies have suggested that this hypothesis is an oversimplification (Fig. 15.4).

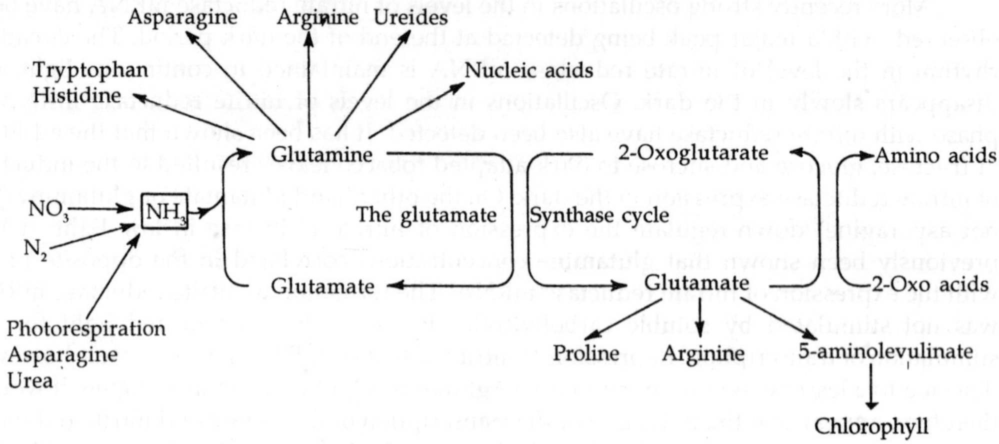

Fig. 15.4 : Ammonia assimilation in higher plants.

The *gln-α, gln-β* and *gln-γ* genes encode the cytosolic α, β and γ polypeptides which are located in the cytoplasm. The DNA sequences show that these three polypeptides have molecular weights of approximately 39kDa and are 85-89% identical. The fourth expressed GS gene *gln δ* encodes a polypeptide of 47 kDa, which has extensions to the amino and carboxy termini of 57 and 16 amino acids, when compared to the three cytosolic GS polypeptides. The N-terminal extension encodes a chloroplast targetting sequence, which is cleaved during transport into the organelle. The resulting chloroplast located GS polypeptide, termed δ is about 42kDa and assembles into an octamer of identical subunits. In contrast, the three cytosolic polypeptides are able to assemble into a range of isoenzymes containing varying proportions of the α, β, and γ polypeptides. The fifth GS gene, *gln-ε* has been identified in a genomic clone, but no evidence for the expression of the gene has yet been obtained.

(a) *gln-α*

This is the least characterized of the four expressed genes, although a high level of mRNA has been demonstrated in dry seeds. During germination, the mRNA remains the most abundant in the cotyledons, radicles and plumules for up to 2 days. After this time, the mRNA declines to barley detectable levels.

(b) *gln-β*

The mRNA for this gene has been detected in all organs of *P. vulgaris*. In roots, *gln-β* mRNA is the major form of GS mRNA and is induced during the early stages or radicle emergence. A similar gene in soybean has been shown to be induced by ammonia, but this is not the case in *P. vulgaris*. During nodulation, the level of *gln-β* mRNA remains constant. Experiments with GUS fusion in *Lotus corniculatus,* indicated that the promoter was expressed in both the cortical and infected regions of the nodule, but as the nodules matured, expression was restricted to the vascular system. The *gln-β* gene is also expressed in leaves, but only at the early stages of development.

(c) *gln-γ*

The *gln-γ* GS gene is highly expressed in the nitrogen-fixing nodule but not in roots and leaves. Low levels of expression have also been detected in the stems, petioles and cotyledons of germinating seeds. Experiments with GUS fusions in *L. corniculatus* has revealed that the *gln-γ* promoter is only expressed in the infected cells of the central nodule tissue. It has been proposed that the release of the *Rhizobium* bacteria may lead to the production of factors specific to the infected cells, which are required for the expression of the *gln-γ* gene in nodules.

(d) *gln-δ*

During leaf development, the chloroplast *gln-δ* gene is expressed 2 days after germination, and after 8 days becomes the major GS transcript. The gene is only expressed at a very low level in dark-grown plants and on transfer to light there is a massive induction of *gln-δ* mRNA synthesis.

Glutamine Synthetase of *Pisum*

In roots, *GS1* is the predominant isoform, although it is also expressed in nodules, *GS3A* and *GS3B* are highly expressed in nodules and also in the cotyledons of germinating seeds. Whereas the *GS3A* and *GS3B* genes have near identical nucleic acid sequences, *GS3A* is expressed at higher levels. The chloroplast-nuclear gene *GS2* encodes the chloroplast enzyme of *GS*, which contains a transit peptide. The presence of mature chloroplasts in dark-adapted pea plants was shown to accelerate the induction of *GS2* mRNA synthesis following transfer of the plants to white light. Levels of the *GS2* mRNA were found to be reduced in the leaves of pea plants that had been grown in elevated CO_2, conditions that would suppress the generation of ammonia from the photorespiratory pathway. The role of photorespiratory ammonia release in the regulation of *GS* is still therefore in doubt.

In an attempt to obtain more information on the control of the different *P. sativum* *GS* genes, Coruzzi and her colleagues have used *GUS* expression in transgenic tobacco plants as an elegant detection method. The promoter for the *GS2* chloroplast gene was shown to direct *GUS* expression in photosynthetically active cells, e.g., pallisade and spongy parenchyma of the leaf, collenchyma of the stem and photosynthetic cotyledons. In contrast, the promoter for cytosolic *GS3A* gene directed expression of *GUS* in the vascular system of the cotyledons, leaves, stems and roots .The expression of the *GS3A* promoter was later shown to be confined to the phloem cells. The promotor conferred a similar pattern of expression in transgenic alfalfa, where expression was also observed, at high levels, in root nodules. The evidence that cytosolic and chloroplastic *GS* are expressed in different cell types, suggests that they play different roles in nitrogen metabolism. The expression of *GS3A* in the phloem cells would indicate that it plays a major role in the synthesis of glutamine required for nitrogen transport, this would be particularly important in the seedling cotyledons and the nitrogen fixing nodule.

Glutamate Synthase Enzyme

The two different forms of glutamate synthase are present in higher plants, one utilizes NADH as a source of reductant and the other ferredoxin–

Glutamine + 2-oxoglutarate + ferredoxin (reduced) → 2-glutamate + ferredoxin (oxidized)

Enzymes can be classified into two types—

(a) Ferredoxin-dependent—The enzyme is a large monomeric protein with a molecular weight of 140-160kDa. Tissue fractionation studies have shown that ferredoxin-dependent glutamate synthase is localized in the chloroplasts of leaves. Using immuno-gold antibody labeling techniques, the enzyme was localized solely in the chloroplast stroma of mesophyll xylem parenchyma. The activity of the enzyme increases during leaf development in the light, a response which may be phytochrome dependent. Despite initial difficulties, data are now available concerning the molecular biology of the enzyme. The first report of the sequence of a gene encoding ferredoxin-dependent glutamate synthase was made by Sakakibara *et. al.* (1991). The cDNA was shown to encode a polypeptide of 1616 amino acids, including a chloroplast transit peptide of 97 amino acids. The enzyme was shown to exhibit a strong similarity to the *E. coli.* NADPH-dependent glutamate synthase protein with 42% of the amino acids residues being identical. The enzymes also contained a short region similar to a potential FMN-binding region of yeast flavocytochrome b_2. Two partial sequences of the gene and protein isolated from tobacco and barley have confirmed the strong similarity with the *E. coli* enzyme.

(b) NADH-dependent—The NADH -dependent enzyme appears to play a major role in the ammonia assimilation in nitrogen fixing nodules. The enzyme is a monomer and has a high molecular weight in the region of 200-225kDa. Antisera raised against alfalfa root nodule glutamate synthase did not detect the presence of a similar protein in roots or leaves. A 7.2kb cDNA clone encoding a 240kDa NADH-dependent glutamate synthase protein has been isolated from alfalfa root nodules. The protein shares significant sequence identity with maize ferredoxin glutamate synthase and with both the large and small subunit of *E. coli* NADH-glutamate synthase. The expression of NADH-glutamate synthase mRNA, protein and enzyme activity was developmentally regulated in the root nodules. A dramatic increase in gene expression occurred coincidentally with the onset of nitrogen fixation in the bacteroid.

SEVERAL MUTANTS WHICH LACK THE ENZYMES OF AMMONIA ASSIMILATION

Ammonia is also liberated in this process at the same rate as CO_2, but is immediately reassimilated via the glutamate synthase cycle. The rate of ammonia production during photo-respiration has been calculated to be ten times the rate of nitrate assimilation. Following an initial suggestion by Somerville & Ogren (1979), it has been possible to isolate mutant plant lines that are deficient in the key enzymes of the photorespiratory nitrogen cycle. Such mutants have been selected on the basis that they grow normally under non-photorespiratory conditions (0.7% CO_2), but show severe stress symptoms and chlorotic lesions on exposure to air. The mutants exhibit remarkably similar properties to plants treated with the GS inhibitors methionine sulphoximine and phosphinothricin. These are characterized by a massive accumulation of ammonia and a reduction in the rate of photosynthetic CO_2 assimilation following exposure to air. The maximum rate of ammonia evolution has been calculated to be 40% of the rate of CO_2 assimilation, and can be taken as a direct measurement of the rate of photorespiration. Freeman *et. al.* (1990) were able to identify three classes of barley mutants lacking chloroplastic *GS2* activity : Class 1 in which the absence of GS2 protein was correlated with low levels of mRNA; class II which had normal or increased levels of GS2 mRNA but very little protein; and class III, which had significant amounts of mRNA and protein.

After exposure to air, the leaves fo the deficient mutants accumulated large quantities of glutamine and small amounts of ammonia; the levels of all other amino acids were greatly reduced. Ten minutes after the transfer to air, the rate of photosynthetic CO_2 assimilation of the mutant leaves had fallen to 10% of the rate of the wild type plant. This loss of photosynthetic capacity was much faster than that seen in the GS deficient mutants and could be reversed by prefeeding alanine or asparagine. Western blot analysis of four of the barley mutants lacking ferredoxin-dependent glutamate synthase indicated that three of the mutants had no detectable cross-reacting material of molecular weight 154 kDa, whereas a fourth (RPr 84/42) had a detectable, but greatly reduced amount of enzyme protein. Using a 1.3kb cDNA fragment specifying the amino terminal portion of barley ferredoxin glutamate synthase, Avilla *et al.* (1993) were able to determine the mRNA levels in the barley mutants. In two of the mutants (RPr 82/9 and RPr 84/82) the probe hybridized to a mRNA similar in mobility (5.7kb) and abundance to that of the wild type. The mutants RPr 82/1 and RPr 84/82, however, exhibited two RNA hybridizing species, one of which was larger than 5.7kb and one smaller. The latter two therefore represent a fourth type of mutant (class IV), in which normal mRNA is replaced by RNA(s) of a different size. Mutants deficient in both enzymes of ammonia assimilation have been obtained by normal genetic crossing. Such mutants lacking both chloroplastic GS and ferredoxin-dependent glutamate synthase are still capable of growing in elevated CO_2.

METABOLISM OF ASPARAGINE

The cotyledons of germinating seeds have proved to be a major source of asparagine synthetase activity and the enzyme has been studied in lupins, soybean and vigna. In maize roots, the enzyme is also able to use ammonia as a substrate which may be of physiological importance in this tissue. The enzyme also plays an important role in root nodules. In leaves, the enzyme has proved very difficult to assay, probably due to the presence of inhibitors. Two classes of asparagine synthetase cDNAs (*AS1* and *AS2*) that encoded homologous but distinct polypeptides with molecular weights of 66.3 and 65.6 kDa were isolated; glutamine binding sites (Met-Cys- Gly- Ile) were detected at the amino terminus of both AS1 and AS2. Northern blot analysis revealed that the level of AS1 mRNA increased in the dark in the leaves of both etiolated seedlings and mature pea plants, in agreement with physiological experiments on the synthesis of asparagine. The light repression of AS1 mRNA synthesis was found to be a phytochrome-mediated response. Both AS1 and AS2 mRNA were shown to accumulate in germinating cotyledons and nitrogen fixing nodules. Asparagine was the first amino acid to be discovered in plants, being isolated from *Asparagus sativus* in 1806. *Asparagus* spears are an important commercial crop, but there is a rapid deterioration following harvest and the levels of asparagine increase markedly in 24 hours. A cDNA clone encoding asparagine synthetase has now been isolated from asparagus spear tips. Levels of mRNA began to increase in the tips 2 hours after harvest and in other sections of the spear after 4 hours.

Breakdown of Asparagine

The nitrogen liberated from the amide group of asparagine has been detected in the endosperm fluid in the form of ammonia, alanine and glutamine. At later stages of maturation, the developing lupin cotyledon is able to metabolized asparagine directly. Similarly asparagine has been shown to be rapidly metabolized in the seed coats of maturing pea seeds giving rise to high concentrations of ammonia in the endosperm fluid. The enzyme (asparaginase) has been purified from *Lupinus polyphyllus* where it was shown to have a molecular weight of 72 kDa with subunits of 38 kDa. Antisera raised against the purified asparaginase cross-reacted with a protein of similar characteristics isolated from the cotyledons and leaves of *L . polyphyllus*, but no reaction was detected with extracts from the roots and nodules following Western blot analysis. The discrepancies in the molecular weight of the putative subunit have now been clarified by Lough *et al.* (1992b) who have isolated a 1.2 kb cDNA clone from *L. arboreus* seeds that codes for a truncated asparaginase protein of 32.8 kDa, indicating that the enzyme comprises two subunits.

A potassium-dependent asparaginase was also detected in the developing seeds of *Vicia faba, Phaseolus multiflorus, Zea mays* and *Hordeum vulgare*. The existence of asparaginase activity in the presence of potassium ions in the testa and cotyledons of pea have been confirmed by Murray and Kennedy (1980) and the enzyme activity has been shown to be reduced by the presence of glutamine in soybean cotyledons. The two forms of asparaginase in seeds have been shown to be immunologically distinct. A potassium-dependent asparaginase has been isolated from the young expanding leaves of pea with a molecular weight of 58 kDa. The enzyme activity has been shown to undergo a complex diurnal variation that is regulated by light.

Transamination of Asparagine

It was argued that asparagine made a small but significant contribution to the synthesis of glycine by the transamination of glyoxylate during the process of photorespiration. Asparagine aminotransferase activity has been detected in high levels in pea leaves, where it is located in the peroxisome, but the enzyme is either absent or present in low activity in other tissues. The enzyme has been purified from pea leaves and it was proposed that one protein catalyzed both serine : Glyoxylate aminotransferase and asparagine; glyoxylate aminotransferase activity. Confirmation that the two activities were carried out by the same protein (and coded for by one gene) was obtained by the isolation of a mutant barley plant that was blocked in the photorespiratory nitrogen cycle and lacked both serine; glyoxylate and asparagine; glyoxylate aminotransferase activity.

ASPARTATE AMINOTRANSFERASES

Glutamate is a major amino donor and is involved in two important reactions—

Aspartate aminotransferase

Glutamate + Oxaloacetate $\rightleftharpoons$ 2-oxoglutarate + Aspartate

Alanine aminotransferase

Glutamate + Pyruvate $\rightleftharpoons$ 2-oxoglutarate + Alanine

Aminotransferases have been detected in plants that can synthesize all the protein amino acids (except proline), provided the correct 2-oxo acid precursor is available. Nitrogen can thus be readily distributed from glutamate, via aspartate and alanine to all amino acids. The reversibility of the reaction allows for sudden changes in demand for key amino acids, although in specific pathways, e.g., the synthesis of glycine in photorespiration, certain enzymes tend only to operate in one direction. These enzymes play significant roles :

A. The rapid transfer of amino groups from glutamate via aspartate to asparagine in nitrogen fixing root nodules. The 2-oxoglutarate molecule cycles to collect more amino groups in the glutamate synthase reaction, and oxaloacetate is synthesized by the action of PEP carboxylase.

B. The malate-oxaloacetate shuttle is often used to transfer reducing power from the mitochondria and chloroplasts to the cytoplasm via the enzyme malate dehydrogenase. Due to the inherent instability of oxaloacetate, the oxoacid may also be transaminated to aspartate to facilitate transport.

C. Following the formation of oxaloacetate, in the PEP carboxylase reaction in C_4 plants, aspartate is transported from the mesophyll cells to the bundle sheath cells in NAD-ME type plants.

In the alfalfa root nodule, two forms of aspartate aminotransferase have been identified, designated AAT-1 and AAT-2. The alfalfa AAT-1 has a native molecular weight of 84 kDa with two subunits of 42kDa. AAT-2 is slightly smaller with a native molecular weight of 80kDa and two subunits of 40 kDa. Antibodies raised against the two enzymes indicate that they are immunologically distinct. During the development of nitrogen fixing root nodules, AAT-2 is the predominant isoenzyme, whereas AAT-1 is the major form in roots. In C_4 plants two forms of aspartate aminotransferase have been characterized in *Panicum* and *Eleusine* species, that display physical characteristics and immunological properties similar to the alfalfa enzymes. A cytosolic isoenzyme is localized in the mesophyll cells and a mitochondrial enzyme is present in the bundle sheath cells. A cDNA clone encoding the alfalfa AAT-2 has now been isolated; the deduced amino acid sequence is 53 and 47% identical to the animal mitochondrial and cytoplasmic enzyme respectively. A N-terminal sequence of 59 amino acids suggests that the enzyme is localized in the plastid. The expression of AAT-2 mRNA in nodules was severalfold higher than the level detected in roots or leaves. Northern and Western blot analysis showed there was a sevenfold increase in AAT-2 mRNA and a comparable increase in enzyme protein during nodule development. A full length cDNA clone encoding aspartate aminotransferase has also been identified in carrot. The deduced amino acid sequence has a 52 and 53% identity with the sequences of the mouse cytoplasmic and mitochondrial enzymes, respectively. Northern blot analysis of the mRNA levels indicated that the gene was expressed throughout the cell culture up to 7 days and was highly expressed in roots but not in leaves.

ASPARTATE KINASE

Experiments with ^{14}C-labeled aspartate have shown that chloroplasts are capable of synthesizing lysine, threonine, isoleucine and homocysteine in light-driven reactions, which can therefore be termed photosynthetic (Fig. 15.5). Subsequent subcellular localization studies have shown that all the enzymes required are present in the chloroplast. The final methylation step involved in the conversion of homocysteine into methionine has been shown to take place in the cytoplasm. It is no coincidence that all the essential amino acids required by animals in their diet are synthesized by plants in the chloroplast. Three isoenzymes of aspartate kinase have been isolated from barley and their presence confirmed by genetic analysis. The first isoenzyme (AKI) is subject to feedback inhibition by threonine alone and is only present in low levels in rapidly growing tissue. The other two isoenzymes (AKII and AKIII) are similar to each other and are both inhibited by lysine. Methionine alone has no direct effect on any of the aspartate kinase isoenzymes but is able to regulate the flow of carbon through the activated form, S-adenosylmethionine. AKII and AKIII are both subject to synergistic inhibition in the presence of S-adenosylmethionine and lysine.

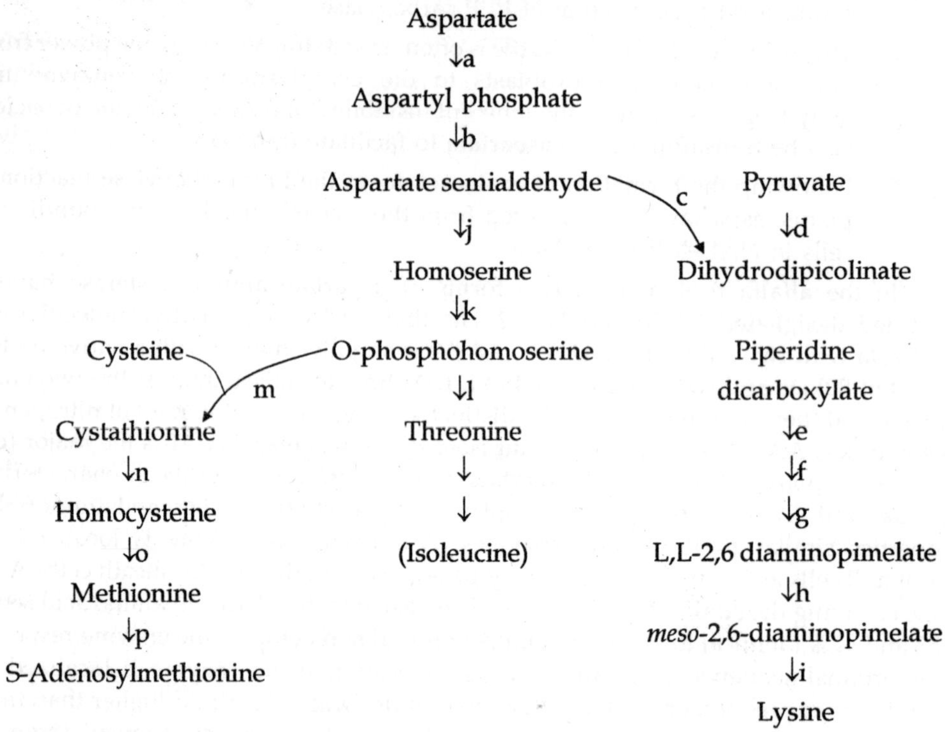

Fig. 15.5 : The metabolic pathways involved in the synthesis of lysine, threonine, methionine and isoleucine like amino acids in higher plants.

In a series of studies on threonine-sensitive homoserine dehydrogenase isolated from carrot tissue culture cells, Wilson et al. (1991) were able to show that the enzyme copurified with aspartate kinase. The amino acid sequence of the 85 kDa polypeptide indicated that both homoserine dehydrogenase and aspartate kinase activities were present on the same protein. A similar copurification of the two enzyme activities was also determined in maize tissue culture cells. Confirmation of the existence of the bifunctional aspartate kinase-homoserine dehydrogenase enzyme in carrot has been obtained by Weisemann & Matthews (1993). Two overlapping cDNA clones encoding the enzyme have been isolated. Together the cDNA clones delineate a 3089bp long sequence encompassing an open reading frame encoding 921 amino acids, including the mature protein and a long transit peptide.

Lysine-sensitive aspartate kinase has not been purified to homogeneity and the gene encoding the protein has not been isolated. Attempts to determine the native molecular weight of the enzyme have encountered difficulties. In carrot tissue cultures, values of 100-225 kDa were obtained, in maize cultures, values varied between 104-225 kDa and 139-150 kDa. Aspartyl phosphate is connected to aspartate semialdehyde by aspartate semialdehyde dehydrogenase. A single form of the enzyme has been isolated from maize tissue cultures. The enzyme does not appear to be regulated by end product amino acids.

SYNTHESIS OF LYSINE

Frisch et al. (1991) proposed that the maize enzyme (dihydrodipicolinate synthase) was a homotetramer with a subunit molecular weight of 38 kDa. However, Dereppe et al., showed that the pea enzyme was a trimer with three subunits of molecular weight 43 kDa. Kampaisal et al. (1987) suggested a ping-pong mechanism for wheat dihydrodipicolinate synthase in which pyruvate binds to the enzyme first, followed by the release of water before aspartate semialdehyde binds to the activated enzyme complex. The data obtained with the maize enzyme were consistent with this mechanism. All the enzymes so far examined have been shown to be extremely sensitive to feedback inhibition by lysine, with I_{50} values in the order of 20-50 µM. cDNA clones encoding dihydrodipicolinate synthase have been isolated from wheat and maize. The maize sequence indicated the presence of a 54 amino acid transit peptide and a subunit molecular weight of 36 kDa, consistent with the chloroplast localization of the enzyme.

Dihydrodipicolinate reductase was initially detected by the ability to catalyze diaminopimelate synthesis from aspartate and pyruvate in extracts of an *E. coli* mutant lacking the reductase. Further evidence of the presence of the enzyme was based on the demonstration of competitive inhibition by dipicolinate. The penultimate step in the synthesis of lysine in bacteria is the epimerization of L-diaminopimelate to *meso*-diaminopimelate.

Meso-diaminopimelate decarboxylase has been shown to be localized solely in the chloroplast. The molecular weight of the enzyme in higher plants varies between 75 kDa for wheat germ and 85 kDa for maize endosperm. Lysine at concentrations up to 1mM has little effect on enzyme activities, a although 20 mM inhibited the enzyme by 60%.

SYNTHESIS OF THREONINE

Distinct threonine-sensitive and threonine-resistant isoenzymes have been isolated in higher plants, the former being localized in the chloroplast and the latter in the cytoplasm. The isoenzyme isolated from maize has been shown to differ in size, subunit composition and antigenic properties. The threonine-sensitive maize enzyme can exist in at least four ligand-induced conformational states, two of which are dimeric and two of which are tetrameric. In carrot, homoserine dehydrogenase activity reversibly converts between a threonine insensitive form in the presence of K^+ and a threonine sensitive form in the presence of threonine. As discussed previously, threonine-sensitive homoserine dehydrogenase and aspartate kinase from carrot were shown to copurify and to be different activities of the same enzyme protein. The full amino acid and nucleic acid sequence of the aspartate kinase-homoserine dehydrogenase has now been determined.

Pea homoserine kinase is also subject to inhibition by several amino-acids including *S*-adenosylmethionine. In contrast, Aarnes (1976) was unable to identify significant effects of amino acids on the activity of the enzyme isolated from barley. Riesmeier et al. (1993) have purified homoserine kinase from wheat germ to homogeneity. The enzyme had a native molecular weight of 75 kDa and a subunit molecular weight of 36 kDa. These data conflict with previous reported molecular weights in the range 120-240 kDa, suggesting that the enzyme can form aggregates. Threonine, methionine, valine and isoleucine were found to have no inhibitory effect on the enzyme at 10 mM. S-adenosylmethionine was shown to inhibit the wheat germ enzyme by 26% at 10 mM, although it is unlikely that this effect would be of any physiological importance. It is possible that plants contain regulated and unregulated forms of homoserine kinase in an analogous way to homoserine dehydrogenase; however, the presence of distinct isoenzymes has not been confirmed. O-phosphohomoserine is converted into threonine in an internal rearrangement, catalyzed by threonine synthase. The enzyme, which is localized in the chloroplast, has been partially purified from radish, pea, barley and *Lemna*. The enzyme has a virtual absolute requirement for S-adenosylmethionine as an allosteric activator. Half maximal activation is obtained at 40-200 μM S-adenosylmethionine. Threosynthase activity is not inhibited by any potential feedback inhibitors, but the enzyme is repressed by methionine supplementation of the growth medium.

SYNTHESIS OF METHIONINE

Cystathionine-γ-synthase catalyzes the combination of O-phosphohomoserine and cysteine to yield cystathionine and inorganic phosphate. The enzyme has not been extensively purified from higher plants, but has been isolated from sugar beet and barley. In *Lemna paucicostata* the enzyme has been shown to be sensitive to a range of potent inhibitors including propargylglycine. Cystathionine-γ-synthase appears to be present in excess in *L. paucicostata* as the enzyme can be inhibited by up to 84% by propargylglycine, before there is a detectable decrease in the rate of methionine synthesis. Although cystathionine-γ-synthase activity *in vitro* is not regulated by any potential feedback inhibitors, the level of the enzyme activity *in vivo* is regulated in *L. paucicostata.*

The addition of 2μM methionine decreased the enzyme activity to 15% of the control value. Conversely, the addition of lysine and threonine which would reduce the flow of carbon into methionine, increased the *in vivo* activity of cystathionine-γ-synthase three fold. Cystathionine is metabolized to homocysteine, pyruvate and ammonia by cystathionine-β-lyase. Mutants of tobacco lacking the enzyme activity require homocysteine or methionine for growth. The enzyme is present in both the chloroplast and cytoplasm of higher plants. Cystathionine-β-lyase has also been purified to homogeneity from the tissue culture cells of *Echinochloa colonum* with a specific activity ten times that of the spinach enzyme. The mature molecular weight of the enzyme is 160 kDa with a subunit molecular weight of 41 kDa. The characteristic flavours and odors of a number of important vegetables are due to the enzymic degradation of nonprotein sulphur- containing amino acids. These enzymes also have the capacity to cleave a C—S bond to form a sulphur-containing product, pyruvate and ammonia by a mechanism analogous to that of cystathionine-β-lyase. The final enzyme in the pathway catalyzes the S-methyltetrahydropteroyltriglutamate-dependent methylation of homocysteine. The enzyme methionine synthase has been detected in a small number of plants including pea leaves and carrot root. Two forms of the enzyme have been partially purified from pea seeds. In *E. coli,* methionine synthase is a cobalamin-dependent enzyme with a molecular weight of 133 kDa. The *metH* gene encoding the enzyme has been isolated and sequenced.

S-adenosylmethionine is used as an important methyl donor for a range of methylation reactions in plants including the synthesis of derivatives of ethanolamine, pectins, chlorophyll, lipids and nucleic acids. The derivative is formed by the combination of methionine and ATP catalyzed by *S*-adenosylmethionine synthetase . Three isoenzymes of *S*-adenosylmethionine synthetase have been isolated from wheat embryos. One of the isoenzymes has been purified to homogeneity and shown to be a dimer with a subunit molecular weight of 84 kDa. The gene (*sam*-1) coding for *S*-adenosylmethionine synthetase has been isolated from *A. thaliana.* High levels of mRNA were expressed in the vascular tissues of the root and stem. Peleman *et al.* (1989) suggested that the *sam*-1 gene may code for the enzyme involved in lignification. The molecular weight of subunits of the *A. thaliana* and carnation enzymes were calculated to be 43 kDa, considerably lower than the value determined for the wheat germ enzyme. By a complex series of analyses, were able to follow the metabolism of the methyl carbon atom, sulfur atom and C_4 group of methionine in *L. paucicostata.* The results showed that the rate of synthesis of methionine from homocysteine is about five times faster than the rate of incorporation of sulfate into homocysteine. In another series of experiments, Giovanelli *et. al.* (1983) showed that the sulfur atom and methyl carbon atom of methionine are recycled independently of the C_4 skeleton. They proposed that S-adenosylmethionine is decarboxylated to yield a C_3 fragment that is used in the synthesis of polyamines. The remainder of the S-adenosylmethionine molecule, S-methylthioadenosine is then metabolized, releasing adenine and formate and finally yielding methionine. In this novel pathway the original methylthio group is retained and the C_4 moiety in the regenerated methionine is derived from the ribosyl moiety of ATP.

REGULATION OF AMINO ACID BIOSYNTHESIS

The various mechanisms known to operate *in vitro* can be summarized as follows—

a. *Lysine* inhibits two isoenzymes of aspartate kinase and the first enzyme unique to its own synthesis, dihydrodipicolinate synthase.

b. *Threonine* inhibits one isoenzyme of aspartate kinase and one isoenzyme of homoserine dehydrogenase.

c. *Methionine* appears to operate through the activated from S-adenosylmethionine. Two isoenzymes of aspartate kinase are synergistically inhibited in the presence of lysine. S-adenosylmethionine is able to activate threonine synthase. There is also evidence that methionine and/or S-adenosylmethionine can repress the synthesis of cystathionine-γ-synthase.

d. *Isoleucine* inhibits the first enzyme unique to its own synthesis, threonine dehydratase.

Giovanelli *et. al.* (1989) proposed that there was no regulation of threonine synthesis along the whole pathway, suggesting that the inhibition of homoserine dehydrogenase and the activation of threonine synthase did not operate *in vivo* (Fig. 15.6).

ALTERED REGULATION IN TRANSGENIC PLANTS

Maize tissue culture cell lines have been isolated following azide mutagenesis, that are resistant to the combined action of lysine and threonine. Three of the cell cultures gave rise to genetically stable plant lines. Similar mutant selection procedures have been successful in barley and tobacco. The mutant lines isolated from all three plant species were shown to contain elevated levels of soluble threonine. In barley leaves, three forms of aspartate kinase activity have been detected following ion exchange chromatography. In the barley mutant R3202 resistant to lysine plus threonine, the AKII peak was nor inhibited by lysine even at the high concentration of 10mM, whereas the AKIII enzyme peak was still inhibited by lysine. Conversely in the mutant R3004, AKIII activity was less sensitive to lysine, with a maximum inhibition detected at 3.5mM lysine compared to 0.36 mM for wild-type barley. Genetic analysis of the two mutants suggested that resistance to lysine plus threonine was controlled by two dominant genes termed *lt1* and *lt2* which are probably the structural genes for AKII and AKIII. A double mutant derived from a cross between R3202 (*lt1*) and R3004 (*lt2*) has been constructed which contained AKII insensitive to lysine inhibition and AKIII with a decreased sensitivity to lysine.

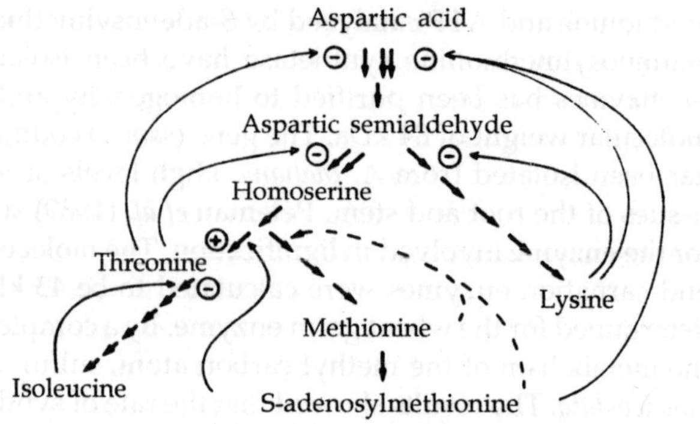

Fig. 15.6 : Regulation of biosynthesis of the aspartate-derived amino acids (–), enzyme inhibition; (+), enzyme activation; enzyme repression.

Aspartate kinase purified from the *Ask1* and *Ask2* mutant lines was less sensitive to inhibition by lysine. Wild-type aspartate kinase activity was inhibited 50% by 10 μM lysine. In contrast, approximately 760 μM lysine was required to inhibit 50% of the aspartate kinase activity of the homozygous mutant *Ask2*. *Ask2* and 25μM lysine was required for 50% inhibition of the heterozygous *Ask1*/ + enzyme. Dotson *et. al.* (1990) suggested that maize aspartate kinase could be a heteromeric enzyme, consisting of two lysine sensitive polypeptides encoded by *Ask1* and *Ask2*. A number of research groups have attempted to select mutant lines that are resistant to the lysine analogue S-(2-aminoethyl)-L-cysteine (AEC) on the grounds that the mutant plants may overproduce lysine. The majority of these experiments have produced plants with a reduced uptake of AEC. However, one mutant of *Nicotiana sylvestris* regenerated from protoplast culture, was resistant to AEC (RAEC-1) and was shown to accumulate lysine. Only 50% of the dihydrodipicolinate synthase activity was inhibited by lysine.

The level of soluble lysine in the leaves of the double mutant was 20-fold higher than the wild type and sometimes rose to 50% of the total soluble amino acid pool. Unfortunately the accumulation of lysine in the leaves caused an abnormal phenotype characterized by a reduced leaf blade surface, absence of stem elongation and sterility. Surprisingly the level of soluble threonine in the double mutant was lower or the same as in the wild type. Dihydrodipicolinate synthase in *E. coli* is considerably less sensitive to feedback inhibition by lysine (I_{50} =1 mM) than the plant enzyme (I_{50}=20 μM). Shaul and Galili (1992) constructed transgenic tobacco plants containing the *E. coli dapA* gene coding for dihydrodipicolinate synthase in the cytoplasm and in the chloroplast. The increase in lysine concentration (up to a maximum of 40-fold), was shown to be correlated with the level of dihydrodipicolinate synthase activity. Compartmentalization of the enzyme inside the chloroplast was found to be essential for lysine overproduction. The highest lysine overproducing plants exhibited abnormal characteristics including the possession of leaves of a mottled yellow/green colour and a lack of apical dominance.

Increase in the soluble level of threonine varied between 20- and 80-fold in transgenic plants containing the enzyme expressed in the chloroplast. Small increases in the levels of lysine and isoleucine was also detected. In contrast to the experiments with dihydrodipicolinate synthase, a chloroplast localization of aspartate kinase was important but not essential for threonine overproduction. It is clear from the experiments with mutants and transgenic plants that it is possible to produce plants with high levels of soluble lysine and threonine but not methionine.

THE AMINO ACIDS OF BRANCHED CHAINS

Threonine dehydratase enzyme is also known as threonine deaminase and catalyzes the first committed step of isoleucine synthesis, forming 2-ketobutyrate and ammonia (Fig. 15.7). Confirmation of the key role of the enzyme has been obtained by Negrutiu *et. al.* (1985) who isolated isoleucine-requiring mutants of tobacco that lacked the enzyme

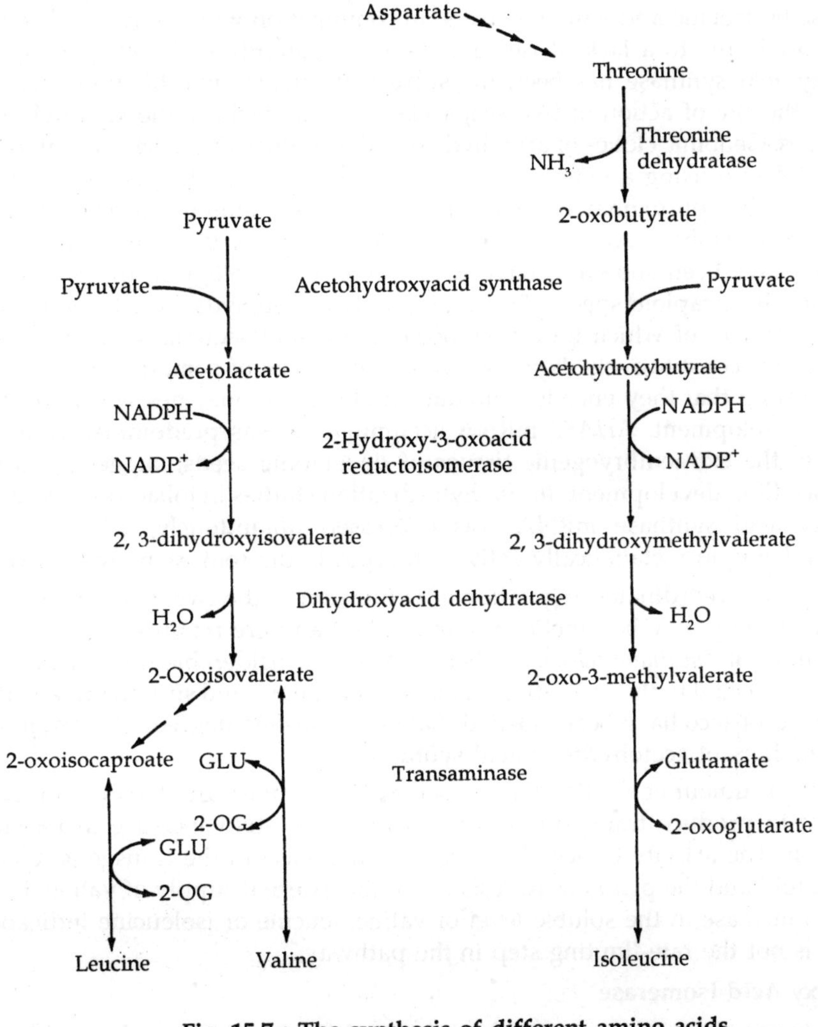

Fig. 15.7 : The synthesis of different amino acids.

activity. Two forms of the enzyme have been isolated from maize, both of which are totally inhibited by low concentrations of the end product isoleucine. In tomato, an isoleucine-sensitive form of threonine dehydratase was shown to predominate in young leaves, but as the leaves aged, an isoleucine insensitive form of the enzyme was shown to be present at high levels. The mature molecular weight of the isoleucine sensitive form was shown to be 370 kDa whereas the insensitive form was 200 kDa. Threonine dehydratase is one of the most abundant soluble proteins in tomato flowers. The gene coding for the 55 kDa polypeptide has been sequenced and shown to contain a plastid transit peptide and eight introns. The expression of threonine dehydratase mRNA was shown to be 50-fold higher in sepals and 500-fold higher in the rest of the flower than in leaves and roots.

Acetohydroxy Acid Synthase

Acetohydroxy acid synthase can either catalyze the synthesis of acetohydroxybutyrate or acetolactate. Control of the pathway is achieved by feedback inhibition of acetohydroxy acid synthase by leucine and valine. Leucine in combination with valine has been shown to inhibit growth due to a lack of isoleucine. The apparently innocent looking enzyme acetohydroxy acid synthase has been the subject of intense investigation, since it was shown to be the site of action of two major classes of herbicides, the sulfonylureas and imidazolinones. Genomic clones of acetohydroxy acid synthase have been isolated from *A. thaliana* and tobacco using a heterologous hybridization probe derived from yeast. The genes proved to be approximately 85% homologous at the amino acid level and neither contained introns. In tobacco, two genes (designated *SurA* and *SurB*) coding for acetohydroxy acid synthase have been shown to have 95% sequence homology in the protein coding regions. In the allotetraploid species *Brassica napus*, five acetohydroxy acid synthase genes have been identified, of which four have been cloned and sequenced. The *AHAS1* and *AHAS3* genes are constitutively expressed in a wide range of somatic and reproductive tissues, suggesting that they encode a fundamental class of enzymes which are vital for growth and development. *AHAS2* mRNA accumulation was predominantly in mature ovules and in the extraembryogenic tissues of developing seeds, suggesting a specific function in seedling development. *In situ* hybridization studies in tobacco showed that the acetohydroxy acid synthase mRNA was expressed ubiquitously, with the greatest accumulation being in metabolically active cell types in the root, stem and flowers.

There was a co-ordinated regulation of expression of the two genes *SurA* and *SurB* suggesting that these may be functionally redundant and are retained as a result of the tetraploid nature of *Nicotiana tabacum*. There has been considerable interest in obtaining plants that are resistant to the toxic effect of the imidazolinone and sulfonylurea herbicides. Mutant lines of tobacco have been isolated that show varying degrees of resistance to the herbicide inhibitors of acetohydroxy acid synthase.

Recently Toumeur et al. (1993) have used a P70 promoter and the *csr-1* mutant gene of *A. thaliana* to produce transgenic tobacco with a 1500-fold increase in resistance to chlorosulfuron. The activity of acetohydroxy acid synthase in the transgenic plants was increased 12-fold and the plants were resistant to the external supply of valine. However, there was no increase in the soluble level of valine, leucine or isoleucine indicating that the enzyme is not the rate-limiting step in the pathway.

Acetohydroxy Acid Isomerase

Three forms of the enzyme have been isolated from spinach, each of which have a native molecular weight of 114 kDa and are made up of two subunits of identical molecular weight (57 kDa) but different isoelectric points. A full length cDNA clone encoding the enzyme has been isolated from spinach which contains a 72 amino acid transit peptide. The amino acid sequence shows only a 23% homology to the yeast and *E. coli* enzymes. The gene encoding acetohydroxy acid isomeroreductase has been over expressed in *E. coli* and large quantities of the enzyme used for a detailed kinetic analysis. The enzyme was shown to have a preference for 2-acetohydroxybutyrate over acetolactate and to operate at high ratios of NADPH/NADP$^+$, conditions which are normally present in illuminated chloroplasts.

Dihydroxyacid Dehydratase

Dihydroxyacid dehydratase carries out the dehydr. ,n and tautomerization of the 2,3-dihydroxycarboxylic acid to the corresponding 2-keto acid. The enzyme contains a [2Fe-2S] cluster which is a novel finding for enzymes of the hydratase class. A mutant of *Datura innoxia* requiring isoleucine and valine for growth has been shown to lack dihydroxyacid dehydratase.

Isopropylmalate Dehydrogenase

The cDNA encodes a 52 kDa protein which has a putative chloroplast transit peptide. Analysis of the sequence of the isopropylmalate dehydrogenase protein, suggested that the enzyme is more similar to bacterial than fungal proteins. The gene encoding the enzyme has also been isolated from potato. A preliminary report has indicated that the final three enzymes of leucine synthesis can be detected in spinach chloroplast extracts.

PROLINE AND ARGININE BIOSYNTHESIS

Proline may either be synthesized directly from glutamate or via the transamination of ornithine (Fig. 15.8). Tomato cells culture in 400 mM NaCl or 25% polyethylene glycol can accumulate proline up to concentrations 300-fold higher than in unstressed conditions. Kinetic studies using ^{15}N-labeling showed that the accumulation of proline was due to a 10-fold increase in the rate of synthesis and a concomitant reduction in the rate of degradation. Proline metabolism has also been shown to play an important role in the process of nitrogen fixation in the legume root nodule. However, Hu *et. al.* (1992) have isolated a cDNA clone from *Vigna aconitifolia* nodules that encodes a bifunctional enzyme Δ'-pyrroline-5-carboxylate synthetase (P5CS) with both γ-glutamyl kinase and glutamate semialdehyde dehydrogenase activity. The P5CS protein was shown to have two distinct domains that had 55.3% and 57.9% overall similarity to the *E. coli ProB* and *ProA* proteins.

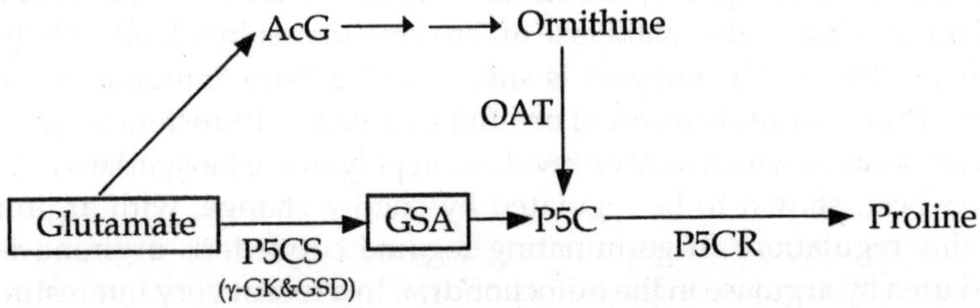

Fig. 15.8 : Proline biosynthesis.

The plant P5CS enzyme protein was isolated from the complemented *E. coli ProBA* mutant and was shown to be 50% inhibited by 5 mM proline; however, the wild type *E. coli* enzyme was 50-fold more sensitive to end-product inhibition.

Northern blot analysis indicated that the P5CS gene was expressed at a greater level in the leaves than the roots and nodules of *V. aconitifolia*. However, the level of mRNA in the roots was shown to be increased by the addition of 200 mM NaCl to the plants. △'-pyrroline-5-carboxylate reductase (P5CR) has been purified from barley and shown to increase in activity following drought stress. The gene encoding the enzyme has been isolated from soybean nodules by the complementation of the *E. coli ProC* mutant. Levels of P5CR mRNA are elevated in the salt stressed plants. The enzyme has been purified to homogeneity by expression of the soybean nodule cDNA in *E. coli* and shown to have a subunit molecular weight of 29 kDa. Subcellular fractionation and Western blot analysis were used to localize P5CR in transgenic tobacco plants synthesizing P5CS constitutionally. Enzyme activity was present in the cytoplasm of the root, nodule and leaf but 15% of the activity was also detected in the chloroplast fraction. As P5CR has previously been detected in the chloroplasts of pea leaves, it would appear that at least two isoenzymes exist and that proline may be synthesized in different subcellular compartments in the leaf and root. Transgenic tobacco plants containing a 50-fold elevated level of P5CR activity did not contain increased concentrations of proline, confirming that the enzyme is not the rate-limiting step in proline synthesis.

Ornithine may be transaminated via either the α or δ amino groups yielding P2C or P5C, respectively. Delauney *et. al.* (1993) have isolated a cDNA encoding ornithine aminotransferase (OAT) by *trans*-complementation of *E. coli ProBA* mutants in the presence of ornithine. Sequence data suggested that the *Vigna aconitifolia* OAT mRNA decreased, whereas those of P5CS increased.

Ornithine and arginine also act as precursors of the important secondary metabolites, polyamines. In addition, citrulline may act as a nitrogen transport compound in non-legume nitrogen fixing symbioses. Arginine is synthesized from glutamate via ornithine and citrulline (Fig. 15.9). Inhibition studies have shown that in soybean cell suspension cultures the acetyl group is rapidly returned from N-acetylornithine to glutamate via ornithine acetyltransferase. N-acetylglutamate kinase has been purified to homogeneity from pea cotyledons and shown to exist in dimeric and tetrameric forms. Arginine was found to strongly inhibit the enzyme activity, this inhibition was relieved by N-acetylglutamate, which activated the kinase enzyme. In an elegant series of tracer studies, Micallef and Shelp (1989) examined the metabolism of [ureido-^{14}C] citrulline and [1-^{14}C] ornithine in the developing cotyledons of soybean. Further metabolism of ornithine requires carbamoylphosphate synthetase an enzyme that is involved in both arginine and pyrimidine biosynthesis. The enzyme is subject to feedback inhibition by UMP which is relieved by ornithine. Multiple forms of ornithine carbamoyltransferase have been detected in higher plants, some of which may be involved in polyamine biosynthesis. Arginosuccinate synthetase has been shown to be regulated by energy change, with arginine acting as a modifier of this regulation. In germinating legume cotyledons arginine is degraded to ornithine and urea by arginase in the mitochondria. In a recent very interesting observation, Ludwig (1993) has shown that *Arabidopsis thaliana* chloroplasts are able to metabolize arginine. The author was able to demonstrate that the chloroplasts contained arginine iminohydratase that yielded citrulline and ammonia. The citrulline may be further metabolized to ornithine and carbamoyl phosphate by ornithine carbamoyltransferase. The carbamoyl phosphate can then be broken down to bicarbonate and ammonia by carbamate kinase. Ludwig (1993) proposed that the interconversion of ornithine—citrulline—arginine may be a novel mechanism of transporting ammonia and CO_2 form the mitochondria to the chloroplast.

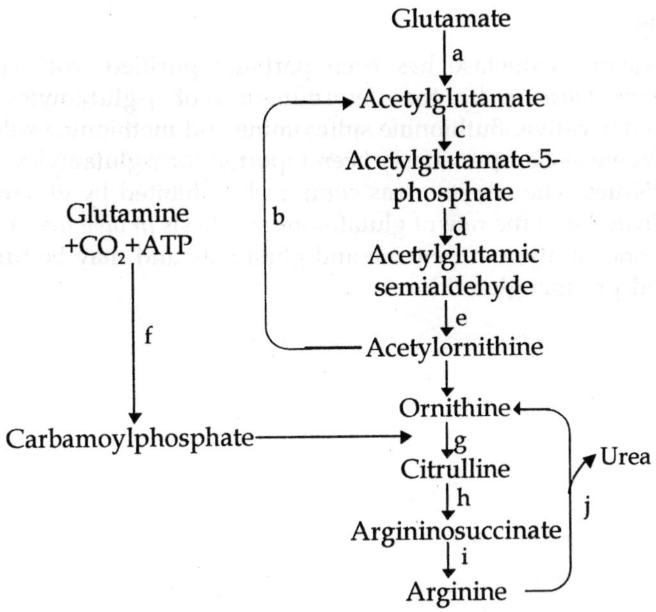

Fig. 15.9 : Arginine synthesis.

SULFUR CONTAINING AMINO ACIDS

Multiple forms of cysteine synthase have been found in *Phaseolus, Brassica napus* leaves and spinach leaves. In a detailed study of cauliflower bud protoplasts, Rolland et al. (1992) demonstrated the presence of four forms by ion exchange chromatography. Two isoenzymes were located in purified proplastids, one in the mitochondria and one in the cytoplasm. The chloroplast cysteine synthase has been purified to homogeneity from spinach. The enzyme consists of two indentical subunits of molecular weight of 310 kDa. Antisera raised against the chloroplast enzyme showed very low cross-reactivity with the cytoplasmic and mitochondrial enzymes. A cDNA clone encoding the cytoplasmic form of cysteine synthase has been isolated from spinach with a molecular weight of 34 kDa. A comparison of the amino acid sequence indicated a 53% identity with the *E. coli* enzyme. The cDNA contained no apparent transit sequence and constitutive expression of mRNA was detected in roots and leaves. A cDNA clone encoding the complete sequence of the chloroplast enzyme has been isolated by Rolland *et. al.* (1993). The 35 kDa polypeptide also contained a 52 amino acid transit peptide. A good primary sequence homology (50%) was demonstrated between the spinach chloroplast cysteine synthase and the A isoenzyme of *E. coli*. Southern blot analysis indicated that the chloroplast enzyme is encoded by multiple genes in spinach. The gene encoding the chloroplast cysteine synthase has also been isolated from spinach by Saito *et. al.* (1993).

Glutathione may exist in the reduced form of GSH or in the oxidized state as a dimer GSSG, where two molecules are linked by a disulfide bond. It has also been proposed that the tripeptide may play a key role in the defence of plants against various environmental stresses, e.g., cold, heat, drought, high light, fungal attack and herbicides. Phytochelatins are polymers of γ-glutamylcysteine, which have been shown to be involved in the chelation of certain toxic metals.

γ-glutamylcysteine synthetase has been partially purified from cell suspension cultures of *Nicotiana tabacum* by the determination of γ-glutamylcysteine as the monobromobimane derivative. Buthionine sulfoximine and methionine sulfoximine were inhibitors of the enzyme as had previously been reported for γ-glutamylcysteine synthase from mammalian tissues. The enzyme was completely inhibited by glutathione (K_i=0.42 mM). The data indicated that the rate of glutathione synthesis *in vivo* may be substantially influenced by the concentration of cysteine and glutamate and may be further feedback regulated by the end product glutathione.

□□□

16

DNA AND RNA VIRUS DISEASES

Viruses are the smallest parasites with the simplest biologic structure—essentially small packaged particles of DNA or RNA that depend on the host cell for their qualities of life. Viruses have special adaptations for entering a cell and for reprogramming its genetic and molecular machinery to produce and release new viruses. Animal viruses are divided into families based on the nature of the nucleic acid (DNA or RNA), the type of capsid, and whether or not an envelope is present. All DNA viruses are double stranded except for the parvoviruses, which have single-stranded DNA. All RNA viruses are single stranded except for the double stranded reoviruses. The genome of RNA viruses can be further described as segmented (consisting of more than one molecule) or nonsegmented (Fig. 16.1).

VIRAL DISEASES : MEDICAL CONSIDERATIONS

Various viruses target nearly all types of tissues, including those of the nervous system (polio and rabies), liver (hepatitis), respiratory tract (influenza, colds), intestine (polio), skin and mucous membranes (herpes, pox), and cells of the immune system (AIDS). Most DNA viruses are assembled and budded off the nucleus, while most RNA viruses multiply in and are released from the cytoplasm. The presence of assembling and completed viruses in the cell gives rise to various telltale disruptions called cytopathic effects. Most cells productively infected with viruses are destroyed, which accounts for the sometimes severe pathology and loss of function.

Viral diseases range from very mild or asymptomatic infections such as colds to severe, deadly syndromes such as rabies and AIDS. Many are so-called "childhood diseases" that occur primarily in the young and are readily transmitted through droplets. Although we tend to consider these infections as self-limited and in-evitable, even measles, mumps, chickenpox, and rubella can manifest severe complications. Many diseases are strictly human in origin, but several such as yellow fever and viral encephalitis are

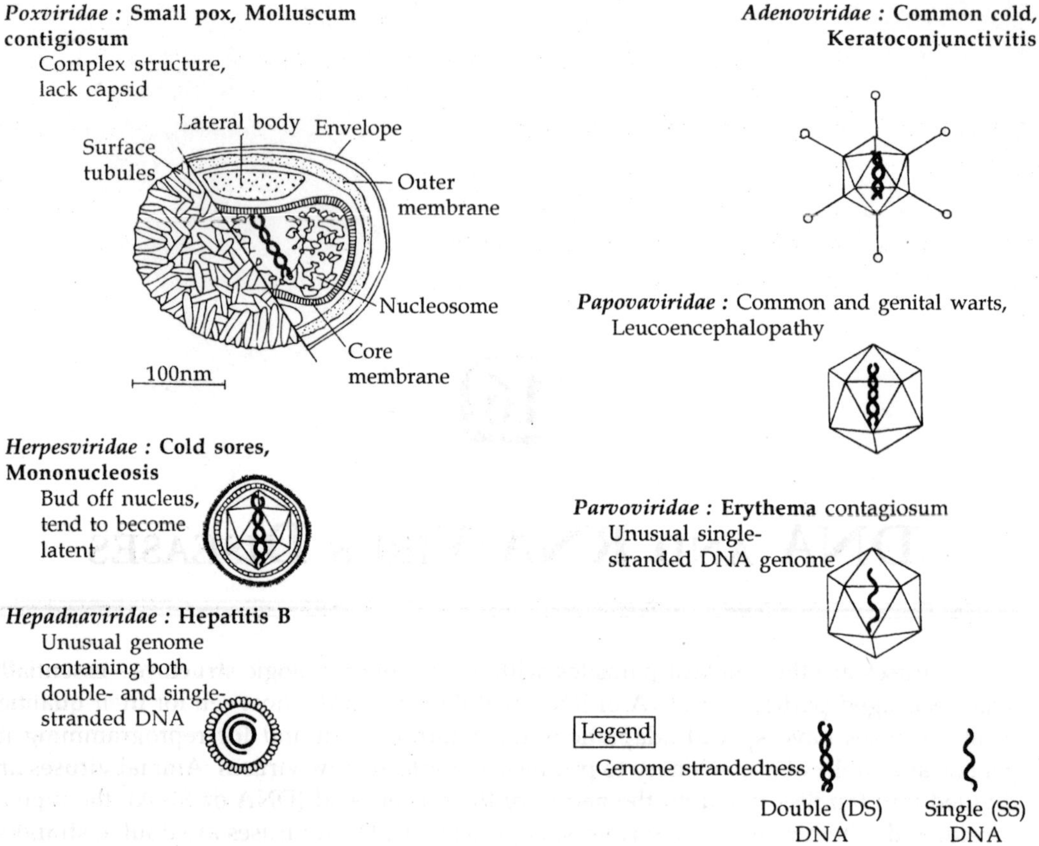

Fig. 16.1 : DNA virus family.

zoonoses transmitted by arthropod vectors. The course of viral diseases starts with invasion at the portal of entry by a few virions and a primary infection. In some cases (influenza, colds), the viruses replicate locally and disrupt the tissue, and in others (rabies, polio), the virus enters the blood and creates complications in tissues far from the initial infection. Common manifestations of virus infections are rashes, fever, muscle aches, respiratory involvement, and swollen glands – symptoms that can be too nonspecific to be of use in diagnosis. Many diseases of unknown etiology are thought to have a viral basis. Protection in viral infections arises from the combined actions of interferon, neutralizing antibodies, and cytotoxic T cells. Due to the nature of the immune response to viruses, infections frequently result in lifelong immunities, and many viruses are adaptable for vaccines.

In some cases, the latent virus alternates between periods of relative inactivity and recurrent infections. Viruses can persist in a more permanent state by splicing their DNA into a site in the host's DNA. This process sometimes transforms the cell into a cancer cell. This potential for oncogenesis is more likely with DNA viruses, though retroviruses, which can convert their RNA to DNA, are also important in cancers. This property of combining with host DNA is a drawback in developing live, attenuated DNA virus vaccines.

Several viruses can cross the placenta from an infected mother to the embryo or fetus. This often causes developmental disturbances and permanent defects in the child that are present at birth (congenital). Among viruses with known teratogenic (Gr. *Teratos*, monster) effects are rubella, cytomegalovirus, and adenovirus. Another risk for infants involves infection at the time of birth, as occurs with hepatitis B and *Herpes simplex*. Viral diseases are diagnosed by a variety of clinical and laboratory methods, depending on the virus. Some can be diagnosed by symptoms alone, while others require isolation in cell or animal culture. Serological testing for antibodies can be of importance in many viral infections. The use of antigen detection assays is on the increase, especially those based on monoclonal antibodies. Great hope is being placed on rapid nucleic acid probes directed against viral DNA or RNA (PCR amplification). These have the distinct advantage of being so sensitive that they could conceivably direct the viral nucleic acid in a single infected cell.

ENVELOPED DNA VIRUSES

Structure and Classification of Pox Viruses

The poxviruses are distinctive because they are the largest and most complex of the animal viruses, they have the largest genome of all viruses, and they multiply in the cytoplasm in well-defined sites called factory areas, which appear as inclusion bodies in infected cells. Among the best known poxviruses are variola (L. *Varius,* varied, mottled) the agent of smallpox, and vaccinia, a closely related virus used in vaccinations. Other members are of interest to the veterinarian for their role in diseases of domestic or wild animals. A common feature of all poxvirus infections is a specificity for the cytoplasm of epidermal cells and subcutaneous connective tissue. In these sites, they produce the typical lesions and also tend to stimulate cell growth than can lead to tumor formation.

It is one of the few infectious agents that now exists only in government laboratories. At one time, smallpox was ranked as one of the deadliest infectious diseases, leaving no civilization untouched. When introduced by explorers into naive populations such as Native Americans and Hawaiians, small pox caused terrible losses and had a hand in destroying these civilizations. Even as recently as 1967, approximately 10 to 15 million cases occurred worldwide, with thousands of fatalities.

Disease manifestations—After crossing the respiratory epithelium, the virus moved into the circulation and multiplied in the local lymph nodes. Viremia was associated with fever, malaise, prostration, and later, a rash that began in the pharynx, spread to the face, and progressed to the extremities. Initially, the rash was macular, evolving in turn to papular, vesicular, and pustular before eventually crusting over, leaving nonpigmented sites pitted with scar tissue. The two principal forms of smallpox were variola minor and variola major. Variola major was 25 times more virulent; death usually occurred early in the illness, as a result of toxemia, shock, and intravascular coagulation.

Diagnosis—Today, smallpox diagnosis is vested with a special importance and cannot be based on clinical observation. Every year since 1977, apparent cases have been reported that must be carefully differentiated from chickenpox, disseminated herpes, vaccinia, monkeypox, and certain nonviral lesions that mimic smallpox. Scanning stained smears of vesicular fluid for cytopathic effects such as cytoplasmic inclusion bodies can provide useful evidence.

Vaccination—As given in more recent times, a single drop of vaccinia virus is punctured into the skin with a double-pronged needle or jetgun. In a successful take, a large, scablike pustule with a reddened periphery develops at the injection site, where it leaves a permanent scar (a pitted emblem that many of us still bear). The vaccine protects people for up to 10 years against variola as well as other poxvirus infections. For the most part, small pox vaccinations have been discontinued not only because smallpox no longer poses a threat, but also due to the slight risk of vaccinia complications, especially in immunodeficient and allergic people.

Other Poxvirus Diseases

An unclassified poxvirus causes a skin disease called Molluscum contagiosum (L. *Molluscus,* soft). This disease is distributed throughout the world, with highest incidence occurring in certain Pacific islands. In endemic regions, it is primarily an infection of children and is transmitted by direct contact and fomites. The moist tropical climate appears to favour skin lesions, which take the form of smooth, round, waxy nodules on the face, trunk, and limbs. In adults, the infection is spread by sexual intercourse, and the lesions occur on the public and genital regions. Lesions can be single or multiple, and they follow a progressive 6 to 36 month course, complicated by itchiness, pain, and self-inoculation. Many mammalian groups host some sort of poxvirus infection. Cowpox, rabbitpox, monkeypox, mousepox, camelpox, buffalopox, and elephantpox are among the types that occur. However, only monkeypox and cowpox appear to be capable of infecting humans symptomatically. Although cowpox is an eruptive cutaneous disease that can develop on cows' udders and teats, cows are not the actual reservoirs of this virus. Other mammals such as rodents, cats, and even zoo animals can carry it. Human infection is rare and usually confined to the hands, although the face and other cutaneous sites can be involved.

Herpes Viruses

Herpesvirus (Gr. *Herpein,* to creep) was named for the tendency of some infections to produce a creeping rash. It is the common name for a large family whose members include *Herpes simplex* 1 and 2, the cause of fever blisters and genital infections; *Herpes zoster*, the cause of chickenpox and shingles; cytomegalovirus (CMV), which affects the salivary glands and other viscera; Epstein-Barr virus (EBV), associated with infection of the lymphoid tissue; and some recently identified viruses (Herpes virus-6 and Herpes virus-7). Prominent features of the family are its tendency toward viral latency and recurrent infections. This behaviour often involves incorporation of viral nucleic acid into the host genome and carries with it the potential for oncogenesis at some time, usually without adverse effect. The herpesviruses are among the larger viruses, with a diameter ranging from about 150 nm to 200 nm. They are enclosed within a loosely fitting envelope that contains glycoprotein spikes. Like other enveloped viruses, herpesviruses are prone to deactivation by organic solvents or detergents and are relatively unstable outside the host's body. The icosahedral capsid houses a core of double-stranded DNA that winds around a proteinaceous spindle in some viruses.

General properties—Humans are susceptible to two varieties of *Herpes simplex* viruses (HSV)—HSV-1, characterized by lesions on the oropharynx, and HSV-2, with lesions on the genitalia, differentiated by several characteristics. Even though laboratory mice, guinea pigs, rabbits, hamsters, and other animals can be experimentally infected, humans appear to be the only natural reservoir for *Herpes simplex* virus.

Epidemiology—Infection occurs in all seasons and among all age groups, and transmission is promoted by direct exposure to secretions containing the virus. The viruses are relatively sensitive to the environment, but can remain infectious in moist secretions on inanimate objects for a few hours. People with active lesions are the most significant source of infection, though a recent study indicates that genital herpes can be transmitted even when no lesions are present. Primary infections tend to be age-specific : HSV-1 frequently occurs in infancy and early childhood, and by adulthood, most people exhibit some serologic evidence of infection. On the other hand, primary infection with HSV-2 occurs most frequently between the ages of 14 and 29, a pattern that reflects the sexual route of transmission. In fact, genital herpes is the third or fourth most common STD in the United States. Instances of HSV-1 genital infections and HSV-2 infections of the oral cavity have recently risen.

Latency and recurrency—In approximately 20% to 50% of primary infections, *Herpes simplex* viruses enter the distal regions of sensory neurons and travel to the dorsal root ganglia. Type 1 HSV enters primarily the trigeminal, or fifth cranial nerve, which has extensive innervations in the oral region. Type 2 HSV usually becomes latent in the ganglion of the lumbosacral spinal nerve trunk. The exact details of this unusual mode of latency are not yet fully understood, but the virus remains inside the neuron in a nonproliferative stage for a variable time. Recurrent infection is triggered by various stimuli such as fever, UV radiation, menstruation, stress, or mechanical injury, all of which reactivate the virus. The virus migrates to the body surface and produces a local skin or membrane lesion, often in the same site as a previous infection.

Spectrum of Herpes

The virus enters cracks or cuts in the membrane surface, and then multiplies in basal and epithelial cells in the immediate vicinity. This is followed by inflammation, edema, cell lysis, degeneration, and a ballooning effect that produces a characteristic thin-walled vesicle. The main disease of *Herpes simplex* viruses are facial herpes (oral, optic, and pharyngeal), genital herpes, neonatal herpes, and disseminated disease.

Type 1 *Herpes simplex*—*Herpes labialis*, otherwise known as fever blisters or coldsores, is the most common recurrent HSV-1 infection. Vesicles usually crop up on the mucocutaneous junction of the lips or on adjacent skin. A few hours of tingling or itching precede the formation of one or more vesicles at the site. Pain is most acute in the earlier stages as vesicles ulcerate; lesions crust over in two or three days and heal completely in a week. Although most primary herpetic infections are mild or asymptomatic, an occasional manifestation known as herpetic *gingivostomatitis*, can occur. This infection of the oropharynx, usually due to HSV-1, strikes young children most frequently. Inflammation of the oral mucosa can involve the gums, tongue, soft palate, and lips, sometimes with ulceration and bleeding. A common complication in adolescents is pharyngitis, a syndrome marked by sore throat, fever, chills, swollen lymph nodes, and difficulty in swallowing. Herpetic keratitis (also called ocular herpes) is an infective inflammation of the eye in which a latent virus travels into the ocular rather than the mandibular branch of the trigeminal nerve. Preliminary symptoms are a gritty feeling in the eye, conjunctivitis, sharp pain, and sensitivity to light. Some patients develop characteristic branched or opaque corneal lesions as well.

Type 2 *Herpes* infections—Type 2 herpes infection usually coincides with sexual maturation and an increase in sexual encounters. **Genital herpes** (*Herpes genitalis*) starts out with malaise, anorexia, fever and bilateral swelling and tenderness in the groin. Then, intensely sensitive vesicles break out in clusters on the genitalia, perineum, and buttocks. The chief symptoms are urethritis, painful urination, cervicitis, itching, and central nervous system involvement. After a period of days to weeks, the vesicles ulcerate and develop a light-gray exudate before healing. Bouts of recurrent genital herpes in either sex are usually less severe than the original infection and are triggered by menstruation, stress, and concurrent bacterial infection.

Herpes of the newborn—However, in the neonate and fetus, HSV infections are very destructive and in some cases can be fatal. Most cases of infection in this group occur when infants are contaminated by the mother's reproductive tract immediately prior to or during birth. Neonatal infections have also been traced to hand transmission from the mother's lesions to the baby. Because type 2 is more often associated with genital infection, it is more frequently involved; however, type 1 infection has similar complications. In infants whose disease is confined to the mouth, skin, or eyes, the mortality rate is 30%, but disease affecting the central nervous system has a 50-80% mortality rate. Very rarely, congenital infections arise when viruses migrate from the infected birth canal to the placenta and fetus.

Miscellaneous *Herpes* infections—When either form of herpes virus enters a break in the skin, a local infection can develop. This route of infection is most often associated with occupational exposure or severely damaged skin. A hazard for health care workers who handle patients or their secretions without hand protection is a disease called herpetic whitlow. Workers in the fields of obstetrics, gynecology, dentistry, and respiratory therapy are probably at greatest risk for contracting it. Whitlows are deepset, usually occur on one finger, and are extremely painful and itchy.

Life-threatening complications—The infection disseminates along nerve pathways to the brain or spinal cord. The effects on the CNS begin with headache and stiff neck, and can progress to mental disturbances and coma. The fatality rate in untreated cases is 70%. Patients with underlying immunodeficiency are more prone to severe, disseminated herpes infection. Of greatest concern are patients receiving organ grafts, cancer patients on immunosuppressive therapy, those with congenital immunodeficiencies, and AIDS patients. For the past 30 years, a considerable amount of evidence has been mounting to implicate Type 2 HSV in cervical carcinoma.

Diagnosis and control—Small, painful, vesiculating lesions on the mucous membranes of the mouth or genitalia, lymphadenopathy, and exudate are typical diagnostic symptoms of *Herpes simplex*. Further diagnostic support is available by examining scrapings from the base of such lesions stained with Giemsa, Wright, or Papanicolaou (Pap) methods. The presence of multinuclear cells, giant cells, and intranuclear eosinophilic inclusion bodies can help establish herpes infection. This method, however, will not distinguish among HSV-1, HSV-2, or other herpes viruses that require more specific subtyping. Laboratory culture and specific tests are essential for diagnosing immunosuppressed and neonatal patients with severe, disseminated herpes infection. A specimen of tissue or fluid is introduced into a primary cell line such as monkey kidney or human embryonic kidney tissue cultures and is then observed for cytopathic effects within 24 to 48 hours. Direct confirmatory tests using fluorescent antibodies or DNA probes can differentiate among HSV-1, HSV-2, and closely related herpesviruses. Serologic analysis is useful for primary infection but is inconclusive for recurrent illness, because only 10% of patients show increased antibody titers. Several agents are available for treatment. Acyclovir (Zovirax) is the most effective therapy developed to date. This drug, in use since the early 1980s, is relatively nontoxic, yet shows high specificity for the *Herpes simplex* virus. Topical medications applied to primary genital and oral lesions cut the length of infection and reduce viral shedding. In addition, systemic therapy is available for more serious complications. The more toxic drugs idoxuridine and trifluridine have been used for topical treatment of herpes keratitis, and vidarabine has been used for complications of disseminated herpes. Over-the-counter coldsore medications containing menthol, camphor, and local anesthetics lessen pain and may protect against secondary bacterial infections, but they probably do not affect the progress of the viral infection. A surprising over-the-counter measure for suppressing cold sores is a tablet or two of the amino acid lysine, taken orally in the earliest phases of recurrence. Although its mechanism is not understood, it can apparently abort the development of the lesions.

Varicella-Zoster Virus

Another herpes virus causes both varicella (L. *Variola*), commonly known as chickenpox, and a recurrent infection called *Herpes zoster* (Gr. *Zoster*, girdle) or shingles (L. *Cingulus*, cinch). Since the same virus causes two forms of disease, it is known by a composite name, varicella-zoster virus (VZV). At one time, the differences in clinical appearance and time separation between varicella and *Herpes zoster* were sufficient to foster the notion that they were caused by different infectious agents. But about 100 years ago, the observation that young children acquired chickenpox from family members afflicted with shingles gave the first hint that the two diseases were related.

Epidemiology—The virus is harbored in the respiratory tract, but it is communicable from both respiratory droplets and the fluid of active skin lesions. The dried scabs are not infectious because the virus is unstable and loses infectivity when exposed to the environment. Patients with shingles are a source of infection for nonimmune children, but the attack rate is significantly lower than with exposure to chickenpox. Only in rare instances will a child develop chickenpox more than once, and even a subclinical case can result in long-term immunity.

Varicella (Chickenpox)—The respiratory epithelium serves as the chief portal of entry and the initial site of viral replication for the chickenpox virus, which is initially without symptoms. Regional lymph nodes and other undetermined body sites probably support secondary multiplication before the virus is detectable in the blood. After an incubation period of 10 to 20 days, the first symptoms to appear are fever and an abundant rash that begins on the scalp, face and trunk, and radiates in sparse crops to the extremities. Skin lesions progress quickly from macules and papules to itchy vesicles that encrust and drop off, usually healing completely but sometimes leaving a tiny pit or scar.

***Herpes zoster* (Shingles)**—In an unknown percentage of individuals, recuperation from varicella is associated with the entry of the varicella-zoster virus into the sensory endings that innervate dermatomes, regions of the skin supplied by the cutaneous branches of nerves, especially the thoracic and trigeminal nerves. Like HSV, VZV migrates toward the central nervous system and becomes established in the sensory ganglia in latent form. In contrast to Herpes simplex, however, zoster generally recurs only once or possibly twice, with a characteristic asymmetric distribution on the skin of the trunk or head. Shingles develops abruptly following reactivation by such stimuli as X-ray treatments, immunosuppressive and other drug therapy, surgery, or developing malignancy. In fact, up to 50% of patients suppressed by radiation therapy, leukemia, or similar malignancies manifest recurrence. The virus is believed to migrate down the ganglion to the skin where multiplication resumes and produces crops of tender skin vesicles typical of varicella, except that they can persist for weeks. Inflammation of the ganglia and the pathways of intercostal nerves can cause pain and tenderness (radiculitis) that can last for several months.

Diagnosis and control—Supportive evidence for a diagnosis of varicella usually comes from a recent history of close contact with a source or with an active case of VZV infection. A diagnosis of shingles is supported by the distribution pattern of skin lesions and a history of varicella. Further laboratory confirmation is achieved by identifying multinucleate giant cells in stained smears prepared from vesicle scrapings. However, unequivocal identification or differentiation of *Herpes simplex* form *Herpes zoster* is best done with fluorescent antibody detection of viral antigen in skin lesions. To prevent secondary bacterial infection that can cause dangerous complications, anaesthetic and antimicrobial ointments are applied. The use of aspirin, though, is contraindicated because of the apparent risk of Reye's syndrome in children and young adults. Drugs that can be therapeutically beneficial for systemic disease are intravenous acyclovir and high dose interferon. Zoster immune globulin (ZIG) and varicella-zoster immune globulin (VZIG) are passive immunotherapies that do not entirely prevent disease but can diminish its complications. Japanese researchers have developed a live, attenuated vaccine that appears to be effective and has recently been approved for use in the United States.

The Cytomegalovirus Group

Another group of herpesviruses, the cytomegaloviruses (Gr. *Cyto,* cell, and *megale,* big, great) (CMV), are named for their tendency to cause a cytopathic effect that produces giant cells and formation of nuclear and cytoplasmic inclusions. These viruses also termed salivary gland virus and cytomegalic inclusion virus, are among the most ubiquitous pathogens of humans.

Epidemiology—Studies in the United States and the United Kingdom revealed infection in 7.5% of newborns, making CMV the most prevalent viral infection in the fetus. Additional surveys that include other countries show that 20% to 100% of women of child-bearing age had CMV antibodies. Other groups with a high carrier rate are IV drug abusers and male homosexuals, many of whom are persistent shedders. Although cytomegalovirus is not noted for being highly communicable, it appears to be transmitted in saliva, respiratory mucus, milk, urine, semen, cervical secretions and feces. Transmission usually involves intimate exposure such as sexual contact, vaginal birth, transplacental infection, blood transfusion, and organ transplantation. As with other herpes viruses, CMV is commonly carried in a latent state in various tissues.

Infection—Three groups that develop a more virulent form of disease are fetuses, newborns, and immunodeficient adults. The incidence of congenital CMV infection is high, occurring in about 20% of pregnancies in which the mother has concurrent infection. Although most infected newborns are born without signs, a certain number exhibit enlarged liver and spleen, jaundice, capillary bleeding, microcephaly, and ocular inflammation. In some cases damage can be so severe and extensive that death follows within a few days or weeks. Babies who survive develop long-term neurological sequelae, including hearing and visual disturbances and even mental retardation. In the United States, this happens to about 5,000 babies a year.

Cytomegalovirus mononucleosis (Gr. *Mono,* one, and *nucleo,* nucleus) is a syndrome characterized by fever and lymphocytosis, somewhat similar to the disease caused by Epstein-Barr virus. Although it can occur in children, it is chiefly an illness of adults. Disseminated cytomegalovirus infection can precipitate a medical crisis. It is a common opportunist of AIDS patients, in whom it produces overwhelming systemic disease, with fever, severe diarrhea, hepatitis, pneumonia, and high mortality. One of its more unfortunate targets in these patients is the retina. It forms spreading lesions that can lead to blindness. Most patients undergoing kidney transplantation and half of those receiving bone marrow develop CMV pneumonitis, hepatitis, myocarditis, meningoencephalitis, hemolytic anemia, and thrombocytopenia.

Diagnosis and control—In neonates, CMV infection must be differentiated from toxoplasmosis, rubella, *Herpes simplex*; in adults, it must be distinguished from Epstein-Barr infection and hepatitis A and B. The cytomegalovirus can be isolated from pathologic specimens taken from virtually all organs as well as from epithelial tissue. Tissues can be stained to observe typical cell enlargement and prominent inclusions in the cytoplasm and nucleus, even though this is not specific to CMV. More specific direct antigen tests involving ELISA and DNA probes are of some benefit. However, virus culture followed by a rapid monoclonal antibody test against the early nuclear antigen is the most effective diagnostic technique. Some antiviral agents give promising results, but most do not modify the course of infection well enough to be clinically useful. Drugs recently approved for therapy are ganciclovir, foscarnet, and hyperimmune CMV immunoglobulin. Interferon appears, at best, to delay excretion of the virus during therapy, but it shows little or no indication of preventing infection or promoting recovery. None of the other anti-herpes drugs is therapeutically active in CMV infections. The development of a vaccine is hampered by the lack of an animal that can be infected with human cytomegalovirus.

Epstein-Barr Virus

Although the lymphatic disease known as infectious mononucleosis was first described more than a century ago, its cause was finally discovered through a series of accidental events starting in 1958, when Michael Burkitt discovered in African children an unusual malignant tumor (Burkitt's lymphoma) that appeared to be infectious. Later, Micheal Epstein and Yvonne Barr cultured a virus from tumors that showed typical herpes virus morphology. Proof that the two diseases had a common cause was completed when a laboratory technician accidentally acquired mononucleosis while working with the Burkitt's lymphoma virus. The Epstein–Barr (EBV) virus shares morphologic and antigenic features with other herpes viruses, and in addition, it contains a circular form of DNA that is readily spliced into the host cell DNA, thus transforming cultured lymphocytes.

Epidemiology—Salivary excretion occurs in about 20% of otherwise normal individuals. Direct oral contact and contamination with saliva are the principal modes of transmission, though fomites and arthropod vectors may play a role. Transmisson by means of blood transfusions and organ transplants is also possible. The nature of infection depends on the patient's age at first exposure, socioeconomic level, geographic region, and genetic predisposition. In less-developed regions of the world, infection rates are generally higher and infection occurs at an earlier age. Children living in east and west central Africa are predisposed to Burkitt's lymphoma, and residents of parts of China and North Africa show high levels of nasopharyngeal carcinoma, another tumor linked to EBV. In industrialized countries, exposure to EBV is usually postponed until adolescence and young adulthood. This delay is generally ascribed to improved sanitation, and less frequent and intimate contact with other children. Since nearly 70% of college-age Americans have never had EBV infection, this population is vulnerable to infectious mononucleosis, often dubbed the "kissing disease" or "mono".

Infection—The epithelium of the oropharynx is the portal of entry for EBV during the primary infection. From this site, the virus moves to the parotid gland, which is the major area of viral replication and the source of viremia. After the initial infection, the virus remains latent in the throat and blood for an extended time, periodically being reactivated by some ill-defined mechanism. The symptoms of infectious mononucleosis are sore throat, high fever, and cervical lymphadenopathy, which develop after a long incubation period (30-50 days). Many patients also have a gray-white exudate in the pharynx, a skin rash, and enlarged spleen and liver. A notable sign of mononucleosis is sudden leukocytosis, consisting initially of infected B cells and later to T-suppressor and cytotoxic T cells. The strong, cell-mediated immune response is decisive in controlling the infection and preventing complications.

Tumors and other complications—Epstein-Barr virus is commonly associated with a B-cell malignancy called Burkitt's lymphoma. This cancer usually develops in the jaw and grossly swells the cheek, though it can also occur in the abdomen. Most cases are reported in central African children four to eight years old, and a small number are seen in North America. The prevalence in Africa may be associated with chronic coinfections with other diseases such as malaria that weaken the immune system and predispose the body to tumors. One theory to explain the neoplastic effect is that persistent EBV infection in young children eventually immortalizes certain B lymphocytes. Unlike Burkitt's lymphoma, nasopharyngeal carcinoma is a malignancy of epithelial cells that occurs in older Chinese and African men. Although little about this disease is understood, high antibody titer to EBV is a consistent finding. It is not known how the virus invades epithelial cells (not its usual target), why the incidence of malignancy is so high in these ethnic groups, or whether dietary carcinogens such as nitrosamines in cured fish are involved. In general, any person with an immune deficiency (especially of T cells) is highly susceptible to Epstein-Barr virus. Organ transplant patients carry a high risk because their immune systems are weakened by measures to control rejection. The oncogenic potential of the virus is another concern, since it is currently believed to cause some types of lymphatic tumors in kidney transplant patients and is commonly isolated from tonsilar, laryngeal, thymic, and certain AIDS-related cancers.

Diagnosis and control—A differential blood count that shows lymphocytosis, neutropenia, and large, atypical lymphocytes with lobulated nuclei and vacuolated, basophilic cytoplasm is suggestive of EBV infection. Serologic assays to detect antibody against the capsid and DNA are helpful, as are direct viral antigen tests using probes and lebeled antibodies. The usual treatments for infectious mononucleosis are nonspecific, directed at symptomatic relief of fever and sore throat. Specific antiviral measures for disseminated and transplant-related disease that have shown some promise are intravenous gamma-globulin, interferon, acyclovir, and monoclonal antibodies. Since Burkitt's lymphoma has usually metastasized by the time of diagnosis, systemic chemotherapy with cyclophosphamide, vincristine, methotrexate, or prednisolone can induce remission.

Other Herpesviruses

The latest addition to the herpesvirus family is human herpesvirus-6 (HHV-6), also known as human B-lymphotropic virus. It was originally isolated from infected lymphocytes and has characteristics similar to EBV, but it is genetically distinct from the other five herpesviruses. Although it can be a dangerous opportunist in leukemia patients, it is primarily involved in a common childhood infection called roseola. This disease, occurring in children between 3 months and 4 years, begins acutely with fever that can spike as high as 105°F (40°C) and cause febrile convulsions. A few days later it is followed by a faint maculopapular rash over the neck, trunk, and buttocks. Because certain herpesviruses are known to be carcinogenic, researchers continue to search for their connections to various unexplained cancers. For example, Kaposi's sarcoma often contains CMV or its antigens, suggesting the possible role of herpesvirus in cancer. It must be emphasized, however, that isolating the virus from the tumor alone is not sufficient proof and that further evidence is required. Scientists of California and New York (1994) carried the link of herpes virus and cancer one step further. They discovered viral DNA in a majority of Kaposi's sarcoma tumors that was not present in non-cancerous tissues from the same patient or other uninfected patients. When the viral DNA was analyzed, it contained fragments of genes identical to herpesviruses. Other mammals are also susceptible to herpesvirus infections. In cases of simian herpesviruses, monkeys can pass the infection to humans, often with fatal results.

Hepadnaviruses

One group of enveloped DNA viruses, called *hepadnaviruses* (Gr. *Hepatos,* liver, plus DNA virus) is quite unlike any other so far discovered. These viruses have never been grown in tissue culture and have an unusual genome containing both double- and single-stranded DNA (Fig. 16.2). Hepadnaviruses show a decided tropism for the liver, where they persist and can be a factor in liver cell carcinoma. One member of the group, hepatitis B virus (HBV), causes a common form of hepatitis, and other members cause hepatitis in woodchucks, ground squirrels, and Peking ducks.

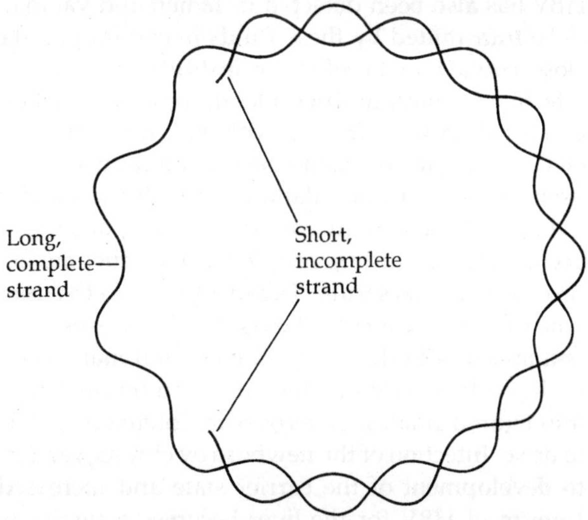

Fig. 16.2 : DNA strand of hepatitis B Virus.

Viral hepatitis—When certain viruses infect the liver, they cause hepatitis, an inflammatory disease marketed by necrosis of hepatocytes and a mononuclear response that swells and disrupts the liver architecture. This pathological change interferes with the liver's excretion of bile pigments such as bilirubin into the intestine. When bilirubin, a greenish-yellow pigment accumulates in the blood and tissues, it cause jaundice, (Fr. *Jaunisse,* yellow) a yellow tinge in the skin and eyes. Hepatitis A virus (HAV) is an RNA enterovirus that is the etiologic agent of infectious hepatitis. In general, HAV disease is far milder, shorter term, and less virulent than the other forms.

Another important agent is hepatitis C virus, a recently discovered RNA virus in the togavirus family that causes most cases of infusion hepatitis. These two viruses have no relationship to hepatitis B virus (HBV), a DNA virus that causes serum hepatitis, which will be covered in this chapter. An unusual virus, hepatitis D or delta agent, is a defective RNA virus that cannot produce infection unless a cell is also infected with HBV. This virus has an unusual nucleocapsid that is similar to viroids and it invades host cells by "borrowing" the outer receptors of HBV. When HBV infection is accompanied by the delta agent, the disease becomes more severe and is more likely to progress to permanent liver damage.

Properties – Because a cell culture system for propagating hepatitis B virus is not yet available, by far the most significant source of information about its morphology and genetics has been obtained from viral fragments, intact virions, and antibodies, which are often abundant in patients' blood. One of these blood-borne pieces, the Dane particle, has been suspected as the infectious virus particle, and modern serologic and electron microscope studies support this view.

Epidemiology – An important factor in the transmission pattern of hepatitis B virus is that it multiplies exclusively in the liver, which continuously seeds the blood with viruses. Electron microscopic studies have revealed up to 10^7 virions per milliliter of infected blood. Even a minute amount of blood (10^{-6} to 10^{-7} ml) can transmit infection. The abundance of circulating virions is so high and the minimal dose so low that such simple practices as sharing a toothbrush or a razor can transmit infection. Over the past 10 years, HBV has also been detected in semen and vaginal secretions, and it is currently believed to be transmitted by these fluids in certain populations. Spread of the virus by means of close contact in families or institutions is also well documented. The virus remains infective for days in dried blood, for months when stored in serum at room temperature, and for decades if frozen. Although it is not inactivated after four hours of exposure to 60°C, boiling for the same period can destroy it. Hepatitis B is an ancient disease that has been found in all populations, though the incidence and risk are higher among people living under crowded conditions, drug addicts, the sexually promiscuous, and certain occupations, including people who conduct medical procedures involving blood or blood products such as serum. The incidence of HBV infection and carriage among drug addicts who routinely share needles is very high, especially if blood is pulled back into the syringe and mixed with the drug. Homosexual males constitute another high-risk group because such practices as anal intercourse can traumatize membranes and permit transfer of virus into injured tissues. Heterosexual intercourse can also spread infection, but is less likely to do so. Infection of the newborn by chronic carrier mothers occurs readily and predisposes to development of the carrier state and increased risk of liver cancer in the child. The tropism of HBV for the liver becomes a significant epidemiologic complication in some people (5–10% of infectees). Persistent or chronic carriers harbor the virus for up to six months, and some continue to shed the virus for many years. Worldwide, 300 million people are estimated to carry the virus, but it is most prevalent in Africa, Asia, and the western Pacific and least prevalent in North America and Europe.

Pathogenesis – Eventually it reaches the liver cells (hepatocytes), where it multiplies and releases viruses into the blood (viremia). After an incubation period of 4 to 24 weeks (7 weeks average), some people experience malaise, fever, chills, anorexia, abdominal discomfort and diarrhea. Surprisingly, the majority of those infected exhibit few overt symptoms and eventually develop an immunity to HBV. In those who do experience disease, the severity of symptoms and the aftermath of hepatic damage vary widely. Fever, jaundice, rashes, and arthritis are common reactions, and a smaller number of patients develop glomerulonephritis and arterial inflammation. In 90% of patients complete liver regeneration and restored function occur. But for reasons that are not entirely clear, a small number of predisposed older persons develop chronic liver disease in the form of necrosis or cirrhosis. The association of HBV with hepatocellular carcinoma is based on these observations : (a) Certain hepatitis B antigens are found in malignant cells and are often detected as integrated components of the host genome; (b) persistent carriers of the virus are more likely to develop this cancer; and (c) young men from areas of the world with a high incidence of hepatitis B (Africa and the Far East) are more frequently affected. This connection is probably a result of infection early in life and the long-term carrier state.

Diagnosis and management – A clinician can usually distinguish hepatitis B from hepatitis A by carefully examining the possible risk factors. Some combination of occupational hazard, drug abuse, age, relatively long incubation period, and insidious onset strongly suggests HBV infection. Serological tests are increasingly important. Testing procedures fall into two categories : Those that detect virus antigen and those that assay for antibodies. Recent developments in RIA and ELISA testing permit detection of the surface antigen very early in the infection. These same tests are essential for screening blood destined for transfusions, semen in sperm banks, and organs intended for transplant. Mothers who are carriers of HBV are highly likely to transmit infection to their newborns at the time of birth. It is not currently known whether transplacental HBV infection occurs. Unfortunately, infection with this virus early in life is a known risk factor for chronic liver disease. In light of the increasing incidence of HBV-positive pregnant women, routine testing of all pregnant women is now strongly supported by the CDC and a majority of obstetricians. If adopted, mandatory hepatitis testing would add to the growing number of procedures used to screen prospective mothers for infection with agents that have especially profound risks for fetuses and/or neonates.

An exciting new treatment for chronic infection is the injection of synthetic interferon, which stops virus multiplication and prevents liver damage in up to 60% of patients. Control measures are currently focused on prevention. Passive immunization with hepatitis B immune globulin (HBIG) gives significant protection to people who have been exposed to the virus through needle puncture, broken blood containers, or skin and mucosal contact with blood. Other groups for whom prophylaxis is highly recommended are the sexual contacts of activity infected people or carriers and neonates born to infected mothers. The first boon to curbing hepatitis B was a vaccine developed and tested in the late 1970s and committed to general use in 1981. Because HBV cannot be cultivated in tissue cultures or embryos, the source of antigen for this first vaccine was the blood of carriers with high levels of the desired surface antigen. The blood is treated to purify and concentrate the antigen and heated to inactivate any intact viruses. Alternate vaccines (Recombivax, Energix) containing the pure surface antigen cloned in yeasts have largely replaced the original vaccine, except in cases of allergy to yeast. Vaccination is a must for medical and dental workers and students, patients receiving multiple transfusions, immunodeficient persons, and cancer patients.

NONENVELOPED DOUBLE STRANDED DNA VIRUSES

The Adenoviruses

During a study seeking the cause of the common cold, a new virus was isolated from *adenoids* (Gr. *Adeno,* gland, and *eidos,* form) that had been surgically excised from young children. It turned out that this virus, termed adenovirus, was not the sole agent of colds but one of several others. It was also the first of about 80 strains of nonenveloped, double-stranded DNA viruses to be discovered and eventually classified as adenoviruses. Of this array, about 30 types are associated with human infection, and the rest are animal pathogens. Besides infecting lymphoid tissue, adenoviruses have a preference for the respiratory and intestinal epithelia and the conjunctiva. Adenoviruses produce aggregations of incomplete and assembled viruses that usually lyse the cell. In cells that do not lyse, the viral DNA may be harbored latently in the nucleus.

Epidemiology—Most cases of conjuctival infection involve preexisting eye damage and contact with contaminated sources such has swimming pools, dusty working places (shipyards and factories), and unsterilized opitical instruments (ocular tonometers used in the glaucoma test). Infection by an adenovirus usually occurs by age 15 in most people, and certain individuals, become chronic respiratory carriers. Although the reason is obscure, military recruits are at greater risk for infection than the general population, and respiratory epidemics are common in armed services installations.

Diseases of adenoviruses—The patient infected with an adenovirus is typically feverish, with acute rhinitis, cough, and inflammation of the pharynx, enlarged cervical lymph nodes, and a macular rash. Certain adenoviral strains produce an acute follicular lesion of the conjunctiva. Usually one eye is affected by a watery exudate, redness, and partial closure.

Acute hemorrhagic cystitis in children has been traced to an adenovirus. This self-limiting illness of four to five days duration is marked by hematuria (blood in the urine), frequent and painful urination with occasional episodes of fever, bedwetting, and suprapubic pain. Because the virus apparently infects the intestinal epithelium and can be isolated from some diarrheic stools, adenovirus is also suspected as an agent of infantile gastroenteritis.

Papovaviruses

Letters from the terms papilloma (L. *Papilla,* pimple, and Gr. *oma,* tumor) polyoma (L. *Poly,* many) and simian vacuolating (L.*Vacuus,* empty) viruses gave rise to the term papovavirus. This family contains two subtypes : The papillomaviruses, which are responsible for human and animal papillomas, and the polyomaviruses, which include several human (BK and JC viruses) and animal pathogens.

Epidemiology of papillomaviruses—Some types are specific for the mucous membranes, while others invade the skin. The appearance and seriousness of the infection vary somewhat from one anatomical region to another. Painless, elevated, rough growths on the fingers and occasionally on other body parts are called common, or seed, warts. These commonly occur in children and young adults. Plantar warts are deep, painful papillomas on the soles of the feet; flat warts are smooth, skin-coloured lesions that develop on the face, trunk, elbows, and knees. A special form of verruca known as genital warts is a prevalent STD and is linked to some types of cancer. Warts are transmissible through direct contact with a wart or contaminated fomites; and they can also spread on the same person by autoinoculation.

Insidious papilloma—Although this disease is not new, case reports have increased to such an extent that it is now regarded as the most common STD in the United States. Studies show that up to 15% of the population are carriers of one of the five types of HPV associated with genital warts. The virus invades the external and internal genital membranes, especially in areas prone to friction in the vagina and the head of the penis. Wart morphology ranges from tiny, flat, inconspicuous bumps to giant, branching, cauliflower-like masses called condylomata acuminata (Gr. *Kondyloma*, knob, and L. *acuminatus*, sharp).

The most disturbing aspect of the sudden increase in HPV infection is its strong association with cancer of the reproductive tract, especially the cervix and penis. Two viral types—16 and 18—can be isolated from a majority of female genital cancers and are known to convert some forms of genital warts into malignant tumors. Susceptiblity to this infection and its potential for cancer can be greatly reduced by early detection and treatment of both the patient and the sexual partner. Such early diagnosis depends on thorough inspection of the genitals and, in women, a Papanicolaou smear to screen for abnormal cervical cells.

Diagnosis and control—The warts caused by papillomaviruses are usually distinctive enough to permit reliable clinical diagnosis without much difficulty. However, a biopsy and histologic examination can help clarify ambiguous cases. A sensitive, DNA probe can detect the HPV type and possibly assess the risk of developing cervical cancer. Although most common warts regress over time, they cause sufficient discomfort and cosmetic concern to require treatment. Treatment strategies for all types of warts include direct chemical application of podophyllin and physical removal of the affected skin or membrane by cauterization, freezing, or laser surgery.

Epidemology of polyomaviruses—Among the most important human polyomaviruses are the JC virus and BK virus. Although many polyomaviruses are fully capable of transforming host cells and inducing tumors in experimental animals, none causes cancer in its natural host. On the basis of serological surveys, it appears that infections by JC virus and BK virus are commonplace throughout the world because the majority of infections are asymptomatic or mild, not a great deal is known about their mode of transmission, portal of entry, or target cells, though urine and respiratory secretions are suspected as possible sources of the viruses. The main complication of infection occur in patients who have cancer or have been immunosuppressed. Progressive multifocal leukoencephalopathy (Gr. *Leuko,* white, *cephalos,* brain, and *pathos* disease) (PML) is an uncommon but generally fatal infection by JC virus that attacks accessory brain cells and gradually demyelinizes certain parts of the cerebrum. Infection with BK viruses, usually associated with renal transplants, causes complications in urinary function. The prospects of finding effective therapy for these diseases are bleak. By the time PML is diagnosed, by the patient is irreversibly immunocompromised and extensive brain damage has already occurred.

NONENVELOPED SINGLE STRANDED DNA VIRUSES

The parvoviruses (L. *Parvus,* small) are unique among the viruses in having single stranded DNA molecules. They are also notable for their extremely small diameter (18-26 nm) and genome size. Parvoviruses are indigenous to and can cause disease in several mammalian groups, for example, distemper in cats, an enteric disease in adult dogs and potentially fatal cardiac infection in puppies.

The most important human parvovirus is B19, the cause of erythema contagiosum, a common infection in children. Often the infection goes unnotices, though the child may have a low grade fever and bright red rash on the cheeks. This same virus can be more dangerous in children with immunodeficiency or sickle cell anaemia, because it destroys red blood stem cells. If a pregnant woman contracts infection and transmits it to the fetus, the severe and often fetal anaemia can result. One type of parvovirus the adeno-associated virus (A-AV) is another example of defective virus that can not replicate in host cell without the function of a helper virus.

ENVELOPED SEGMENTED SINGLE STRANDED RNA VIRUSES

Influenza

Influenza viruses are spherical particles, size 80-120 nm and are covered with a lipoprotein envelope that is studded with glycoprotein spikes acquired during viral maturation (Fig. 16.3). The two glycoproteins that comprise the spikes of the envelope and contribute to virulence are hemagglutinin (HA) and neuraminidase (NA). Because hemagglutinin spikes combine with a specific carbohydrate molecule found in all eukaryotic cell membranes, the virus has the capacity to bind and clump a variety of animal cells. The name *hemagglutinin* is derived from its particular agglutinating action on red blood cells. The property of erythrocyte agglutination is the basis for viral assays that have been used to identify several antigenic types. Hemagglutinin contributes to infectivity by binding to host cell receptors of the respiratory mucosa, a process that facilitates viral penetration. Neuraminidase has the principal effect of hydrolyzing the protective mucous coating of the respiratory tract, assisting in viral budding and release, keeping viruses from sticking together, and participating in host cell fusion.

The genome of the influenza virus is known for its extreme variability. The influenza virus genome has the unique property of being divided into eight separate segments that can reassort in various ways during viral assembly. Constant changes in antigenic expression create new influenza strains and give rise to epidemiologic patterns known as *antigenic shift* and *antigenic drift.* Antigenic shift is the sudden emergence of a new viral strain due to recombination of the genome segments from different viruses.

Epidemiology—Influenza is an acute, highly contagious respiratory illness afflicting people of all ages, and marked by seasonal regularity and pandemics at predictable intervals. So convinced were the ancients that this disease was influenced by a celestial alignment of the stars that one of its first names was an Italian derivative of *un influenza di freddo,* alluding to a vague and mysterious influence. This name has remained and is now contracted to the vernacular "flu." Periodic outbreaks of respiratory illness that were probably influenza have been noted throughout recorded history. Of the three types of influenza viruses, the more virulent type A is most commonly associated with human disease.

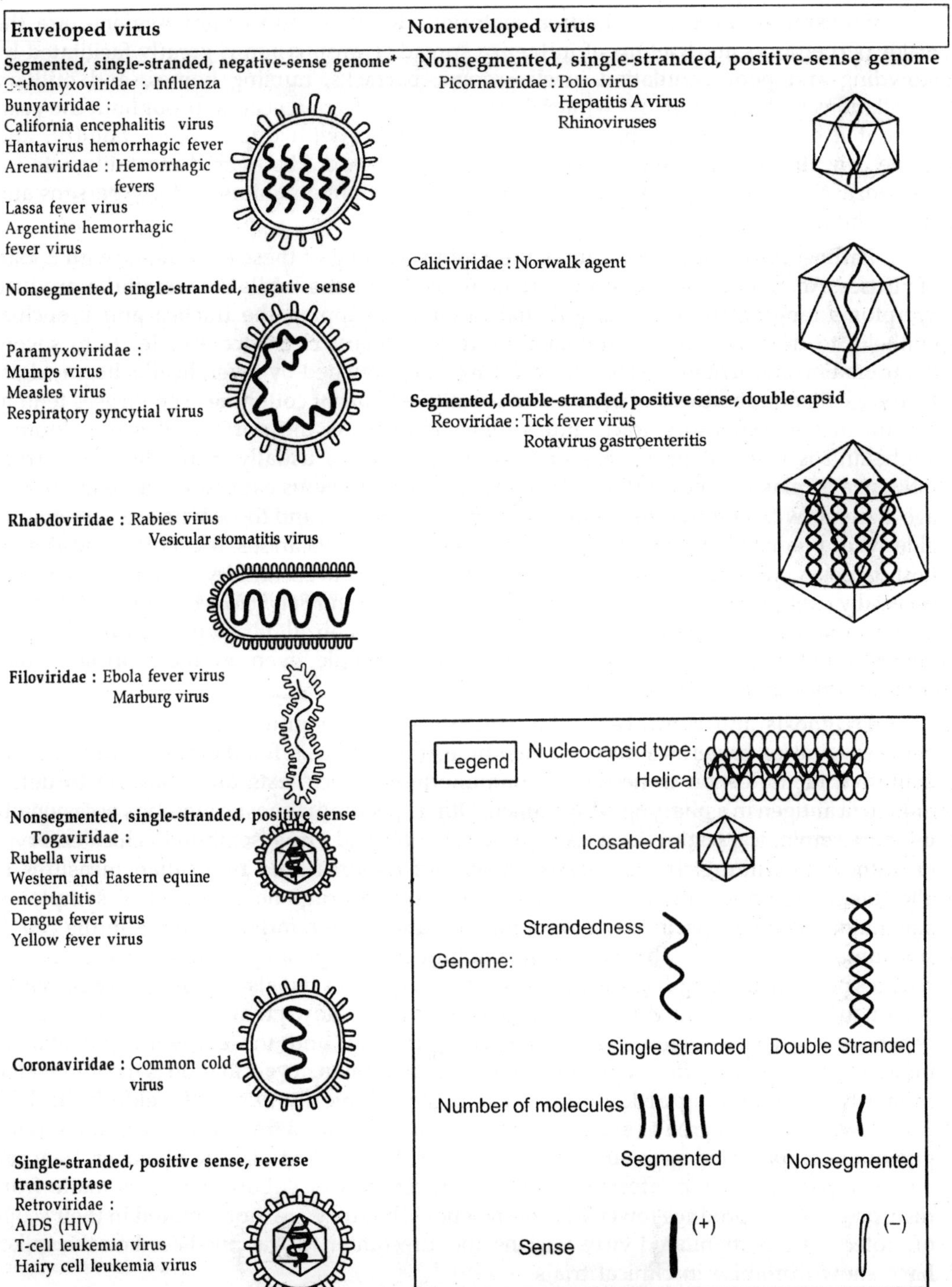

Fig. 16.3 : RNA viruses family.

Transmission—Young children, notorious for their nonchalant hygiene, are the major perpetrators and victims of influenza spread. Transmission is greatly facilitated by crowding and poor ventilation in classrooms, barracks, nursing homes, and military installations in the late fall and winter. In the tropics, infection occurs throughout the year, without a seasonal distribution. Occupational contact with ducks, other poultry, and swine constitutes a special high-risk category for disease. The overall mortality rate for influenza A is about 0.1% of cases. Deaths are most common among elderly persons and small children.

Pathogenesis—Infection causes the rapid shedding of these cells along with a load of viruses. About onehalf of infections remain asymptomatic, while the remainder produce symptoms ranging from inconsequential to lethal, stripping the trachea and bronchial epithelia to the basal layer eliminates protective ciliary clearance and leads to severe inflammation and irritation. The illness is further aggravated by fever, headache, myalgia, pharyngeal pain, shertness of breath and repetitive bouts of coughing. The viruses tend to remain in the respiratory tract, and viremia is rare. As the normal ciliated columnar epithelium is restored in a week or two, the symptoms usually abate. Recovery from influenza is related to general health, immunity from previous exposure, vaccination, and age. Patients with emphysema or cardiopulmonary disease, and those who are very young, elderly, or pregnant, and more susceptible to serious complications. Most often, weakened host defenses and immune function predispose patients to secondary bacterial infections, especially pneumonia caused by *Streptococcus pneumoniae* and *Staphylococcus aureus*. Although less common that bacterial pneumonia, primary influenzal virus pneumonia has the ominous reputation for causing rapid death of people even in their prime. Other complications are bronchitis, meningitis, and neuritis.

Diagnosis and control—An initial Gram stain of sputum can often differentiate between viral or bacterial infection, thereby forming a basis for therapy. If rapid, more definitive identification is necessary, immunofluorescence tests are available to detect influenza antigen in a pharyngeal specimen. Other specific diagnostic methods occasionally used are serological testing for a rising antibody titer and virus isolation in chick embryos or kidney cell culture. In most cases, treatment requires little more than measures to alleviate symptoms. Fluids, bed rest, and nonaspirin drugs help to control symptoms. Amantadine, which specifically blocks the uncoating of the influenza virus in the cell, is sometimes prescribed to abate the course of influenza, especially in more severe cases. It reduces not only the length of the disease and its symptoms but also the spread of the virus. Amantadine is given orally, though recent studies have shown that aerosol sprays may be beneficial as well. Given the lack of specific therapy and the epidemic potential of influenza, the need for effective prophylaxis is great. Influenza vaccination is recommended primarily for certain highrisk groups, such as the chronically ill and the elderly, and for people with a high degree of exposure to the public. The standard vaccines contain viruses grown in embryonated eggs and have an overall effectiveness of 70%; however, they can be accompanied by side effects of chills and fever. Newer subunit vaccines made from disrupted viruses produce fewer adverse reactions, but they are more limited in their scope of protection. An attenuated virus vaccine and a recombinant vaccine based on the spikes have shown promise in clinical trials.

Bunyaviruses and Arenaviruses

Between them, the bunyavirus (from Bunyamwera, a African area) and arenavirus

(L. *Arena,* sand) groups contain hundreds of viruses whose normal hosts are arthropods or rodents. Although none of the viruses are natural parasites of humans, they can be transmitted zoonotically; occasionally leading to rampant epidemics in human populations. Bunyaviruses are discussed in a later section along with other arboviruses because they are transmitted primarily by insects and ticks. Some of the more important bunyavirus diseases are California encephalitis, spread by mosquitos; Rift Valley fever, an African disease transmitted by sand flies; and Korean hemorrhagic fever associated with hantavirus and transmitted by rodents. Four major arenavirus diseases are Lassa fever, found in parts of Africa; Argentine hemorrhagic fever, typified by severe weakening of the vascular bed, hemorrhage, and shock; Bolivian hemorrhagic fever; and lymphocytic choriomeningitis, a widely distributed infection of the brain and meninges. The viruses are closely associated with a rodent host and are continuously shed by the rodent throughout its lifetime. Transmission of the virus to humans is through aerosols and direct contact with the animal or its excreta. These diseases vary in symptomology and severity from mild fever and malaise to complications such as hemorrhage, renal failure, cardiac damage, and shock syndrome.

ENVELOPED NON SEGMENTED SINGLE STRANDED RNA VIRUSES

Paramyxoviruses

The important human paramyxoviruses are *Paramyxovirus* (parainfluenza and mumps viruses), *Morbillivirus* (measles virus), and *Pneumovirus* (respiratory syncytial virus), all of which are readily transmitted through respiratory droplets. The envelope of a paramyxovirus possesses HN spikes and F glycoprotein spikes that allow it to infect neighboring cells by a novel mechanism. First, the cell membrane of an infected cell is modified by insertion of the spikes. The HN spikes immediately bind an uninfected neighboring cell, and in the presence of F spikes, the two cells permanently fuse. A chain reaction of multiple cell fusions then produces a *syncytium* (Gr. *Syn,* together, and *kytos,* cell) or multinucleate giant cell, with cytoplasmic inclusion bodies.

Epidemiology—One type of infection by *Paramyxovirus,* called parainfluenza, is as widespread as influenza but usually more benign. Primary infection with parainfluenza virus occurs throughout the year, with seasonal peaks in late fall and winter. Droplets and respiratory secretions effectively disseminate the virus into the air and onto fomites. Inhalation or inoculation of the mucous membranes by contaminated hands are the usual routes of transmission. Parainfluenzal respiratory disease is seen most frequently in children, most of whom have been infected by the age of six. Because protective transplacental immunity is frequently missing, babies in their first year are particularly susceptible, and they develop more severe symptoms. Characteristic symptoms of parainfluenza are minor upper respiratory disease (a cold), bronchitis, bronchopneumonia, and laryngotracheobronchitis (croup). *Croup* (Scot. *Kropan,* to cry a loud) causes labored and noisy breathing accompanied by a hoarse cough and is most common in infants and young children.

Diagnosis and control—Presenting symptoms typical of a cold are often sufficient to presume respiratory infection of viral origin. Determining the actual viral agent, however, is difficult and usually unnecessary. Whereas in older children and adults infection is usually self-limited and benign; primary infection in infants can be severe enough to be life-threatening. So far no specific chemotherapy is available, but supportive treatment with immune serum globulin or interferon can be of benefit.

Mumps

Another infection caused by *Paramyxovirus* is mumps. So distinctive are the pathologic features of this disease that Hippocrates clearly characterized it several hundred years B.C. as a self-limited, mildly epidemic illness associated with painful swelling at the angle of the jaw. Also called epidemic parotitis, this infection typically targets the parotid salivary glands, but it is not limited to this region.

Epidemiology—It is communicated primarily through salivary and respiratory secretions, though it is less contagious than measles or chickenpox viruses. Infection occurs worldwide, with epidemic increases in the late winter and early spring in temperate climates. High rates of infection arise among crowded populations or communities with poor herd immunity. Most cases occur in children under the age of 15, and as many as 40% are subclinical. Because lasting immunity follows any form of mumps infection, no long-term carrier reservoir exists in the population. After an average incubation period of two to three weeks, symptoms of fever, nasal discharge, muscle pain, and malaise develop. These may be followed by inflammation of the salivary glands (especially the parotids), producing the classic gopherlike swelling of the cheeks on one or both sides. This swelling of the gland and its duct can induce considerable discomfort, especially when the gland is stimulated. Viral multiplication in salivary glands is followed by viremia and invasion of other organs, especially glands. Such diverse structures as the testes, ovaries, thyroid gland, pancreas, meninges, heart, and kidney can be involved.

Complications—The resultant syndrome of orchitis and epididymitis may be rather painful, but no permanent damage occurs. The popular belief that mumps readily causes sterilization of the postpubertal male is still held, despite medical evidence to the contrary. Perhaps this notion has been reinforced by the excruciating pain experienced during testicular infection, by the tenderness that continues afterward, and by the partial atrophy of one testis that occurs in about half the cases. Permanent sterility due to mumps is very rare. In mumps pancreatitis, the virus replicates in beta cells and pancreatic epithelial cells. Viral meningitis, characterized by fever, headache, nausea, vomiting, and stiff neck, is rather common in mumps. It appears 2 to 10 days after the onset of parotitis, lasts for 3 to 5 days, and then dissipates, leaving few or no adverse side effects. On rare occasions, loss of hearing develops abruptly in the course of parotitis, usually affecting only one side. Unfortunately, because the virus replicates in the organ of Corti and causes inflammation and irreversible injury, deafness is usually permanent. Any case of mumps, even if it develops on one side only or if it is subclinical leads to permanent immunity.

Diagnosis and control—Mumps can be tentatively diagnosed in a child with swollen parotid glands and a known exposure two or three weeks previously. Since parotitis is not always present and the incubation period can range from 7 to 23 days, a practical diagnostic alternative is to perform a direct fluorescent test for viral antigen or an ELISA test on the serum. The general pathology of mumps is mild enough that symptomatic treatment to relieve fever, dehydration, and pain is usually adequate. A live, attenuated mumps vaccine given routinely as part of the MMR vaccine at 15 months of age is a powerful and effective control agent. A separate single vaccine is available for adults who require protection.

Measles

Measles, an acute disease caused by *Morbillivirus,* is also known as red measles (Dutch *Maselen,* spotted) and rubeola (L. *Ruber,* red). Unfortunately, German measles (rubella), an entirely unrelated viral infection, is also called rubeola in some countries. Despite a tendency to think of measles as a mild childhood illness, several recent outbreaks have reminded us that measles can kill babies and young children, and that it is the seventh most frequent cause of death worldwide.

Epidemiology—Measles is one of the most contagious infectious diseases, transmitted principally by direct contact with respiratory aerosols. Epidemic spread is favoured by crowding, low levels of herd immunity, a prevalence of nonimmune children, malnutrition, and inadequate medical care. Occasional outbreaks of measles have been linked to the lack of immunization in children or the failure of a single dose of vaccine in many children. The most recent statistics show that measles incidence is at an all-time low. There is no reservoir other than humans, and although a person is infectious during the periods of incubation, prodromium, and skin rash, the virus is not carried for any length of time during convalescence. Only relatively large, dense populations of susceptible individuals can sustain the continuous chain necessary for transmission.

Pathogenesis—The incubation period of nearly to weeks ends with symptoms of sore throat, dry cough, headache, conjunctivitis, lymphadenitis, and fever. In a short time, unusual oral lesions called *Koplik's spots* appear as a prelude to the characteristic red maculopapular *exanthem* that erupts on the head and then progresses to the trunk and extremities, until most of the body is covered. Although in most instances, measles is a self-limited infection, complications can be sufficiently severe to cause death in about 1 in 500 children. Among the complications are laryngitis, bronchopneumonia, bronchitis, and bacterial secondary infections such as otitis media and sinusitis. Children afflicted with leukemia or thymic deficiency are especially predisposed to pneumonia due to their lack of the natural T-cell defense. Undernourished children frequently have severe diarrhea and abdominal discomfort that adds to their debilitation.

Diagnosis and control—Clinical characteristics such as dry cough, sore throat, conjunctivitis, lymphadenopathy, fever, and especially the appearance of Koplik's spots and a rash are presumptive of measles. Because specific therapy for measles is not available, treatment relies on reducing fever, suppressing cough, and replacing lost fluid. Complications require additional remedies to relieve neurologic and respiratory symptoms and to sustain nutrient, electrolyte, and fluid levels. Antibiotics are sometimes given for bacterial complications, and large doses of immune globulin can be therapeutic. Vaccination is the most practical, economical and enduring strategy for combating measles. The attenuated viral vaccine is administered by subcutaneous injection and achieves immunity that persists for about 20 years. It must be stressed that since the virus is live, it can cause an atypical infection sometimes accompanied by a rash and fever. Measles immunization is recommended for all healthy children at the age of 15 months (MMR vaccine, with mumps and rubella), and a booster is given prior to entering school. A single antigen vaccine (Meruvax) is also available for older patients who require protection against measles alone.

Respiratory Syncytial Virus

Pathogenesis—As its name indicates, respiratory syncytial virus (RSV), also called *Pneumovirus,* infects the respiratory tract and produces giant multinucleate cells. Outbreaks of droplet-spread RSV disease occur regularly throughout the world, with peak incidence in the winter and early spring. Children six months of age or younger are especially susceptible to serious disease of the respiratory tract. The chance of a youngster being hospitalized for RSV disease before reaching one year of age in estimated to be 5 per 1,000 live births, making RSV the most prevalent cause of respiratory infection in this age group. An estimated total of 91,000 children are hospitalized with RSV infection each year in the United States and approximately 4,500 die from it. The mortality rate is highest for children who have complications like prematurity, congenital disease, and immunodeficiency.

The first symptoms of primary infection are fever that lasts for three days, rhinitis, pharyngitis, and otitis. Infections of the bronchial tree and lung parenchyma gives rise to symptoms of croup that include acute bouts of coughing, wheezing, *dyspnea,* and abnormal breathing sounds (rales). In severe cases compromised breathing can lead to fatalities. Adults and older children with RSV infection can experience cough and nasal congestion but are frequently asymptomatic.

Diagnosis and control—The afflicted child is conspicuously ill, with signs typical of pneumonia and bronchiole inflammation. In order to begin the correct treatment, a positive identification of RSV must be made as rapidly as possible. The best diagnostic procedures are those that demonstrate the viral antigen directly from specimens. The sample must be fresh (the virus is very fragile), and it must contain sufficient infected cells. Direct and indirect fluorescent staining, ELISA testing, and DNA probes are useful methods for evaluating a specimen's viral content. A highly beneficial new treatment is the administration of RSV immunoglobulin. This serum is obtained from people with high RSV antibody titers, and it greatly reduces complications and the need for hospitalization. An antiviral drug ribavirin (virazole) is an alternate therapy administered in aerosol inhaled form. This treatment is very costly and does have some adverse side effects. Other supportive measures include drugs to reduce fever, assisting pulmonary ventilation, and treating secondary bacterial infection if present. In view of the potentially life-threatening nature of RSV infection in youngsters, vaccination seems an attractive alternative.

Rhabdoviruses (Rabies)

The most conspicuous rhabdovirus (Gr. *Rhabdos,* rod) is the rabies (L. *Rabidus,* rage) virus, genus *Lyssavirus* (Gr. *Lyssa,* madness). The particles of this virus have a distinctive bulletlike appearance, round on one end and flat on the other. Additional features are a helical nucleocapsid and spikes that protrude through the envelope. The family contains approximately 60 different viruses, but only the rabies virus and rarely, some other mammalian lyssaviruses, affect humans.

Epidemiology—Rabies is a slow, progressive zoonotic disease characterized by fatal meningoencephalitis. It is distributed nearly worldwide, all though 34 countries have remained rabies free by practicing regorous animal control. The primary wild reservoirs of the virus are mammals such as wild canines, skunks, raccoons, badgers, wild cats, and bats. These animals can spread the infection to domestic dogs and cats, and both wild and domestic mammals can spread the disease to humans through bites, scratches, and inhalation of droplets. The annual worldwide total for human rabies is estimated at about 30,000 cases, but fewer than 5-10 of these cases occur in the United States. Rabies incidence in wild reservoirs have been on the increase, while domestic dog rabies has declined. So the end papers for statistics on the reported incidence of animals and human rabies in the United States. The most common wild animal reservoir host has changed from foxes in 1966 skunks in 1977 and to raccoons in 1993. Regional differences in the dominant reservoir also occur. Bats, skunks, and bobcats are the most common carriers of rabies in California, while raccoons are the predominant carriers in the East, and coyotes dominate in Texas.

Pathogenesis—Infection with rabies virus typically begins when an infected animal's saliva enters a puncture site. Occasionally the virus is inhaled or inoculated orally. The virus then gradually enters sensory nerve endings or the neuromuscular junction and advances toward the sensory ganglia, spinal cord, and brain. Viral multiplication throughout the brain is eventually followed by migration to such diverse sites as the eye, heart, skin, and oral cavity. The infection cycle is completed when the virus replicates in the salivary glands and is shed into the saliva. Clinical rabies proceeds through the identifiable stages of incubation, prodromium, acute neurologic phase (encephalitis), coma, and death.

Clinical studies—The average incubation period of rabies is one to two months with extremes of a week to more than a year. Its length depends upon the wound site, its severity, and the inoculation dose. The incubation period is shorter in facial, scalp, or neck wounds because of greater proximity to the brain. The prodromal phase begins with fever, anorexia, nausea, vomiting, headache, fatigue, and other nonspecific symptoms. Some patients continue to experience pain, burning, prickling, or tingling sensations at the wound site. In the form of rabies termed *furious,* the first acute signs of neurological involvement are periods of agitation, disorientation, seizures, and twitching. Spasms in the neck and pharyngeal muscles lead to severe pain upon swallowing. As a result, attempts to swallow or even the sight of liquids bring on *hydrophobia.* Throughout this phase, the patient is fully coherent and alert. If afflicted with the *dumb* form of rabies, a patient is not hyperactive but paralyzed, disoriented, and stuporous. Ultimately both forms progress to the coma phase, resulting in death from cardiac or respiratory arrest. Until recently, humans were never known to survive rabies.

Diagnosis and control—When symptoms appear after a rabid animal attack, the cause-and-effect relationship is vivid, and the disease is readily diagnosed. But the diagnosis can be obscured when contact with an infected animal is not clearly defined or when symptoms are absent or delayed. Anxiety, agitation, and depression can pose as a psychoneurosis; the mental state and pharyngeal spasms resemble tetanus; and encephalitis with convulsions and paralysis mimics a number of other viral infections. Often the disease is diagnosed at autopsy. The laboratory criteria most diagnostic of rabies are intracellular inclusions (Negri bodies) in infected nervous tissue, isolation and identification of rabies virus in saliva or brain tissue, and demonstration of rabies virus antigens in specimens of the brain, serum, cerebrospinal fluid.

A bite from a wild or stray animal demands prompt action, including assessment of the animal, meticulous care of the wound, and a specific treatment regimen. A wild mammal, especially a skunk, raccoon, fox, or coyote that bites without provocation, is presumed to be rabid, and treatment of the patient is immediately commenced. If the animal is captured, brain samples and other tissue are examined for verification of rabies. Apparently healthy domestic animals are observed closely for signs of disease and sometimes quarantined. Following an animal bite, the wound should be scrupulously washed with soap or detergent and water, followed by debridement and application of an antiseptic such as alcohol or peroxide. Rabies is one of the few infectious disease for which a combination of passive and active post exposure immunization is therapeutically indicated and successful. Initially the wound is infused with human rabies immune globulin (HRIG) to impede the spread of the virus, and globulin is also injected intramuscularly to provide immediate systemic protection.

Measures that effectively prevent and limit rabies are preexposure vaccination and animal control. Three intramuscular injections of HDCV are recommended for high-risk groups such as veterinarians, animal handlers, laboratory personnel, and travelers to endemic areas. Control measures such as vaccination of domestic animals, elimination of strays, and strict quarantine practices have helped reduce the virus reservoir. Several countries have tested a new genetically engineered oral vaccine made with a vaccinia virus that carries the gene for the rabies virus surface antigen. The vaccine is incorporated into bait placed in the habitats of wild reservoir species such as skunks and raccoons.

NONENVELOPED SINGLE STRANDED RNA VIRUSES

Poliovirus and Poliomyelitis

Poliomyelitis (Gr. *Polios,* gray, *myelos,* medulla, and *itis,* inflammation) (polio) is an acute enteroviral infection of the spinal cord that can cause neuromuscular paralysis. Because it often affects small children, it is also called infantile paralysis. No civilization or culture has escaped the devastation of polio. Drawings made by ancient civilizations hint at its long history, and such famous persons as Sir Walter Scott and Franklin Delano Roosevelt suffered its complications. Those of us who experienced the epidemics at the middle of this century will never forget the lasting images of polio; crippling paralysis, braces, the iron lung.

Epidemiology—The protective capsid and lack of an envelope confer chemical stability and resistance to acid, bile, and detergents. By this means the virus survives the gastric environment and other assaults to factors that contribute to its transmission. Sporadic cases of polio can break out at any time of the year, but its incidence is more pronounced during the summer and fall in temperate zones. The virus is passed within the population through food, water, hands and objects contaminated with feces. Transmission via mechanical vectors such as flies sometimes occurs, but respiratory droplets are only rarely involved. The prevalence of polio varies according to region. The polio rate in 26 Asian, African, and European countries is rapidly declining due to intensive immunization campaigns.

Pathogenesis—The precise site of viral multiplication is still unclear, but the likeliest candidates are mucosal epithelia or lymphoid tissue such as tonsils and Peyer's patches. Multiplication of the virus results in large numbers of them being shed into the throat and feces, and some of them leaking into the blood. Following an incubation period of one to two weeks, the infection takes one of four courses : (a) Subclinical infection, (b) minor disease, (c) aseptic, nonparalytic meningitis, and (d) paralytic disease.

A small number of persons develop mild nonspecific symptoms of fever, headache, nausea, sore throat, and myalgia. If the viremia persists, viruses can be carried to the central nervous system through its blood supply and cross the blood-brain barrier. The virus then spreads along specific pathways in the spinal cord and brain. Being *neurotropic,* the virus infiltrates the motor neurons of the anterior horn of the spinal cord, though it can also attack spinal ganglia, cranial nerves, and motor nuclei.

Paralytic disease—Depending on the level of damage to motor neurons, paralysis of the muscles of the legs, abdomen, back, intercostals, diaphragm, pectoral girdle, and bladder can result. In rare cases of bulbar poliomyelitis, the brain stem, medulla, or even cranial nerves are affected. This leads to loss of control of cardiorespiratory regulatory centers, palate, pharynx, and vocal cords. Some protection is afforded by antibodies to poliovirus in the intestine and tonsils and in the serum.

Diagnosis and control—Polio is mainly suspected when epidemics of neuromuscular disease occur in the summer in temperate climates. Poliovirus can usually be isolated by inoculating cell cultures with stool or throat washings in the early part of the disease. The stage of the patient's infection can also be demonstrated by testing serum samples for the type and amount of antibody. In the absence of specific therapy, treatment of polio rests largely on alleviating pain and suffering. During the acute phase, muscle spasm, headache, and associated discomfort can be alleviated by pain-relieving drugs. Respiratory failure may require artificial ventilation maintenance. Since reflex swallowing and coughing can be impaired, drainage or mechanical suction of mucous and secretions is necessary to prevent choking. Impaired ciliary and cough reflexes predispose the patient to secondary pulmonary infection that can require antibiotic treatment.

The two forms of vaccine currently in use are inactivated poliovirus vaccine (IPV), known as the Salk vaccine, and oral poliovirus vaccine (OPV), known as the Sabin vaccine. Both are prepared from animal cell cultures and are trivalent (combinations of the three serotypes). Both vaccines are effective, but one may be favoured over the other under certain circumstances. In some tropical countries where exposure of infants to wild virus is common, both vaccines may be given. The Sabin vaccine is favoured in the United States because it is easily administered by mouth, but it is not free of medical complications. It contains an attenuated virus that can multiply in vaccinated people and be spread to others. In very rare instances, the attenuated virus reverts to a neurovirulent strain that causes disease rather than protects against it. Numerous instances of paralytic polio have occurred among children with hypogammaglobulinemia who have been mistakenly vaccinated.

Nonpolio Enteroviruses

Such viruses as coxsackieviruses A and B, *echoviruses,* and nonpolio enteroviruses, are similar to the poliovirus in many of their epidemiologic and infectious characteristics. The incidence of infection is highest from late spring to early summer in temperate climates and is most frequent in infants and persons living under unhygienic circumstances.

Pathogenesis—Symptoms are usually mild and self-limited, and last only a few days. The initial phase of infection is intestinal, after which viruses enter the lymph and blood, and disseminate to other organs. The outcome of this infection largely depends on the organ affected.

Complications—All ages, however, are susceptible to the "common cold syndrome" of enteroviruses. *Pleurodynia* is an acute disease characterized by recurrent sharp, sudden intercostal or abdominal pain accompanied by fever and sore throat. Although nonpolio enteroviruses are less virulent than the polioviruses, rare cases of coxsackievirus and echovirus paralysis, aseptic meningitis and encephalitis occur. Even in severe childhood cases involving seizures, ataxia, coma, and other central nervous system symptoms, recovery is usually complete. Eruptive skin rashes (exanthems) that resemble the rubella rash are other manifestations of enterovirus infection. Coxsackievirus can cause a peculiar pattern of lesions on the hands, feet, and oral mucosa (hand-foot-mouth disease). Rashes are accompanied by fever, headache, and muscle pain. Acute hemorrhagic conjunctivitis is an abrupt inflammation associated with subconjunctival bleeding, serous discharge, painful swelling, and sensitivity to light. Spread of the virus to the heart in infants can cause extensive damage to the myocardium, leading to heart failure and death in nearly half of the cases.

Hepatitis A Virus and Infectious Hepatitis

One enterovirus that reacts primarily with the intestinal tract is the hepatitis A virus (HAV; enterovirus 72), the cause of infectious, or short-term, hepatitis. Otherwise, the two viruses are different in almost every respect. The hepatitis A virus is a cubical picornavirus relatively resistant to heat but sensitive to formalin, chlorine, and ultraviolet radiation. There appears to be only one major serotype of this virus.

Epidemiology of hepatitis A—In general, the disease is associated with deficient personal hygiene and lack of public health measures. In countries with inadequate sewage control, most outbreaks are associated with fecally contaminated water and food. The United States has a yearly incidence of 20,000-25,000 cases. Most of these result from close institutional contact, unhygienic food handling, eating shellfish, sexual transmission, or travel to other countries. Occasionally, hepatitis A can be spread by blood or blood products. In developing countries, children are the most common targets, because exposure to the virus tends to occur early in life, while in North America and Europe, more cases appear in adults.

Pathogenesis—Swallowed hepatitis A virus multiplies in the small intestine during an incubation period of two to six weeks. The virus is then shed in the feces for about two weeks, and at some point, it enters the blood and is carried to the liver. Most infections are either subclinical or accompanied by vague, flulike symptoms. In more overt cases, the presenting symptoms are loss of appetite, nausea, diarrhea, fever, pain and discomfort in the region of the liver, and darkened urine. Jaundice is present in only about one in 10 cases. Occasionally, hepatitis A occurs as a fulminating disease and causes liver damage, but this is quite rare.

Diagnosis and control—Certain liver enzymes are elevated, and leukopenia is common. Diagnosis is aided by detection of anti-HAV IgM antibodies produced early in the infection and tests to identify HA antigen or virus directly in stool samples. There is no specific treatment for hepatitis A once the symptoms begin. Patients receiving immune serum globulin early in the disease usually experience milder symptoms. Prevention of hepatitis A is based primarily on prophylactic immunization with pooled immune serum globulin. Potential globulin recipients are travelers and armed forces personnel planning to enter endemic areas, contacts of known cases, and occupants of day-care centers and other institutions during epidemics. An inactivated viral vaccine is currently approved and an oral vaccine, based on an attenuated strain of the virus is in development. Control of this disease can be improved by sewage treatment, hygienic food handling and preparation, and adequate cooking of shellfish.

Human Rhinovirus (HRV)

The rhinoviruses (Gr. *Rhinos,* nose) are a large group of picornaviruses associated with the common cold. Although the majority of characteristics are shared with other picornaviruses, two distinctions set rhinoviruses apart. First, they are sensitive to acidic environments such as that of the stomach, and second, their optimum temperature of multiplication is not normal body temperature, but 33°C, the average temperature in the human nose. Virologists have subjected rhinovirus (type 14) to a detailed structural analysis. The results of this study provided a striking three-dimensional view of its molecular surface, and also explained why immunity to the rhinovirus has been so elusive. The capsid subunits are of two types : Protuberances (knobs), which are antigenically diverse among the rhinoviruses, and indentations (pockets), which exist only in two forms. Since the antigens on the surface are the only ones accessible to the immune system, a successful vaccine would have to contain hundreds of different antigens, which is not practical. Unfortunately the recessed antigens are too inaccessible for either immune surveillance or antibody fit.

Epidemiology—Epidemics caused by a single type arise on occasion, but usually many strains circulate in the population at one time. As mutations increase and herd immunity to a given type is established, newer types predominate. Children are the most successful disseminators of colds, and often transmit the virus to the rest of a family. Infected respiratory membranes shed the virus for several days to weeks. People acquire infection from contaminated hands and fomites, and to a lesser extent from droplet nuclei. Although other animals have rhinoviruses, interspecies infections have never been reported. After an incubation period of one to three days, the patient can experience some combination of headache, chills, fatigue, sore, throat, cough, a mild nasal drainage, and atypical pneumonia.

Control—The classic therapy is to force fluids, and relieve symptoms with various cold remedies and cough syrups that contain nasal decongestants, alcohol, antihistamines, and analgesics. The actual effectiveness of most of these remedies is rather questionable. Owing to the extreme diversity of rhinoviruses, prevention of colds through vaccination will probably never be a medical reality. Some protection is afforded by such simple measures as handwashing and care in handling nasal secretions.

NONENVELOPED DOUBLE STRANDED RNA VIRUSES

Reoviruses have an unusual double-stranded RNA genome and both an inner and outer capsid. Two of the best studied viruses of the group are *Rotavirus* and *Reovirus.* Named for its wheel-shaped capsomer. Rotavirus (L. *Rota,* wheel) is significant cause of diarrhea in new born humans, calves and piglets. Because the viruses are transmitted via fecally contaminated food, water, disease is prevalent in areas of the world with poor sanitation. Globally, rotavirus is primary viral cause of mortality and morbidity resulting from diarrhea, accounting for nearly 50% of all the cases. Babies from 6 to 24 months of age lacking maternal antibodies have the greatest risk for fatal disease. These children present with symptoms of watery diarrhea, fever, vomiting and dehydration. In the United States, rotavirus infection is relatively common, but its course is generally mild. A live, attenuated oral vaccine was tested in Venezuela with some success and will be approved in the near future for children at risk.

Reovirus (respiratory enteric orphan virus) is not considered a significant human pathogen. Adults who were voluntarily inoculated with the virus developed symptoms of the comman cold. The virus has been isolated from the feces of children with enteritis and from people suffering from an upper respiratory infection and rash, but most infections are asymptomatic. Although a link of biliary meningeal, hepatic and renal syndromes has been suggested, a causal relationship has never been proved.

OTHER ENVELOPED VIRUSES

Coronaviruses

These viruses are common in domesticated animals and are responsible for epidemic respiratory, enteric, and neurologic diseases in pigs, dogs, cats, and poultry. Thus far, two types of Human coronaviruses have been characterized. One of these (HCV) is an etiologic agent of the common cold. The same virus is also thought to cause some forms of viral pneumonia and myocarditis. The other human virus is very closely associated with the intestine and may have a role in some human enteric infections.

Rubella

There are several important members, including *Rubivirus,* the agent of rubella, and certain arboviruses. Rubella, or German measles, was first recognized as a distinct clinical entity in the mid-eighteenth century, and was considered a benign childhood disease with mild prodromal symptoms and a brief rash. Then, in 1941, the Australian physician Norman Gregg called attention to the *teratogenic* aspect of rubella virus infection. He discovered that cataracts often developed in neonates born to mothers who had contracted rubella in the first trimester.

Epidemiology—Rubella is an endemic disease with worldwide distribution. Infection is initiated much like the paramyxoviruses through contact with respiratory secretions and occasionally urine from infected individuals. The virus is shed during the prodromal phase and up to a week after the rash appears. Because the virus is only moderately communicable, close living conditions are required for its spread. Although epidemics and pandemics of rubella once regularly occurred in six- to nine- year cycles, the introduction of vaccination has essentially stopped this pattern in the United States. Most cases are reported among adolescents and young adults in military training camps, colleges, and summer camps.

Pathogenesis and transmission—Two clinical forms of rubella can be distinguished: Postnatal infection, which develops in children or adults, and congenital (prenatal) infection of the fetus, which is expressed in the newborn as various types of birth defects. During an incubation period of two to three weeks, the virus multiplies in the respiratory epithelium, infiltrates local lymphoid tissue, and enters the bloodstream. Despite widespread invasion of organs, half of infections are asymptomatic. Early symptoms include malaise, mild fever, sore throat, and lymphadenopathy. The rash of pink macules and papules first appears on the face and progresses down the trunk and toward the extremities, advancing and resolving in about three days. Adult rubella is often accompanied by joint inflammation and pain rather than a rash.

Transmission of the rubella virus to a fetus in utero can result in a serious complication called congenital rubella. The mother is able to transmit the virus even if she is asymptomatic. Fetal injury appears to come from disruption in the development of the germ layers. The severity of damage varies according to the time of infection. It is generally accepted that infection in the first trimester is most likely to induce miscarriage or multiple permanent defects in the newborn such as cardiac abnormalities, ocular lesions, deafness, and mental and physical retardation. Less drastic sequelae that usually resolve in time are anemia, hepatitis, pneumonia, carditis, and bone infection.

Diagnosis and control—Rubella mimics other diseases and is often asymptomatic, so it should never be diagnosed on clinical grounds alone. The confirmatory methods of choice are serologic testing and virus isolation systems. Diagnosis of fetal infection by examining cells and amniotic fluid obtained by amniocentesis is a promising technique that awaits further evaluation. IgM tests can determine recent infection, and a rising titer is a clear indicator of continuing rubella infection. Several serological tests for IgM are used, including complement fixation, ELISA, fluorescent assay, and hemagglutination. Postnatal rubella is generally benign and requires only symptomatic treatment. Because no specific therapy for rubella is available, most control efforts are directed at maintaining herd immunity with attenuated rubella virus vaccine. This vaccine is usually given to children in the combined form (the MMR vaccination) once at 15 months of age.

Arboviruses

Vertebrates are hosts to more than four hundred viruses transmitted primarily by arthropods. For simplicity's sake, these viruses are lumped together in a loose grouping called the arboviruses (arthropod-borne viruses). The major arboviruses pathogenic to humans are togaviruses (*Alphavirus* and *Flavivirus*), some bunyaviruses (*Bunyavirus* and *Phlebovirus*), and reoviruses (*Orbivirus*). The chief vectors are bloodsucking arthropods such as mosquitos, ticks, flies, and gnats. Most types of illness caused by these viruses are mild, undifferentiated fevers, though some cause severe encephalitides (plural of encephalitis) and life-threatening hemorrhagic fevers. Although these viruses have been assigned taxonomic names, they are more often known by their common names, which are based on geographical location and primary clinical profile.

Epidemiology—The vectors and viruses tend to be clustered in the tropics and subtropics, but many temperate zones report periodic epidemics. A given arbovirus type may have very restricted distribution, even to a single isolated region, but some range over several continents, and others can spread along with their vectors. As one would expect, the activity and distribution of arboviruses are closely tied to the ecology of the vectors. Factors that weigh most heavily are the longevity of the arthropod, the availability of food and breeding sites, and climatic influences such as temperature and humidity. When vectors take a blood meal to finish their reproductive cycle, they acquire the virus and harbor it for varying time periods. Infections show a peak incidence when the arthropod is actively feeding and reproducing, usually from late spring through early fall. Warm-blooded vertebrates also maintain the virus during the cold and dry seasons. Humans can serve as dead-end, accidental hosts, as in equine encephalitis and Colorado tick fever, or they can be a maintenance reservoir, as in dengue fever and yellow fever. Arboviral diseases have a great impact on humans. Although exact statistics are unavailable, it is accepted knowledge that Millions of people acquire infections each year and thousands of them die. The uncertain nature of host and viral cycles frequently results in sudden, unexpected epidemics, sometimes with previously unreported viruses.

General characters—Although the arboviruses have many inherently interesting characteristics, our coverage here will be limited to : (a) The chief manifestations of disease in humans, (b) aspects of the most important North American viruses, and (c) methods of diagnosis, treatment, and control. One type of disease elicited by arboviruses is an acute, undifferentiated fever, often accompanied by rash. These infections, typified by dengue fever and Colorado tick fever, are usually mild and self-limited, and leave no sequelae. Prominent symptoms are fever up to 40°C prostration, headache, myalgia, orbital pain, muscle aches, and joint stiffness. About midway through the illness, a maculopapular or petechial rash can erupt over the trunk and limbs. When the brain, meninges, and spinal cord are invaded, symptoms of viral encephalitis occur. The most common American types are western equine, eastern equine, St. Louis, and California encephalitis. The viruses cycle between wild animals (primarily birds) and mosquitos or ticks, and humans are not usually reservoir hosts. The disease begins with an arthropod bite, the release of the virus into the tissues, and its replication in nearby lymphatics. Prolonged viremia establishes the virus in the brain, where inflammation can cause swelling and damage to various nuclei and nerve tracts as well as to the meninges. Symptoms are extremely variable, and can include coma, convulsions, paralysis, tremor, rigidity, loss of coordination, palsies, memory deficits, changes in speech and personality, and heart disorders. In some cases survivors experience some degree of permanent brain damage.

Colorado tick fever (CTF) is the most common tick-borne viral fever in the United States. Restricted in its distribution to the Rocky Mountain states, it occurs sporadically during the spring and summer months, corresponding to the time of greatest tick activity and human forays into the wilderness.

Western equine encephalitis (WEE) occurs sporadically in the western United States and Canada, first in horses and later in humans. The mosquito that carries the virus emerges in the early summer when irrigation begins in rural areas and breeding sites are abundant. The disease is extremely dangerous to infants and small children, with a case fatality rate of 3% to 7%.

Eastern equine encephalitis (EEE) is endemic to an area along the east coast of North America and Canada. The usual pattern is sporadic cases, but occasional epidemics can occur in humans and horses. High periods of rainfall in the late summer increase the chance of an outbreak, and disease usually appears first in horses and caged birds. A disease known commonly as California encephalitis is caused by two different viral strains. The California strain occurs occasionally in the western states and has little impact on humans. The LaCrosse strain is widely distributed in the eastern United States and Canada, and is a prevalent cause of viral encephalitis in North America. St. Louis encephalitis (SLE) is the most common of all American viral encephalitides. Cases appear throughout North and South America, but epidemics occur most often in the midwestern and southern states. Inapparent infection is very common, and the total number of cases is probably thousands of times greater than the two hundred or so reported each year. People over the age of 55 are most susceptible to life-threatening complications.

Hemorrhagic fevers—Certain arboviruses, principally the yellow fever and dengue fever viruses, can disrupt the vascular bed to such an extent that sudden localized bleeding erupts in the tissues, leading to shock and even death. The exact mechanisms of pathology are obscure, but somehow the virus causes capillary fragility and disrupts the blood clotting system. These hemorrhagic syndromes, which are usually attended by fever and pain, are caused by a variety of viruses, are carried by a variety of vectors, and are distributed globally. Reservoir animals are usually small mammals, although yellow fever and dengue fever can be harbored in the human population. As a result of intensive studies begun in the 1920s, more is known about yellow fever than any other arboviral disease. Although this disease was at one time cosmopolitan in its distribution, mosquito control measures have eliminated it in many countries, including the United States. Two patterns of transmission occur in nature. One is an urban cycle between humans and the mosquito *Aedes aegypti,* which reproduces in standing water in cities. The other is a sylvan cycle, maintained between forest monkeys and jungle species of mosquitos. Most cases in the western hemisphere occur in the jungles of Brazil, Peru, and Colombia during the rainy season. Infection begins acutely with fever, headache, and muscle pain. While the disease is benign in 80-90% of cases, some patients experience oral hemorrhage, nosebleed, vomiting, jaundice, and liver and kidney damage.

Dengue fever is caused by a flavivirus and carried by *Aedes* mosquitos. Although mild infection is the usual pattern, a form called dengue hemorrhagic shock syndrome can be lethal. This disease is also called "breakbone fever" because of the severe pain it induces in the muscles and joints. The illness is endemic to Southeast Asia and India, and several epidemics have occurred in Central America, the Caribbean, and Mexico. Concern over the possible spread of the disease to the United States had led the CDC to survey mosquito populations in states along the southern border. They discovered a recently introduced species, the Asian tiger mosquito (*Aedes albopictus*) that maintains breeding colonies in several states. Because this mosquito is a potential vector of dengue fever virus, it is feared that contact of these mosquitos with infected humans could establish the virus in these mosquito populations.

Diagnosis and control—The patient's history of travel or contact with vectors, along with serum analysis, are highly supportive of a diagnosis. Treatment of the various arbovirus disease relies completely on support measures to control fever, convulsions, dehydration, shock, and edema. The most reliable arbovirus vaccine is a live, attenuated yellow fever vaccine that provides relatively long-lasting immunity. Vaccination is a requirement for travelers in the tropics and can be administered to entire populations during epidemics. Live, attenuated vaccines for WEE and EEE are administered to laboratory workers, veterinarians, ranchers, and horses. Most of the control safeguards for arbovirus disease are aimed at the arthropod vectors. Mosquito abatement by eliminating breeding sites and broadcasting insecticides have been highly effective in a restricted urban setting, but if the vectors are widely dispersed through a woodland habitat, these techniques are of little value. In the case of sylvan vectors, the only real preventive is to restrict human contact with the vector through insect repellents and protective clothing.

Retroviruses

The most familiar of these viruses, called *lentiviruses,* is the human immunodeficiency virus (HIV) that causes acquired immune deficiency syndrome (AIDS). HIV is only one of a whole array of retroviruses associated with cancer and other diseases. Judging from what is currently known about these viruses, they will continue to influence medicine, research, politics, and economics for decades to come. An outstanding characteristics of retroviruses is an unusual enzyme called reverse transcriptase, which can convert a single-stranded RNA genome into double-stranded viral DNA. Not only can this newly formed DNA chronically infect host cells, but it also can be incorporated into the host genome as a provirus that can be passed on to progeny cells. These viruses do not follow a central thesis of biology, which maintains that the direction of flow of information and function is from DNA to RNA. Previously, all biologic entities, including other viruses, fulfilled this expectation, but this virus has a radical strategy. It comes already equipped with an enzyme that works in the opposite direction by catalyzing the replication of double-stranded DNA from single-stranded RNA. Because these viruses reverse the order of replication, they are termed *retroviruses*. The association of *retroviruses* with their hosts can be so intimate that viral genes are permanently integrated into the host genome. In fact, as the technology of DNA probes for detecting retroviral genes is employed, it becomes increasingly evident that retroviruses may be integral parts of chromosomes.

Taxonomy—Most retroviruses isolated to date cause slow infections that lead to leukemia, tumors, anaemia, immune dysfunction, and even neurological diseases in birds and mammals. The first human retrovirus (discovered in 1978) was the T-cell leukemia virus I (HTLV I). This was followed by the isolation of HTLV II and HTLV III. Types I and II are associated with human leukemia or lymphoma, while type III is the agent of AIDS. Since it was determined that HTLV III, lymphadenopathy virus (LAV), and AIDS retrovirus (ARV) are all different names for the same virus, the unified name HIV 1 (human immunodeficiency virus) was assigned to it.

Structure—The outermost component is a lipid envelope with transmembrane glycoprotein spikes and knobs that mediate viral adsorption to the host cell. Although a few retroviruses are nonspecific, most can infect only host cells that present the required receptors.

Acquired Immune Deficiency Syndrome (AIDS)

The explosion of AIDS on the world scene in the early 1980s has altered society's collective view of infectious diseases. At no time in history has a concerted research effort uncovered so much information about a microbe and a disease in so short a time. Never before has so much money been spent on treatment, vaccine development, education, and public health for one disease. To reconstruct the discovery of AIDS, we must go back to 1979 when physicians in Los Angeles, San Francisco, and New York City began to see clusters of patients who were all young males with one or more of a complex of symptoms : Severe pneumonia caused by *Pneumocystis carinii* (ordinarily a harmless fungus), a rare vascular cancer called *Kaposi's sarcoma;* sudden weight loss; swollen lymph nodes; and general loss of immune function. Another mysterious finding was that all of these young men were homosexuals. Early hypotheses attempted to explain the disease as a consequence of the homosexual life-style or a result of immune suppression by chronic drug abuse or infections. Soon, however, cases were reported in nonhomosexual patients who had been transfused with blood or blood products. Eventually, virologists at the Pasteur Institute in France isolated a novel retrovirus. Once HIV was isolated from patients' tissues, researchers were hampered by the inability to grow sufficient amounts of the virus for further study and by the lack of accurate tests for differentiating this virus from other retroviruses. Finally in 1984, a functioning line of T cells that remained alive after infection with the new virus became available. This was soon followed by the creation of sensitive tests for detecting the virus and the antibody against it.

There are no conclusive answers to the question of where and how HIV originated. Initially, it was hypothesized that the virus originated in central Africa from a simian retrovirus carried by green monkeys. Although this virus is benign in monkeys, it could develop high virulence if transmitted to humans. This theory was called into question when Japanese researchers showed that HIV 1 is much too genetically different from the simian virus to be a mutated form of it. It now appears that the closest simian relative of HIV appears to be a chimpanzee virus that is very similar to HIV 2. Serological examination of stored blood samples provides tantalizing hints at the history of HIV. HIV 2 antibodies were present in samples taken in Africa as early as 1966, and convincing evidence from Europe has linked HIV 1 to HIV-like diseases occurring in the late 1950s. The recent discovery of an AIDS-like retrovirus among healthy Amazonian Indians has given credence to the idea that HIV and its relatives have existed for a long time in remote human populations.

Epidemiology—Estimates of the number of individuals currently infected with the virus range from 10 to 20 million worldwide, with approximately one million (the range is 800,000 to 2 million) in the United States. It is thought that most of these people have not yet begun to show symptoms because they are in the latent phase of the disease. The number of diagnosed AIDS cases has increased continuously since AIDS first became a notifiable disease at the national level in 1984. It took seven years for the first 100,000 cases to appear, but only two years for the next 100,000 cases. The disease has continued this epidemic pattern, although it is recently showing signs of slowing. Despite new therapies that have lengthened the lives of many patients, the overall death rate is still around 55%. Although all people appear to be susceptible to the AIDS virus, epidemiologic statistics show that, overall, it's more common among males than females, and that ethnically, whites account for the most cases, followed by blacks, Hispanics, and other groups. The average adult patient is about 35 years of age. If one observes specific sociologic groups, however, the sexual, racial, and age statistics can be quite different.

Transmission—It is now clear that the mode of HIV transmission is exclusively through two forms of contact : Sexual intercourse and transfer of blood or blood products. Each of these can be divided into specific high-risk groups that include persons who become infected through a particular life-style or behaviour, or through medical accidents. The mode of transmission is rather similar to that of hepatitis B virus, except that the AIDS virus does not survive for as long outside the host, and it is far more sensitive to heat and disinfectants. In general, AIDS is spread only by direct and rather specific routes involving intimate contact with an infectious dose. An infection can take place only if the virus crosses the body's epithelial barriers into the fluid compartments. Since the blood of HIV-infected people often harbors high levels of free virus and infected leukocytes, any form of intimate contact involving transfer of blood (trauma, injection) can be a potential source of infection. Semen and vaginal secretions also harbor the virus within white blood cells, thus they are significant factors in sexual transmission. The virus can be isolated from urine, tears, sweat, and saliva, but these fluids contain too few viruses (less than 1 virus per cc) to be a source of infection.

Myths and misconceptions—So horrifying is the prospect of acquiring a lethal disease like AIDS that inaccurate notions and misconceptions rather than established facts have sometimes abounded. Rumors about the role of bloodsuking insects have been common; however, insects like mosquitos do not serve as biological vectors of the AIDS virus, and the chance of the insect harboring and delivering an infectious dose mechanically is infinitesimally small. At first it was mistakenly believed that airborne droplets, shaking hands, handling fomites, and sharing public facilities, food, and swimming pools were possible ways to become infected. However, in-depth studies have indicated that even within families, infection of healthy family members has occurred only after known exposure to blood. One peculiar fear that surfaced during the early stages of the AIDS epidemic was that donating blood could expose the donor to AIDS, but obviously, blood banks use new sterile equipment for each donor. And although kissing is a form of intimate contact, no documented cases have been traced to it.

AIDS : overview—In 1993, the CDC redefined the case definition of AIDS, based on changing epidemiologic patterns and medical information. AIDS is defined as a severe immunodeficiency disease arising from infection with HIV and accompanied by some of the following symptoms; life-threatening opportunistic infections, persistent fever, unusual cancers, chronically swollen lymph nodes, weight loss, diarrhea, and neurological disorders. This definition also includes many patients suffering from AIDS-related complex (ARC), once considered a stage in development of the disease. At some point during the infection, the patient tests positive for antibodies to HIV. Seropositivity means only that an infection with HIV has taken place, but does not indicate that the disease exists.

Order of prevalence in homosexuals—Still account for the majority of AIDS cases (56%), though the rate of infection appears to be declining. This demographic group has been so hard hit apparently due to certain types of homosexual practices. Multiple anonymous sexual encounters were once relatively common among gay men frequenting bars and bathhouses. Anal sex, which is known to lacerate the rectal mucosa, could certainly provide entry of viruses from semen into the blood. Studies have shown that the passive anal partner in these encounters is the more likely of the two to become infected.

Order of prevalence in drug abusers—In large metropolitan areas such as New York City, as many as 60% of intravenous drug addicts can be HIV carries. Needle sharing is most common in the "shooting galleries" of large eastern cities and is especially risky for cocaine users whose rate of drug injection can be more frequent. Infection from contaminated needles is growing more rapidly than for any other mode of transmission (24% of cases), and is another significant factor in the spread of AIDS to the heterosexual population. A fairly large segment of IV users are also homosexual men.

Order of prevalence of heterosexuals of HIV carriers—Individuals in this group include the partners of drug addicts and bisexual men, prostitutes, and spouses of people who have received transfusions or blood factors. The virus spreads readily in either direction, and multiple sexual encounters increase the chance of infection, although even a single encounter with an AIDS patient has been known to transmit the virus. The overall rate of infection is the general heterosexual population has increased dramatically in the past few years in adolescent and young adult women.

Order of Prevalence of blood transfusion—Since routine screening of donated blood for antibodies to the AIDS virus was adopted in 1985, transfusions are no longer considered a serious risk. Before then, approximately 2% of AIDS cases arose in patients receiving multiple transfusions, and most new-transfusion related cases can be traced to transfusions given prior to 1985. Since there can be a lag period of a few weeks to nearly three years before antibodies appear in an infected person, it is remotely possible to be infected through donated blood (estimated at 1 in 500,000 transfusions). As added precautions, potential donors are thoroughly screened, and people anticipating surgery are encouraged to store their own blood in advance. A new technique of sterilizing blood with a laser beam shows some promise, but it has not been perfected.

Order of prevalence in coagulation disorder replacement—For several years in this country, replacement coagulation products (factor VIII and others) were derived from human plasma and filtered to remove cellular microbial contaminants, but not viruses. Before this route of HIV transmission was discovered, several hundred hemophiliacs became infected. Now replacement infusions are heat-treated to kill any contained viruses, a process that has essentially eliminated risk for this group of people.

Order of prevalence in unknown risk factors—This is the most troublesome category because it opens up questions about some other, unknown route of spread. When the CDC has carefully followed up such cases, a majority of the patients have admitted to prior contact with a prostitute or a history of other STDs. Factors such as patient denial, unavailability, death, or uncooperativeness make it impossible to explain every case.

Primary effects of AIDS—Over the long period of chronic infection, the infected cells play host to the virus in either a latent or an active phase. Cells with active viruses release large quantities into the circulation and are killed in the process. The death of T cells and other white blood cells results in extreme leukopenia and loss of essential T4-memory clones and stem cells. Evidence indicates that the viruses also cause the formation of giant T-cell and other cell syncytia, which allow the spread of viruses directly from cell to cell, followed by mass destruction of the syncytia. As the dose of viruses entering the system increases, a vicious cycle occurs, involving a greater infection rate and death of cells, and greater viral burden. Because the T4 lymphocytes have multiple cytotoxic, helper, and inducer functions, their destruction or disability paves the way for invasion by opportunistic agents and cancer. When the T4 cell levels fall below 200 cells/mm^3 of blood, symptoms of AIDS appear. Targets of the AIDS virus are not limited to the immune system. The central nervous system is involved in a large proportion of cases. It is believed that infected macrophages cross the blood-brain barrier and release viruses, which then invade nervous tissues. Studies have indicated that some of the viral envelope proteins can have a direct toxic effect on the brain's glial cells and other cells. Other research has shown that some peripheral nerves become demyelinated and the brain becomes inflamed.

Secondary effects of AIDS—Among the earliest symptoms of AIDS are debilitating, potentially fatal, and often multiple opportunistic infections. In general, the infections involve fungi, protozoa, viruses, or bacteria that establish a foothold in the absence of adequate cytotoxic immunity. The most frequent sites of attack are the respiratory, digestive, and nervous systems. *Pneumocystis carinii* pneumonia (PCP), caused by a wide-spread fungus, is the leading cause of morbidity and morality in AIDS patients. Its symptoms include fever, cough, shortness of breath, thickening of the lung epithelium, and fluid infiltration that compromises respiration. The course of infection can be acute and rapid, or slow and progressive. Fungi most frequently involved in AIDS complications are *Cryptococcus neoformans,* a cause of meningitis; *Candida albicans,* which attacks the oral, pharyngeal, and esophageal mucosa; and *Aspergillus* species, which infect the lungs. Common herpesviruses also afflict the AIDS patient. Cytomegalovirus produces a disseminated pneumonia, and can infect the brain, eyes, and liver as well. *Herpes simplex* and *Herpes zoster* can reactivate to produce severe ulceration of the skin, lips, genitalia, and anus. Infections with bacteria such as *Mycobacterium* are particularly problematic and severe. Tuberculosis in AIDS patients is now at epidemic levels.

The predisposition of AIDS patients for tumors is due to the loss of natural cancer-killing T cells along with the intrusion of microbes that can cause cancer. Kaposi's sarcoma (KS) accounts for most cancers seen clinically. This sarcoma, which is apparently caused by an unidentified herpesvirus, is a nodular purple lesion that develops from endothelial cells in blood vessels of the skin, intestine, and mucous membranes. Other symptoms that accompany KS are fever, lymphadenopathy, weight loss, and diarrhea. Complications are hemorrhage, perforation, and intestinal obstruction.

AIDS patients experience many nonspecific, disease-related symptoms. Pronounced wasting of body mass that accompanies weight loss, diarrhea, and poor nutrient absorption is a common complaint. Protracted fever, fatigue, sore throat, and night sweats are significant and debilitating. Some of the most virulent complications are neurological. Lesions occur in the brain, meninges, spinal column, and peripheral nerves.

Prevalence of coinfection—One relationship that seems certain is the effect of coinfections, especially STDs. Genital ulcers such as those present in chlamydia, syphilis, and warts increase susceptibility to AIDS because breaks in the mucous membrane promote the shedding and entrance of the virus. AIDS patients frequently have concurrent infections with Epstein-Barr virus (mononucleosis), CMV, syphilis, tuberculosis, hepatitis B, and a recently discovered mycoplasma. The severe immune suppression accompanying AIDS appears to enhance the pathogenicity and virulence of both participants and escalate damage.

Diagnosis and control—When these signs are coupled with a patient's history of predisposing activities and a positive serological test for AIDS antibodies, the diagnosis is confirmed. The enzyme-linked immunosorbent assay for AIDS antibodies (ELISA) is highly sensitive, but variations in the quality of testing and the patient's condition can lead to false-positive results in a small number of tests. An alternate latex agglutination test that is simple, completed in five minutes, and 99% accurate has recently been marketed. Positive antibody readings call for the more specific follow-up test known as the Western blot, which detects several different anti-HIV antibodies and can usually rule out a false-positive result. False-negative results occur in as many as one in four patients who are indeed infected with HIV. Such silent infections contribute considerably to the unpredictable nature of latency and how to test for it. Because an infected person has viruses in his blood from the very earliest stages, a more reliable diagnostic approach would be direct discovery of the AIDS virus, its genes, or antigenes in blood and other specimens. This form of testing could eliminate the false tests inherent in antibody analysis and be a particular boon in monitoring transfusion blood. So far, there is no cure for AIDS. None of the therapies do more than prolong life or diminish symptoms. Treatment for AIDS patients focuses on supportive care and drugs to control both opportunistic and HIV infections. Drugs used to control PCP are pentamidine, taken in oral or inhaled form; trimethoprim-sulphamethoxazole and leucovorin. Drugs that have some control over CMV infections are ganciclovir and foscarnet. Disseminated fungal infections are treated with flucomazole, and Kaposi's sarcoma is relieved by alpha interferon and human chorionic gonadotropin.

Drugs to inhibit infection and replication by AIDS virus are beings actively sought and tested. More than 100 drugs are in development or clinical trials. The only class of drugs formally approved by the Food and Drug Administration for HIV therapy are the synthetic nucleotide analogs azidothymidine (AZT), dedeoxyinosine (ddI), dedeoxycytidine (ddC), and dedeoxythymidine (ddT). They interrupt the HIV multiplication cycle by being incorportated into the DNA molecule by reverse transcriptase. Since these drugs lack all of the correct binding sites for further DNA synthesis, this terminates viral replication and the viral cycle. Recent studies show that ddI is more effective than AZT in delaying the onset of symptoms in HIV positive patients. It is currently recommended as a primary or combined therapy in this group. Other considerations are their relative high cost (AZT costs $2,200 for one year of treatment), toxic side effects, and development of viral resistance. A sampling of other strategies being pursued include the development of drugs : (a) To prevent HIV adsorption (CD4), (b) to stop virus fusion with the host cell (Peptide T22), (c) to block the integration and replication of viral DNA into the host cell, (d) to block translation of the viral RNA (ribozymes), (e) to inhibit maturation (protease inhibitors), and (f) to block virus budding (interferon). Several clinical trials are underway for protease inhibitors and ribozyme therapy. Protease inhibitors block the action of a unique HIV enzyme required for the final assembly and maturation of the virus. In their urgency for a cure, many AIDS patients are willing to try untested but potentially useful therapies. Some patients have acquired drugs from countries with less stringent drug policies than the United States. To ease this tremendous demand for alternate treatments, the FDA has relaxed its usual regulations to allow for personal use of experimental AIDS drugs among patients. Another serious problem with all AIDS drugs is the tendency for the virus to develop resistance. Hundreds of virus strains exist. This is largely due to the high rate of mutation the virus experiences and its lack of a proofreading function for DNA synthesis. The drugs actually have the effect of selecting those strains that mutate into resistant forms.

□□□

17

Antimicrobial Drugs : Their Mechanism of Action

In one ancient Chinese folk remedy, a fermented soybean curd was applied to skin infections. The Greeks used wine and plant resins (myrrh and frankincense), rotting wood, and various mineral salts to treat diseases. Some folk medicines were occasionally effective, but most of them were probably witches' brews that either had no effect or were even harmful. It is interesting that the Greek word *pharmakeutikos* originally meant the practice of witchcraft. These ancient remedies were handed down from generation to generation, but it was not until the Middle Ages that a specific disease was first treated with a specific chemical. An enormous breakthrough in the science of drug therapy came with the proof of the germ theory to infection by Robert Koch. This allowed disease treatment to focus on a particular microbe, which in turn opened the way for Paul Ehrlich to formulate the first theoretical concepts in chemotherapy in the late 1800s. Ehrlich had observed that specific dyes often affixed themselves to specific microorganisms and not to animal tissue. This led to the profound idea that if a drug was properly selective in its actions, it would zero in and destroy a microbial target and leave human cells unaffected. He decided to test this theory in treating syphilis. First, he worked out the structure of an arsenic-based drug that was very toxic to the spirochete of syphilis but, unfortunately, to humans as well. Ehrlich gradually and systematically altered this parent molecule, creating numerous derivatives. Finally, on the 606th try, he arrived at a compound he called *salvarsan*. Another pathfinder in early drug research was Gerhard Domagk, whose discoveries in the 1930s launched a breakthrough in therapy that marked the true beginning of broad-scale usage of antimicrobial drugs. Experimenting with numerous synthetic dyes, Domagk showed that the red dye prontosil was active against certain bacterial infections in animals, even though it was inactive against that same infectious agent in a test tube. It turned out that prontosil was chemically changed by the body into an entirely different compound, with specific activity against bacteria. This new substance was sulfonamide – the first *sulfa* drug. In a short time, the structure of this durg was determined, and it became possible to synthesize it on a wide scale and to develop scores of other sulfonamide drugs. The discovery of antibiotics dramatically demonstrates how developments in science and medicine often occur through a combination of accident, persistence, collaboration, and vision. In the London laboratory of Alexander Fleming in 1928, a plate of *Staphylococcus aureus* became contaminated with the mold *Penicillium notatum*. Observing these plates, Fleming noted that the colonies of *Staphylococcus* were evidently being destroyed by some activity of the nearby *Penicillium* colonies. Struck by this curious phenomenon he extracted from the fungus a compound, he called penicillin, and showed that it was responsible for the inhibitory effects. Fleming lost interest in penicillin and nothing was done with it for a decade, until the pressures of impending war spurred Howard Florey and Ernst Chain to develop methods for industrial production of penicillin in England. Clinical trials of penicillin, conducted in 1941, ultimately proved its effectiveness, and cultures of the mold were brought to the United States for an even larger-scale effort. When penicillin was made available to the world's population, it seemed a godsend at first, but in time, due to extreme overuse and mis-understanding of its capabilities, it also became the model for one of the most serious drug problems—namely, drug resistance. Fortunately, by the 1950s the pharmaceutical industry had entered an era of drug research and development that soon made penicillin only one of a large assortment of antimicrobial drugs.

PRINCIPLES OF ANTIMICROBIAL THERAPY

Early death or severe lifelong debilitation from scarlet fever, diphtheria, tuberculosis, meningitis, and many other bacterial diseases was a fearsome yet undeniable fact of life to most of the world's population. The introduction of modern drugs to control infections in the 1930s began a medical revolution that has added significantly to the lifespan and health of humans. It is no wonder that for many years, antibiotics in particular were regarded as the miracle cure-all for infectious disease. In later dicussions, we will evaluate this misconception in light of the shortcomings of chemotherapy. Although antimicrobic drugs have greatly reduced the incidence of certain infections, they have definitely not eradicated infectious disease and probably never will. In fact, in many parts of the world, mortality rates from infectious diseases are as high as before the arrival of antimicrobic drugs. Any chemical uses in treatment, relief, or prophylaxis (Gr. *Prophylassein,* to keep guard before) of disease is defined as a chemotherapeutic drug or agent. When chemotherapeutic drugs are given as a means to control infection, the practice is termed antimicrobial chemotherapy (Gr. *Chemieia,* chemistry, and *therapeia,* service to the sick). Antimicrobial drugs are a special of compounds capable even in high dilutions of destroying or inhibiting microorganisms. The origin of modern antimicrobial drug is varied. Some, called antibiotics, are substances produced by the natural metabolic processes of microorganisms that can inhibit or destroy other microorganisms. Other antimicrobial substances, termed synthetic drugs, are derived in the laboratory from dyes or other organic compounds. Although separation into these two categories has been traditional, they tend to overlap, because many antibiotics are now chemically altered in the laboratory *(semisynthetic)*. Antimicrobial drugs vary in their scope of activity. The so-called narrow-spectrum agents are effective against a limited array of different microbial types. Examples are bacitracin, an antibiotic whose inhibitory effects extend mainly to certain Gram-positive bacteria or griseofulvin, which is used chiefly in fungal skin infections. Broad-spectrum agents are active against a wider range of different microbes. The targets of antibiotics in the tetracycline group, for example, are a variety of Gram-positive and Gram-negative bacteria, rickettsias, and mycoplasmas. The concept of spectrum of activity has also become blurred to some extent, because an agent can be induced by artificial means to change its spectrum.

Drugs, Host and Microbe Interactions

To put it simply, the drug should destroy the infectious agent without harming the host's cells (Fig. 17.1). This process can be visualized more tangibly as a series of stages : (a) The drug is administered to the host via a designated route – primarily by mouth (oral), by injection into a vein (intravenous), by injection into a muscle (intramuscular), by applying to the skin surface (topical), and occasionally, by injection into the skin or body cavity. (b) The drug is dissolved in body fluids. (c) The drug is delivered to the infected area (extracellular or intracellular). (d) The drug destroys the infectious agent or inhibits its growth. (e) The drug is eventually excreted or broken down by the host's organs, ideally without harming them. Among the factors that affect the outcome of drug therapy are those that aid or block drug action and those that influence the behaviour of microbe or host tissues.

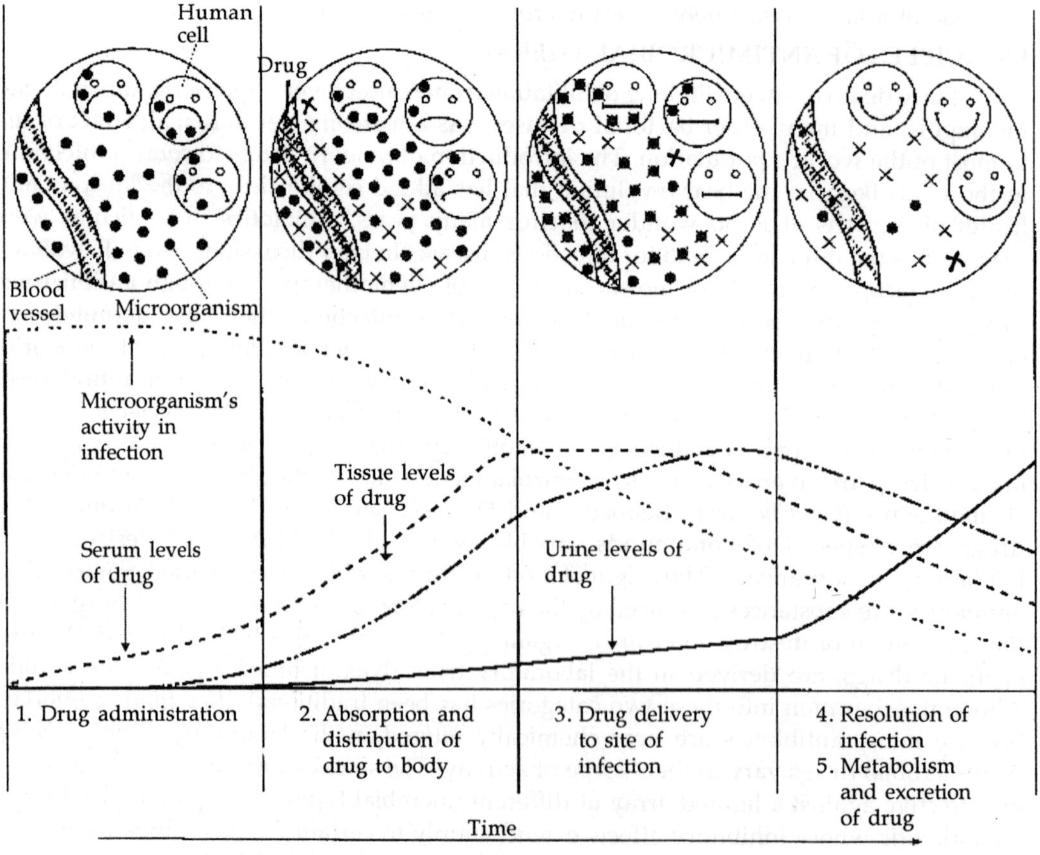

Fig. 17.1 : Interaction among host, drugs and microorganisms.

Antibiotics, after all, are common metabolic products of aerobic spore-forming bacteria and fungi. By inhibiting the growth of other microorganisms in the same habitat (antagonism), antibiotic producers presumably enjoy less competition for nutrients and space. The greatest numbers of antibiotics are derived from bacteria in the genera *Streptomyces* and *Bacillus,* and molds in the genera *Penicillium* and *Cephalosporium.* Chemists have sought to decipher the structure of the more useful antibiotics in order to produce related semisynthetic compounds with specialized features.

Most of the drugs used in chemotherapy interfere with the function of enzymes required to synthesize or assemble macromolecules, or they destroy structures already formed in the cell. Preferably, such drugs should be selectively toxic, which means they produce adverse effects on microbial cells without simultaneously damaging host tissues. This concept of selective toxicity is central to chemotherapy, and the best drugs are those that block the actions or synthesis of molecules in microorganisms but not in vertebrate cells. Examples of drugs with selective toxicity are those that block the synthesis of the cell wall in bacteria (penicillins). They have low toxicity and few direct effects on human cells because human cells lack a wall and are thus neutral to this action of the antibiotic. Among the most toxic to human cells are drugs that act upon a common structure such as the cell membrane.

Mechanism of Action of Drugs

Microbicidal drugs lyse and kill microorganisms by inflicting direct damage upon specific cellular targets. Microbistatic drugs interfere with the machinery in the cell required for cell division, and thus inhibit reproduction. Drugs that inhibit growth are not considered directly responsible for subsequent microbial death; their primary importance is to keep the microbes at bay and prevent their spread, thus allowing the host defenses an opportunity to destroy and remove the infectious agent (Fig. 17.2). Antimicrobial drugs

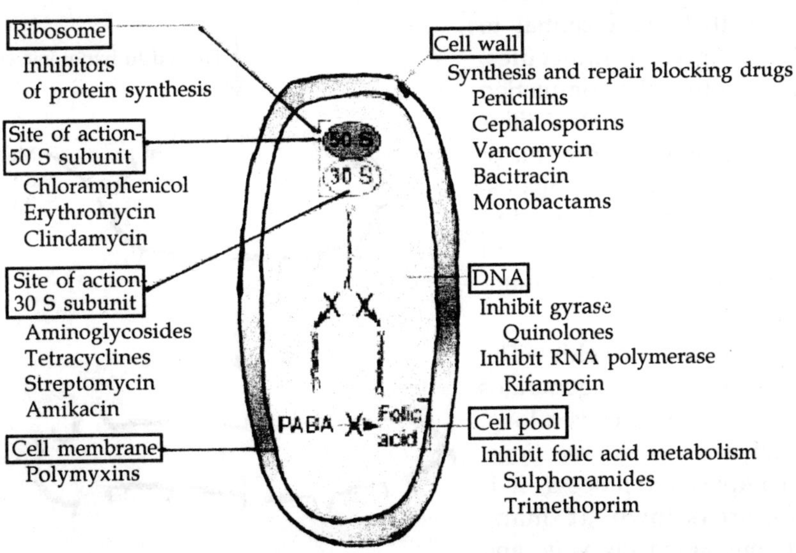

Fig. 17.2 : Action of antimicrobial drugs.

function specifically in one of the following ways : (a) They inhibit cell wall synthesis (b) they inhibit nucleic acid synthesis; (c) they inhibit protein synthesis; or (d) they interfere with the function of the cell membrane. These categories are not completely discrete, and some effects can overlap. It is worth noting that much knowledge of cellular structure and function has emerged as an added bonus from research on the effects of drugs.

Effects on bacterial cell wall—The cell walls of most bacteria contain a rigid girdle of peptidoglycan. This structure, which is many layers thick in Gram-positive species and quite thin in Gram-negative ones, protects the cell against rupture from hypotonic environments. Cells actively engaged in enlargement or binary fission must constantly synthesize new peptidoglycan and transport it to its proper place in the cell envelope. Drugs such as penicillins and cephalosporins react with one or more of the enzymes required to complete this process, causing the cell to develop weak points at growth sites and to become osmotically fragile (Fig. 17.3). Antibiotics that produce this effect are considered bactericidal, because the weakened cell is subject to lysis. It is essential to note that most of these antibiotics are active only in young, actively growing cells, since old, inactive, or dormant cells do not synthesize peptidoglycan. Cycloserine inhibits the formation of the basic peptidoglycan subunits and vancomycin hinders the elongation of the peptidoglycan. The beta-lactams (penicillins and cephalosporins) bind to peptidases that are essential to cross-link the glycan molecules, thereby interrupting the completion of the cell wall (Fig. 17.4). Penicillins that do not penetrate the outer membrane are less effective against Gram-negative becteria, but broader-spectrum penicillins and cephalosporins can destroy Gram-negative species.

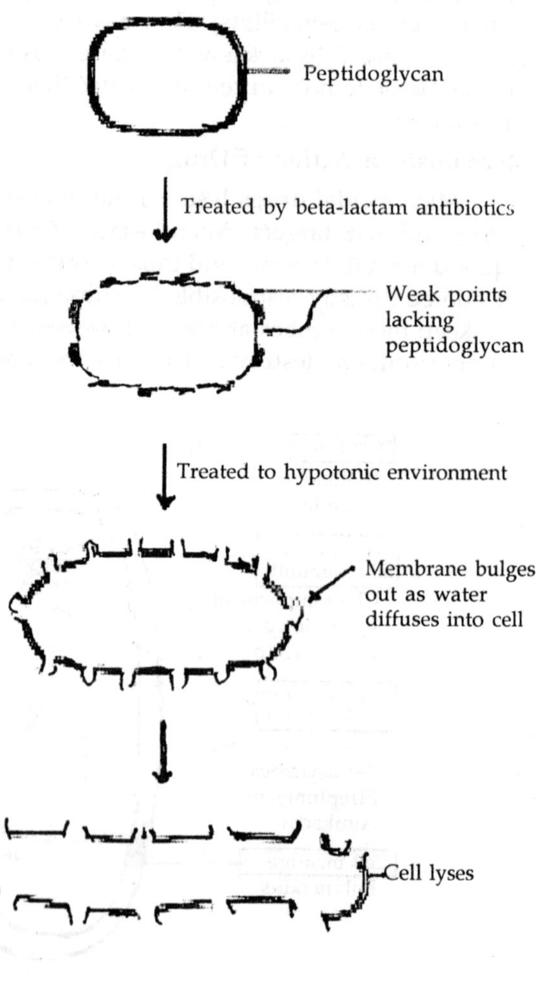

Fig. 17.3 : Effect of cephalosporins on cell.

Effects on nucleic acids—The metabolic pathway that generates DNA and RNA molecules is a long, enzyme-catalyzed series of reactions. Like any complicated process, it is subject to breakdown at many different points along the way, and inhibition at any point in the sequence can block subsequent events. Antimicrobial drugs interfere with nucleic acid synthesis by blocking synthesis of nucleotides, inhibiting replication, or stopping transcription. Because functioning DNA and RNA are required for proper translation as well, the effects on protein metabolism can be far-reaching. One of the more extensively studied modes of action is that of the sulphonamides (sulpha drugs). Because these synthetic drugs interfere with an essential metabolic process in bacteria, they represent a model for competitive inhibition. They act as structural or metabolic analogs that mimic the natural substrate of an enzyme and via for its active site. In practice, sulpha drugs are very similar to the natural metabolic compound PABA (para-aminobenzoic acid) required by bacteria to synthesize folic acid. Folic acid, in turn, is a component of the coenzyme tetrahydrofolic acid that participates in the synthesis of purines and certain amino acids. A sulphonamide molecule has extreme affinity for the PABA site on the enzyme that synthesizes folic acid, thus it can successfully complete in a "chemical race" with PABA for the same sites (Fig. 17.5). Sulphonamides ultimately cause an inadequate supply of folic acid for purine production, which invariably halts nucleic acid synthesis and prevents bacterial cells from multiplying. Sulphonamides are valuable in therapy because they inhibit bacteria and certain protozoa, but not mammalian cells. This one-sided inhibition stems from a basic nutritional difference between humans and microorganisms. Although humans require folic acid for nucleic acid synthesis as much as bacteria do, humans cannot synthesize folic acid; it is an essential nutrient (vitamin) that must come

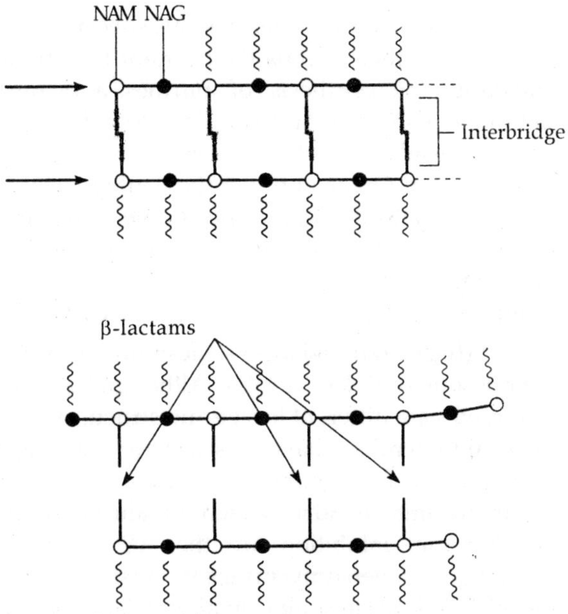

Fig. 17.4 : Action of β-lactam antibiotics on bacterial cell wall.

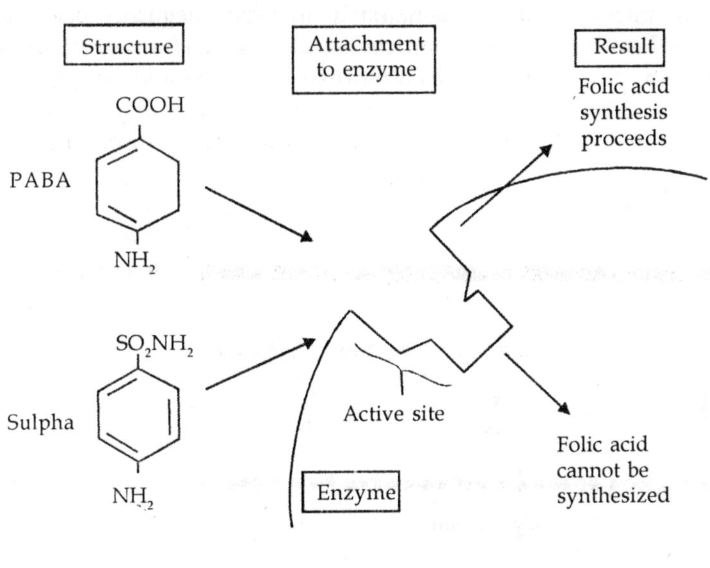

Fig. 17.5 : Inhibition of nucleic acid synthesis.

from the diet. Human cells lack this special enzymatic system for incorporating PABA into folic acid, so their metabolism cannot be inhibited by sulfa drugs. Blocking this enzyme interrupts the synthesis of purines and pyrimidines. Trimethoprim is often given simultaneously with sulfonamides to achieve a synergistic effect. In pharmacology, this refers to an additive effect achieved by two drugs working together, thus requiring a lower dose of each. Other antimicrobics inhibit DNA synthesis. Chloroquine (an antimalarial drug) binds and cross-links the double helix. The newer broad-spectrum quinolones inhibit DNA unwinding enzymes or gyrases, thereby stopping DNA transcription. Antiviral drugs that are analogs of purines and pyrimidines (including idoxuridine and acyclovir) insert in the viral nucleic acid and block further replication.

Effects on translation—Most inhibitors of translation, or protein synthesis, react with the ribosome-mRNA complex. Although human cells also have ribosomes, the ribosomes of eukaryotes are different in size and structure from those of prokaryotes, so for the most part, these antimicrobics have a selective action against bacteria. One potential therapeutic consequence of drugs that bind to the prokaryotic ribosome is the damage they can do to eukaryotic mitochondria, which contain a prokaryotic type of ribosome. Two possible targets of ribosomal inhibition are the 30S subunit and the 50S subunit. Aminoglycosides (streptomycin, gentamycin) insert on sites on the 30S subunit and cause the misreading of the mRNA, leading to abnormal proteins. Tetracyclines block the attachment of tRNA on the A acceptor site and effectively stop further synthesis. Other antibiotics attach to sites on the 50S subunit in a way that prevents the formation of peptide bonds (chloramphenicol) or inhibits translocation of the subunit during translation (erythromycin).

Inhibition of cell membrane functions—The antibiotic classes that damage cell membranes have specificity for a particular microbial group, based on differences in the types of lipids in their cell membranes.

Polymyxins interact with membrane phospholipids, distort the cell surface, and cause leakage of proteins and nitrogen bases, particularly in Gram-negative bacteria (Fig. 17.6). The polyene antifungal antibiotics (amphotericin B and nystatin) form complexes with the sterols on fungal membranes, which causes abnormal openings and seepage of small ions. Unfortunately, this selectivity is not exacting, and the universal presence of membranes in microbial and animal cells alike means that most of these antibiotics can be quite toxic to humans.

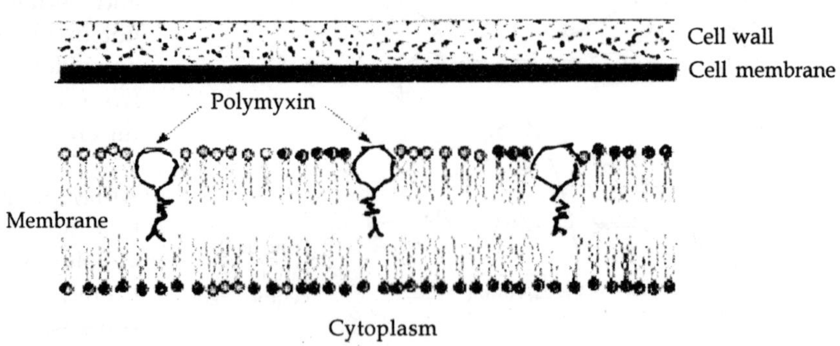

Fig. 17.6 : Polymyxin action on cell membrane.

Microbial Resistance

The wide-scale use of antimicrobics soon led to microbial drug resistance, an adaptive response in which microorganisms become able to tolerate an amount of drug that would ordinarily be inhibitory. The development of mechanisms for circumventing or inactivating antimicrobic drugs is due largely to genetic versatility and adaptability of microbial populations. The property of drug resistance can be intrinsic as well as acquired. Intrinsic drug resistance exists when the natural structure or physiology of a species or group is protective, not acquired through specific genetic changes. In our context, the term *drug resistance* will mean resistance acquired by microbes that had formerly been sensitive to the drug.

Development—Chromosomal drug resistance usually results from spontaneous random mutations in bacterial populations. The chance that such a mutation will be advantageous is minimal, and the chance that it will confer resistance to a specific drug is lower still. Nevertheless, given the huge numbers of microorganisms in any population and the constant rate of mutation, such mutations do occur. The end result varies from slight changes in microbial sensitivity, which can be overcome by larger doses of the drug, to compete loss of sensitivity. Resistance associated with intermicrobial transfer originates from plasmids called *resistance factors,* or R factors, that are transferred through conjugation, transformation, or transduction. Studies have shown that plasmids encoded with drug resistance, like mutations occurring on chromosomes, are naturally present in microorganisms before they have been exposed to the drug. Such traits are "lying in wait" for an opportunity to be expressed and to confer adaptability on the species. Many bacteria also maintain transposable drug resistance sequences (transposons) that are duplicated and inserted from one plasmid to another or from a plasmid to the chromosome. Chromosomal genes and plasmids containing codes for drug resistance are faithfully replicated and inherited by all subsequent progeny. In general, a microorganism loses its sensitivity to a drug by expressing genes that stop the action of the drug. Gene expression takes the form of : (a) Synthesis of enzymes that inactivate the drug, (b) decrease in cell permeability and uptake of the drug, (c) change in the number or affinity of the drug receptor sites, or (d) modification of an essential metabolic pathway. Some bacteria can become resistant indirectly by lapsing into dormancy, or, in the case of penicillin, by converting to a cell-wall-deficient form (L form) that penicillin cannot affect.

Microbes inactivate drugs by producing enzymes that permanently alter drug structure. One example, bacterial exoenzymes called beta-lactamases, hydrolyze the *beta-lactam* ring structure of some penicillins and cephalosporins. Two beta-lactamases—*penicillinase* and *cephalosporinase* – disrupt the structure of certain penicillin or cephalosporin molecules. So many strains of *Staphylococcus aureus* produce penicillinase that regular penicillin is rarely a possible therapeutic choice. Now that some strains of *Neisseria gonorrheae,* called PPNG, have also acquired penicillinase, alternate drugs are required to treat gonorrhea. A large number of other Gram-negative species are inherently resistant to some of the penicillins and cephalosporins because of naturally occurring beta-lactamases. The resistance of some bacteria can be due to a mechanism that prevents the drug from entering the cell and acting on its target. For example, the outer membrane of the cell wall of certain Gram-negative bacteria is a natural blockade for some of the penicillin drugs. Resistance to the tetracyclines can arise from plasmid-encoded proteins that pump the drug out of the cell. Resistance to the aminoglycoside antibiotics is a special case in which microbial cells have lost the capacity to transport the drug intracellularly.

Because most drugs act on a specific target such as protein, RNA, DNA, or membrane structure, microbes can circumvent drugs by altering the nature of this target. In bacteria resistant to rifampcin and streptomycin, the structure of key proteins has been altered so that these antibiotics can no longer bind. Erythromycin and clindamycin resistance is associated with an alteration on the 50 S ribosomal binding site. The types of penicillin resistance in *Streptococcus pneumoniae* and methicillin resistance to *Staphylococcus aureus* are related to an alteration in the binding proteins in the cell wall. Fungi can become resistant by decreasing their synthesis of ergosterol, the principal receptor for certain antifungal drugs. The action of antimetabolites can be circumvented if a microbe develops an alternative metabolic pathway or enzyme (Fig. 17.7). Sulphonamide and trimethoprim resistance develops when microbes deviate from the usual patterns of folic acid synthesis. Fungi can acquire resistance to flucytosine by completely shutting off certain metabolic activites.

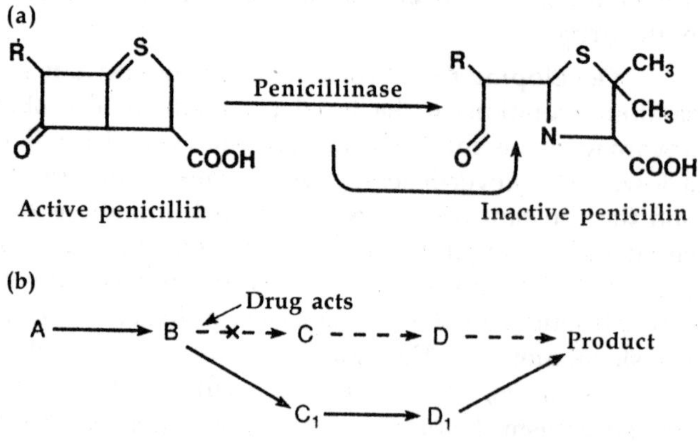

Fig. 17.7 : Mechanism of acquired drug resistance.

Natural Selection and Drug Resistance

So far, we have been considering drug resistance at the cellular and molecular levels, but its full impact is felt only if this resistance occurs throughout the cell population. Let us examine how this might happen and its longer-term therapeutic consequences. Recall that any large population of microbes is likely to contain a few individual cells that are already drug resistant because of prior mutations or transfer of plasmids. As long as the drug is not present in the habitat, the numbers of these resistant forms will remain low because they have no particular growth advantage. But if the population is subsequently exposed to this drug, sensitive individuals are inhibited or destroyed, and resistant forms survive and proliferate. During subsequent population growth, all offspring of these genetically altered microbes will inherit this drug resistance. In time, the replacement population will have a preponderance of the drug-resistantal forms and can even become completely resistant. In ecological terms, the environmental factors (in this case, the drug) has put selection pressure on the population, thus the more "fit" microbe (the drug-resistant one) survived, and the population has evolved to a condition of drug resistance. Natural selection for drug-resistant forms is apparently a common phenomenon.

Modern Concept of Drug Design

Hundreds of investigators began the laborious task of screening samples from soil, dust, muddy lake sediments, rivers, estuaries, oceans, plant surfaces, compost heaps,

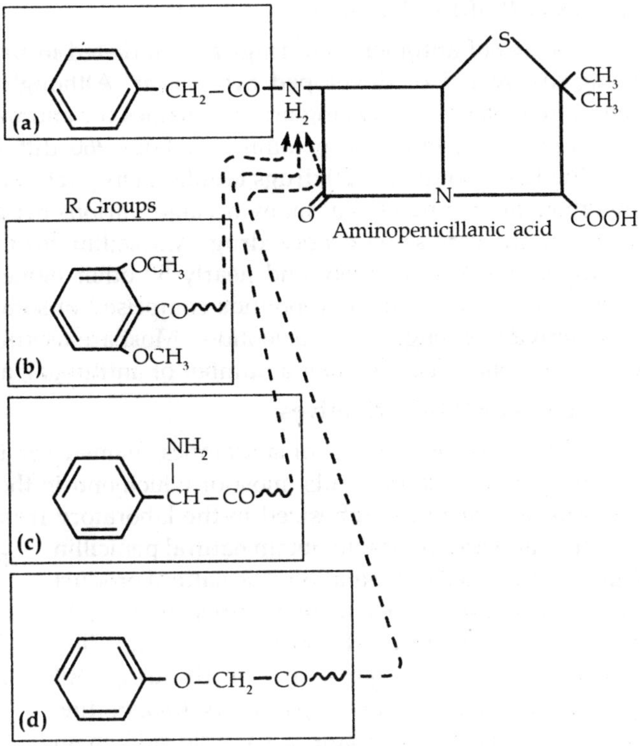

Fig. 17.8 : Synthesis of penicillins.

sewage, skin, and even the hair and skin of animals for antibiotic-producing bacteria and fungi. This intense effort has paid off over the past 50 years, because more than 10,000 antibiotics were eventually discovered. Finding a new antimicrobic substance is only a first step. The complete pathway of drug development from discovery to therapy takes at least nine years at a cost of several billion dollars. Antibiotics are products of fermentation pathways that occur in many spore-forming bacteria and fungi. The true role of antibiotics in the lives of these microbes continues to be somewhat mysterious, although the evolutionary preservation of genes for antibiotic production does point to an important function for them. Some experts theorize that antibiotic-releasing microorganisms can inhibit or destroy nearby competitors or predators, while others feel that these compounds play a part in spore formation. Whatever benefit the microbes derive, the existence of these compounds has been extremely profitable for humans. Every year, the pharmaceutical industry farms vast quantities of microorganisms and harvests their products to treat diseases caused by other microorganisms. Another approach in the drug quest is that of organic synthesis—the chemical manipulation of molecules by adding or removing functional groups. Drugs produced in this way are designed to have advantages over other, related drugs. With the semisynthetic method, a natural product of the microorganism is joined with various preselected functional groups. The antibiotic is reduced to its basic molecular framework (called the nucleus), and to this nucleus specially selected side chains (R groups) are added. A case in point is the metamorphosis of the semisynthetic penicillins. The nucleus is an inactive penicillin derivative called aminopenicillanic acid, which has an opening on the number 6 carbon for addition of R groups. A particular carboxyllic acid (R group) added to this nucleus can "fine-tune" the penicillin, giving it special characteristics. For instance, some R groups will make the product resistant to penicillinase (methicillin), some confer a broader activity spectrum (ampicillin), and others make the product acid-resistant (penicillin V). Cephalosporins and tetracyclines have likewise been subjected to semisynthetic conversions. The potential for using bioengineering techniques to design drugs seems almost limitless, and indeed, several drugs have already been produced by manipulating the genes of antibiotic producers (Fig. 17.8).

ANTIMICROBIAL DRUGS

Scores of antimicrobial drugs are marketed in the United States, and new ones are being discovered or developed every year. Although the medical and pharmaceutical literature contains a dizzying array of names for antimicrobics, most of them are variants of a small number of drug families. About 260 different antimicrobial drug currently classified are grouped in 20 drugs families. Drug reference books may give the impression that there are ten times that many because various drug companies assign different trade names to the very same generic drug. Ampicillin, for instance, is sold as Ampen, Amcill, Omnipen, PenA, Principen, and nearly 50 other names. Most antimicrobics are organic cyclic (ring or aromatic) compounds comprised of some combination of peptides, amino acid derivatives, sugars, or nucleotides. Most are useful in controlling bacterial infections, though we shall also consider a number of antifungal, antiviral, and antiprotozoan drugs.

Penicillins and Its Derivatives

The penicillin group of antibiotics, named for the parent compound, is a large, diverse group of compounds, most of which end in the suffix-*cillin.* Although penicillins could be completely synthesized in the laboratory from simple raw materials, it is more practical and economical to obtain natural penicillin through microbial fermentation. The natural product can then be used either in unmodified form or to make semisynthetic derivatives. *Penicillium chrysogenum* is the major source of the drug. All penicillins consist of three parts : A thiazolidine ring, a beta-lactam ring, and a variable side chain that dictates its microbicidal activity (Fig. 17.9).

Penicillins G and V are the most important natural forms. Penicillin G is considered the drug of choice for infections by known sensitive, Gram-positive cocci (most streptococci, pneumococci) and some Gram-negative bacteria (meningococci and the spirochete of syphilis). Penicillin V is given for many of the same types of infectious agents. It is one of a very few antibiotics approved for antimicrobic prophylaxis and is currently suggested as a daily therapy for children with sickle-cell anemia who are threatened by systemic pneumococcal infections. Certain semisynthetic penicillins such as ampicillin, carbenicillin, and amoxicillin have broader spectra and thus can be used to treat infections by Gram-negative enteric rods. Penicillinase-resistant penicillins such as methicillin, nafcillin, and cloxacillin are useful in treating infections caused by some penicillinase-producing bacteria. Mezlocillin and azlocillin have such an extended spectrum that they can be substituted for combinations of antibiotics.

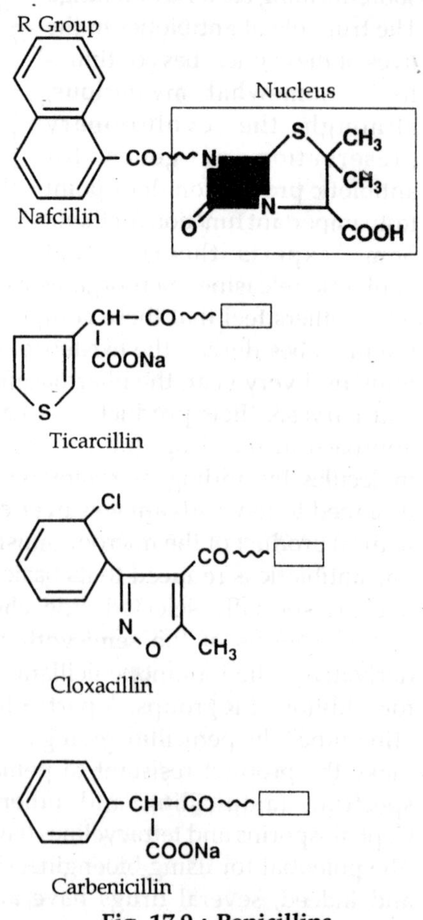

Fig. 17.9 : Penicillins.

The Cephalosporins

The cephalosporins are a comparatively new group of antibiotics that currently account for a majority of all antibiotics administered. The first compounds in this group were isolated in the late 1940s from the mold *Cephalosporium acremonium.* However, it was not until 20 years later that drug researchers discovered a mutant strain of this species that could be used in drug manufacture. Cephalosporins are similar to penicillins in their beta-lactam structure that can be synthetically altered (Fig. 17.10), and in their mode of action. The generic names of these compounds are often recognized by the presence of the root "cef" in their names.

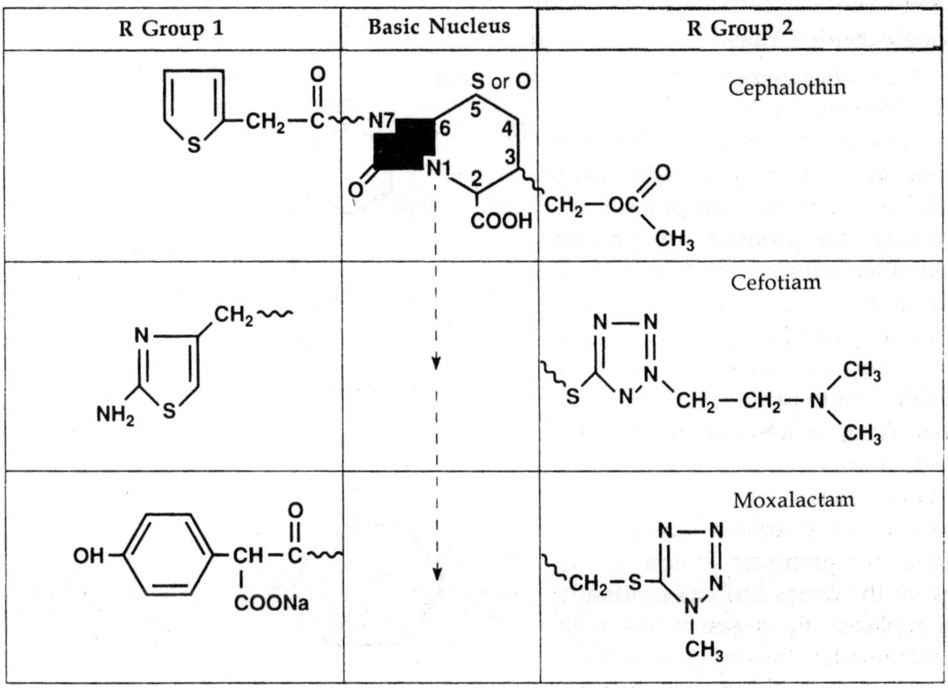

Fig. 17.10 : Cephalosporins.

The cephalosporins are versatile. They are relatively broad spectrum, resistant to penicillinases, and cause fewer allergic reactions than penicillins. Although some cephalosporins are given orally, many are poorly absorbed from the intestine and must be administered parenterally (Gr. *Para,* beyond, and *enteron,* intestine) by injection into a muscle or intravenously. Three generations of cephalosporins exist, based upon their antibacterial activity. First-generation cephalosporins such as cephalothin and cefazolin are most effective against Gram-positive cocci and are mildly active against a few Gram-negative bacteria. Second-generation forms include cefaclor and cefonacid, which are more effective than the first-generation forms in treating infections by certain Gram-negative bacteria such as *Enterobacter, Proteus,* and *Haemophilus.* Third-generation cephalosporins such as cephalexin (Keflex) and cefotaxime, are broad spectrum with especially well-developed activity against enteric bacteria that produce beta-lactamases. Ceftriaxone is a new semisynthetic drug that shows great promise in treating difficult Gram-positive and Gram-negative infections and anaerobes.

β-lactam Antibiotics

One of the newer antibiotics, Imipenem, has proved to be an excellent broad-spectrum drug for infections with Gram-positive and Gram-negative aerobic and anaerobic pathogens. Not only is it active in very small concentrations, but also it can be taken by mouth and appears to have low human toxicity. Aztreonam, isolated from the purple pigmented bacterium *Chromobacterium violaceum,* is a newer narrow-spectrum drug with special effectiveness in pneumonia, septicemia, and urinary tract infections by *Pseudomonas* and other Gram-negative aerobic bacilli.

Aminoglycoside Drugs

Antibiotics composed of two or more amino sugars and an aminocyclitol (6-carbon) ring are referred to as aminoglycosides (Fig. 17.11). These complex compounds are exclusively the products of various species of soil *actinomycetes* (Gr. *Actinos,* ray, and *myles,* fungus) in the genera *Streptomyces* and *Micromonospora.* As a group, aminoglycosides have a relatively broad antimicrobial spectrum by inhibiting protein synthesis. They are especially useful in treating infections caused by aerobic Gram-negative rods and certain Gram-positive bacteria. Streptomycin is among the oldest of the drugs and has gradually been replaced by newer forms with less mammalian toxicity. It is still the antibiotic of choice for treating bubonic plague and tularemia, and is considered a good antituberculosis agent. Gentamycin is less toxic and is widely administered for infections caused by Gram-negative rods (*Echerichia coli, Proteus, Pseudomonas, Salmonella,* and *Shigella).* Two relatively new aminoglycosides, tobramycin and amikacin, are also used for Gram-negative bacillary infections and have largely replaced kanamycin. A relative of the aminoglycosides, spectinomycin, is used to treat infections with penicillin-resistant *Neisseria gonorrheae.*

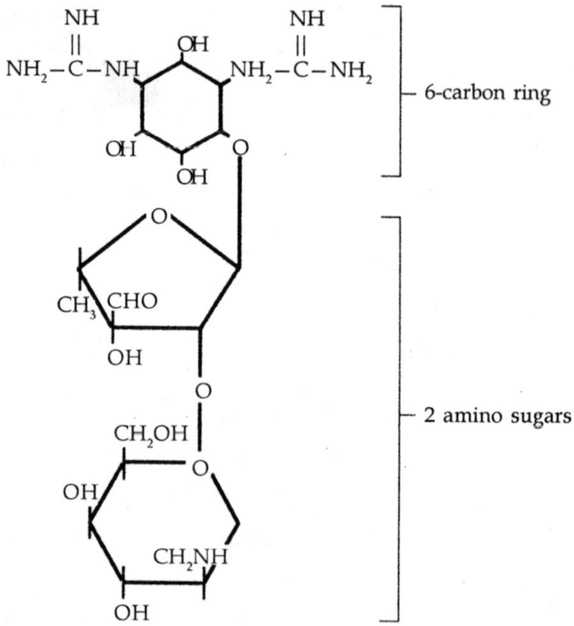

Fig.17. 11 : Streptomycin.

Tetracyclines

In 1948, a yellow colony of *Streptomyces* isolated from a soil sample gave off a substance, aureomycin, with strong antimicrobic properties. This antibiotic was used to synthesize its relatives terramycin and tetracycline. These natural parent compounds and semisynthetic derivatives are called collectively, the tetracyclines (Fig. 17.12). They are effective broad-spectrum drugs that can bind to prokaryotic ribosomes and block protein synthesis. The scope of microorganisms inhibited by tetracyclines includes Gram-positive and Gram-negative rods and cocci, aerobic and anaerobic bacteria, mycoplasmas, rickettsias, and spirochetes. Tetracycline compounds such as doxycycline and minocycline are administered orally to treat several sexually transmitted diseases, Rocky Mountain spotted fever, typhus, *Mycoplasma* pneumonia, cholera, leptospirosis, acne, and even some protozoan infections. Although generic tetracycline is low in cost and easy to administer, its side effects—namely, gastrointestinal disruption and deposition in hard tissues—can limit its use.

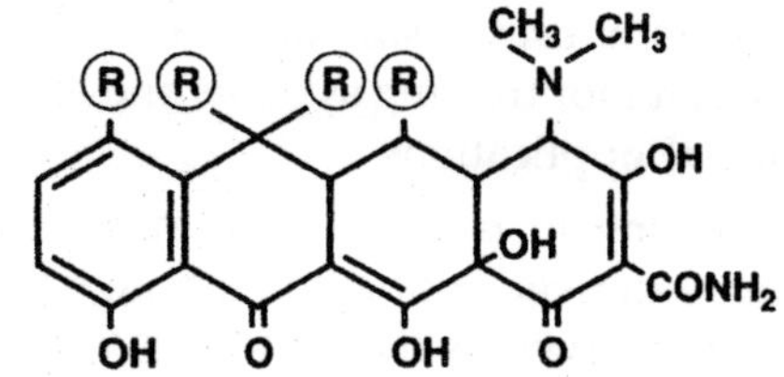

(a) Tetracyclines

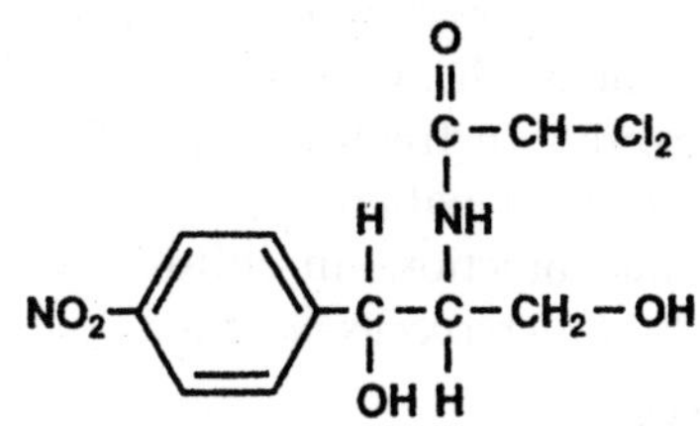

(b) Chloramphenicol

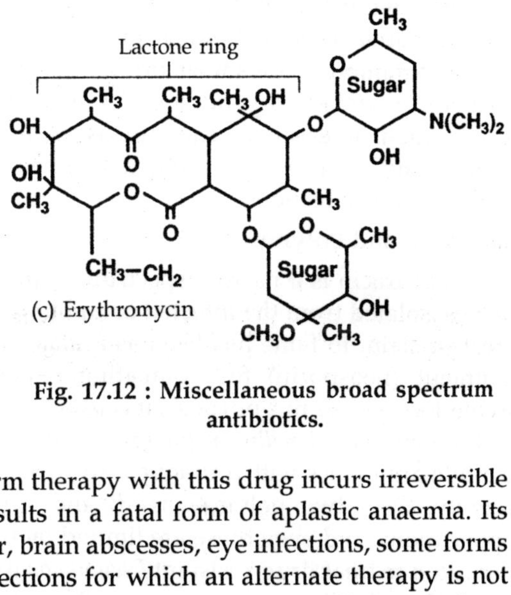

(c) Erythromycin

Fig. 17.12 : Miscellaneous broad spectrum antibiotics.

Chloramphenicol

Originally isolated in the late 1940s from *Streptomyces venezuelae,* chloramphenicol is a potent broad-spectrum antibiotic with a unique nitrobenzene structure (Fig. 17.12). Its primary effect on cells is to inhibit protein synthesis by blocking peptide bond formation on the ribosomes. It is one type of antibiotic that is no longer derived from the natural source but is entirely synthesized through chemical processes. Although this drug is fully as broad spectrum as the tetracyclines, evidence of its extreme toxicity began to mount within a short time after its introduction, and by the mid-1950s, restrictions were placed on its use. One out of 24,000—40,000 people undergoing long-term therapy with this drug incurs irreversible damage to the bone marrow that usually results in a fatal form of aplastic anaemia. Its administration is now limited to typhoid fever, brain abscesses, eye infections, some forms of meningitis, and certain life-threatening infections for which an alternate therapy is not available. Chloramphenicol should never be given in large doses repeatedly over a long time period, and the patient's blood must be monitored during therapy.

Erythromycin, Clindamycin, Vancomycin

Erythromycin is a macrolide antibiotic first isolated in 1952 from a strain of *Streptomyces* isolated from a Philippine soil sample. Its structure consists of a large lactone ring with sugars attached (Fig. 17.12). This drug is relatively broad spectrum and of fairly low toxicity. Its mode of action is to block protein synthesis by attaching to the ribosome. It is administered orally as the drug of choice for *Mycoplasma* pneumonia, legionellosis, *Chlamydia* infections, pertussis, and diphtheria, and as a prophylactic drug prior to intestinal surgery.

It also offers a useful substitute for dealing with penicillin-resistant streptococci and gonococci, and for treating syphilis and acne. Erythromycin is the only macrolide antibiotic with clinical applications.

Clindamycin is a relatively new antibiotic synthesized from lincomycin, another *Streptomyces*-derived antibiotic. Although both antibiotics have relatively broad antibacterial spectra, clindamycin is the more active of the two. The tendency of clindamycin to cause adverse reactions in the gastrointestinal tract limits it applications to : (a) Serious infections in the large intestine and abdomen due to anaerobic bacteria that are unresponsive to other antibiotics (*Bacteroides* and *Clostridium*), (b) infections with penicillin-resistant staphylococci, and (c) acne medications applied to the skin. Vancomycin is a narrow-spectrum antibiotic most effective in treating staphylococcal infections in cases of penicillin and methicillin resistance or in patients with an allergy to penicillins. It has also been chosen to treat *Clostridium* infections in children and endocarditis (infection of the lining of the heart) caused by *Enterococcus faecalis.*

Refampin

The dark red antibiotic has a somewhat limited spectrum because the molecule is too large to pass through the cell envelope of many Gram-negative bacilli. It is, however, very effective in treating infections by several Gram-positive rods and cocci, and a few Gram-negative bacteria. It must enter the cell in order to block transcription and inhibit cell growth. Rifampin figures most prominently in treating mycobacterial infections, especially tuberculosis and leprosy, but it is usually given in combination with other drugs to prevent development of resistance. Rifampin is recommended for prohylaxis in *Neisseria meningitidis* carriers and their contacts, and it is occasionally used to treat *Legionella, Brucella,* and *Staphylococcus* infections.

Bacitracin and Polymyxin

Bacitracin is a narrow-spectrum peptide antibiotic produced by a strain of *Bacillus subtilis* isolated from the infected compound fracture of a girl named Tracy. Since then, its greatest claim to fame has been as a major ingredient in a common drugstore antibiotic ointment (neosporin) for combating superficial skin infections by streptococci and staphylococci. For this purpose, it is usually combined with neomycin (an aminoglycoside) and polymyxin. *Bacillus polymyxa* is the source of the polymyxins, narrow-spectrum peptide antibiotics with a unique fatty acid component that contributes to their detergent activity. Only two polymyxins—B and E (also known as colistin)—have any routine applications, and even these are limited by their toxicity to the kidney. Either drug can be indicated to treat drug-resistant *Pseudomonas aeruginosa* and severe urinary tract infections caused by other Gram-negative rods. Polymyxin B alone is commonly added to topical skin ointments.

Sulphonamides, Trimethoprim, and Sulfones

The very first modern antimicrobic drugs were the sulphonamides, or sulfa drugs, named for para-aminobenzenesulphonamide (Fig. 17.13). Although thousands of sulphonamides have been formulated, only a few have gained any importance in chemotherapy. Because of its solubility, sulphisoxazole is the best agent for treating shigellosis, acute urinary tract infections, and certain protozoan infections. Silver sulphadiazine ointment and solution are prescribed for treatment of burns and eye infections. In many cases, sulphamethoxazole is given in combination with trimethoprim (Septra, Bactrim) to take advantage of the synergistic effect of the two drugs. This combination is one of the primary treatments for *Pneumocystis carinii* pneumonia (PCP) in AIDS patients. Sulphones are compounds chemically related to the sulphonamides but lack their broad-spectrum effects. This does not diminish their importance as key drugs in treating leprosy.

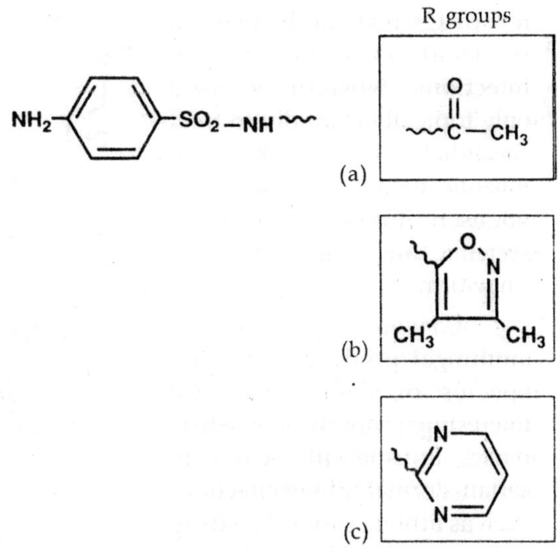

Fig. 17.13 : Sulphonamides (a) sulfacetamide (b) sulfadiazine (c) sulfioxazole.

Isoniazid (INH) has been in use since 1952 to treat tuberculosis. It is bactericidal to *Mycobacterium tuberculosis,* but only against growing cells. Oral doses are indicated for both active tuberculosis and prophylaxis in cases of a positive TB test. Ethambutol, a closely related compound, is effective in treating the early stages of tuberculosis. Nitrofurantoin, a derivative of furan sugars, is a fairly broad-spectrum synthetic drug most effective in Gram-negative urinary tract infections because it is excreted by the kidneys and is concentrated in the urine. Much excitement has been generated by the new class of synthetic drugs chemically related to quinine called fluoroquinolones. These drugs exhibit several ideal traits, including potency and broad spectrum. Even in minimal concentrations, quinolones inhibit nearly every Gram-positive and Gram-negative bacterial species tested. In addition, they are readily absorbed from the intestine and less subject to microbial resistance than other drugs. The principal quinolones, norfloxacin and ciprofloxacin, have been successful in therapy for urinary tract infections, sexually transmitted diseases, gastrointestinal infections, osteomyelitis, respiratory infections, and soft tissue infections.

ANTIFUNGAL DRUGS

Since the cells of fungi are eukaryotic, they present special problems in chemotherapy. For one, the great majority of chemotherapeutic drugs useful in treating bacterial infections are generally infective in combating fungal infections, and for another, the similarities between fungal and human cells often mean that drugs toxic to fungal cells are capable of harming human tissues. A small battery of agents with special antifungal properties has been developed for treating systemic and superficial fungal infections. Four main drug groups currently in use are the macrolide polyene antibiotics, griseofulvin, the synthetic azoles and flucytosine (Fig. 17.14).

Macrolide polyenes, represented by amphotericin B (named for its acidic and basic—amphoteric—properties) and nystatin (for New York State, where it was discovered), have a structure that mimics the lipids in some cell membranes. Both compounds were originally isolated from species of *Streptomyces.* Amphotericin B (Fungizone) is by far the most versatile and effective of all antifungals. Not only does it work on most fungal infections, including skin and mucous membrane lesions caused by *Candida albicans,* but one of the few drugs that can be injected to treat systemic fungal infections. Nystatin is used only topically or orally to treat candidiasis of the skin and mucous membranes, but it is not useful for subcutaneous or systemic fungal infections or for ringworm.

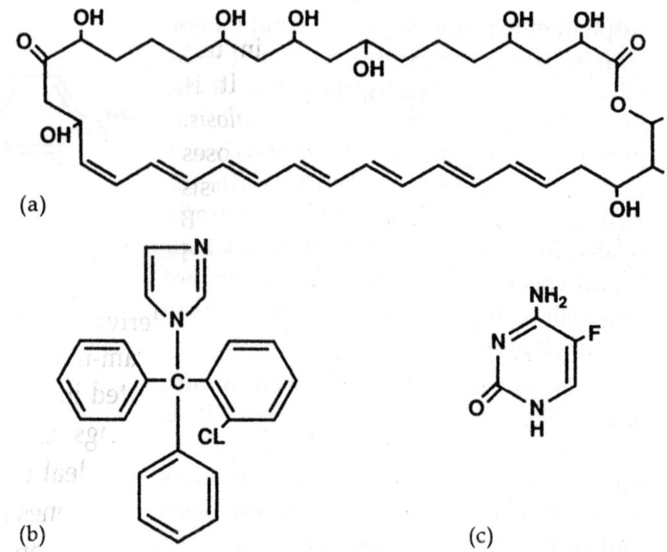

Fig. 17.14 : Antifungal drugs (a) polyenes (b) clotrimazole (c) flucytosine.

Griseofulvin is an antifungal product of another species of *Penicillium*. An interesting property of this drug makes it especially active in certain dermatophyte infections such as athlete's foot. The drug is deposited in the epidermis, nails, and hair, where it inhibits fungal growth. Becausec complete eradication requires several months and griseofulvin is relatively nephrotoxic, this therapy is given in only the most extreme cases. The *azoles* are broad-spectrum antifungal agents with a complex ringed structure. The most effective drugs are : Ketoconazole, which is used orally and topically for cutaneous mycoses, vaginal and oral candidiasis, and some systemic mycoses; and clotrimazole and miconazole, which are applied in creams and solutions to infections in the skin, mouth, and vagina. Flucytosine is an analog of cytosine that was first developed for tumor therapy in mammals. Although not an effective anticancer drug, it turned out to be useful in combating fungi. Its best features are its rapid absorption after oral therapy and its property of entering the blood and cerebrospinal fluid. Alone, it can be used to treat *Cryptococcus* meningitis and candidiasis. Now that many fungi are resistant to flucytosine, it must be combined with amphotericin B to effectively treat other systemic mycoses.

ANTIPARASITIC DRUGS

The enormous diversity among protozoan and helminth parasites and their corresponding therapies reaches far beyond the scope of this textbook; however, a few of the more common drugs will be surveyed here. Presently, a small number of approved and experimental drugs are used to treat malaria, leishmaniasis, trypanosomiasis, amebic dysentery, and helminth infections, but the need for new and better drugs has spurred considerable research in this area.

Quinines & its Derivatives

Quinine, extracted from the bark of the *Cinchona* tree, was the principal treatment for malaria for hundreds of years. After World War II, however, it was replaced by the synthesized quinolines, mainly chloroquine and primaquine, which had less toxicity to humans. Since there are several species of *Plasmodium* (the malaria parasite) and many stages in its life cycle, no single drug is universally effective for every species and stage, and each drug is restricted in application. For instance, primaquine eliminates the liver phase of infection, and chloroquine suppresses acute attacks associated with infection of red blood cells. The drugs are taken together for malarial prophylaxis and cure. Although quinine chemotherapy was abandoned for a time because of its toxicity, the development of chloroquine-resistant *Plasmodium* in South America and Southeast Asia restored quinine to a role in treating drug-resistant infections.

Miscellaneous Drugs

A widely used amebicide, metronidazole (Flagyl), is effective in treating mild and severe intestinal infections, and hepatic disease caused by *Entamoeba histolytica.* Given orally, it also has applications for infections by *Giardia lamblia* and *Trichomonas vaginalis.* Another antidote for amebic infections is dihydroemetine, a drug whose action is to induce emesis (vomiting), which helps eliminate the pathogen.

ANTIHELMINTHIC DRUGS

Flukes, tapeworms, and roundworms are much larger parasites than other microorganisms and have greater similarities to human physiology, so drugs that merely block their reproduction are usually not successful in eradicating the adult worms. The most effective drugs immobilize, disintegrate, or inhibit the metabolism of the worms. Mebendazole and thiabendazole are broad-spectrum antiparasitic drugs used in several roundworm and tapeworm intestinal infestations. Working locally within the intestine, these drugs inhibit the function of the cellular microtubules of adult worms, thereby interfering with their glucose utilization and disabling them. Egg and larval stages are also killed by these drugs. The compounds pyrantel and piperazine paralyze the muscles of intestinal roundworms. Niclosamide destroys the scolex and the adjoining proglottids of tapeworms, thereby loosening the worm's holdfast. In these forms of therapy, the worms are unable to maintain their grip on the intestinal wall and are expelled along with the feces by the normal peristaltic action of the bowel.

ANTIVIRAL DRUGS

Traditionally, viral infections have been prevented by vaccination, but useful vaccines have not been developed for many viruses. Recent epidemics of AIDS, genital warts, and measles, not to mention the common cold, continue to emphasize the need for effective antiviral drugs. The major limitations of the small number of existing antiviral agents are that most of them have an extremely narrow spectrum and act only on viruses inside the cell. Most compounds halt the viral multiplication cycle by : (a) Barring complete penetration of the virus into the host cell, (b) blocking the transcription and translation of viral molecules, or (c) preventing the maturation of viral particles. Although antiviral drugs protect uninfected cells by keeping viruses from being synthesized and released, most are unable to destroy extracellular viruses or those in a latent state. The mode of action of several antiviral agents is to mimic the structure of nucleotides and compete for sites on replicating DNA. The incorporation of these synthetic nucleotides inhibits further DNA synthesis. Acyclovir (Zovirax) is a synthetic purine compound that blocks DNA synthesis in a small group of viruses, particularly the *Herpes simplex* viruses (HSV). In the topical form, it is most effective in controlling the primary attack of facial or genital herpes. Intravenous or oral acyclovir therapy has proved successful in reducing the severity of primary and recurrent genital HSV episodes. A relative called Fancicloviris used to treat shingles and chickenpox caused by the *Herpes zoster* virus. *Idoxuridine* is an analog of thymine that inhibits the synthesis of DNA in certain viruses; at present, it is prescribed only for eye infections by *Herpes simplex*. An analog of adenine, *vidarabine,* is also effective against the herpesviruses. Ribavirin is a guanine analog used in aerosol form to treat life-threatening infections by the respiratory syncytial virus (RSV) in infants. Azidothymidine (AZT or Retrovir) is a thymine analog used exclusively to treat AIDS patients. This drug is specific for HIV by preventing the natural action of the viral reverse transcriptase and blocking further DNA synthesis and viral replication. It is indicated for patients with full-blown AIDS and for those in the earlier phases of disease. Although it is not a cure, AZT does slow the course of the disease is most instances. Other approved anti-AIDS drugs that have a similar action are didanosine (ddI), zalcitabine (ddC), and stavudine (d4T). *Amantadine* and its relative, rimantidine, are amines restricted almost exclusively to treating infections by influenza A virus. Because their action appears to inhibit the uncoating of the viral RNA, these drugs must be given rather early in an infection. Amantadine is also used prophylactically to reduce the incidence of influenza A infections in the elderly. At one time, great hope was held for the antiviral application of interferon. Interferon is a carbohydrate-containing protein produced naturally by infected animal cells. In humans, interferon is produced primarily by fibroblasts and leukocytes in response to various infectious stimuli. The discovery of interferon's potential application in viral infections and cancer therapy prompted a tremendous push for research on its biological effects. We now know that it is a versatile part of animal host defenses, having a broad spectrum of activities and great import in natural immunities. The first investigations of interferon's antiviral activity were limited by the extremely minute quantities that could be extracted from human blood. As techniques in recombinant DNA technology made it possible to produce larger quantities of interferon, extensive clinical trials to test its effectiveness in viral infections and cancer were undertaken, with inconclusive and somewhat disappointing results. Some major conclusions regarding interferon therapy can be summarized as follows : (a) It reduces the time of healing and some of the complications in certain infections (mainly of herpesviruses) but does not cure them. (b) It can prevent some symptoms of cold and papillomaviruses (warts). (c) It is somewhat toxic if given in large doses. (d) It apparently slows the progress of a few cancers, including bone cancer and cervical cancer, and certain leukemias and lymphomas.

HOST-DRUG REACTIONS

Although selective antimicrobial toxicity is the ideal constantly being sought, chemotherapy by its very nature involves contact with foreign chemicals that can harm human tissues. In fact, estimates indicate that at least 5% of all persons taking an antimicrobic drug experience some type of serious adverse reaction to it. The major side effects of drugs fall into one of three categories : Direct damage to tissues through toxicity, allergic reactions, and disruption in the balance of normal microbial flora. The damage incurred by antimicrobial drugs can be short-term and reversible or permanent, and it ranges in severity from cosmetics to lethal.

Drug Toxicity

Certain drugs adversely affect the following organs : The liver (hepatotoxic drugs), kidneys (nephrotoxic drugs), gastrointestinal tract, cardiovascular system and blood-forming tissue, nervous system (neurotoxic drugs), respiratory tract, skin, bones, and teeth. Because the liver is responsible for metabolizing and detoxifying foreign chemicals in the blood, it can be damaged by a drug or its metabolic products. Injury to liver cells can result in enzymatic abnormalities, fatty liver deposits, and hepatitis. A longterm effect can be liver failure. The kidney, too, excretes drugs and their metabolites in its work as a blood filter. Some drugs irritate the nephron tubules, creating changes that interfere with their filtration abilities. The most common complaint associated with oral antimicrobial therapy is diarrhea, which can progress to severe intestinal irritation or colitis. Although some drugs directly irritate the intestinal lining, the usual gastrointestinal complaints are caused by disruption of the intestinal microflora. Many drugs given for parasitic infections are toxic to the heart, causing irregular heart beats, and even cardiac arrest in extreme cases. Chloramphenicol can severely depress blood-forming cells in the bone marrow, resulting in either a reversible or permanent (fatal) anaemia. Some drugs hemolyze the red blood cells, others reduce white blood cell counts, and still others damage platelets or interfere with their formation, thereby inhibiting blood clotting.

Others, such as aminoglycosides, damage nerves (very commonly, the 8th cranial nerve), leading to dizziness, deafness, or motor and sensory disturbances. When drugs block the transmission of impulses to the diaphragm, respiratory failure can result. The skin is a frequent target of drug-induced side effects. The skin response can be a symptom of drug allergy or a direct toxic effect. Some drugs interact with sunlight to cause photodermatitis, a skin inflammation. Tetracyclines are contraindicated (not advisable) for children from birth to eight years of age because they bind to the enamel of the teeth, creating a permanent gray to brown discolouration. Pregnant women should avoid tetracyclines because they cross the placenta and can be deposited in the developing fetal bones and teeth.

Allergic Reactions

This reaction occurs because the drug acts as an antigen (a foreign material capable of stimulating the immune system) and stimulates an allergic response. This response can be provoked by the intact drug molecule or by substances that develop from the body's metabolic alteration of the drug. In the case of penicillin, for instance, it is not the penicillin molecule itself that causes the allergic response, but a major product, *benzylpenicilloyl*. Allergic reactions have been reported for every major type of antimicrobic drug, but the penicillins account for the greatest number of antimicrobic allergies, followed by the sulfonamides. People who are allergic to drug become sensitized to it during the first contact, usually without symptoms. Once the immune system is sensitized, a second exposure to the drug can lead to a reaction such as a skin rash (hives), respiratory inflammation, and rarely, anaphylaxis, an acute, overwhelming allergic response that develops rapidly and can be fatal.

Inhibition of Growth of Microflora

Most normal, healthy body surfaces, such as the skin, large intestine, outer openings of the urogenital tract, and oral cavity, provide numerous habitats for a virtual "garden" of microorganisms. These normal colonists or residents, called the flora or microflora, consist mostly of harmless or beneficial bacteria, but some can be potential pathogens. If a broad-spectrum antimicrobic is introduced into the host to treat infection, it will destroy microbes regardless of their roles in the ecologic balance, affecting not only the targeted infectious agent, but also many others in sites far removed from the original infection. In some cases, the result of this therapy is the destruction of beneficial resident species and the subsequent survival and overgrowth of opportunistic residents or contaminants. This complication is called a superinfection. Some common examples demonstrate how a disturbance in microbial flora leads to replacement flora and superinfection. A broad-spectrum cephalosporin used to treat a urinary tract infection by *Escherichia coli* will cure the infection but it will also kill off the lactobacilli in the vagina that normally maintain a protective acidic environment there. The drug has no effect however, on *Candida albicans,* a yeast that also resides in normal vaginas. Released from the inhibitory pH normally provided by lactobacilli, the yeasts proliferate and cause an infection. *Candida* can cause similar superinfections of the oropharynx (thrush) and the large intestine. Oral therapy with tetracyclines, clindamycin, and broad-spectrum penicillins and cephalosporins is associated with a serious and potentially fatal condition known as *antibiotic-associated colitis* (pseudomembranous colitis). This condition is due to the overgrowth in the bowel of *Clostridium difficile,* a spore-forming bacterium that is resistant to the antibiotic. It invades the intestinal lining and releases toxins that induce diarrhea, fever, and abdominal pain.

CRITERIA FOR DRUG SELECTION

Before actual antimicrobial therapy can begin, it is important that at least three factors be known : (a) The nature of the microorganism causing the infection; (b) the degree of the microorganism's susceptibility to various drugs; and (c) the overall medical condition of the patient.

Investigation of Causal Agent

It is especially important that such specimens be taken before the antimicrobic drug is given, because the drug can eliminate the infectious agent. Direct examination of Gram-stained specimens of body fluids, sputum, or stool is still a valuable method for rapidly detecting and perhaps even identifying bacteria or fungi. A doctor often begins the initial therapy on the basis of such immediate findings. The choice of drug will be based on past experience with drugs that are known to be effective against the microbe; this is called the "informed best guess". For instance, if a sore throat appears to be caused by *Streptococcus pyogenes,* the physician might prescribe penicillin, because this species seems to be universally sensitive to it so far. If the infectious agent is not or cannot be isolated, epidemiological statistics may be required to predict the most likely agent in a given infection. For example, *Haemophilus influenza* accounts for the majority of cases of meningitis in children, followed by *Streptococcus pneumoniae* and *Neisseria meningitidis.*

Testing for Drug Susceptibility

Testing is essential in those groups of bacteria commenly showing resistance, primarily *Staphylococcus* species, *Neisseria gonorrhoeae,* certain streptococci (*S. pneumoniae* and *Enterococcus faecalis*), and the aerobic Gram-negative enteric bacilli. However, not all infectious agents require antimicrobial sensitivity testing. Drug testing in fungal or protozoal infections is difficult and is often unnecessary. And because certain groups, such as group A streptococci and all anaerobes (except *Bacteroides*), have so far been uniformly susceptible to penicillin G, testing may not be necessary unless the patient is allergic to penicillin. Selection of a proper antimicrobial agent begins by demonstrating the *in vitro* activity of several drugs against the infectious agent by means of standardized methods. In general, these tests involve exposing a pure culture of the isolated bacterium to several different drugs and observing it for the effects of the drugs on growth. The *Kirby-Bauer* technique is an agar diffusion test that provides useful semiquantitative or qualitative data on antimicrobic susceptiblity.

In this test, the surface of a plate of special medium is covered completely with the isolated test bacterium, and small discs containing a premeasured amount of antimicrobic are dispensed onto the bacterial lawn. After 18 to 24 hours of incubation at 37°C, the zone of inhibition surrounding the discs is measured and compared with a standard for each drug. The profile of antimicrobic sensitivity, or antibiogram, provides data for drug selection. An advantage of the Kirby-Bauer procedure is that many drugs can be tested simultaneously in a single plate; however, it is less effective for anaerobic, highly fastidious, or slow-growing bacteria. More sensitive and quantitative results can be obtained with tube dilution tests. First the antimicrobic is diluted serially in containers of broth, and then each tube is inoculated with a small uniform sample of pure culture. After incubation for 18-24 hours, the tubes are examined for growth (turbidity). The smallest concentration of drug that visibly inhibits growth is called the minimum inhibitory concentration, or MIC. The MIC is useful in determining the smallest effective dosage of a drug and in providing a comparative index against other antimicrobics.

The physician must also take a careful history of the patient to discover any preexisting medical conditions that will influence the activity of the drug or the response of the patient. A history of allergy to a certain class of drugs should preclude the administration of that drug. Underlying liver or kidney disease will ordinarily necessitate the modification or drug therapy, because these organs play such an important part in metabolizing or excreting the drug. Infants, the elderly, and pregnant women also necessitate special precautions. Because drugs can interact with one another, the patient's current intake of other drugs must be carefully scrutinized. Drug incompatibilities can result in increased toxicity or failure of one or more of the drugs. For example, the combination of aminoglycosides and cephalosporins increases nephrotoxic effects; antacids reduce the absorption of isoniazid; and the interaction of tetracycline or rifampin with oral contraceptives can abolish the contraceptive's effect. Some drugs act synergistically so that reduced doses of each can be used in combined therapy.

Consider the case of an elderly alcoholic patient with pneumonia caused by *Serratia marcescens* and complicated by diminished liver and kidney function. All drugs must be given parenterally because of prior damage to the gastrointestinal lining and poor absorption. Drug tests show that the infectious agent is sensitive to third-generation cephalosporins, gentamycin, amikacin, and azlocillin. The patient's history shows previous allergy to the penicillins, so these would be ruled out. Drug interactions occur between alcohol and the cephalosporins, which are also associated with serious bleeding in elderly patients, so this is perhaps not a good choice. The aminoglycosides are nephrotoxic and poorly cleared by damaged kidneys. On the surface, it would appear that none of the drugs is a good choice, but closer examination reveals that amikacin is less nephrotoxic than gentamycin. In the case of a cancer patient with severe systemic *Candida tropicalis* infection, there will be fewer criteria to weigh. Intravenous amphotericin B alone or in combination with flucytosine is about the only possibility, despite the drug's nephrotoxicity and other possible adverse side effects. In a life-threatening situation, where a dangerous chemotherapy is perhaps the only chance for survival, the choices are reduced and the priorities are different.

18

FERMENTATION AND SYMBIOTIC INTERACTIONS IN ORGANISMS

FERMENTATION

The introductory reactions whereby a variety of alternate sugars and other substrates are metabolized to intermediates that can enter these central metabolic routes are also described. Individual organisms may use the fructose bisphosphate aldolase, phosphoketolase, or ketogluconate aldolase pathways or variations of them. Additional fermentative products may be derived from pyruvate or other intermediates (Fig. 18.1). Oxidation-reduction reactions play a major role in the fermentative metabolism of carbohydrates. It is, therefore, advisable to perform an oxidation-reduction or O–R balance on the products formed. The O – R balance provides an indication as to whether the formed products balance with regard to their oxidized or reduced states. It may not be possible to balance the hydrogen and oxygen of the substrate directly because hydrations or dehydrations may occur as intermediary steps in the fermentation pathway. However, a simple way to achieve the same end is to calculate an oxidation value for each compound and compare the total amount of oxidized and reduced products. If the ratio of oxidized products to reduced products is close to the theoretical value of 1.0, this provides further indication that the products are in balance. The oxidation value of a compound is determined from the number of oxygen atoms less one-half the number of hydrogen atoms. For glucose, which has 6 oxygen atoms and 12 hydrogen atoms, the oxidation value is 0. Thus, glucose is referred to as a neutral compound, and to be equally balanced the fermentation products should contain equivalent amounts of oxidized and reduced products. Another aspect of fermentation balances is the C_1 balance. The amount of expected C_1 product is calculated from the amounts of those products for which CO_2 or formate is expected as an accompanying product. For example, if a C_2 compound such as ethanol or acetate is among the final products, an equal amount of CO_2 will be expected, since ethanol is derived from pyruvate by decarboxylation.

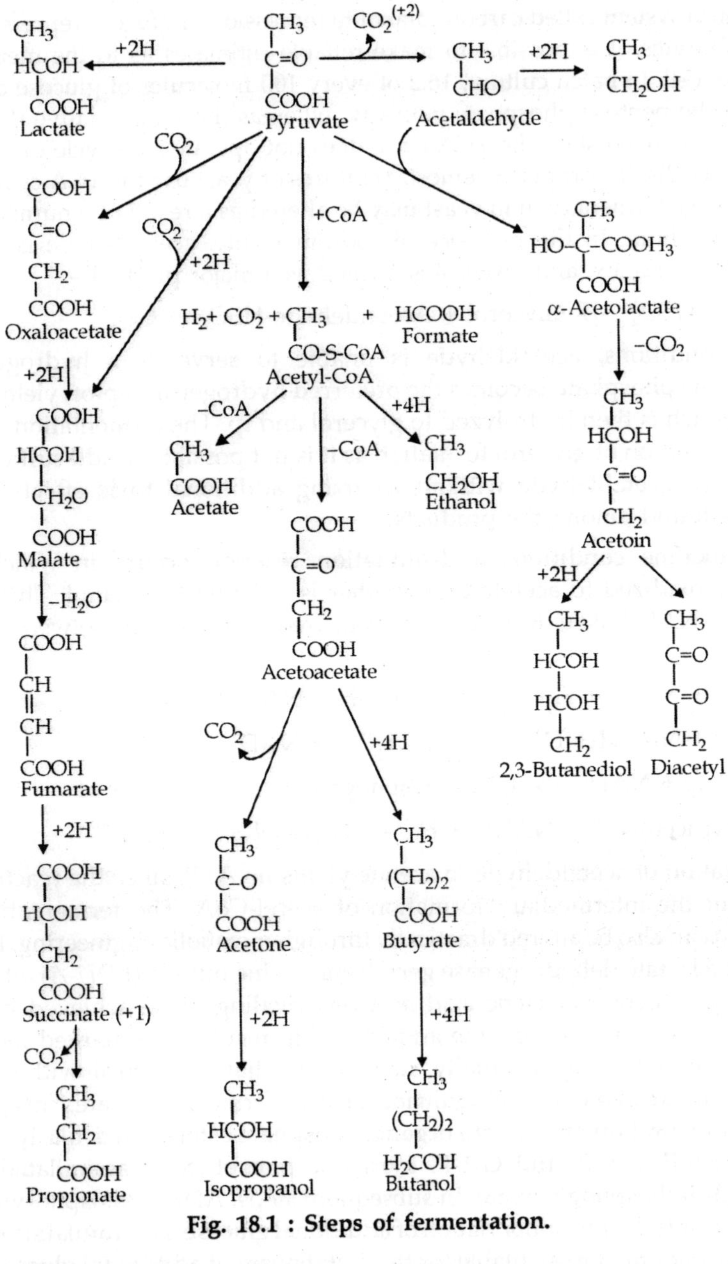

Fig. 18.1 : Steps of fermentation.

FERMENTATION BY YEAST

The yeast *Saccharomyces cerevisiae* species uses the EMP pathway of glucose metabolism under the conditions of neutral or slightly acid pH and an anaerobic environment. The major products formed under these conditions are carbon dioxide and ethanol. However, certain facets of the alcoholic fermentation of yeast bear additional consideration. When yeast cells are grown in high concentrations of glucose, the cells function under a regulatory control system called carbon catabolite repression or glucose repression. By using ^{13}C-labeled substrates, it is possible to make reliable estimates as to the metabolic fluxes occurring in the cell. In batch culture, 16.2 of every 100 molecules of glucose consumed by the cells enters the pentose phosphate pathway, whereas the same relative flux is 44.2 per 100 molecules in a chemostat. The TCA cycle does not operate as a cycle in batch cultures of growing yeast cells. In contrast, chemostat cultures of yeast use the TCA cycle in its intact form. The ethanolic fermentation in yeast may be altered as a result of a number of changes in the culture medium. In the presence of sodium sulfite, acetaldehyde is trapped as a bisulfite addition complex, and glycerol is formed as a major product–

$$\text{Glucose} + HSO_3^- \rightarrow \text{Glycerol} + \text{Acetaldehyde-}HSO_3^- + CO_2$$

Under these conditions, acetaldehyde is unable to serve as a hydrogen acceptor. Dihydroxyacetone phosphate becomes the preferred hydrogen acceptor, yielding glycerol-3-phosphate, which is then hydrolyzed to glycerol and ip. The fermentation is not shifted entirely in the direction of glycerol formation as it is not possible to add sufficient bisulfite to bind all of the acetaldehyde without incurring additional toxic effects. Thus, some ethanol will be found among the products.

Under alkaline conditions a dismutation reaction occurs in which 1 mol of acetaldehyde is oxidized to acetate and another is reduced to ethanol. This sequence is catalyzed by two NAD-linked dehydrogenases and requires a balance between the following reactions–

$$CH_3CHO + NAD^+ + H_2O \rightarrow CH_3COOH + NADH + H^+$$

$$CH_3CHO + NADH + H^+ \rightarrow CH_3CH_2OH + NAD^+$$

$$\text{GA-3-P} + \text{ip} + NAD^+ \rightarrow \text{1,3-diphosphoglycerate} + NADH + H^+$$

$$\text{Dihydroxyacetone-P} + NADH + H^+ \rightarrow \text{Glycerol-3-P} + NAD^+$$

The oxidation of acetaldehyde to acetate yields no ATP, since the reaction proceeds directly without the intermediary formation of acetyl-CoA. The fermentation products formed by yeast can also be altered drastically through metabolic engineering. For example, introduction of a lactate dehydrogenase gene from bovine muscle (*LDH-A*) into S. *cerevisiae* engenders the production of lactic acid at levels rivaling those achieved by lactic acid bacteria. In cell-free yeast extracts the addition of ip results in a marked increase in the fermentation rate. The rate eventually subsides to that of control without added ip. Addition of second quality of ip will again increase the fermentation rate. This phenomenon results from incorporation of ip a into organic phosphate esters, particularly fructose-1, 6-bisphosphate, F-6-P, G-6-P, and G-1-P. Inorganic phosphate is assimilated during the formation of 1,3-diphosphoglycerate. In subsequent steps, ADP is phosphorylated to ATP which in turn permits the phosphorylation of additional glucose. Accumulation of phosphate esters limits the amount of ip available for the metabolism of additional glucose. This effect can be counteracted by the addition of arsenate ion, which results in the formation of phosphoglyceryl arsenate. This compound hydrolyzes nonenzymatically, preventing the accumulation of phosphorylated intermediates.

During the usual procedure of preparing cell-free extracts of yeast, the enzyme ATPase may be inactivated. Preparations in which ATPase is active, or to which exogenous ATPase is added, do not exhaust the ip, and phosphate esters do not accumulate. Under these conditions, fermentation continues at a constant rate. In a resting yeast cell suspension that is fermenting glucose, the introduction of oxygen results in the cessation of ethanol formation. This phenomenon, termed the Pasteur effect, has been studied extensively in attempts to elucidate the mechanism involved. Growing yeast cells do not show a noticeable Pasteur effect. Some of the difficulty in studying alcoholic fermentation in yeast results from the fact that several isozymes of alcohol dehydrogenase (ADH) are present, and the activities of these and other enzymes concerned with carbohydrate metabolism are partitioned between the cytoplasm, the mitochondrial membrane, and the mitochondrial cytosol. The key enzyme of fermentative alcohol production (ADH1) reduces acetaldehyde to ethanol in the presence of NADH and serves to maintain the redox balance of glucose fermentation occurring in the cytoplasm of yeast cells. During ethanol oxidation, ADH2 catalyzes the formation of acetaldehyde. Oxidative utilization of ethanol in the mitochondrion involves ADH3. The inner mitochondrial membrane is impermeable to NAD^+ or NADH. Yeast mutants lacking ADH1 may regenerate cytoplasmic NAD^+ by reducing DHAP to glycerol-3-phosphate. Glycerol-3-phosphate can enter the mitochondrial membrane, where it is reoxidized to DHAP by an FAD-linked glycerol-3-phosphate dehydrogenase. The electrons generated in this oxidation are transferred to the respiratory chain. For this reason, ADH1-deficient cells are unable to grow anaerobically. As a result of the release of ip from glycerol-3-phosphate, glycerol appears as an end product. However, yeast deficient in at least four of the known ADH isozymes still produces up to one-third of the theoretical maximum yield of ethanol from glucose. This may result from the presence of additional forms of ADH. The published genome for *S. cerevisiae* identified at least seven genes that could potentially code for members of the ADH family. Glycerol is a major fermentation product, but acetaldehyde and acetate are also produced.

PRODUCTION OF LACTIC ACID BY FERMENTATION

Those species that ferment glucose primarily to lactic acid are homofermentative and those that produce a mixture of other products along with lactic acid are heterofermentative. Some homofermentative species can form a wider variety of products if the conditions of the fermentations are altered. At alkaline pH the amounts of formate, acetate, and ethanol are increased at the expense of lactate. As discussed earlier, isotope-labeling studies are helpful in determining the major pathway used in the formation of end products. *Lactobacillus casei, L. pentosus,* and *S. faecalis* were shown to produce ^{14}C-methyl-labeled lactate from glucose-1-^{14}C with 50% dilution and ^{14}C-carboxyl-labeled lactate from glucose-3,4-^{14}C without dilution of the specific activity. These labeling patterns fit the distribution of carbon atoms expected from utilization of the EMP pathway (Fig. 18.2). In *L. lactis* an operon has been shown to code for phosphofructokinase, pyruvate kinase, and lactate dehydrogenase, three enzymes crucial to the operation of the EMP pathway. Distribution of ^{14}C-labeled substrates indicates that more than 90% of the carbon in fermented sugars is converted into metabolic end products, usually lactic acid (Fig. 18.2). True homolactic acid bacteria metabolize glucose via the EMP pathway while those that are truly heterofermentative presumably follow a hexose monophosphate pathway. However, some data suggest that a combination of both pathways may be operative in certain organisms.

In the oxidative step in the triose phosphate pathway, NAD^+ is reduced. The electrode potential of this system ($NAD^+/NADH_2$) is –0.28 V. The electrode potential of the lactate dehydrogenase reaction is –0.18 V. Since this is at a higher potential, reduced NAD

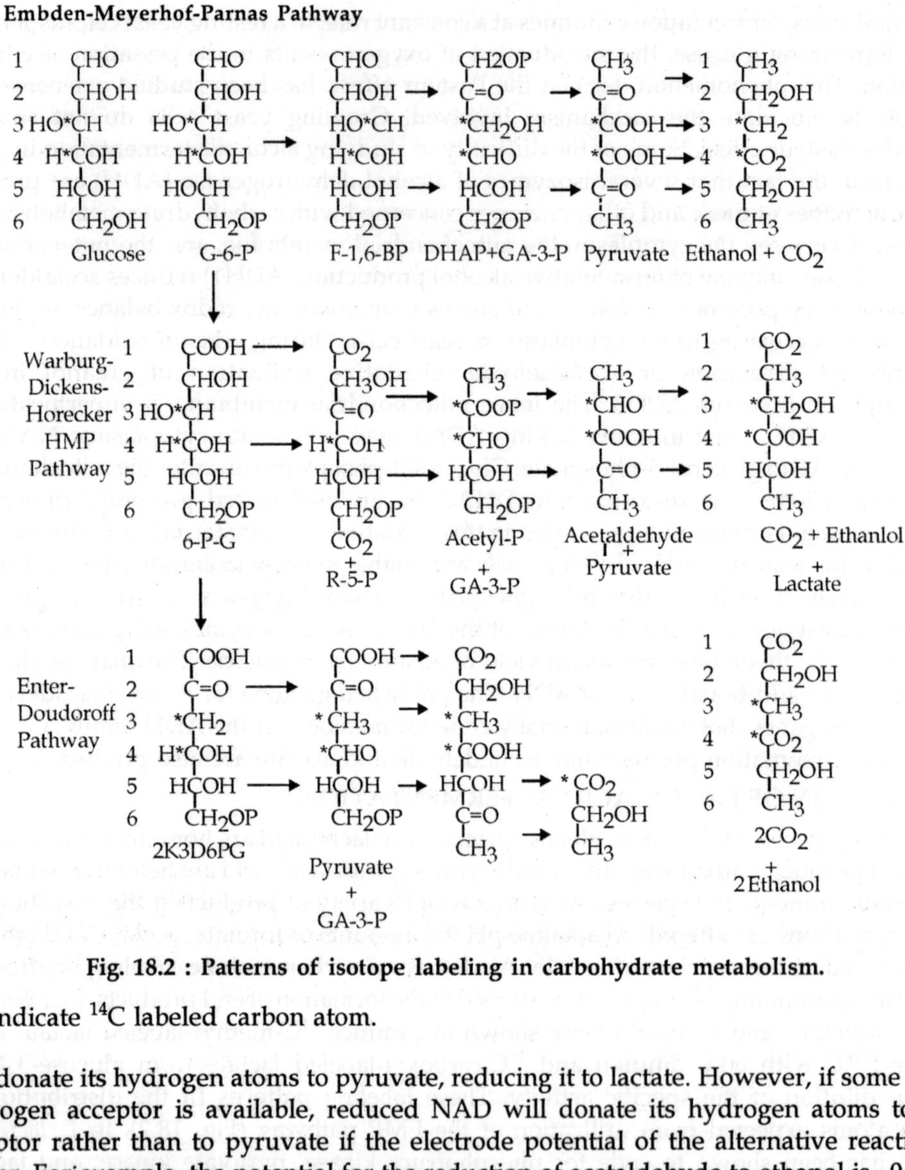

Fig. 18.2 : Patterns of isotope labeling in carbohydrate metabolism.

* C indicate ^{14}C labeled carbon atom.

will donate its hydrogen atoms to pyruvate, reducing it to lactate. However, if some other hydrogen acceptor is available, reduced NAD will donate its hydrogen atoms to this acceptor rather than to pyruvate if the electrode potential of the alternative reaction is higher. For example, the potential for the reduction of acetaldehyde to ethanol is –0.07 V. If acetaldehyde is formed, then NADH will donate its hydrogen atoms to this system. Under the usual conditions of lactic acid fermentation, acetaldehyde does not appear to be formed in any appreciable quantity. However, by alteration of the pH, the fermentation may be diverted to a variety of products. Since the streptococci appear to utilize the EMP pathway, the diversion of fermentation must occur at a step after pyruvate formation rather than by metabolizing hexose through an alternative pathway. Ethanol, when produced, is formed via reduction of acetyl phosphate–

$$CH_3COOPO_3H_2 + 2\ (2H) \rightarrow CH_3CH_2OH + ip$$

Because of a difference in the pyruvate cleaving systems, formate, rather than carbon dioxide is the major one carbon product. Carbon dioxide is formed in small amounts by some homolactic organisms under conditions in which the fermentation is diverted from the production of lactate. Homofermentative organisms, such as *S. faecalis,* have been shown to possess high levels of glucose-6-phosphate dehydrogenase and 6-phosphogluconate dehydrogenase. However, a homofermentative pattern is maintained through regulatory interrelationship between the EMP and HMP pathways (Fig. 18.3). The specific activation of lactate dehydrogenase (LDH) by FDP is characteristic of *S. bovis* and many other lactic acid-producing bacteria. The level of LDH, which is dependent on FDP for activity, decreases as fermentation becomes heterolactic. In *S. faecalis* a single NADP-linked 6PGD is observed when glucose is the primary energy source. Gluconate-adapted cells produce two 6-PGDs : One specific for NADP and the other specific for NAD. The NADP-linked enzyme is insensitive to FDP inhibition but is inhibited by ATP and other nucleotides. The G-6-P isomerase is inhibited by 6-PG. When there is a cellular demand for ATP, these regulatory activities direct glucose carbon through the energy-producing EMP pathway. Inhibition of G-6-P isomerase by 6-PG prevents the accumulation of F-6-P. Alteration of the fermentative pattern of these organisms by change in pH or other environmental influences may result from the disruption of these regulatory functions. In *S. mutans,* triose phosphates (glyceraldehyde-3-phosphate or dihydroxyacetone phosphate) strongly inhibit pyruvate-formate lyase (PFL), the first step in the dismutation sequence.

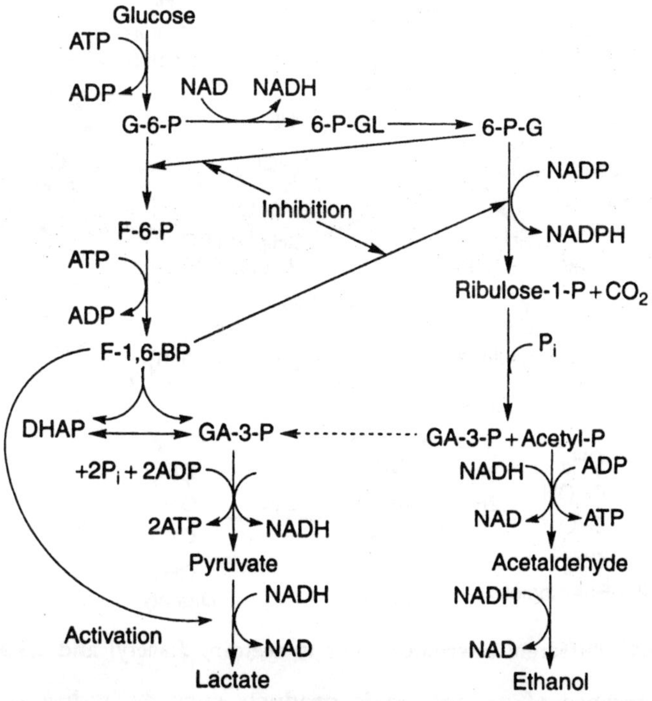

Fig. 18.3 : Regulation of metabolic pathways in *Streptococcus faecalis.*

Leuconostoc and *Lactobacillus*, produce acetoin, diacetyl, and 2,3-butanediol are responsible for the characteristic and desirable flavour in certain dairy products such as butter. In certain other fermentations, the presence of these compounds is undesirable, and it is of importance to be able to minimize their production (Fig. 18.4). With acetate as substrate, acetyl-CoA is formed, directly from acetate without the intermediary formation of pyruvate. *Lactococcus diacetylactis, Lactobacillus casei*, and other lactate-producing organisms that require lipoic acid for growth activate acetate to acetyl phosphate and then from acetyl-CoA via acetate kinase and phosphotransacetylase. Therefore, in media devoid of lipoic acid, these organisms cannot form acetyl-CoA from pyruvate. In a lipoate-free medium containing glucose and acetate, acetoin is formed via both acetyl-CoA-C_2-TPP condensation and C_2-C_3 condensation routes. In *L. lactis* the gene for α-acetolactate decarboxylase (*aldB*) is clustered with the genes for the branched-chain amino acids. This enzyme appears to play a dual role in this organism by regulating valine and leucine biosynthesis as well as catalyzing the second step of the pathway to acetoin and 2,3-butanediol. During growth in milk, PFL and lactate dehydrogenase are important to sustain growth of *L. lactis* and to regenerate reduced cofactors under anaerobic conditions. The gene encoding PFL has been cloned and characterized. The enzyme is activated via free-radical formation and is degraded in the presence of oxygen. Under anaerobic conditions, the carbon flux pyruvate is distributed mainly between the two competing enzymes : Lactate dehydrogenase (LDH) and pyruvate-formate lyase (PFL)

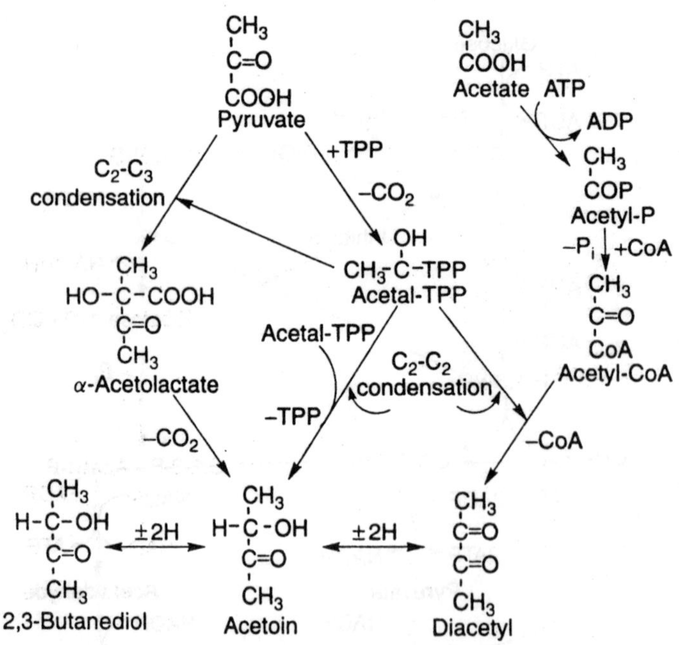

Fig. 18.4 : Different alternate routes to acetoin, diacetyl and 2,3-butanediol.

In the presence of oxygen, toxic products such as hydrogen peroxide may be generated by the autooxidation of reduced FAD, and hydroxyl radicals may be formed by the interaction of this compound with superoxide. Anaerobic organisms that tolerate the presence of oxygen are usually able to do so by virtue of the pressence of enzymes that protect them from these toxic compounds. Some fermentative organisms such as *L. plantarum* are aerotolerant simply because they do not interact with molecular oxygen. High intracellular concentrations of Mn^{2+} also aid these organisms in scavenging superoxide anion. *Streptococcus pyogenes* and other lactic acid bacteria produce NADH oxidase, a unique flavoprotein that catalyzes the four-electron reduction of oxygen to water. This enzyme may actually allow certain fermentative organisms to use molecular oxygen to advantage in producing energy. Under anaerobic conditions the conversion of pyruvate to lactate or a mixture of formate, acetate, and ethanol serves to oxidize NADH back to NAD^+. In the presence of oxygen, flavin-containing enzymes catalyze the oxidation of NADH by oxygen. Although *S. mutans* depends on glycolysis and prefers an anaerobic environment, the presence of oxygen induces the formation of NADH oxidase and superoxide dismutase. The presence of free FAD enhances the activity of Nox-1 but not that of Nox 2. *S. mutans* also produces a peroxidase (AhpC) encoded by *ahpC,* a gene located upstream of the *nox-1* gene. This peroxidase enzyme complements Nox-1 by reducing the peroxide to water and emulating the action of Nox-2. However, Nox-2 appears to play an important role in aerobic energy metabolism in oxygen-tolerant *S. mutans,* whereas Nox-1 serves a negligible role despite the face that it is the only enzyme produced when the organism is grown under anaerobic conditions.

The organism also produces another antioxidant system, encoded by the *dpr* gene that is independent of the Nox-1-AhpC system and can compensate for deletions of both *nox-1* and *ahpC* in mutant strains. The product of the *dpr* gene, Dpr, is an iron-binding protein that presumably contributes to oxygen tolerance by maintaining a low level of iron. Most lactic acid bacteria do not produce cytochromes because they are unable to synthesize heme. However, when grown in the presence of hematin, *Enterococcus faecalis* produces functional cytochromes that yield ATP via oxidative phosphorylation. An *E. faecalis* gene cluster, *cydABCD,* encodes a functional cytochrome *bd*-terminal oxidase that is produced when the cells are grown in the presence of heme. The ability to form cytochromes seems to be limited to *E. faecalis* and a few strains of *L. lactis.*

FERMENTATION BY *CLOSTRIDIUM* FOR BUTYRIC ACID AND OTHER SOLVENT PRODUCTION

Clostridium acetobutylicum utilizes the EMP pathway for glucose catabolism with the formation of ethanol, carbon dioxide, hydrogen, acetone, isopropanol, butyrate, and butanol from pyruvate via acetyl-CoA. During growth, the organism first forms acetate and butyrate (acidogenic phase). As the pH drops and the culture enters the stationary phase, there is a metabolic shift to solvent production (solvetogenic phase). The pivotal reaction in these fermentations is the formation of acetyl-CoA from pyruvate. Acetyl-CoA can then undergo condensation to form acetoacetate, which may be reduced to butyrate and butanol or cleaved via decarboxylation to acetone Acetone may be further reduced to isopropanol. Acetyl-CoA may also be reduced to acetaldehyde and ethanol. With other species such as *Bacillus acetoethylicum* or *B. polymyxa,* lactate or 2,3-butanediol may be produce other solvents. In balancing the fermentation products formed in these complex fermentations, it is not difficult to account for the carbon in the products formed. The carbon atoms balance, but there is a deficit of some reduced product. In most clostridial fermentations there is a compensatory formation of small amounts of several reduced products. Production of these compounds appears to be necessary because the condensation of 2 mol of acetyl-

CoA to form acetoacetyl-CoA and the subsequent conversion to acetone via acetoacetate decarboxylase results in a sharp decrease in the number of hydrogen acceptors available. Acetyl-CoA can act as a hydrogen acceptor, giving rise to ethanol, and acetone can be further reduced is isopropanol.

Generally, small amounts of either oxidized or reduced products will be formed to provide for a balance of the oxidized and reduced products. Formate is produced in relatively few of these fermentations. Although carbon dioxide and hydrogen gas are produced as in the fermentation of Gram-negative organisms, in clostridia and other anaerobes hydrogen is formed through a pyruvate : Ferredoxin (Fd) oxidoreductase without the intermediary production of formate. The reduced ferredoxin is converted to hydrogen by hydrogenase—

$$\text{Pyruvate} + \text{Fd} \rightarrow \text{Acetyl-CoA} + CO_2\ \text{FdH} + H^+$$

$$\text{FdH} + H^+ \rightarrow H_2 + \text{Fd}$$

The flow of electrons from NADH to FdH to H_2 explains why these organisms produce large quantities of H_2 and do not consume H_2. An enzyme that contains two Fe_4S_4 clusters catalyzes the oxidation of pyruvate to acetyl-CoA, carbon dioxide, and reduced ferredoxin. Ferredoxin also contains two Fe_4S_4 clusters arranged in approximately two-fold symmetry within the ferredoxin protein. Hydrogenases that result in the production of hydrogen gas contain iron only enzymes. The *C. pasteurianum* enzyme contains five distinct iron-sulfur clusters covalently attached to the protein. One Fe_2S_2 cluster and three Fe_4S_4 clusters transfer electrons to the hydrogen center, the presumed catalytic site of the enzyme. The hydrogen center is composed of a six-iron cluster and a two-iron subcluster bound to carbon monoxide and/or cyanide (Fig. 18.5). Production of acetic acid from glucose by *C. thermoaceticum* is efficient in that 3 molecules of acetate are produced per molecule of glucose. It is particularly significant that the synthesis of acetate from carbon dioxide by this organism was the first demonstration of total synthesis of an organic compound from carbon dioxide by a heterotrophic organism. Bacteria that can form an organic compound from carbon dioxide and hydrogen contain hydrogenase, an enzyme that converts hydrogen to two protons and two electrons. These electrons provide the necessary reductive capacity for the utilization of carbon dioxide. The overall

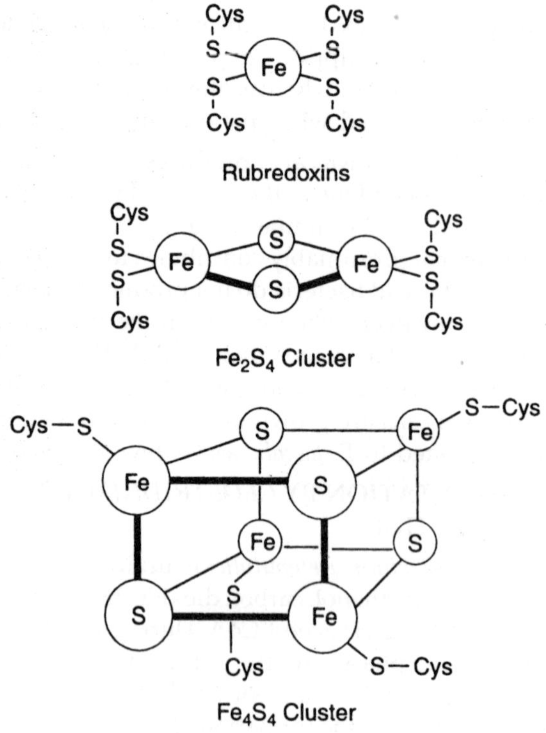

Fig. 18.5 : Hypothetical arrangements of the iron (Fe) and sulphur (S) atoms in rubredoxin, ferredoxin, and hydrogenase systems.

reaction involves participation of ferredoxin as a reduced electron carrier and the enzymes hydrogenase, carbon monoxide dehydrogenase, and methylenetetrahydrofolate reductase.

PRODUCTION OF MIXED ACIDS BY FERMENTATION

The production of acetate and formate as major products is notable. As much as 85% of the glucose fermented by *E. coli* is metabolized via the EMP pathway. Other pathways must contribute to the products to some extent, however. As opposed to clostridial fermentations in which acetyl-CoA, carbon dioxide, and dihydrogen usually arise without format as an intermediate, formate is consistently found as a product of sugar metabolism by enteric bacteria. A CoA-dependent pyruvate-formate lyase (Pfl) encoded by the *pfl* gene initiates the sequence.

$$CH_3COCOOH + CoASH \leftrightarrow CH_3COSCoA + HCOOH$$

Phosphotransacetylase converts acetyl-CoA to acetyl phosphate–

$$CH_3COSCoA + ip \leftrightarrow CH_3COOPO_3H_2 + CoASH$$

This enzyme results in the production of acetate and generates a major portion of the ATP produced during anaerobic growth. The formate-hydrogen lyase (FHL) system converts formate to H_2 and CO_2. The FHL system actually consists of two enzymes : Formate dehydrogenase-H (FDH-H), which yields carbon dioxide and a reduced carrier, and a hydrogenase-3, which produces H_2 from the reduced carrier. Induction of FDH-H (encoded by *fdhF*) requires the presence of formate, molybdate, the absence of electron acceptors, such as oxygen or nitrate, and acidic pH. Expression of *fdhF* requires an alternate σ factor (NtrA) and an upstream activating sequence (UAS). Expression of *pfl* is dependent on phosphoglucoisomerase and phosphofructokinase activities.

Production of ethanol occurs via the reduction of acetyl-CoA to ethanol through a CoA-linked acetaldehyde dehydrogenase coupled to an NAD-linked alcohol dehydrogenase. Acetoin is formed by *Enterobacter aerogenes* but not by *E. coli*. A variety of other organisms (e.g., *Serratia, Erwinia*) also form acetoin or its oxidation (diacetyl) or reduction (2,3-butanediol) products. In Gram-negative organisms, thiamine pyrophosphate is involved as a cofactor and α-acetolactate is produced as an intermediate in acetoin formation. Acetate induces the production of the acetolactate-forming enzyme, acetolactate decarboxylase, and diacetyl, reductase. Diacetyl reductase also functions as an acetoin dehydrogenase in *E. aerogenes* and serves as a regulator of the balance between acetoin and 2,3-butanediol. In *E. coli* the direct formation of diacetyl from 2-hydroxyethyl-TPP and acetyl-CoA has been demonstrated. Diacetyl is reduced to acetoin by an $NADP^+$ –specific diacetyl reductase that does not act on acetoin. Two genes, *ilvBN* (coding for acetohydroxy acid synthase) and *alsS* (coding for α-acetolactate synthase), are involved in producing acetolactate from pyruvate in *B. subtilis*. Acetolactate can be converted to acetoin by spontaneous decarboxylation at low pH or via the action of acetolactate decarboxylase (encoded by *alsD*). Acetoin serves as a carbon storage compound. It is reimported from the growth medium and utilized during the stationary phase when other carbon sources have been depleted. There appears to be more than one pathway of acetoin utilization in *B. subtilis*. Three genes form an operon, *acuABC*, genes that encode acetoin utilization enzymes. Another system, the *acoABCL* operon, encoding the multicomponent acetoin dehydrogenase enzyme complex, appears to constitute the main route of acetoin catabolism in this organism.

More oxidized substrate resulted in the production of more neutral product in the form of acetate. The redox level of the substrate is important in governing the ratio of oxidized to reduced products. For example, sugar alcohols, which are more reduced than the corresponding hexoses, must yield a higher proportion of more reduced fermentation products to achieve hydrogen balance. Fermentation of citrate (oxidation value : +3) yields sufficient oxidized products in the form of HCOOH (oxidation value : +1) and carbon dioxide (oxidation value : +2) to balance the oxidation/reduction equation. The fermentation of citrate by *Klebsiella pneumoniae* begins with cleavage by citrate lyase to form acetate and oxaloacetate. Oxaloacetate is decarboxylated by the oxaloacetate decarboxylase Na^+ pump. Part of the energy from this reaction is used for citrate transport. The formed pyruvate is degraded by pyruvate-formate lyase to acetyl-CoA and formate. Phosphotransacetylase forms acetyl phosphate from acetyl-CoA an ip. Transfer of high-energy phosphate from acetyl phosphate to ADP generates ATP, leaving acetate as a final end product. A portion of the formate is acted upon by formate-hydrogen lyase, yielding CO_2 and H_2. An NADP-reducing hydrogenase converts H_2 to NADPH + H^+, which is available for biosynthetic reactions. Utilization of glycerol (oxidation value : –1) involves transport into the cell via a facilitator encoded by *glpF*. Glycerol is phosphorylated by glycerol kinase (encoded by *glpK*) trapping it inside the cell as glycerol-3-phosphate (G-3-P). Under anaerobic conditions a dehydrogenase encoded by the *glpACB* operon yields dihydroxyacetone phosphate, which can then be metabolized via the triose phosphate pathway. In the presence of oxygen, an aerobic dehydrogenase encoded by the *glpD* operon accomplishes the same end. The *fnr* gene product FNR, a pleiotropic activator of genes involved in anaerobic respiration, regulates the utilization of glycerol, G-3-P, and glycerophosphate diesters.

PRODUCTION OF PROPIONIC ACID BY FERMENTATION

Clostridium propionicum, Bacteroides ruminicola, and *Peptostreptococcus* can remove water from lactate to form acrylate with subsequent reduction to propionate. Coenzyme A and lactyl CoA dehydrase mediate the reaction.

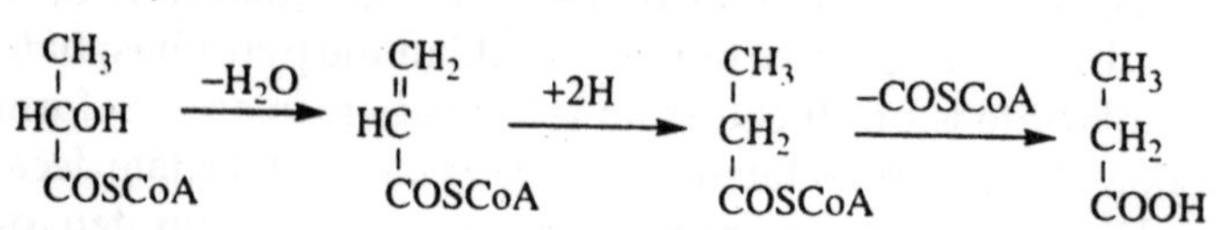

Lactyl-CoA Acrylyl-CoA Propionyl-CoA Propionate

The details of the pathway from glucose to propionate, acetate, and CO_2 in propionibacteria are shown in Figure 18.6. A major factor in the elucidation of this route of propionate formation was the demonstration of transcarboxylation by methylmalonyl-oxaloacetate transcarboxylase. This multimeric transcarboxylase catalyzes the reversible transfer of a carboxyl group from methylmalonyl CoA to pyruvate to form propionyl-CoA and oxaloacetate. The transferred carboxyl group is never released or exchanged with the CO_2; nor it is exchanged with the CO_2 dissolved in the medium. In this reaction, biotin plays an important catalytic role. Cobalamin (vitamin B_{12}) also serves as a cofactor in the formation of methylmalonyl-CoA from succinyl-CoA in the reaction sequence.

Under appropriate conditions and in the presence of the requisite enzymes, yields of 11 mol of ATP per 3 mol of glucose can be achieved :

$$3\ \text{Glucose} + 4\text{Pip} + 11\text{ADP} \longrightarrow 2\ \text{Acetate} + 4\ \text{Propionate} + 2CO_2 + 11\text{ATP} + 4\text{Pip}$$

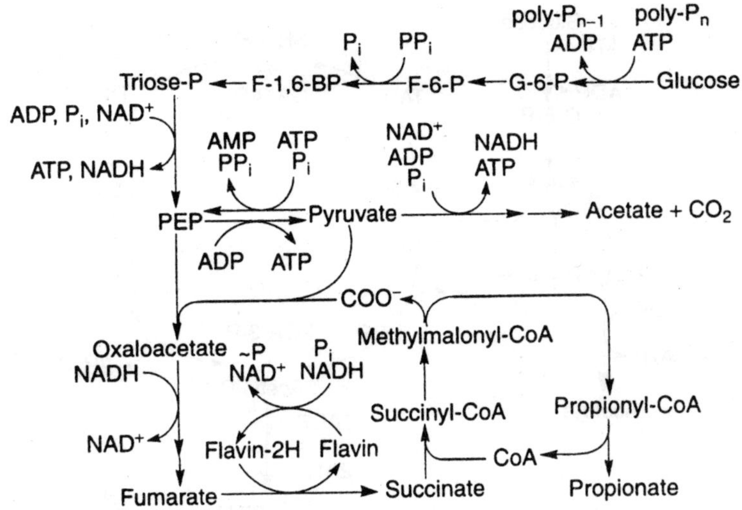

Fig. 18.6 : Production of propionate, acetate, and CO_2 by *Propionibacterium*.

The participation of polyphosphate and Pip also explains the high cell yields observed in the growth of propionibacteria. Demonstration of reactions in which the energy inherent in polyphosphate and Pip may be utilized and not wasted through hydrolysis has far-reaching implications in other systems, since many biological reactions yield Pip, or polyphosphate as production. Some examples are as follows—

Glucokinase : Glucose + $polyP_n \longrightarrow$ G-6-P + $polyP_{n-1}$

Phosphofructokinase : F-6-P + Pip $\longrightarrow$ F-1,6-BP + ip

Carboxytransphosphorylase : PEP + CO_2 + ip $\longrightarrow$ Oxaloacetate + Pip

PRODUCTION OF ACETIC ACID BY FERMENTATION

Gluconobacter suboxydans lacks a functional TCA cycle, although all of the enzymes of this cycle except succinate dehydrogenase are present. This organism cannot ferment glucose or other carbohydrates since the EMP and Entner-Doudoroff pathways are not present. A modified pentose cycle is used for the metabolism of glucose under aerobic conditions (Fig. 18.7). *Acetobacter* species have a functional TCA cycle. The acetyl phosphate resulting from the C_2-C_3 cleavage of pentose is oxidized to CO_2 and energy is generated via oxidative phosphorylation. A characteristic activity of *Acetobacter* and *Gluconobacter* is the oxidation of ethanol to acetic acid. Ethanol oxidation occurs via two membrane-associated dehydrogenases : Alcohol dehydrogenase and acetaldehyde dehydrogenase—

$$CH_3CH_2OH \longrightarrow CH_3CHO + 2H \longrightarrow CH_3\,COOH + 2H$$

Alcohol is first produced by yeast fermentation converting glucose to CO_2 and alcohol. The second stage uses *A. aceti* to convert the ethanol to acetic acid, with the reducing equivalents being transferred to oxygen via the respiratory chain as described above. The process requires carefully controlled conditions to achieve the maximum yield of acetate. Acetate is a major product of carbohydrate metabolism in *B. subtilis*. Conversion

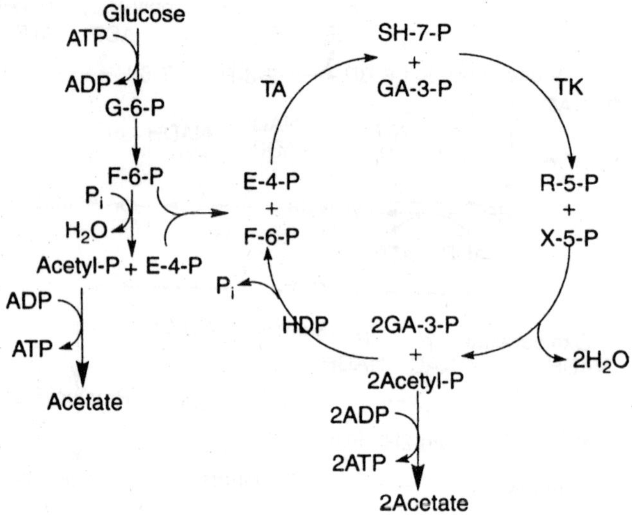

Fig. 18.7 : Pentose cycle (modified) for *Acetobacter* and *Gluconobacter*. SH-7-P, sedoheptulose-7-phosphate; GA-3-P, glyceraldehyde-3-phosphate; HDP, hexose diphosphate; X-5-P, xylulose-5-phosphate; R-5-P, Ribose-5-phosphate; TA-Transaldolase; TK; Transketolase; E-4-P-Erythrose-4-Phosphate.

of pyruvate to acetate occurs primarily during exponential growth. During the stationary phase, acetate is generated via butanediol. Overacidification of the medium due to pyruvate and acetate accumulation during exponential growth is apparently alleviated by acetoin production under aerobic conditions or 2,3-butanediol under anaerobic conditions. These compounds can be reutilized in the stationary phase. The phosphotransacetylase (*pta*) gene, together with the acetate kinase (*ackA*) gene, encode the enzymes that catalyze the conversion of acetyl-CoA to acetate via acetyl~P.

MICROBIAL SYMBIOSIS

Symbiotic relationships may provide protection, nutrients or other benefits among microorganisms. From an evolutionary standpoint, such associations are directed towards meeting the challenge of the environment by adapting to existence in continuous close association with other forms of life. This biological phenomenon is known as symbiosis. The associations vary widely. On the basis of their degree of intimacy, these may be—

A. Ectosymbioses—The microbe remains external to the cells of its host. The term host is sometimes used for larger of two symbionts.

B. Endosymbioses —The microbe grows within the cells of its host.

On the basis of relative advantage occurring to each partner, the associations are classified as follows—

a. Neutralism—The host remains unaffected. Some persons do not recognise it under symbiosis.

b. Mutualism—Both partners benefit from the association.

c. Parasitism—One partner benefits, but the second gains nothing and often suffers mild or severe damage.

ROLES OF SYMBIOTIC RELATIONS

Endo-as well as ectosymbionts living in body cavities of animals are protected from adverse environmental conditions. These habitats protect the symbionts from dessication and in case of warm-blooded hosts, from extremes of temperature. The association may provide one partner with a position that is favourable with respect to the supply of nutrients. For example, marine ciliates attached on the body surfaces of crustaceans, where the hosts respiratory and feeding currents assure the microbe of a constant supply of food. Other examples are favourable position provided by marine invertebrates to their photosynthetic algal symbionts. Other cases are the thick conical protuberances on the mantle of the tridacnid clams (family–tridacnidae). Each protuberance contains one of more lenslike structures—the hyaline organs made up of transparent cells. Each hyaline organ is surrounded by a dense mass of algae.

Bioluminescence is widespread in animals as jellyfish, earthworms, fireflies, squid and fish. The emission of light appears to be a recognition device, promoting schooling, mating or attraction of prey. This is the most common function. The provision of nutrients may be indirect, as in the case of fungi infecting plant roots, where they increase the water-absorbing capacity of root system. Generally, the association is direct, where the symbiont furnishes one or more essential nutrients to the partner.

RHIZOSPHERE AND PHYLLOSPHERE

Rhizosphere is the region of the soil immediately surrounding the roots of a plant. Some authors include the root surface-rhizoplane also under rhizosphere. From an operational viewpoint rhizosphere can be defined as the region extending a few millimeters from the root surface in which the microbial population of soil is influenced by the chemical activities of plant. The major effect is quantitative : The numbers of microorganisms in the rhizosphere usually is more than in the neighbouring soil. There is also qualitative effect : Short Gram-negative rods predominate in the rhizosphere, while Gram-positive rods and coccoid forms are less numerous in the rhizosphere than elsewhere in the soil. Root exudates in the rhizosphere are rich in soluble sugars, amino acids, vitamins and other factors that serve as nutrients for microbes in the rhizosphere. Similar associations exist in the phyllosphere or more precisely, the phylloplane, the surfaces of living leaves. The surface harbours a variety of microflora. The leaf leachates are the nutrients for microbes. There are both, quantitative and qualitative effects in the phylloplane.

MYCORRHIZAE

The roots of higher plants are infected by fungi. The composite root-fungus structure is called mycorrhiza. With few exceptions, mycorrhizas are not species specific. There is invasion of root by soil fungus. The hyphae penetrate the root cells by haustoria and develop intracellularly. In VAM (vesicular-arbuscular mycorrhiza) the fungus forms intracellular branching structures called arbuscules. A single species of pine may be associated with any of the 40 different fungi. The fungal components is unable to use complex polysaccharides. The auxins excreted by the fungi induce a dramatic flow of carbohydrate from the leaves to the roots. The plants are said to be helped through facilitation of absorption of water and minerals from soil. Thus pines absorb 2 to 3 times more phosphorous, nitrogen and potassium when mycorrhizae are present than when they are absent.

ROOT NODULE BACTERIA AND LEGUMES

It is known since long time that fertility of agricultural land can be maintained by crop rotation. Cultivation of leguminous plant could restore the declining levels of land productivity due to cultivation of single crop for several years. The bacteria are Gram-negative motile rods that are classified in the genus *Rhizobium*. Their numbers are very variable, depending upon the nature of the soil and on its previous agricultural treatment. The soil under nonleguminous crop, such as wheat may have fewer than 10 *Rhizobium* cells per gram, the same soil will contain between 10 and 10^7 cells of *Rhizobium* following the cultivation of legume crop. The ability of legume plants to stimulate the growth of *Rhizobium* in the soil extends as far as 10 to 20 mm from the roots. The effect is highly specific. The nature of substance responsible for the stimulation is not yet very well known. The infection of root by *Rhizobium* resembles that by pathogens. Nevertheless, this association is not usually considered as parasitism. The plant gains much more from the association. This is thus taken as mutualism.

ZOOCHLORELLAE AND ZOOXANTHELLAE

These are the endosymbionts of protozoa. The green algae are present in freshwater forms of Ciliophora and Rhizopoda, whereas in marine forms there are generally yellow or brown algae. The two types of algae are called zoochlorellae and zooxanthellae respectively. Each protozoan cell harbors from 50 to several hundred algae. The algae resist digestion by the host. Alga converts CO_2 to organic products through photosynthesis liberating O_2. The animal uses the O_2 for respiration and produces CO_2 as by-product.

CYANELLAE

These are blue-green bacterial endosymbionts of protozoa. They are found in a few genera of freshwater protozoa as *Cyanophora, Peliaina* and *Paulinella*. *Paulinella* contains from one to six cyanellae. The cyanellae were discovered by Pascher in 1929.

LICHENS

These are examples of symbioses of algae with fungi. A lichen is a composite organism, consisting of a specific fungus, usually an ascomycete, living in association with one or sometimes two species of algae or blue-green bacteria. The symbionts form a vegetative body, or thallus. There are three morphological types of lichen thalli : Crustose, adhering closely to the substrates (rock, tree bark); foliose, leaflike and more loosely attached to substrate; and the fruticose, forming pendulous strands or upright stalks. The bulk of thallus is made up of fungal hyphae. The partners may be easily separated and grown in pure cultures. The algal partner may be modified by lichenization. Thus, certain filementous blue-green bacteria do not form normal filaments in the thallus, but each cell is separated and surrounded by fungal hyphae. The association of fungus with alga in a lichen is not specific. Algal symbionts belong to 26 genera : 17 of green algae; 1 of yellow-green algae and 8 of blue-green algae. One green algal genus, *Trebouxia* is found in more than half of the described lichens. Most fungi are ascomycetes, and a few are deuteromycetes and basidiomycetes. The lichen may remain viable in dry state for months. They scavenge minerals by producing lichen acids, that dissolve as well as chelate the minerals. Chelation, the process of binding metal atoms to organic ligands, plays an important role in the solubilization and uptake of minerals by lichens. Both partners are able of free-living existence. The association is, therefore, of mutual benefit only in very special ecological situations, i. e., in environments where nutrients are extremely scarce and where extremes of wetting and drying occur. Under such conditions, the fungus benefits from alga for its source of organic nutrients. The contribution of the fungus is less clear.

PARASITISM

The bdellovibrios, of which *Bdellovibrio bacteriovorus* is the type species are very small, Gram-negative bacteria bearing a single polar flagellum. They parasitise and kill other Gram-negative bacteria multiplying inside the host's cell wall. It begins in a violent collision with a host cell. The cells move at high speed (100 cell lengths per second). Thus a host cell many times its size is carried a considerably distance by the momentum of impact. They are unique group of Gram-negative bacteria, very small in size (0.3 to 0.45 μm in diameter), with a sheathed flagellum, obligate aerobes, and obligate parasites of other Gram-negative bacteria. They have been found in soil samples from many parts of the world, as well as in sewage. Host-dependent species can be cultivated in a host-free medium containing an extract of host cells.

BACTERIAL ENDOSYMBIONTS OF PROTOZOA

Bacteria have been described inside amebae, flagellates, ciliates and sporozoa. None of them could be cultivated outside its host. Bacterial contribution to host are unknown. In one case, it is said to provide amino acids and other growth factors to the host. The most extensively studied symbiosis is that between *Paramecium aurelia* and its *endosymbiont, Kappa.* The Kappa is the genetic element in the cytoplasms of *Paramecium aurelia.* It is a bacterium like endosymbiont which divides by binary fission in the cytoplasm of paramecium and is responsible for liberation of toxic particles in killer strains of paramecium. There are two strains of paramecium. Killers which are immune to toxic substance and sensitives, to which these substances are lethal. Kappa contains DNA and can undergo mutations. In addition to these, a number of other bacterial endosymbionts have been found in killer stocks of *P. aurelia* isolated from nature. These have also been designated by Greek letters. One of them, called alpha has affinities with *Cytophaga.* The endosymbionts perhaps synthesize the folic acid required by paramecium.

ECTOSYMBIOSES OF PROTOZOA WITH INSECTS : THE INTESTINAL FLAGELLATES OF WOOD-EATING TERMITES AND COCKROACHES

The cellulose and lignin components of woody tissues of trees can not be utilised by most animals, since they lack the enzymes for degradation of these polymers. Nevertheless, many insects obtain their food from wood through an ectosymbiotic association with cellulose and lignin-digesting microorganisms. All wood-eating termites and cockroaches harbor flagellated protozoa (Polymastigotes and Hypermastigotes) in their gut. The flagellates digest cellulose which insects can not. The flagellates, in turn, are hosts to extracellular spirochetes and to intracellular bacteria, and perhaps some, if not all, of the cellulases may come from intracellular symbionts. Nitrogen fixation is also reported in gut of termites. N_2 fixing bacteria may occur free in the gut or as intracellular symbionts of the flagellates. Encystment of flagellates is regulated by hormones of the cockroach. Hatching of eggs coincides with the peak of the molting season, and cyst formation in protozoan is induced by the molting hormone, ecdysone.

ENDOSYMBIOSES OF FUNGI AND BACTERIA WITH INSECTS

Symbionts are never present in insects that have a nutritionally complete diet, but are present in all insects that have a nutritionally deficient diet during their developmental stages. Thus no carnivorous insect has symbionts, whereas those living on blood or plant sap all contain symbionts. The main function of symbiont is to provide one or more growth factor to the insect host that is absent from its diet. There are some exceptions, however, as in mosquitoes. The endosymbionts are both bacteria and yeasts. In *Rhodnius,* symbiont is an actinomycete of the genus *Nocardia*. Other symbionts are Gram-negative rods or coryneforms. The specialised cells which contain symbionts are known as mycetocytes (for fungi) or bacteriocytes (for bacteria). In cockroaches, symbionts provide the host with some essential amino acids. In some insects they appear to help the host in the breakdown of nitrogenous waste products (uric acid, urea, xanthine etc.).

Ruminant gut provides a more obvious example of mutualism. They use plant cellulose as the major carbohydrate source of their diet. However, their normal gut can not digest cellulose. Their digestive tract contains no less than four successive stomachs. They have developed a special region for cellulose digestion. This region is the rumen, that is essentially vast incubation chamber teeming with bacteria and protozoa. In cow, this resembles a large fermentation vat, about 100 litres, into which masticated plant materials enter for further digestion by large number of anaerobic bacteria and protozoa. The mutualistic microbes hydrolyse cellulose and other complex plant polysaccharides to their component monosaccharides which are then fermented to simple fatty acids (acetic, propionic, butyric) and gases (methane, carbondioxide). The fatty acids are absorbed through the wall of rumen into the bloodstream for use as carbon and energy source. The gases are passed out of rumen at frequent intervals. The microbial population of the rumen grows rapidly. These microbial cells pass out of the rumen along with undigested plant material into the stomach. These cells are destroyed and digested in the stomach through proteases (the rumen produces no digestive enzymes), in normal way to provide essential amino acids and vitamins etc., required for the growth of animal. The cellulose-digesting bacteria of rumen are all strict anaerobes. The species include *Bacteroids succinogenes, Ruminococcus flavofaciens, R. albus* and *Botryovibrio fibrisolvens*. The great bulk of bacterial population, however, is noncellulolytic. Many of the rumen bacteria including some of the cellulolytic species are able of digesting starch, proteins and lipids.

ECTOSYMBIOSES OF MICROBES WITH BIRDS

The honey guides, a group of birds belong to the genus *Indicator*. They are found in India and Africa. They guide honey badgers, as well as humans, to the nests of wild bees, where they wait for their followers to break open the hive. The honey guide feeds on the remnants of exposed honeycomb, after the badger or human has left. The birds do not have enzymes for digesting beeswax. They harbor in their intestines two microbes – a bacterium, *Micrococcus cerolyticus* and a yeast, *Candida albicans,* which carry out digestion.

THE NORMAL FLORA OF HUMAN BODY

Both, skin and mucous membranes are exposed to external environment. These become contaminated at different stages and establish a normal flora. They are harmless unless under certain conditions. The throat and mouth support a variety of microbes including most of the eubacterial groups. Gram-positive cocci (micrococci, pneumococci, *Streptococcus salivarius)* are common throat inhabitants. There are also Gram-negative rods of genera *Bacteroids* and *Spirillum*. Gram-positive rods include mainly lactobacilli and corynebacteria. Spirochetes (*Treponema dentium*), yeasts *(Candida albicans)* and actinomycetes are also common, in mouth. On the skin, there are present mainly corynebacteria, micrococci, nonhemolytic streptococci and mycobacteria. The stomach and small intestine are unsuitable environments for microbial development. Large intestine harbors a resident flora. In adults intestines, *Bacteroides* spp., *E. coli, Streptococcus faecalis* are common.

HUMAN MOUTH AND GUT

The normal flora has a high species diversity, being composed chiefly of Gram-positive cocci, rods and filaments, but aerobic and anaerobic Gram-negative cocci and rods may also be found. These microbes live in neutralism in mouth without causing any harm. However, under right conditions they may become pathogenic and cause dental diseases. One such disease is caries, the localised dissolution of tooth enamel by acid. The major organism involved in caries is *Streptococcus mutans* which utilises dietary sucrose in two ways : To make extracellular polysaccharides which aid in its adherence to tooth enamel; and in fermentation, producing lactic acid. Thus there are set up localised sites of attack on the enamel. Caries, unlike other bacterial diseases, is the result of complex intractions between many different organisms living with and interacting with *Streptococcus.*

The acids in the stomach prevent the growth and cause the death of microbes. Some acid-tolerant bacteria can colonise the stomach lining, but their role is unknown. The concentration of microbes increases from the small to the large intestines and to the bowels. The gut in fact is like a continuous culture system which an entry part in the mouth and an exit part at the anus. Most of these microorganisms could be now grown in the laboratory. Most of these were thought to be facultative anaerobes like *E. coli.* However, it appears that most gut microbes are very strict anaerobes, being killed by even the small amounts of oxygen used in the laboratory sub-culture. Special methods are used for their isolation. The germ-free animals have poorly developed defense mechanisms, have an abnormal gut and suffer increased nutritional needs compared with normal animals. Some of these effects can also be produced in humans who are administered wide-spectrum antibiotics. Majority of gut flora is killed. We have already described the more obvious example of mutualism in the gut of ruminants.

PLANT ROOT

The roots, however, are surrounded by the region, rhizosphere where nutrients for microbes are available in excretion products from the root. Environmental conditions in rhizosphere do not fluctuate much as in the phyllosphere (for leaf). The number of microbes is greater in rhizosphere than in the soil away from the influence of roots. However, they are shown to play important role in antagonising the growth of potential root pathogens. A more organised relationship is with the fungi, resulting into the mycorrhizae. These are said to confer an advantage to plants growing in poor soils where they aid in the absorption of nutrients, especially phosphates, by increasing the area of absorption. Two contrasting relationships between bacteria and plant roots examplify the spectrum of interactions that we find in nature. *Rhizobium* and *Agrobacterium* are closely related Gram-negative bacteria (they show very close DNA homology). Both infect roots to cause abnormal localised growths. The relationship between *Rhizobium* and leguminous plants is a good example of a well developed mutualism. Each root nodule may contain as many as 10^9 bacteria. In contrast, *Agrobacterium* gives an example of parasitism. It forms crown gall tumors. The bacteria enter through a would and move between its cells that are stimulated to rapid division. The bacteria have a large plasmid, Ti (tumor inducing) plasmid, part of which is transferred to the plant cell, and incorporated and expressed in the plant nuclear DNA.

❒❒❒

19

IMMUNOTHERAPY AND IMMUNE ASSAYS

Greater understanding of the functions of B and T cells and antibodies has led to valuable techniques for manipulating and monitoring the immune system. It is now regular practice to achieve active immunity with vaccines against common infectious diseases such as polio and diphtheria that once caused untold sickness and death. In a similar vein, immediate protection against rabies and tetanus can be delivered to persons at risk. Methods for testing the immune status of an individual are another outgrowth of this technology. It is routine medical practice to test the blood of patients to diagnose HIV, syphilis, hepatitis B, and rubella infections. Both separately and in combination, these methods have made sweeping contributions to individual and community health, treating disease, and controlling the spread of disease. Many hopes for the survival of humankind lie in our ability to harness the amazing workings of the immune system.

METHODS OF IMMUNITY MANIPULATION

A knowledge of the immune system and its responses to antigens has provided extremely valuable biomedical applications in two major areas : (a) Use of antiserum and vaccination to provide artificial protection against disease, and (b) diagnosis of disease through immunologic testing. Methods that actively or passively immunize people are widely used in disease prevention and treatment. When a patient is given preformed antibodies, the immunization is a form of passive therapy. When a patient is vaccinated with a microbe or its antigens, the immunization is considered active and primarily preventive.

Artificial Passive Immunity

The first attempts at passive immunization involved the transfusion of horse serum containing antitoxins to prevent tetanus and to treat patients exposed to diphtheria. Since then, antisera from animals have been replaced with products of human origin that function with various degrees of specificity. Immune serum globulin (ISG), sometimes called gamma globulin, contains immunoglobulin extracted from the pooled blood of at least 1,000 human donors. Each lot of serum presents a broad cross section of IgG (and some IgM) antibodies, and the content varies from lot to lot. The method of processing ISG concentrates the antibodies to increase potency and eliminates potential pathogens. It is a treatment of choice in preventing measles and hepatitis A, and in replacing antibodies in immunodeficient patients. Most forms of ISG are injected intramuscularly to minimize adverse reactions, and the protection it provides lasts two to three months. A preparation called specific immune globulin (SIG) is derived from a more defined group of donors. Companies that prepare SIG obtain serum from patients who are convalescing and in a hyperimmune state after such infections as pertussis, rabies, tetanus, chickenpox, and hepatitis B. These globulins are preferable to ISG because they contain higher titers of specific antibodies obtained from a smaller pool or patients. Although useful for prophylaxis in persons who have been exposed or may be exposed to infectious agents, these sera are often limited in availability. When a human immune globulin is not available, antisera and antitoxins of animal origin can be used. Sera produced in horses are available for diphtheria, botulism, spider and snake bites. Unfortunately, the presence of horse antigens can stimulate allergies such as serum sickness or anaphylaxis. Most passive immunization involves the administration of antibodies, but occasionally T cells are also given.

Artificial Active Immunity

Active immunity can be conferred artificially by vaccination—exposing a person to material that is antigenic but not pathogenic. The discovery of vaccination was one of the farthest reaching and most important developments in medical science. It profoundly reduced the prevalence and impact of many infectious diseases that were once common and often deadly.

Undoubtedly, the earliest crude attempts involved bringing a susceptible person into contact with a diseased person or animal. The first recorded attempt at immunization occurred in sixth century China. It consisted of drying and grinding up smallpox scabs and blowing them with a straw into the nostrils of vulnerable family members. By the tenth century, this practice had changed to the deliberate inoculation of dried pus from the smallpox pustules of one patient into the arm of a healthy person, a technique later called variolation. This method was used in parts of the Far East for centuries before Lady Mary Montagu brought it to England in 1721. Although the principles of the technique had some merit, unfortunately, many recipients and their contacts died of smallpox. This vividly demonstrates, a cardinal rule for a workable vaccine : It must contain an antigen that will provide protection but not cause the disease. Variolation was so controversial that any English practitioner caught doing it was charged with a felony. Eventually, this human experimentations paved the way for the first really effective vaccine, developed by the English physician Edward Jenner in 1796. Jenner conducted the first scientifically controlled study, one that had a tremendous impact on the advance of medicine. His work gave rise to the words vaccine and vaccination (L. *Vacca,* cow), which now apply to any immunity obtained by inoculation with selected antigens. Jenner was inspired by the case of a farmer, Benjamin Jesty, who had been infected by cowpox (a related virus that afflicts cows but causes only a mild infection in humans) and was thereafter free of smallpox. Jesty even went so far as to inoculate his family with cowpox, which protected them as well. In his initial test of this cross-resistance, Jenner prepared material from human cowpox lesions and inoculated a young boy. When challenged two months later with an injection of crusts from a smallpox patients, the boy proved immune. Jenner's discovery—that a less pathogenic agent could confer protection against a more pathogenic one—is especially remarkable in view of the fact that microscopy was still in its infancy and the nature of viruses was unknown. At first, the use of the vaccine was regarded with some fear and skepticism. When his method proved successful and word of its significance spread, it was eventually adopted in many other countries. Other historical developments in vaccination included using heat-killed bacteria in vaccines for typhoid fever, cholera, and plague, and techniques for using neutralized toxins for diphtheria and tetanus.

Vaccine preparation—In natural immunity, an infectious agent stimulates appropriate B and T lymphocytes, and creates memory clones. In artificial active immunity, the objective is to obtain this same response with a modified version of the microbe or its components. A safe and effective vaccine should mimic the natural protective response, not cause a serious infection or other disease, have long-lasting effects in a few doses, and be easy to administer. Most vaccine preparations contain one of the following antigenic stimulants : (a) Killed whole cells or inactivated viruses, (b) live, attenuated cells or viruses, (c) parts of cells or viruses, or (d) genetically engineered microbes or microbial parts.

Large, complex antigens such as intact cells or viruses are very effective immunogens. Depending on the vaccine, these are either killed or attenuated. Killed, whole vaccines are prepared by cultivating the desired strain or strains of a bacterium or virus and treating them with formalin, radiation, or some other agent that does not destroy antigenicity. One type of vaccine for the bacterial diseases pertussis and typhoid fever are of this type.

Salk polio vaccine and rabies vaccine contain killed whole viruses. Because the microbe does not multiply, killed vaccines often require a larger dose and more boosters to be effective. Although one might think that vaccines containing dead microorganisms would be quite safe, they are not without adverse side effects. A number of vaccines are prepared from live, attenuated (L. *Attenuare*, to thin) microbes. Attnuation is any process that substantially lessens or negates the virulence of viruses or bacteria. It is usually achieved by modifying the growth conditions or manipulating microbial genes in a way that eliminates virulence factors. Attenuation methods include long-term cultivation, selection of mutant strains that grow at colder temperatures (cold mutants), passage of the microbe through unnatural hosts or tissue culture, and removal of virulence genes. The vaccine for tuberculosis (BCG) was obtained after 13 years of subculturing the agent of bovine tuberculosis. Vaccines for measles, mumps, polio (Sabin), and rubella contain live, nonvirulent viruses. The advantages that favour live preparations are : (a) Viable microorganisms can multiply and produce infection like the natural organism; (b) they confer long-lasting protection; and (c) they usually require fewer doses and boosters than other types of vaccines. Disadvantages of using live microbes in vaccines are that they require special storage facilities, can be transmitted to other people, and can mutate back to a virulent strain.

If the exact antigenic determinants that stimulate immunity are known, it is possible to produce a vaccine based on a selected part of a microorganism. These are called subunit, or acellular, vaccines. Examples of subunit antigens are the capsules of the pneumococcus and meningococcus, the protein suface antigen of anthrax, and the surface receptors of hepatitis B virus. An alternate BCG vaccine containing only purified proteins is currently being tested. A special type of vaccine is the toxoid, which consists of a purified exotoxin that has been chemically denatured. By eliciting the production of antitoxins that can neutralize the natural toxin, toxoid vaccines provide protection against toxinoses such as diphtheria and tetanus.

One reason many infectious diseases lack practical vaccines is that a suitable antigen preparation has yet to be developed. Currently, much attention is being focussed on newer strategies for vaccine preparation that employ antigen synthesis, recombinant DNA and gene cloning technology, and other laboratory manipulations. When the exact composition of an antigenic determinant is known, it is possible to synthesize it. This permits preservation of antigenicity while greatly increasing antigen purity and concentration. The malaria vaccine currently being tested in areas of South America and Africa is composed of three synthetic peptides from the parasite. These methods are particularly effective in designing vaccines for obligate parasites that are difficult or expensive to culture, such as the syphilis spirochete or the malaria parasite.

This technology provides a means of isolating the genes that encode various microbial antigens, inserting them into plasmid vectors, and cloning them in appropriate hosts. The outcome of recombination can be varied as desired. For instance, the cloning host can be stimulated to synthesize and secrete a protein product (antigen), which is then harvested and purified. This is how one type of hepatitis B vaccine is made, and it is the basis for several AIDS vaccines undergoing clinical trails. Antigens from the agent of syphilis, *Schistosoma,* and influenza have been similarly isolated and cloned, and are currently being considered as potential vaccine material. Another ingenious technique using genetic recombination has been nicknamed the *Trojan horse* vaccine. The term derives from an ancient legend in which the Greeks sneaked soldiers into the fortress of their Trojan enemies by hiding them inside a large, mobile wooden horse. In the microbial equivalent, genetic material from a selected infectious agent is inserted into a live carrier microbe that is nonpathogenic. In theory, the recombinant microbe will multiply and express the foreign genes, and the vaccine recipient will be immunized against the microbial antigens. *Vaccinia,* the virus originally used to vaccinate for smallpox, and adenoviruses have proved practical agents for this technique. Vaccine researchers are also pursuing DNA-based vaccines that contain isolated DNA molecules encoding microbial antigens. Injecting this preparation appears to induce protective antibodies. This promising strategy is currently being tested for influenza, AIDS, and genital warts. Other diseases for which vaccines are being engineered and tested are gonorrhea, syphilis, rotavirus, salmonellosis, shigellosis, and gastric ulcers. Vaccine effectiveness relies, in part, on the production of antibodies that closely fit the natural antigen. Realizing that such reactions are very much like a molecular jigsaw puzzle, researchers have proposed an entirely new concept in vaccines. The *anti-idiotypic vaccine* is based on the principle that the antigen binding (variable) region, or *idiotype* (Gr. *Idios,* own peculiar) of a given antibody (A) can be antigenic to a genetically different recipient and can cause that recipient's immune system to produce antibodies (B; also called anti-idiotypic antibodies) specific for the variable region on antibody A. The purpose for making anti-idiotypic antibodies is that they will display an identical configuration as the desired antigen and can be used in vaccines. The antibodies it stimulates in the vaccine recipient will be able to react with the natural antigen. This method avoids giving a microbial antigen, thus reducing the potential for dangerous side effects. In addition, the exact nature of the microbial antigen need not be known in order for a workable vaccine to be developed.

Administration and Effects of Vaccines

Oral vaccines are available for only two diseases, but have some distinct advantages. An oral dose of a vaccine can stimulate protection (IgA) on the mucous membrane of the portal of entry. Oral vaccines are also easier to give, more readily accepted, and well tolerated. An influenza vaccine given intranasally is another method that shows promise. Some vaccines require the addition of a special binding substance, or *adjuvant* (L. *Adjuvare,* to help). An adjuvant is any compound that enhances immunogenicity and prolongs antigen retention at the injection site. The adjuvant precipitates the antigen and holds it in the tissues so that it will be released gradually. This presumably facilitates contact with macrophages and lymphocytes. Common adjuvants are alum (aluminum hydroxide salts), Freund's adjuvant and beeswax. Vaccines must go through many years of trials in experimental animals and humans before they are licensed for general use. Even after they have been approved, like all therapeutic products, they are not without complications. The most common of these are local reactions at the injection site, fever, allergies, and other adverse reactions. Relatively rare reactions are panencephalitis (from measles vaccine), back-mutation to a virulent strain (from polio vaccine), disease due to contamination with dangerous viruses or chemicals, and neurological effects of unknown cause (from pertussis and swine flu vaccines). Some patients experience allergic reactions to the medium (eggs or tissue culture), rather than to vaccine antigens. Professionals involved in giving vaccinations must understand their inherent risks but also realize that the risks from the infectious disease almost always outweigh the chance of an adverse vaccine reaction.

VACCINATION

Vaccination confers long-lasting, sometimes lifetime, protection in the individual, but an equally important effect is to protect the public health. It is an effective method of establishing herd immunity in the population. According to this concept, individuals immune to a communicable infectious disease will not be carriers, which reduces the occurrence of that microbe; therefore, the larger the number of immune individuals in a population (herd), the less likely that an unimmunized member of the population will encounter the agent. In effect, collective immunity through mass immunization confers indirect protection on the nonimmune. Vaccination is recommended for all typical childhood diseases for which a vaccine is available and for people in certain special circumstances. Not only are some vaccines mixtures of antigens, but in certain cases several vaccines are also administered simultaneously. Examples are military recruits who receive as many as 15 injections within a few minutes and children who receive boosters for DPT and polio at the same time they receive the MMR vaccine. Experts doubt that immune interference is a significant problem in these instances, and the mixed vaccines are carefully balanced to prevent this eventuality. The main problem with simultaneous administration is that side effects can be amplified.

SEROLOGICAL AND IMMUNE TESTS

The antibodies formed during an immune reaction are important in combating infection, but they hold additional practical value. Characteristics of antibodies can reveal the history of a patient's contact with microorganisms or other antigens. This is the underlying basis of serological testing. Serology is the branch of immunology that traditionally deals with *in vitro* diagnostic testing of serum. Serological testing is based on the familiar concept that antibodies have extreme specificity for antigens, so that when a particular antigen is exposed to its specific antibody, it will fit like a hand in a glove. The ability to visualize this interaction by some means provides a powerful tool for detecting, identifying, and quantifying antibodies—or for that matter, antigens. The scheme works both ways, depending on the situation. One can detect or identify an unknown antibody, using a known antigen, or if the antigen is unknown, one can use an antibody of known specificity to help detect or identify it. Modern serological testing has grown into a field that tests more than just serum. Urine, cerebrospinal fluid, whole tissues, and saliva can also be used to determine the immunologic background of patients. These and other immune tests are helpful in confirming a suspected diagnosis or in screening a certain population for disease.

General features—The strategies of immunologic tests are diverse, and they underline some of the brilliant and imaginative ways that antibodies and antigens can be used as tools. We will summarize them under the headings of agglutination, precipitation, immunodiffusion, complement fixation, fluorescent antibody tests, and immunoassay tests. First we will overview the general characteristics of immune testing, and then look at each type separately. The most effective serological tests have a high degree of specificity and sensitivity. Specificity is the property of a test to focus upon only a certain antibody or antigen and not to react with unrelated or distantly related ones. Sensitivity means that the test can detect even very small amounts of antibodies or antigens that are the targets of the test.

Antigen-antibody interactions—The primary basis of most tests is the binding of an antibody to a specific molecular site on an antigen. Because this reaction cannot be readily seen without an electron microscope, tests involve some type of endpoint reaction visible to the naked eye or with regular magnification that tells whether the result is positive or negative. In the case of large antigens such as cells, Ab binds to Ag and creates large clumps or aggregates that are visible macroscopically or microscopically. Smaller Ag-Ab complexes that do not result in readily observable changes will require special indicators in order to be visualized. Endpoints are often revealed by dyes or fluorescent reagents that can tag molecules of interest. Similarly, radioactive isotopes incorporated into antigens or antibodies constitute sensitive tracers that are detectable with photographic film. An antigen-antibody reaction can be used to read a titer. Knowing the relative quantity of antibodies permits different samples to be standardized and compared. To determine titer, a sample is serially diluted in tubes or a multiple-welled microtiter plate and mixed with antigen. Titer is expressed as the highest dilution of serum that produces a visible reaction with an antigen.

Agglutination and precipitation reactions—The essential differences between agglutination and precipitation are in size, solubility, and location of the antigen. In agglutination, the antigens are whole cells such as red blood cells or bacteria with determinant groups on the surface. In precipitation, the antigen is a soluble molecule. In both instances, when Ag and Ab are optimally combined so that neither is in excess, one antigen is interlinked by several antibodies to form an insoluble, three-dimensional aggregate so large that it can not remain suspended and it settles out.

Agglutination is discernible because the antibodies called *agglutinins* cross-link the antigens or *agglutinogens* to form visible clumps. Agglutination tests are performed routinely by blood banks to determine ABO and Rh (Rhesus) blood types in preparation for transfusions. In this test, antisera containing antibodies against the blood group antigens on red blood cells are mixed with a sample of blood and read for the presence or absence of clumping. The *Widal test* is an example of a tube agglutination test for diagnosing salmonelloses and undulant fever.

The VDRL (Venereal Disease Research Lab) is a common test for antibodies to syphilis. Since the reagents for the test include a heterophile antigen (cardiolipin), false positives can occur. The cold agglutinin test, named for antibodies that react only at lower temperatures (4°-20° C), was developed to diagnose *Mycoplasma* pneumonia. The *Weil-Felix reaction* is an agglutination test sometimes used in diagnosing rickettsial infections. In some tests, special agglutinogens have been prepared by affixing antigen to the surface of an inert particle. In *latex agglutination* tests, the inert particles are tiny latex beads. Kits using latex beads are available for assaying pregnancy hormone in the urine, identifying *Candida* yeasts and bacteria (staphylococci, streptococci, and gonococci), and diagnosing rheumatoid arthritis. In *viral hemagglutination* testing, agglutinogen is a red blood cell that reacts naturally with certain viral antigens. The RBCs are mixed with a known virus and a patient's serum of unknown content. The test is interpreted differently than other agglutination tests because it is based on a competition between the RBCs and the antibodies for the virus antigens. If the patient's serum does not contain antibodies specific to the virus, the virus reacts with the RBCs instead and agglutinates them. Thus, agglutination here indicates no antibodies and a seronegative result. On the other hand, if the specific antibodies for the virus are present, they attach to the virus particles, the RBCs remain free, and there is no agglutination. As performed in the wells of microtiter plates, the difference is apparent to the naked eye.

In precipitation reactions, the soluble antigen *(precipitinogen)* is precipitated (made insoluble), by an antibody *(precipitin)*. This reaction is observable in a test tube where antiserum has been carefully laid over an antigen solution. At the point of contact, a cloudy or opaque zone forms. Although precipitation is a useful detection tool, the precipitates are so easily disrupted in liquid media that most precipitation reactions are carried out in agar gels. These substrates are sufficiently soft to allow the reactants (Ab and Ag) to freely diffuse, yet firm enough to hold the Ag-Ab precipitate in place. One technique with applications in microbial identification and diagnosis of disease is the double diffusion (Ouchterlony) method. It is called double diffusion because it involves diffusion of both antigens and antibodies. The test is performed by punching a pattern of small wells into an agar medium and filling them with test antigens and antibodies. A band forming between two wells indicates that antibodies from one well have met and reacted with antigens from the other well. Variations on this technique provide several results : (a) Antigen placed in the center well can be used to assess the content of sera. (b) Antiserum placed in the center well can help identify antigens placed in the outer wells. (c) Identification reactions between the contents of the outer wells can be determined.

Immunoelectrophoresis constitutes yet another refinement of diffusion and precipitation in agar. With this method, a serum sample is first electrophoresed to separate the serum proteins as previously shown. After antibodies that can react with these serum samples are placed in a trough parallel to the direction of migration, reaction arcs specific for blood proteins develop (Fig. 19.1). This text is widely used to detect disorders in the production of antibodies. *Counterimmunoelectrophoresis,* which uses an electric current to speed up the migration of antibody and antigen, is a newer technique for identifying bacterial and viral antigens in blood.

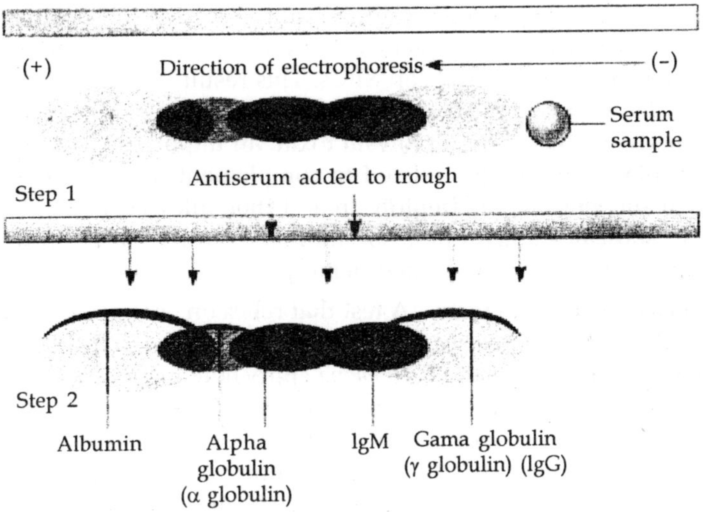

Fig. 19.1 : Immunoelectrophoresis of normal human serum.

The western blot—This test is a counterpart of the Southern blot test for identifying DNA. It is a very specific and sensitive way to identify or verify a particular (protein antibody or antigen) in a sample. First, the test material is electrophoresed in a gel, which separates out particular bands. The gel is then transferred to a special blotter that binds the reactants in place. The blot is developed by incubating it with a solution of antibody or antigen that has been labelled with radioactive, fluorescent, or luminescent labels. Sites of specific binding will appear as a pattern of bands that can be compared to known positive and negative samples. This is currently the verification test for people who are antibody-positive for HIV in the ELISA test, because it tests more types of antibodies and is less subject to misinterpretation. The technique has significant applications for detecting microbes and their antigens in specimens.

Complement fixation—An antibody that requires (fixes) complement to complete the lysis of its antigenic target cell is a lysin or cytolysin. When lysins act in conjunction with the intrinsic complement system on red blood cells, the cells hemolyze (lyse and release their hemoglobin). This lysin-mediated hemolysis is the basis of a group of tests called complement fixation, or CF. CF testing uses four components : Antibody, antigen, complement, and sensitized sheep red blood cells, and it is conducted in two stages. In the first stage, the test antigen is allowed to react with the test antibody in the absence of complement. If the Ab-Ag are specific for each other, they form complexes. To this mixture, purified complement proteins from guinea pig blood are added. If antibody and antigen have complexed during the previous step, they attach, or fix, the complement to them, thus preventing it from participating in further reactions. The extent of this complement fixation is determined in the second stage, sheep RBCs are mixed with sheep-specific lysins. This will produce an indicator complex that can also fix complement. Contents of the stage 1 tube are mixed with the stage 2 tube and observed for hemolysis which can be observed with the naked eye as a clearing of the solution. If hemolysis does not occur, it means that the complement was used up by the first stage Ab-Ag complex and that the unknown antigen or antibody was indeed present. This result is considered positive. If hemolysis does occur, it means that unfixed complement from tube 1 reacted with the RBC complex instead, thereby causing lysis of the sheep RBCs. This result is negative for the antigen or antibody that was the target of the test. The antistreptolysin O (ASO) titer test measures the levels of antibody against the streptolysin toxin, an important hemolysin of group A streptococci. It employs a technique related to complement fixation. A serum sample is exposed to known suspensions of streptolysin and then allowed to incubate with RBCs. Lack of hemolysis indicates antistreptolysin antibodies in the patient's serum that have neutralized the streptolysin and prevented hemolysis.

Miscellaneous serological tests—A test that relies on changes in cellular activity as seen microscopically is the *Treponema pallidum immobilization* (TPI) test for syphilis. The impairment or loss of motility of the *Treponema* spirochete in the presence of test serum and complement indicates that the serum contains anti-*Treponema pallidum* antibodies. In *toxin neutralization* tests, a test serum is incubated with the microbe that produces the toxin. Sereotyping is an antigen-antibody technique for identifying, classifying, and subgrouping certain bacteria into categories called serotypes, using antisera for cell antigens such as the capsule, flagellum, and cell wall. It is widely used in typing *Salmonella* species and strains, and is the basis for the numerous serotypes of streptococci. The Quellung test that identifies serotypes of the pneumococcus involves a precipitation reaction in which antibodies react with the capsular polysaccharide. Although the reaction makes the capsule seem to swell, it is actually creating a zone of Ab-Ag complex on the cell's surface.

Fluorescent antibodies and immunofluorescence testing—This property of fluorescence has found numerous applications in diagnostic immunology. The fundamental tool in immunofluorescence testing is a fluorescent antibody—a monoclonal antibody labeled by a fluorescent dye (fluorochrome). In *direct testing,* an unknown test specimen or antigen is fixed to a slide and exposed to a fluorescent antibody solution of known composition. If the antibodies are complementary to antigens in the material, they will bind to it. After the slide is rinsed to remove unattached antibodies, it is observed with the fluorescent microscope. Fluorescing cells or specks indicate the presence of Ab-Ag complexes and a positive result. These tests are valuable for identifying and locating antigens on the surfaces of cells or in tissues, and in identifying the diseases agents of syphilis, gonorrhea, chlamydiosis, whooping cough, Legionnaires' disease, plague, trichomoniasis, meningitis, and listeriosis. In *indirect testing* methods, the fluorescent antibodies are *anti-isotypic* antibodies made to react with the Fc region of another antibody. In this scheme, an antigen of known character is combined with a test serum of unknown antibody content. To visualize whether the serum contains antibodies that have affixed to the antigen, the fluorescent antibody solution made to react with the unknown antibody is applied and rinsed off. A positive test shows fluorescing aggregates or cells, indicating that the fluorescent antibodies have combined with the unlabeled antibodies. In a negative test, no fluorescent complexes will appear.

Immunoassays

The elegant tools of the microbiologist and immunologist are being used increasingly in athletics, criminology, government, and business to test for trace amounts of substances such as hormones, metabolites, and drugs. But traditional techniques in serology are not refined enough to detect a few molecules of these chemicals. Extremly sensitive alternate methods that permit rapid and accurate measurement of trace antigen or antibody are called immunoassays. Examples of the technology for detecting an antigen or antibody in minute quantities include radioactive isotope labels, enzyme labels, and sensitive electronic sensors.

Radioimmunoassay (RIA) – Antibodies or antigens labeled with a radioactive isotope can be used to pinpoint minute amounts of a corresponding antigen or antibody. Although very complex in practice, these assays compare the amount of radioactivity present in a sample before and after incubation with a known, labeled antigen or antibody. The labeled substance completes with its natural, nonlabeled partner for a reaction site. Large amounts of a bound radioactive component indicate that the unknown test substances was not present. The amount of radioactivity is measured with an isotope counter or a photographic emulsion (autoradiograph).

Enzyme-linked immunosorbent assay [ELISA] – The ELISA test contains an enzyme-antibody complex that can be used as a colour tracer for antigen-antibody reactions. The enzymes used most often are horseradish peroxidase and alkaline phosphatase, both of which release a dye (chromogen) when exposed to their substrate. This technique also relies on a solid support such as a plastic microtiter plate that can *adsorb* (attract on its surface) the reactants. As with immunofluorescence, this technique is applicable for both direct and indirect assays. In *direct ELISA* or sandwich tests, a known antibody is adsorbed to the bottom of a well and incubated with a solution containing unknown antigen. After rinsing off excess unbound components, an enzyme-antibody indicator that can react with the antigen is added. If antigen is present, it will attract the indicator-antibody and hold it in place. Next, the substrate to the enzyme is placed in the wells and incubated. Enzymes affixed to the antigen will hydrolyze the substrate and release a coloured dye. Thus, any colour developing in the wells is a positive results. Lack of colour means that the antigen was not present and that the subsequent rinsing removed the enzyme-antibody complex. The *indirect ELISA* test can detect antibodies in a serum sample. As with other indirect tests, the final positive reaction is achieved by means of an antibody-antibody reaction. Like the direct ELISA, the indicator antibody is complexed to an enzyme that produces a colour change with positive serum samples. The starting reactant is a known antigen that is adsorbed to the surface of a well. To this, an unknown serum is added. After rinsing, an enzyme-Ab reagent that can react with the test antibody is placed in the well. The substrate to the enzyme is then added, and the wells are scanned for colour changes.

Colour development indicates that all the components reacted and that the antibody was present in the patient's serum. This is the common screening test for the antibodies to HIV (AIDS virus), various rickettsial species, *Salmonella*, the cholera vibrio, and *Helicobacter*, a cause of gastric ulcers. Since false positives can occur, a verification test may be necessary (such as Western blot for HIV). A newer technology uses electronic monitors that directly read out antibody-antigen reactions. Without belaboring the technical aspects, these systems contain computer chips that sense the minute changes in electric current given off when an antibody binds to antigen. The potential for sensitivity is extreme—it is thought that amounts as small as 12 molecules of a substance can be detected in a sample. In another procedure, antibody-substrate molecules are incubated with sample and then exposed to the enzyme alkaline phosphatase.

Tests of T and B Cells

So far we have concentrated on tests that identify antigens and antibodies in samples, but techniques also exist that differentiate between B cells and T cells, and can quantify subjects of each. Information on the types and number of lymphocytes in blood and other samples is a common way to evaluate immune dyfunctions such as those in AIDS, immunodeficiencies, and cancer. A simple method for identifying T cells is to mix them with untreated sheep red blood cells. Receptors on the T cell bind the RBCs into a flowerlike cluster called a *rosette formation.* Rosetting can also occur in B cells if one uses Ig-coated bovine RBCs or mouse erythrocytes. Indirect fluorescent techniques have been developed that can differentiate not only between T cells and B cells but also can subgroup them. These subgroup tests utilize monoclonal antibodies produced in response to specific cell markers. B-cell tests categorize different stages in B-cell development and are very useful in characterizing B-cell cancers. Tests that can help differentiate the CD4, CD8, and other T-cell subsets are important in monitoring AIDS and other immunodeficiency diseases.

***In Vivo* Testing**

A classic example of one such technique is the *tuberculin test,* which uses a small amount of purified protein (PPD) from *Mycobacterium tuberculosis* injected into the skin. The appearance of a red, raised lesion can indicate previous exposure to tuberculosis. In practice, *in vivo* tests employ principles similar to serological tests, except in this case an antigen or an antibody is introduced into a patient to elicit some sort of visible reaction. Like the tuberculin test, some of these diagnostic skin tests are useful for evaluating infections due to fungi.

□□□

20

FUNGAL DISEASES

FUNGAL PATHOGENS

Molds and yeasts are so widely distributed in air, dust, fomites, and even among the normal flora that humans are incessantly exposed to them. The fact that the planet's surface is totally dusted with spores led W.B. Cooke to christen it "our moldy earth." Fortunately, because of the relative resistance of humans and the comparatively nonpathogenic nature of fungi, most exposures do not lead to overt infection. Of the estimated 100,000 fungal species, only about 300 have been linked to disease in animals, though among plants, fungi are the most common and destructive of all pathogens. Human mycotic disease, or mycosis. (Gr. *Mykos,* fungus, and *osis,* disease) is associated with true fungal pathogens that exhibit some degree of virulence, or with opportunistic pathogens that take advantage of defective resistance. A true fungal pathogen is a species that can invade and grow in a healthy, noncompromised animal host. This behaviour is contrary to the metabolism and adaptation of fungi, most of which are inhibited by the relatively high temperature and low oxygen tensions of a warm-blooded animal's body. But a small number of fungi have acquired the morphological and physiological adaptations required to survive and grow in this habitat. By far their most striking adaptation is a switch from hyphal cells typical of the saprobic phase to yeast cells typical of the parasitic phase. This biphasic characteristic of the life cycle is termed *thermal dimorphism* because it is initiated by changing temperature. An opportunistic fungal pathogen is different from a true pathogen in several ways.

An opportunist has weak to nonexistent invasiveness or virulence, and the host's defenses must be impaired to some degree for the microbe to gain a foothold. Although some species show both mycelial and yeast stages in their life cycles, thermal dimorphism is an exception. Opportunistic pathogens vary in their manifestation from superficial and benign colonizations to deep, chronic systemic diseases that is rapidly fatal. Mycoses due to opportunists are an increasingly serious problem. Some fungal pathogens exist in a category between true pathogens and opportunists. These species are not inherently invasive but can grow when inoculated into the skin wounds or abrasions of healthy people. Examples are *Sporothrix,* the agent of a subcutaneous infection, and the *dermatophytes* (Gr. *Dermos,* skin, and *phyte,* plant), which cause ringworm and athlete's foot.

Epidemiology

Notable exceptions are some dermatophyte and *Candida* species that naturally inhabit the human body and are transmissible. For the remainder of the pathogens, exposure and infection occur when a human happens upon fungal spores in the environment (usually air, dust, or soil). Unlike the opportunists, true fungal pathogens are distributed in a predictable pattern that coincides with the pathogen's adaptation to the specific climate, soil, or other factors of a relatively restricted geographic region. The incidence of all fungal infections has traditionally been difficult to measure because none of the reportable infections monitored by the CDC are caused by fungi. Dermatophytoses are probably the most prevalent, and it is thought that at least 90% of all humans will acquire ringworm or althlete's foot at least once during their lifetimes. Estimates provided by routine skin testing indicate that millions of people have experienced true mycoses, though most cases probably go undiagnosed or misdiagnosed. Epidemics of mycotic infections can occur following mass exposure to a common source. Memorable incidents include spelunkers who were stricken by histoplasmosis after exploring bat caves, and South African mine workers who rubbed against contaminated wood planks and came down with sporotrichosis. Coccidioidomycosis is a particular hazard of construction workers and people living in the paths of aberrant windstorms. Transmissible fungal infections such as dermatophytoses are readily passed by means of shared personal articles, public facilities, swimming pools, gymnasiums, and contact with infected animals.

Pathogenesis

Fungi enter the body mainly via respiratory, mucous and cutaneous routes. In general, the agents of primary mycoses have a respiratory portal (spores inhaled from the air); subcutaneous agents enter through inoculated skin (trauma); and cutaneous and superficial agents enter through contamination of the skin surface. Spores, hyphal elements, and yeasts can all be infectious, but spores are most often involved due to their durability and abundance. Spore size is a factor in respiratory infections, because smaller spores are likely to be inhaled more deeply into the respiratory tract, and a large infecting dose increase the severity of disease. Thermal dimorphism greatly increases virulence by enabling fungi to tolerate the relatively high temperatures and low O_2 tensions of the body. Fungi in the yeast form can be more invasive because they grow more rapidly and spread through tissues and blood, while those producing hyphae tend to localize along the course of blood vessels and lymphatics. Specific factors contributing to fungal virulence are the subject of much research. Toxinlike substances have been isolated from several species, but their method of damaging the tissues remains obscure. The human body is extremely resistant to establishment of fungi. Among its numerous antifungal defenses are the normal integrity of the skin, mucous membranes, and respiratory cilia, but the most important defenses are cell-mediated immunity, phagocytosis, and the inflammatory reaction.

Diagnosis and Control

Accurate and speedy diagnosis is especially critical to the immunocompromised patient who must have prompt antifungal chemotherapy. For example, a patient with systemic *Candida* infection can die if the infection is not detected and treated within five to seven days. Since therapy can vary among the pathogenic fungi, identification to the species level is often necessary. A suitable specimen can be obtained from sputum, skin scrapings, skin biopsies, cerebrospinal fluid, blood, tissue exudates, urine, or vaginal samples, as determined by the patient's symptoms. Routine laboratory procedures include isolation, microscopic and macroscopic examination, histologic stains, serology, and animal inoculation. Because culture of the sample can require several days, the immediate direct examination of fresh samples is recommended. Wet mounts can be prepared by mixing a small portion of the sample on a slide with saline, water, or potassium hydroxide to clear the specimen of background debris. The relatively large size and uniqe appearance of many fungi help them stand out. Large round of oval budding cells are evidence of yeasts, while thick branching strands suggest hyphae.

Isolating the fungal pathogen requires planting the specimen onto three or four types of solid media such as Sabouraud's dextrose agar, mycosel agar, inhibitory mold medium, and brain-heart-infusion agar. These media can be modified by adding one of the following agents : Blood to increase the growth of fastidious species, chloramphenicol and gentamycin to inhibit bacteria, or cyclohexamide to slow the growth of fungal contaminants. Cultures are incubated at room temperature, at 30 °C, and at 37 °C, and observed daily for growth, which can require several weeks in some species. Gross colonial morphology, such as the colour and texture of the colony's surface and underside, can be very distinctive. Initial identification of the pathogen can be followed by confirmatory physiological, antigen, and DNA-based testing. *In vitro* tests to detect antifungal antibodies in serum can be a useful diagnostic tool in some infections. *In vivo* skin testing for delayed hypersensitivity to fungal antigens is mostly used to trace epidemiologic patterns. It does not verify ongoing infection, and considerable cross reactivity exists among *Histoplasma, Blastomyces,* and *Coccidioides.* Although immunization for most fungal diseases is not considered feasible at this time, T-cell infusions can be used successfully to treat children with candidiasis. Vaccines for coccidioidomycosis and histoplasmosis are being prepared for clinical trials. However, few specific prevention measures exist for fungal infections. Traditional methods are based on taxonomic group, location of infection, and type of pathogen.

SYSTEMATIC PATHOGENS

They are restricted to certain endemic regions of the world. Infection occurs when soil or other matter containing the fungal conidia is disturbed and the spores are inhaled into the lower respiratory tract. In the lungs, the spores germinate into yeasts or yeastlike cells and produce an asymptomatic or mild primary pulmonary infection (PPI) that parallels tuberculosis. In a small number of hosts, this infection becomes systemic and creates severe, chronic lesions. In a few cases, spores are inoculated into the skin where they form localized granulomatous lesions.

Histoplasmosis

The most common true pathogen is *Histoplasma capsulatum* (Gr. *Histo,* tissue, *plasm* shape, and, L. *Capsula,* small box), the cause of histoplasmosis. This disease has probably afflicted humans since antiquity, though it was not described until 1905 by Dr. Samuel Darling. Through the years, it has been known by various synonyms—Darling's disease, Ohio Valley fever, and reticuloendotheliosis. The impact of histoplasmosis throughout history is not really known.

Epidemiology—*Histoplasma capsulatum* is typically dimorphic. Growth on media below 35 °C is characterized by a white or brown, hairlike mycelium, and growth at 37 °C on blood agar produces a white, smooth colony. *H. capsulatum* is discontinuously distributed on all continents except Australia, though its areas of greatest endemicity are the eastern and central regions of the United States (the Ohio Valley). This fungus appears to grow most abundantly in moist soils high in nitrogen content, especially those supplemented by bird and bat *guano* (Sp. *Huanu,* dung). A useful tool for determining the distribution of *H. capsulatum* is to inject a fungal extract called histoplasmin into the skin and monitor for allergic reactions. Application of this test has verified the extremely widespread distribution of the fungus. In high prevalence areas such as southern Ohio, Illinois, Missouri, Kentucky, Tennessee, Michigan, Georgia, and Arkansas, 80% to 90% of the population show signs of prior infection. The spores of the fungus are probably dispersed by the wind and, to a lesser extent, animals. The most striking outbreaks of histoplasmosis occur when concentrations of spores have been dislodged by humans working in parks, bird roosting areas, and old buildings. Poeple of both sexes and all ages incur infection, but adult males experience the majority of cases.

Pathogenesis—Histoplasmosis presents the most formidable array of manifestations of any mycosis. It can be benign or severe, acute or chronic, and it can show pulmonary, systemic, or cutaneous lesions. Inhaling a small dose of microconidia into the deep recesses of the lung establishes a primary pulmonary infection that is usually asymptomatic. Some people experience mild symptoms such as aches, pains, and coughing, but a few develop more severe symptoms, including fever, night sweats, and weight loss. The most serious chronic form of histoplasmosis occurs in patients with defective cell-mediated immunity. The fungus spreads from the primary site of infection, often hidden within macrophages. In children, this can lead to hepatosplenomegaly, anemia, circulatory collapse, and death. Adults with systemic disease can acquire lesions in the brain, intestine, adrenal gland, heart, liver, spleen, lymph nodes, bone marrow, and skin. Persistant colonization of patients with emphysema and bronchitis causes *chronic pulmonary histoplasmosis,* a complication that has signs and symptoms similar to those of tuberculosis.

Diagnosis and control—Usually it appears as spherical, "fish-eye" yeasts intracellularly in macrophages, and occasionally as free yeasts in samples of sputum and cerebrospinal fluid. An essential confirmatory step is to isolate the agent and demonstrate dimorphism, but this step can require up to 12 weeks. Complement fixation and immunodiffusion serological tests can support a diagnosis by showing a rising antibody titer. The histoplasmin test does not indicate concurrent infection and is not useful in diagnosis. Undetected or mild cases of histoplasmosis resolve without medical management, but chronic or disseminated disease calls for systemic chemotherapy. The principal drug, amphotericin B, is administered in daily intravenous doses for a few days to a few weeks. Under some circumstances, ketoconazole or other azoles are the drugs of choice.

Coccidioidomycosis

Coccidioides immitis is the etiologic agent of coccidioidomycosis. Although the fungus has probably lived in soil for millions of years, human encounters with it are relatively recent and coincide with the encroachment of humans into its habitat.

Epidemiology—At 25°C, it forms a moist, white to brown colony with abundant, branching, septate hyphae. These hyphae fragment into thick-walled, blocky arthroconidia at maturity. On special media incubated at 37 °C–40°C, an arthrospore germinates into the parasitic phase, a small, spherical cell (spherule). This structure swells into a giant sporangium that cleaves internally to form numerous endospores that look like bacterial endospores but have none of their resistance traits. *Coccidioides immitis* occurs endemically in various natural reservoirs and casually in areas where it has been carried by wind and animals. Conditions favoring its settlement in a given habitat include high carbon and salt content, and a semiarid, relatively hot climate. The fungus has been isolated in such regions from soils, plants, and a large number of vertebrates. The natural history of *C. immitis* follows a cyclic pattern—a period of dormancy in winter and spring, followed by growth in summer and fall. Skin testing has disclosed that highest incidence of coccidioidomycosis is in the southwestern United States, though it also occurs in Mexico and parts of Central and South America. Especially concentrated reservoirs exist in the San Joaquin Valley of California and in southern Arizona. Occasional epidemics are reported in conjunctions with soil agitation by humans. Archeologists digging in ancient dwellings of Native Americans have been common victims of the disease. A highly unusual outbreak of coccidioidomycosis was traced to the 1994 earthquake in Northridge, California. The CDC reported over 200 cases and one death over a two-month period following the quake and its aftershocks. Clouds of dust bearing loosened spores were given off by landslides in the mountains north of the epicenter.

Pathogenesis—In the lung they convert to spherules, which swell, sporulate, burst, and release spores that continue the cycle. In 60% of patients this primary pulmonary infection is inapparent, while in the other 40%, it is accompanied by coldlike symptoms such as fever, chest pain, cough, headaches, and malaise. In uncomplicated cases, complete recovery and lifelong immunity are the rule. In about five out of a thousand cases, the primary infection does not resolve, and progresses with varied consequences. Chronic progressive pulmonary disease is manifested by nodular growths called *fungomas* and cavity formation in the lungs that compromise respiration. Dissemination of the endospores into major organs occasionally takes place in people with impaired cell-mediated immunity.

Diagnosis and control—Diagnosis of coccidioidomycosis is straightforward when the highly distinctive spherules are found in sputum, spinal fluid, and biopsies. This finding is further supported if the typical mycelium and spores are isolated on Sabouraud's agar and if spherules are induced to form. Newer specific antigen tests have been a great boon in identifying and differentiating *Coccidioides* from other fungi. So great is the potential for contamination and infection that cultures are grown in closed tubes or bottles, and are opened in a biological containment hood or killed prior to inspection. Immunodiffusion and latex agglutination tests on serum samples are excellent screens for detecting early infection. The majority of coccidioidomycosis patients do not require treatment. However, in people with disseminated disease, amphotericin B is administered intravenously. The azole drugs can have some benefit, but they may require a longer term of therapy. Minimizing contact with the fungus in its natural habitat has been of some value.

Blastomycosis

Blastomyces dermatitidis (Gr. *Blastos,* germ, *myces,* fungus, *dermato,* skin, and *itis,* inflammation) the cause of blastomycosis, is another fungal pathogen endemic to the United States. Alternate names for this disease are Gilchrist's disease, Chicago disease, and North American blastomycosis. The dimorphic morphology of *Blastomyces* follows the pattern of other pathogens. Colonies of the saprobic phase are uniformly white to tan, with a thin, septate mycelium and simple ovoid conidia. Temperature-induced conversion results in a wrinkled, creamy-white colony that yields large, heavy-walled yeasts with buds nearly as large as the mother cell. Studies indicate that *Blastomyces dermatitidis* inhabits areas high in organic matter such as forest soil, decaying wood, animal manure, and abandoned buildings, but it has been difficult to isolate with regularity. Its life cycle features dormancy during the warmer, dryer times of the year and growth and sporulation during the colder, wetter seasons. In general, blastomycosis occurs from southern Canada to southern Louisiana and from Minnesota to Georgia. Cases have also been reported in Central America, South America, Africa, and the Middle East. Humans, dogs, cats, and horses are the chief targets of infections, which is usually acquired by inhaling conidia-laden dust from living quarters, farm building, or forest litter.

Pathogenesis—The primary portal of energy of *B. dermatitidis* is the respiratory tract, though it also may enter through accidental inoculation. Inhaling only 10 to 100 conidia is enough to initiate infection. As conidia convert to yeasts and multiply, they encounter macrophages in the lung. A large proportion of primary pulmonary infections are probably symptomatic to some degree. Mild disease is accompanied by cough, chest pain, hoarseness, and fever. More severe, chronic blastomycosis can progress to the lungs, skin, and numerous other organs. Abscesses and tumorlike nodules developing in the lungs are often mistaken for cancer, compared to other fungal diseases, the chronic cutaneous form of blastomycosis is rather common. It frequently begins on the face, hand, wrist, or leg as a subcutaneous nodule that erodes to the skin surface. Yeasts disseminating into the bone induce symptoms of arthritis and osteomyelitis. Involvement of the central nervous system can bring on headache, convulsions, coma, and mental confusion. *Blastomyces* can also spread to the reproductive tract, spleen, liver, and kidneys.

Diagnosis and control—Cultures prepared on selective and enriched media and incubated at room temperature can require several weeks to develop. Further confirmatory tests are done to verify dimorphism and the presence of antibodies through complement fixation and ELISA tests. Most skin tests for *Blastomyces* are not useful for general testing because of the frequency of false-positive and false-negative results. Although disseminated infections were once nearly 100% fatal, modern drugs have greatly improved the prognosis. The drug of choice for severe disseminated and cutaneous disease is amphotericin B, and milder cases typically respond to several months of dihydroxystilbamidine therapy.

Paracoccidioidomycosis

Paracoccidioides brasiliensis causes paracoccidioidomycosis, also known as paracoccidioidal granuloma and South American blastomycosis. Due to its relatively restricted distribution, this disease is the least common of the primary mycoses. *Paracoccidioides* forms a small, nondescript colony with scanty, undistinctive spores at room temperature, but it develops an unusual yeast from at 37°C. The natural history of *Paracoccidioides* has not yet been completely clarified. It has been isolated from the cool, humid soils of tropical and semitropical regions of South and Central America , particularly Brazil, Colombia, Venezuela, Argentina, and Paraguay. People most often afflicted with paracoccidioidomycosis are rural agricultural workers and plant harvesters. A possible animal role in transmission has been postulated but not demonstrated. Other factors believed to influence disease are altered physiology, poor nutrition, and impaired host resistance. Infection is initiated when fungal spores enter the lungs or, occasionally, are inoculated into the skin. Most infections are probably benign, self-limited events that go completely unnoticed. In the minority of patients who do incur progressive systemic disease, the lungs, skin and mucous membranes, and lymphatic organs are frequently involved. Paracoccidioidomycosis is diagnosed by standard procedures previously described for the other mycoses in this chapter. Closely scrutinizing fresh or stained clinical specimens for the distinctive yeast phase is important, and a cultural follow-up with conversion media is essential. Serological testing can be instrumental in diagnosing and monitoring the course of infection.

SUBCUTANEOUS MYCOSES

Such infections are termed subcutaneous because they involve tissues within and just below the skin. Most species in this group are greatly inhibited by the higher temperatures of the blood and viscera, and only rarely do they disseminate. Nevertheless, these diseases are progressive and can destroy the skin and associated structures. Mycoses in this category are sporotrichosis, chromoblastomycosis, phaeohyphomycosis, and mycetoma.

Sporotrichosis

The cause of sporotrichosis, *Sporothrix schenckii,* is a very common saprobic fungus that decomposes vegetative matter in soil and humus, and exhibits both mycelial and yeast phases. *Sporothrix* resides in warm, temperature, and moist areas of the tropics and semitropics, though the incidence of sporotrichosis is highest in Africa, Australia, Mexico, and Latin America. A prick by a rose thorn is the classic origin of the cutaneous form of infection, and has given rise to its common name. Most episodes of human infection follow contact with thorns, wood, sphagnum moss, bare roots, bark, and other vegetation. Horticuturists, gardeners, farmers, and basket weavers incur the majority of cases. Overt disease is uncommon unless a person has inadequate immune defenses or is exposed to a large inoculum or a virulent strain.

Pathogenesis, diagnosis and control – In lymphocutaneous sporotrichosis, the fungus grows at the site of penetration and develops a small, hard, nontender nodule within a few days to months. This subsequently enlarges, becomes necrotic, breaks through to the skin surface, and drains. Infection often progresses along the regional lymphatic channels, leaving a chain of lesions at various stages. If left untreated, this condition persists for several years, though it is contained by the regional lymph nodes and does not spread. When *Sporothrix* conidia are drawn into the lungs, primary pulmonary sporotrichosis can result. To enhance its appearance, enzymes that clear the specimen and special stains that highlight the tiny, cigar-shaped yeasts can be applied. Demonstration of the typical macroscopic and microscopic morphology on artificial media confirm the diagnosis.

Reliable serological tests are available but are usually not necessary. However, the sporotrichin skin test can be used to determine prior infection. An older but effective drug for sporotrichosis is potassium iodide, given orally in milk or applied topically to open lesions. Amphotericin B and flucytosine can assist in unresponsive cases. The fungus cannot withstand heat, thus local applications of heat packs to lesions have helped resolve some infections.

Chromoblastomycosis and Phaeohyphomycosis

Chromoblastomycosis is a progressive subcutaneous mycosis characterized by highly visible *verrucous* lesions. The principal etiologic agents are a collection of widespread soil saprobes containing large amounts of dark pigments. *Fonsecaea pedrosoi, Phialophora verrucosa,* and *Cladosporium carrionii* are among the most common causative species. Phaeohyphomycosis, a closely related infection, differs primarily in the causative species and the appearance of the infectious agent in tissue. The fungi associated with these infections are of low inherent virulence, and none exhibits thermal dimorphism. Infection occurs when body surfaces, especially the legs and feet, are penetrated by soiled vegetation or inanimate objects. Chromoblastomycosis occurs throughout the world, with peak incidence in the American subtropics and tropics. Most vulnerable to infection are men with rural occupations who go barefoot. After a very long (2- to 3-year) incubation period, small, a coloured, warty plaque, ulcer, or papule appears. Unfortunately, the patient can scratch the nodules, which aggravates the spread of infection on the body and provokes secondary bacterial infections. Chromoblastomycosis is frequently confused with cancer, syphilis, yaws, and blastomycosis. Diagnosis is based on the clinical appearance of the lesions, microscopic examination of biopsied lesions, and the results of culture. Therapy for the condition includes heat, drugs, surgical removal of early nodules, and amputation in advanced cases.

Combined topical amphotericin B and thiabendazole or systemic flucytosine have shown some success in arresting the disease. The infectious diseases termed phaeohyphomycoses are among strangest oddities of medical science. The etiologic agents are soil fungi or plant pathogens with brown-pigmented mycelia. Victims of the disease are usually highly compromised patients who have become inoculated with these fungi, which are so widespread in household and medical environments that contact is inevitable. The disease process can be extremely deforming, and the microbe is sometimes hard to identify. Some genera of fungi commonly involved in these mycoses are *Alternaria, Aureobasidium, Curvularia, Dreschlera, Exophiala, Phialophora,* and *Wangiella.*

Mycetoma

It is also called *madura foot* for the Indian region where it was first described. About half of all mycetomas are caused by fungi in the genera *Pseudallescheria* or *Madurella,* though filamentous bacteria called actinomycetes are often etiologically implicated. Mycetomas are endemic to equatorial Africa, Mexico, Latin America, and the Mediterranean, and cases occur sporadically in the United States. Infection begins when bare skin is pierced by a thorn, sliver, leaf, or other type of sharp plant debris. First to appear is a localized abscess in the subcutaneous tissues, which gradually swells and drains.

CUTANEOUS MYCOSES

Common terms used in reference to these disease are ringworm, because they tend to develop in circular, scaly patches, and tinea. About 39 species in the genera *Trichophyton,*

Microsporum, and *Epidermophyton* are involved in various dermatophytoses. The causative agent of a given type of ringworm differs from person to person and from place to place, and is not restricted to a particular genus or species.

Dermatophytes

Various species exhibit unique macroconidia, microconidia, and unusual types of hyphae. In general, *Trichophyton* produces thin-walled, smooth macroconidia and numerous microconidia. *Microsporum* produces thick-walled, rough macroconidia and sparser microconidia; and *Epidermophyton* has ovoid, smooth, clustered macroconidia and no microconidia.

Epidemiology—Important factors that promote infection are the hardiness of the dermatophyte spores (they can last for years on fomites); hot, sweaty, chafed body parts; and intimate contact. Most infections exhibit a long incubation period (months), followed by localized inflammation and allergic reactions to the fungal proteins. As a general rule, infections acquired from animals and soil cause more severe reactions, and infections eliciting stronger immune reactions are resolved faster. Dermatophytic fungi exist throughout the world. Although some species are endemic, they tend to spread rapidly to nonendemic regions through travel. Dermatophytoses endure today as a serious health concern, not because they are life-threatening, but because of the extreme discomfort, stress, pain, and unsightliness they cause.

Atlas—In the following section the dermatophytoses are organized according to the area of body they affect, their mode of acquisition, and their pathologic appearance. This mycosis results from the fungal invasion of the scalp and the hair of the head, eyebrows, and eyelashes. Very common in children, tinea capitis is acquired from other children and adults or from domestic animals. Manifestations range from small, scaly patches (gray patch), to a severe inflammatory reaction (kerion), to destruction of the hair follicle and permanent hair loss. Unless infected hairs are controlled, reinfection is common.

Ringworm of beard and body—This tinea, also called "barber's itch", afflicts the chin and beard of adult males. Although once a common aftereffect of unhygienic barbering, it is now contracted mainly from animals. This extremely prevalent infection of humans can appear nearly anywhere on the body's glabrous (smooth and bare) skin. The principal sources are other humans, animals, and soil, and it is transmitted primarily by direct contact and fomites (clothing, bedding). The infection usually appears as one or more scaly reddish rings on the trunk, hip, arm, neck, or face. The ringed pattern is formed when the infection radiates from the original site of invasion into the surrounding skin. Depending on the causal species and the health and hygiene of the patient, lesions vary from mild and diffuse to florid and pustular.

Ringworm of groin and foot—Sometimes known as "jock itch", crural ringworm occurs mainly in males on the groin, perianal skin, scrotum, and occasionally, the penis. The fungus thrives under conditions of moisture and humidity created by profuse sweating or tropical climates. It is transmitted primarily from human to human, and is pervasive among athletes and persons living in close situations. Tinea pedis is known by a variety of synonyms, including athlete's foot and jungle rot. The disease is clearly connected to wearing shoes, since it is uncommon is cultures where the people customarily go barefoot. Conditions that encase that feet in a closed, warm, moist environment increase the possibility of infection. Because tinea pedis is a known hazard in shared facilities such as shower stalls, public floors, and locker rooms, many public facilities have rules against bare feet.

Ringworm of hand and nail—Infection of the hand by dermatophytes is nearly always associated with con-current infection of the foot. Lesions usually occur on the fingers and palms of one hand and they vary from white and patchy to deep and fissured. Fingernails and toenails, being masses of keratin, are often sites for persistent fungus colonization. The first symptoms are usually superficial white patches in the nail bed. A more invasive form causes thickening, distortion, and darkening of the nail.

Occasionally, the presenting symptoms are so dramatic and suggestive that no further testing is necessary, but in most cases, direct microscopic examination and culturing are needed. Diagnosis of tinea of the scalp is sometimes aided by a Wood's light, a special long-wave ultraviolet lamp that causes infected hairs to fluoresce with a greenish glow. Samples of hair, skin scrapings, and nail debris treated with heated KOH show a thin, branching fungal mycelium if infection is present. Ringworm therapy is based on the knowledge that the dermatophyte is feeding on dead epidermal tissues. These regions undergo constant replacement from living cells deep in the epidermis, so if multiplication of the fungus can be blocked, the fungus will eventually be sloughed off along with the skin or nail. Unfortunately, this takes time. By far the most satisfactory choice for therapy is a topical antifungal agent. Over-the-counter ointments containing imidazoles, or thiabendazine are applied regularly for several weeks. Some drugs work by speeding up loss of the outer skin layer. Intractable infections can be treated with griseofulvin, but placing a patient on this hepatotoxic and nephrotoxic drug for up to two years is probably too risky in most cases.

SUPERFICIAL MYCOSES

Tinea versicolor is caused by the yeast *Malassezia furfur,* a normal inhabitant of human skin that feeds on the high oil content of the skin glands. Even though this yeast is very common, in some people its growth elicits mild, chronic scaling, and interferes with production of pigment by melanocytes. The trunk, face, and limbs take on a mottled appearance. The disease is most pronounced in young people living in the tropics who are frequently exposed to the sun. Other skin conditions in which *M. furfur* is implicated are folliculitis, psoriasis, and seborrheic dermatitis. *Piedras* are marked by a tenacious, coloured concretion forming on the outside surface of hair shafts. In white piedra, caused by *Trichosporon beigelii,* a white to yellow adherent mass develops on the shaft on scalp, pubic, or axillary hair. At times, the mass is invaded secondarily by brightly coloured bacterial contaminants, with startling results. Black piedra, caused by *Piedraia hortae,* is characterized by dark-brown to black gritty nodules, mainly on scalp hairs.

OPPORTUNISTIC MYCOSES

The prevailing opportunistic pathogens of humans are the yeasts *Candida* and *Cryptococcus, Pneumocystic,* and a small number of filamentous fungi, primarily *Aspergillus* and certain zygomycetes.

Candidiasis

Candida albicans, an extremely widespread yeast, is the major cause of candidiasis. Manifestations of infection run the gamut from short-lived, superficial skin irritations to overwhelming, fatal systemic diseases. Microscopically, *C. albicans* has budding cells of varying size that can be elongated and attached in a pseudomycelium.

Epidemiology—A healthy person with an unimpaired immune system can hold it in check, but the risk of invasion increases with extreme youth, pregnancy, use of birth control or antimicrobial drugs, nutritional and organic disease, immunodeficiency, the presence of invasive devices, and trauma. Any situation that maintains the yeast in contact with moist skin provides an avenue for infection. Although candidiasis is usually endogenous and not contagious, it can be spread in nurseries or through surgery, childbirth, and sexual contact.

Pathogenesis—Mucous membranes most frequently involved are the oral cavity and vagina. Thrush is a white, adherent, patchy infection affecting the membranes of the mouth, gums, cheeks, or throat, usually in newborn infants and elderly, debilitated patients. Vulvovaginal candidiasis (VC), known more commonly as yeast infection, has widespread occurrence in adult women, especially if taking antibiotics or oral contraceptives, diabetic, or pregnant. These conditions are thought to increase the glucose content of vaginal secretions, which favours candidal overgrowth and disrupts the balance of normal flora. Binding, impervious undergarments (pantyhose, nylon panties) are probably contributing factors. The chief symptoms of VC are a yellow to white milky discharge, inflammation, painful ulceration, and itching. Of all the areas of the gastrointestial tract, *Candida* most often infects the esophagus and the anus. Esophageal condidiasis, which afflicts 70% of AIDS patients, causes painful, bleeding ulcerations; perforations; nausea, and vomiting.

People whose occupations require their hands or feet to be constantly immersed in water are at risk for finger and nail invasion. *Intertriginous* infection occurs in moist areas of the body where skin rubs against skin, as beneath the breasts, in the armpit, and between folds of the groin. Cutaneous candidiasis can also complicate burns and produce a scaldlike rash on the skin of neonates. Candidal blood infection in patients chronically weakened by surgery, bone marrow transplants, advanced cancer, or intravenous drug addiction is usually followed by systemic invasion. Although the mechanisms of virulence are not well understood, the presence of *C. albicans* in the blood is such a serious assault that it causes more human mortalities than any other fungal pathogen. Principal targets of systemic infections are the urinary tract, endocardium, and brain. Patients with valvular disease of the heart or indwelling prosthetic devices are vulnerable to candidal endocarditis, usually caused by other species.

Diagnosis and control—Specimens are cultured on standard media incubated at 30 °C. Identification is complicated by the numerous species of *Candida* and other look-alike yeasts, though confirmatory evidence of *C. albicans* can be obtained by the germ tube test, the presence of chlamydospores, and multiple panel systems that test for biochemical characteristics. Therapy for superficial mucocutaneous infection consists of topical antifungal agents (azoles and polyenes). Amphotericin B with or without flucytosine is the only choice available for systemic infections. Recurrent bouts of vulvovaginitis are managed by topical ketoconazole or miconazole ointments, now available as over-the-counter drugs.

Cryptococcosis

This yeast has a spherical to ovoid shape, with small, constricted buds and a large capsule that is important in its pathogenesis. Its role as an opportunist in supported by evidence that healthy humans have strong resistance to it and that frank infection occurs primarily in debilitated patients.

Epidemiology—It is prevalent in urban areas where pigeons congregate, and it proliferates in the high nitrogen content of droppings that accumulate on pigeon roosts. Masses of dried yeast cells are readily scattered into the air and dust. The species is sporadically isolated from dairy products, fruits, and the healthy human body (tonsils and skin), where it is considered a contaminant. The highest rates of cryptococcosis occur among patients with AIDS. It causes a severe form of meningitis, which affects at least 10% of them and is frequently fatal. Other conditions that predispose to infection are steroid treatment, diabetes, cancer, and lupus.

Pathogenesis—A few patients with pulmonary cryptococcosis develop fever, cough, and nodules in their lungs. The escape of the yeasts into the blood is intensified by weakened host defenses and attended by severe complications. Sites for which *Cryptococcus* show an extreme affinity are the brain and meninges. The tumorlike masses formed in these locations can cause headache, mental changes, coma, paralysis, eye disturbances, and seizures.

Diagnosis and control—Colony development on blood-enriched or selective media requires two to three days. Two quick screening tests that presumptively differentiate *C. neoformans* from the seven other cryptococcal species are the urease and phenoloxidase tests. Confirmatory results include a negative nitrate assimilation, pigmentation on birdseed agar, and fluorescent antibody tests. Cryptococcal antigen can be detected in a specimen by means of serological tests, and DNA probes can make a positive genetic identification.

Pneumocystis Pneumonia

Pneumocystis carinii pneumonia is the most frequent opportunistic infection in AIDS patients, most of whom will develop one or more episodes during their lifetimes. Because this unicellular microbe has characteristics of both protozoa and fungi, its taxonomic status has been somewhat uncertain. Until recently it was considered a type of protozoan because of its cystlike and trophozoite-like phases. Unlike most of the human fungal pathogens, little is known about the life cycle or epidemiology of *Pneumocystis*. Evidently the parasite is a common, relatively harmless resident of the upper respiratory tract. Contact with the agent is so widespread that in some populations a majority of people show serological evidence of infection. Until the AIDS epidemic, symptomatic infections by this organism were very rare, occurring only among the elderly or premature patients that were severely debilitated or malnourished. *Pneumocystis* is probably spread in droplet from between humans. In people with intact immune defenses, it is usually held in check by lung phagocytes and lymphocytes, but in those with deficient immune systems, it multiplies intracellularly and extracellularly. The massive numbers of parasites adhere tenaciously to the lung pneumocytes and cause an inflammatory condition. The linings of the lung alveoli slough off and a foamy exudate builds up. Since human strains of the species have not yet been successfully cultured, PCP must be diagnosed by symptoms and direct examination of lung secretions and tissue.

Aspergillosis

The estimated 600 species are widely dispersed in dust and air and can be isolated from vegetation, food, and compost heaps. Of the eight species involved in human diseases, the thermophilic fungus *A. fumigatus* accounts for the most cases. Aspergillosis is almost always an opportunistic infection, lately posing a serious threat to AIDS, leukemia, and organ transplant patients. It is also involved in allergies and toxicoses. *Aspergillus* infections usually occur in the lungs. People breathing clouds of conidia from the air of granaries, barns, and silos are at greatest risk. In heavily exposed healthy people, the spores simply germinate in the lungs and form *fungus balls.* Similar benign, noninvasive infections occur in colonization of the sinuses, ear canals, eyelids, and conjunctiva. In the more invasive form of aspergillosis, the fungus produces a necrotic pneumonia and disseminates to the brain, heart, skin, and a wide range of other organs. An obvious branching, septate mycelium with characteristics conidial heads in a specimen is presumptive evidence of aspergillosis. A culture of the specimen that produces abundant colonies and growth at 37 °C will rule out its being only an air contaminant.

Mucormycosis

Their large, prolific sporangia release multitudes of lightweight spores that literally powder the living quarters of humans and usually do little harm besides rotting fruits and vegetables. But an increasing number of critically ill patients are contracting a disease called *mucormycosis.* The genera most often involved in mucormycoses are *Rhizopus, Absidia,* and *Mucor.* Underlying debilities that seem to increase the risk for mucormycosis are leukemia, burns, and malnutrition. Uncontrolled diabetes is a special risk because it can lead to reduced pH in the blood and tissue fluids that enhances invasion of the fungus. The specific pathology depends on the portal of entry of the fungus, which can be nasal, pulmonary, cutaneous, or gastrointestinal. In rhinocerebral mucormycosis, infection begins in the nasal passages or pharynx, and then grows in the blood vessels and up through the palate into the eyes and brain. It progresses so swiftly that death can occur in a few days. Invasion of the lungs occurs most often in leukemics, and gastrointestinal mucormycosis is a complication in children suffering from protein malnutrition.

Miscellaneous Mycoses

Although it is still relatively unusual to become infected by common airborne fungi, these infections are currently on the rise. Geotrichosis is a rare mycosis caused by *Geotrichum candidum,* a mold commonly found in soil, in dairy products, and on the human body. The mold is not highly aggressive and is primarily involved in secondary infection in the lungs of tubercular or highly immunosuppressed patients. Species of *Fusarium* a common soil inhabitant and plant pathogen, occasionally infects the eyes, toenails, and burned skin. Mechanical introduction of the fungus into the eye (for example, by contaminated contact lenses) can induce a severe mycotic ulcer. It can also cause a patchy infection of the nail bed somewhat like tinea unguium. *Fusarium* colonization of patients with widespread burns has occasionally led to mortalities. Unusual opportunistic infections have been reported for other commonplace fungi such as *Penicillium, Paecilomyces,* and the red yeast *Rhodotorula.*

ACTINOMYCETES

The precedent for including these decidedly prokaryotic organisms with the eukaryotic fungi was established by a century-old investigation that showed a superficial resemblance between them. This resemblance includes the formation of spores by some actinomycetes, the production of a branching, fragmented mycelium, and the chronic, granulomatous, disfiguring diseases they cause. In time, it became evident that actinomycetes were not fungi or even ancestors of fungi as had once been thought. In structure and physiology, the actinomycetes are filamentous, Gram-positive bacteria related to the mycobacteria.

Actinomycosis

Actinomycosis is an endogenous infection of the cervicofacial, thoracic, or abdominal regions by species of *Actinomyces* living normally in the human oral cavity, tonsils, and intestine. One form of disease, lumpy jaw, was once a common complication of tooth extraction, poor oral hygiene, and rampant dental caries. In cervicofacial involvement, the bacteria enter a damaged area of the oral mucous membrane and begin to multiply there. Diagnostic signs are swollen, tender nodules in the neck or jaw that give off a discharge containing macroscopic (1-2 mm) sulfur granules. In most cases, infection remains localized, but bone invasion and systemic spread can occur in people with poor health. Thoracic actinomycosis is a necrotizing lung disorder that can project through the chest wall and ribs. Abdominal actinomycosis is a complication of burst appendices, gunshot wounds, ulcers, or intestinal damage. Uterine actinomycosis has been increasingly reported in women using intrauterine contraceptive devices. Compelling evidence indicates that oral actinomyces play a strategic role in the development of plaque and dental caries. Studies of the oral environment show that *A. viscosus* and certain oral streptococci are the first microbial colonists of the tooth surface.

Nocardiosis

Nocardia is a genus of bacteria widely distributed in the soil. Most species are not infectious, but *N. brasiliensis* is a primary pulmonary pathogen, and *N. asteroides* and *N. caviae* are opportunists. Nocardioses fall into the categories of pulmonary, cutaneous, or subcutaneous infection. Most cases in the United States are reported in patients with deficient immunity, but a few occur in normal individuals. Pulmonary nocardiosis is a form of bacterial pneumonia with pathology and symptoms similar to tuberculosis. The lung develops abscesses and nodules, and can consolidate. Often the lesions extend to the pleura and chest wall and disseminate to the brain, kidneys, and skin.

REFERENCES

Adams, M. W. W. and Stiefel E. I. (1998). Biological hydrogen production : not so elementary. *Science* **282** : 1842-43.

Agrios, G. N. (1978). *Plant Pathology,* Academic, New York.

Albright, R. A. and Matthews, B. W. (1998). How Cro and Lambda-repressor distinguish between operators : the structural basis underlying a genetic switch. *Proc. Natl. Acad. Sci. USA* **95** : 3431-36.

Alexander, M. (1977). *Introduction to Soil Microbiology. Wiley,* New York.

Alexopoulos, C. J. and Mims C. W. (1979). *Introductory Mycology.* Wiley, New Yrok.

Avery, O. T., MacLeod, C. M. and McCarty, M. (1994). Studies on the chemical nature of the substance inducing transformation of pneumococcal types. Induction of transformation by a deoxyribonucleic acid fraction isolated from pneumococcus type III. *J. Exp. Med.* **79** : 137.

Ayala, F. J. and Kiger, J. A. Jr. (1980). *Modern Genetics.* Benjamin/Cummings, Menlo Park, Calif.

Baas, P. D. (1985). DNA replication of single-stranded *Escherichia coli* DNA phages. *Biochim. Biophys. Acta* **825** : 111-39.

Baumler, A. J., Tsolis, R. M. and Heffron F. (1996). The *lpf* fimbrial operon mediates adhesion of *Salmonella typhimurium* to murine Peyer's patches. *Proc. Natl. Acad. Sci. USA* **93** : 279-83.

Birge, E. A. (1981). *Bacteriophage and Bacteriophage Genetics.* Springer-Verlag, New York.

Birge, E. A. (2000). *Bacterial and Bacteriophage Genetics.* Springer-Verlag, New Yrok.

Bock, A. (2000). Fermentation. In : *Encyclopedia of Microbiology,* J. Lederberg (ed.), Vol. 20, Academic Press, San Diego.

Bulloch, W. (1938). *The History of Bacteriology.* Oxford, London.

Chattoraj, D. K. (2002). Control of plasmid DNA replication by interons : no longer paradoxical. *Mol. Microbiol.* **37** : 467–76.

Cheryan, M. (2000). Acetic acid production. In : *Encyclopedia of Microbiology,* J. Lederberg (ed.), Vol. 1, Academic Press, San Diego.

Cheryan, M. (2000). Acetic acid production. In : *Encyclopedia of Microbiology,* J. Lederberg (ed.), Vol. 1, Academic Press, San Diego.

Christie, P. J. and Vogel, J. P. (2000). Bacterial type IV secretion : conjugation systems adapted to deliver effector molecules to host cells. *Trends Microbiol.* **8** : 354-60.

Clark, Paul F. (1961). *Pioneer Microbiologists of America,* Wisconsin, Madison.

Clewell, D. B. (ed.). (1993). *Bacterial Conjugation.* Plenum Press, New York.

Collins, C. H., Allwood, M. C., Bloomfield, S. F. and Fox A. (eds.) (1981). *Disinfectants : Their Use and Evaluaton of Effectivenss.* Academic, New York.

Covacci, A., Telford J. L., Del Guidice, G., Parsonnet, J. and Rappuoli, R. (1999). *Helicobacter pylori* virulence and genetic geography. *Science* **284** : 1328-33.

Cox, M. M. (1991). The RecA protein as a recombinational repair system. *Mol. Microbiol.* **5** : 1295-99.

Cox, M. M., Goodman, M. F., Kreuzer, K. N., Sherratt, D. J., Sandler, S. J. and Marians K. J. (2000). The importance of repairing stalled replication forks. *Nature* **404** : 37-41.

Das, A., and Ljungdahl. L. G. (2000). Acetogenesis and acetogenic bacteria. In : *Encyclopedia of Microbiology,* J. Lederberg (ed.), Vol. 1, Academic Press, San Diego.

Davies, D. D. (ed.) (1987). *The Biochemistry of Plants.* vol. 2, Academic Press, New York.

Doelle, H. W. (1975). *Bacterial Metabolism* Academic, New York.

Dowling, H. F. (1977). *Fighting Infection, Conquests of the Twentieth Century.* Harvard Press, Cambridge, Mass.

Dubnau, D., and Provedi, R. (2000). Internalizing DNA. *Res. Microbiol.* **151** : 475-80.

Evans, A. S. (ed.) (1982). *Viral Infections Of Humans.* Plenum, New York.

Favero, M. S. (1985). Sterilization, Disinfection, and Antisepsis in the Hospital. In : *Manual of Clinical Microbiology.* Lennette, E. H., et al. (eds.). Washington, American Society for Microbiology.

Finegold, S. M. and Martin, W. J. (1982). *Diagnostic Microbiology,* Mosby, St. Louis.

Fox, G. E., Stackebrandt, E., Hespell, R. B., Gibson, J., Maniloff, J., Dyer, T. A., Wolfe, R. S., Blach, W. E., Tanner, R. S., Magrum, L. J., Zablen, L. B., Blakemore, R., Gupta, R., Bonen, L., Lewis, B. J., Stahl, D. A., Luehrsen, K. R., Chen, K. N. and Woese, C. R.: (1980). The phylogency of prokaryotes. Science, **209** : 457-63.

Franklin, D. J., and Snow, G. A. (1981). *Biochemistry of Antimicrobial Action.* Chapman and Hall Ltd., London.

Frieg, A. (1981). The Genus *Bacillus.* In : sect Pathogens. *The Prokaryotes : A handbook on Habitats, Isolation, and Identification of Bacteria.* Starr, M. P., Stolp, H., Truper, H. G., Balows, A. and Schlegel, H. G. (eds.). vol. II, Springer-Verlag, New York.

Frost, L. S., Ippen-Ihler, K. and Skurray, R. A. (1994). Analysis fo the sequence and gene products of the transfer region of the F sex factor. *Microbiol. Rev.* **58** : 162-210.

Fu, Y. and Galan, J. E. (1999). A *Salmonella* protein antagonizes *Rac-I* and *Cd-42* to mediate host cell recovery after bacterial invasion. *Nature* **401** : 293-97.

Fuller, J. G. (1974). Fever! *The Hunt for a New Killer, Virus. Reader's Digest*, New York.

Galan, J. E. and Zhou, Y. (2000). Striking a balance : modulation of action cytokeleton by *Salmonella. Proc. Natl . Acad. Sci. USA* **97** : 8754-61.

Gallon, J. R. and Chaplin, A. E. (1987). *An Introduction to Nitrogen Fixation*. Cassell, London.

Garcia-del Portillo, F. and Finlay, B. B. (1995). Targeting of *Salmonella typhimurium* to vesicles containing lysosomal membrane glycoproteins bypasses compartments with mannose 6-phosphate receptors. *J. Cell Biol.* **129** : 81-97.

Garrod, L. P., Lambert, H. P., O'Grady F. (eds.) (1981). *Antibiotics and Chemotherapy.* Churchill Livingstone, Edinburgh,

Gerhardt, P., Murray, R. G., Costilow, R. N., Nester, E. W., Wood, W. A., Krieg, N. R. and Phillips, G. B. (eds) (1981). Manual of Methods for General Bacteriology. American Society for Microbiology, Washington.

Gherna, R. L. (1981). Preservation, In : *Manual of Methods for General Bacteriology*, Gerhardt, P., Murray, R. G. E., Costilo, R. N., Jester, E. W., Wood, W. A., Krieg, N. R. and Phillips, G. B. (eds.). American Society for Microbiology, Washington, D. C.

Ghosh, B. K. (ed.) (1981). Organization of Prokaryotic Cell Membranes. CRC Press, Boca Ration, Florida.

Gibson, C. M., Mallett Claiborne, T. C. A. and Caparon, M.G. (2000). Contribution of NADH oxidase to aerobic metabolism of *Streptococcus pyogenes. J. Bacteriol.* **182** : 448-55.

Gombert, A. K., DosSantos, M. M., Christensen, B. and Nielsen, J. (2001). Network identification and flux quantification in the central metabolism of *Saccharomyces cerevisiae* under different conditions of glucose repression. *J. Bacteriol.* **183** : 1441-51.

Gottesman, M. (1999). Bacteriophage lambda : the untold study. *J. Mol. Biol.* **293** : 177-80.

Gottschalk, G. (1979). *Bacterial Metabolism* Springer-Verlag, New York.

Goupil-Feuillerat, N. (2000). Dual role of a-acetolactate decarboxylase in *Lactococcus lactis* subsp. lactis. *J. Bacteriol.* **179** : 6285-93.

Gray, T. R. G. and Parkinson, D. (eds.) (1968). *The Ecology of Soil Bacteria : An International Symposium.* University of Toronto Press, Toronto.

Grell, Karl G. (1973). *Protozoology*. Springer-Verlag, Heidelberg.

Gresshoff, P. M., Roth, E. L. Stacey, G. and Newton, W.E. (eds.), *Nitrogen Fixation : Achievements and Objectives.* Chapman & Hall, New York.

Griffith, F. (1928). The significance of pneumococcal types. *J. Hyg.* **27** : 113.

H. Rennenberg, Brunold, C. H., DeKok, L. J. and Stulen, I. (eds.), *Sulfur Nutrition and Sulfur Assimilation in Higher Plants.* SBP Academic, The Hague.

Hackstadt, T., Scidmore, M. A. and Rockey, D. D. (1995). Lipid metabolism in *Chlamydia trachomatis*-infected cells : directed trafficiking of Golgi-derived sphingolipids to the chlamydial inclusion. *Proc. Natl. Acad. Sci USA* **92** : 877-881.

Hanegraaf, P. P. F., Stouthamer, A. H . and Kooijman, S. A. L. M. (2000). A mathematical model for yeast respiro-fermentative physiology. *Yeast* **16** : 423-37.

Harrmann, K. M. and Somerville R. L. (eds.) (1983). *Amino Acids : Biosynthesis and Genetic Regulation.* Addision-Wesley, Reading, MA.

Hattori, Tsutomu (1973). *Microbial Life in the Soil : An Introduction.* Marcel Dekker, New York.

Hugo, W. B. and Russell, A. D. (eds.) (1977). *Pharmaceutical Microbiolgy.* Blackwell Scientific, Oxford.

Kandler, O. (1982). Cell wall structures and their phylogenetic implications. *Syst. Appl. Microbiol.* **3** : 149–160.

Karam, J. D. (1994). *Bacteriophage T4.* ASM Press, Washington DC.

Kennedy, A. K., Gudhathakurta, A., Kleckner, N. and Hainford, D. B. (1998). Tn*10* transposition via a DNA hairpin intermediate. *Cell* **95** : 125-34.

Kornberg, A. and Baker, T. (1992). *DNA Replicaton.* W. H. Freeman and Co., San Francisco.

Krieg, N. R. (1981). Enrichment and Isolation, In : *Manual of Methods for General Bacteriology,* Gerhardt, P., Murray, R. G. E., Costilow, R. N., Nester, E. W., Wood, W. A., Krieg, N. R. and Phillips G. B. (eds.). American Society for Microbiology, Washington. D.C.

Krieg, N. R. (ed.) (1984). *Bergey's Manual of Systematic Bacteriology.* vol. 1, Williams & Wilkins, Baltimore.

Krieg, N. R. (ed.) (1984). *Bergey's Manual of Systematic Bacteriology.* vol. 1., Williams & Wilkins, Baltimore.

Kubori, T., Matsushima, Y., Nakamura, D., Uralil, J., Lara-Tejero, M., Sukhan, A., Galan, J. and Aizawa S.-I. (1998). Supramolecular structure of the *Salmonella typhimurium* type III protein secretion system. *Science* **280** : 602-05.

Lara-Tejero, M. and Galan, J. E. (2000). A bacterial toxin that controls cell cycle progression as a deoxyribonuclease I-like protein. *Science* **290** : 254-57.

Lawley, T.D., Burland, V. and Taylor, D.E. (2000). Analysis of the complete nucleotide sequence of the tetracycline-resistance transposon Tn*10*. *Plasmid* **43** : 235-39.

Lea, P. J. and Leegood, R. C. (eds.), *Plant Biochemistry and Molecular Biology.*

Lechevalier, H. and Solotorovsky, M. (1965). *Three Centres of Microbiology.* McGrawHill, New York.

Lennette, E. H., Balows, A. W., Hausler, J. Jr.,Truant, J. P. and Shadomy, H. J. (eds.) (1985). *Manual of Clinical Microbiology.* American Society for Microbiology, Washington DC.

Lewin, B. (1983). *Genes.* Wiley, New York.

Linton, A. H. (1982). *Microbes, Man and animals.* John Wiley, New York.

Marra, G., and Schar, P. (1999). Recognition of DNA alterations by the mismatch repair system. *Biochem. J.* **338** : 1-13.

Marx, J. L. (1984). Strong new candidate for AIDS agent. *Science.* **224** : 475-477.

Matth, E. H. (ed.) (1978). *Standard Methods for Examination of Dairy Products.* American Public Health Association, Washington DC.

Mazel, D., Dychinco, B., Webb, V.A. and Davies, J. (1998). A distinctive class of interon in the *Vibrio cholerae* genome. *Science* **280** : 605-08.

Mengal, K. and Pilbeam, D. J., (eds.), *Nitrogen Metabolism of Plants.* Clarendon Press, Oxford.

Mengel, K. and Pilbeam, D. J. (eds). *Nitrogen Metabolism of Plants.* Clarendon Press, Oxford.

Mengel, K. and Pilbeam, D. J. (eds.), *Nitrogen Metabolism in Plants.* Clarendon Press, Oxford.

Michael, B., Smith, J. N., Swift, S., Heffron, F. and Ahmer, B. M. (2001). SdiA of *Salmonella enterica* is a LuxR homolog that detects mixed microbial communities. *J. Bacteriol.* **183** : 5733-42.

Miflin, B. J. (ed.) *The Biochemistry of Plants.* Vol. 5, Academic Press, New York.

Miller, B. M. and Litsky W. (1976). *Industrial Microbiology.* McGraw-Hill, New York.

Miller, J. H. E. (1999). *Bacterial Genetic Systems. Methods in Enzymology.* Academic Press, San Diego.

Mitchell, R. (1978). *Water Pollution Microbiology.* Wiley, New York.

Moat, A. G. (1979). *Microbial Physiology* Wiley, New York.

Moore-Landecker, E. (1982). *The Foundamentals of Fungi.* Prentice-Hall, Englewood Cliffs, New Jersey.

Mossel, D. A. A. (1982). *Microbiology of Foods.* Faculty of Veterinary Medicine, University of Utrecht, The Netherlands.

Muller, F., (ed.), *Chemistry and Biochemistry of Flavoenzymes.* Vol. 3, CRC Press, Boca Raton.

Murray, N. E. (2000). Type I restriction systems : sophisticated molecular machines (a legacy of Bertani and Weigle). *Microbiol. Biol. Rev.* **64** : 412-34.

Nair, R. V. (1999). Regulation of the *sol* locus genes for butanol and acetone formation in *Clostridium acetobutylicum* ATCC824 by a putative transcriptional repressor. *J. Bacteriol.* **181** : 319-30.

O'Loughlin, E. V. and Robins-Browne, R. M. (2001). Effect of Shiga toxin and Shiga-like toxins on eukaryotic cells. *Microbe Infect.* **3** : 49-57.

Orth, k., Palmer, L. E., BaO, Z. Q., Stewart, S., Rudolph, A. E., Bliska, J. B. and Dixon, J. B. (1999). Inhibition of the mitogen-activated protein kinase superfamily by a *Yersinia* effector. *Science* **285** : 1920-23.

Paley, L. G. and Aspinall, D. (eds.), *Physiology and Biochemistry of Drought Resistance in Plants.* Academic Press, Sydney.

Pesci, E. C., Milbank, J. B., Pearson, J. P., McKnight, S., Kende, A., Greenberg, E. P. and Iglewski, B. H. (1999). Quinolone signaling in the cell-to-cell communication systems of *Pseudomonas aeruginosa. Proc. Natl. Acad. Sci. USA* **96** : 1129-34.

Peters, J. W., Lanzilotta, W. N., Lemon, B. J. and Seefelt, L. C. (1998). X-ray crystal structure of the Fe-only hydrogenase (Cpl) from *Clostridium pasterianum* to 18 angstrom resolution. *Science* **282** : 1853-58.

Pickett-Heaps, J. D. (1975). *Green Algae* : Structure, Reproduction and Evolution in Selected Genera. Sinauer Associates, Sunderland, Mass.

Poulson, J. E., Romeo, J. T. and Conn, E. E. (eds), *Plant Nitrogen Metabolism.* Plenum Press, New York.

Preer, J. R., Jr., Preer, L. B., and Jurand, A. (1974). *Kappa and other endosymbionts in Paramecium aurelia,* Bacteriol Rev. **38** : 113-163.

Presecan-Siedel, E., Galinier, A., Longin, R. (1999). Catabolite regulation of the *pta* gene as part of carbon flow pathway in *Bacillus subtilis. J. Bacteriol.* **181** : 6889-97.

Rangarajan, S., Woodgate, R. and Goodman, M. F. (1999). A phenotype for enigmatic DNA polymerase II : a pivotal role for pol II in replication restart in UV-irradiated *Escherichia coli. Proc. Natl. Acad. Sci. USA* **96** : 9224–29.

Reed, G. (ed.) (1982). *Prescott and Dunn's Industrial Microbiology.* AVI Publishing Co., Inc., Westport, Conn.

Reinberg, D., Zipursky, S. L., Weisbeek, P., Brown, D. and Hurwitz, J. (1983). Studies on the phi X17 gene A protein-mediated termination of leading strand DNA synthesis. *J. Biol. Chem.* **258** : 529-37.

Rheinheimer, G. (1980). *Aquatic Microbiology.* Wiley, New York.

Rogers, L. J. and Gallon, J. R. (eds.) *Biochemistry of the Algae and Cyanobacteria.* Oxford University Press, Oxford.

Rosche, W. A. and Foster, P. L. (2000). Mutation under stress : adaptive mutation in *Escherichia coli.* In : *Bacterial Stress Responses.* G. Storz and R. Hengge-Aronis (eds.), ASM Press, Washington D.C., pp. 239-48.

Rose, A. A. (ed.) (1982). *Fermented Foods.* vol. 7 : *Academic,* New York.

Rose, A. H. (1976). *Chemical Microbiology,* Plenum, New Yrok.

Ross, I. K. (1979). *Biology of the Fungi.* McGraw-Hill, New York.

Russell, A. D., Hugo, W. B. and Ayliffe, G. A. J. (eds.) (1982). *Principles and Practice of Disinfection.* Blackwell Scientific Publications, Oxford.

Rutledge, R. G., Oueellet, T., Hattori, J. and Miki, B. L. (1991) *Mol. Gen. Genet* : 229 : 31-40.

Sauer, F. G., Futterer, K., Pickner, J. S., Dodson, K. W., Hultgren S. J. and Waksman G. (1999). Structural basis of chaperone basis of chaperone function and pilus biogenesis. *Science* **285** : 1058-66.

Schmidt, G. D. and Roberts, L. R. (1981). *Foundations of Parasitology.* Mosby St. Louis.

Shatursky, O., Heuck, A. P. , Shepard, L. A., Rossjohn, J., Parker, M. W. Johnson, A. E. and Tweten, R. K. (1999). The mechanism of membrane insertion for a cholesterol-dependent cytolysin : a novel paradigm for pore-forming toxins. *Cell* **99** : 293-99.

Sieburth, J. M. (1975). *Microbial Seascapes.* University Park Press, Baltimore.

Sieburth, J. M. (1979). *Sea Microbes.* Oxford University Press, New York.

Singh, B. K., Flore, H. E. and Shannon, J. C. (eds.), *Biosynthesis and Molecular Regulation of Amino Acids in Plants.* American Society of Plant Physiologists, Maryland.

Singh, B. K., Flores, H.E. and Shannon, J. C. (eds), *Biosynthesis and Molecular Regulation of Amino Acids in Plants.* American Society of Plant Physiologists, Maryland.

Slack, John M., and Irvin, S. (1978). *Bacteria and Human Disease,* Year Book, Chicago.

Sleigh, M. A. (1973). *The Biology of the Protozoa.* American Elsevier, New York.

Smith, D. C. (1973). *The Lichen Symbiosis.* Oxford, London.

Sneath, P. H. A. (1978). Classification of microorganisms, In : *Essays in Microbiology,* Norris, J. r. and Richmond, M. H. (eds.) John Wiley, Chichester, United Kingdom.

Sprent, J. I. and Sprent, P. (1990). *Nitrogen Fixing Organisms : Pure and Applied Aspects.* Chapman & Hall, London.

Starr, M. P., H. Stolp, H. G. Truper, A. Balows and Schlegel, H. G. (eds.) (1981). *The Prokaryotes : A Hand book of Habitats, Isolation, and Identification of Bacteria.* Springer-Verlag, New York.

Steuber, J., Bott, M. and Dimroth, P. (1999). A membrane-bound $NAD(P)^+$ -reducing hydrogenase provides reduced pyridine nucleotides during citrate fementation by *Klebsiella pneumoniae. J. Bacteriol.* **181** : 241-45.

Stewry, P. R. (ed.), *Barley : Genetics Biochemistry, Molecular Biology and Biotechnology.* C.A.B. International, Wallingford.

Tang, M. P., Pham-Shen, X., Taylor, J. S., O'Donnell, M.,Woodgate, R. and Goodman, M. F. (2000). Roles of *E. coli* DNA polymerases IV and V in lesion-targeted and untargeted SOS mutagenesis. *Nature* **404** : 1014-18.

Trainor, F. R. (1978). *Introductory Phycology.* Wiley, New York.

Truper, H. G. and Kramer, J. (1981). Principles of characterization and identification of prokaryotes, In : *The Prokaryotes : A Handbook on Habitats, Isolation, and Identification of Bacteria,* Starr, M. P., Stolp, H., Truper, H. G. Balows, A. and Schelegel, H. G. (eds.). Springer-Verlag, New York.

Van Iterson, G.,den Dooren de Jong, L., Kluyver, A. J., Van Neil, C. B. and Brock Martinus Beijerinck T. D. (1984). *His Life and Work.* Science Tech, Madison.

Walker, G. M. (1998). *Yeast Physiology and Biotechnology.* John Wiley & Sons, Chichester, UK.

Watson, J. D. (1976). *Molecular Biology of the Gene.* W. A. Benjamin, Inc., New York.

Whittaker, R. H. (1969). New concepts of kingdoms of organisms, *Science* **163** : 150-160.

Willetts, N. and Wilkins, B. (1984). Processing of plasmid DNA during becterial conjugation. *Microbiol. Rev.* **48** : 24-41.

Winstedlt, L., Franskenberg, L., Hederstedt, L. and Wachenfeldt C. von. (2000). *Enterococcus faecalis* V583 contains a cytochrome *bd*-type respiratory oxidase. *J. Bacteriol.* **182** : 3863-66.

Wray, J. L. (ed.), *Inducible Plant Proteins.* Cambridge University Press, Cambridge.

Wray, J. L. (ed.), *Inducible Plant Proteins.* Cambridge University Press, Cambridge.

Wray, J. L. and Kinghorn, J. R. (eds.), *Molecular and Genetic Aspects of Nitrate Assimilation.* Oxford University Press, Oxford.

Xavier, R. J. and Podolsky, D. K. (2000). How to get along—friendly microbes in a hostile world. *Science* **289** : 1483-84.

Yamaguchi, M. and Baker, T. A. (1998). An ATP-ADP switch in MuB controls progression of the Mu transposition pathway. *EMBO J.* **17** : 5509-18.

Yamamoto, Y., Higuchi, M., Poole, L. B. and Kamio, Y. (2000). Role of the *dpr* product in oxygen tolerance in *Streptococcus mutans. J. Bacteriol.* **182** : 3740–47.

Zumbihl, R., Andor, A., Jacobi, C. A., Ruckdeschel, K., Rouot, B. and Heesemann, J. (1999). The cytotoxin YopT of *Yersinia enterocolitica* induces modification and cellular redistribution of the small GTP-binding protein RhoA. *J. Biol. Chem.* **274** : 289-93.

□□□

INDEX

I

K

L

M

N

❑❑❑